# HIDDEN COSTS OF
# ENERGY

## UNPRICED CONSEQUENCES OF ENERGY PRODUCTION AND USE

Committee on Health, Environmental, and Other External Costs
and Benefits of Energy Production and Consumption

Board on Environmental Studies and Toxicology

Division on Earth and Life Studies

Board on Energy and Environmental Systems

Division on Engineering and Physical Sciences

Board on Science, Technology, and Economic Policy

Policy and Global Affairs Division

NATIONAL RESEARCH COUNCIL
*OF THE NATIONAL ACADEMIES*

THE NATIONAL ACADEMIES PRESS
Washington, D.C.
**www.nap.edu**

THE NATIONAL ACADEMIES PRESS   500 Fifth Street, NW   Washington, DC 20001

NOTICE: The project that is the subject of this report was approved by the Governing Board of the National Research Council, whose members are drawn from the councils of the National Academy of Sciences, the National Academy of Engineering, and the Institute of Medicine. The members of the committee responsible for the report were chosen for their special competences and with regard for appropriate balance.

This project was supported by Contract No. TOS-08-038 between the National Academy of Sciences and the U.S. Department of the Treasury. Any opinions, findings, conclusions, or recommendations expressed in this publication are those of the author(s) and do not necessarily reflect the view of the organizations or agencies that provided support for this project.

International Standard Book Number-13: 978-0-309-14640-1 (Book)
International Standard Book Number-10: 0-309-14640-2 (Book)
International Standard Book Number-13: 978-0-309-14641-8 (PDF)
International Standard Book Number-10: 0-309-14641-0 (PDF)
Library of Congress Control Number: 2010925089

Additional copies of this report are available from

The National Academies Press
500 Fifth Street, NW
Box 285
Washington, DC 20055

800-624-6242
202-334-3313 (in the Washington metropolitan area)
http://www.nap.edu

# THE NATIONAL ACADEMIES
*Advisers to the Nation on Science, Engineering, and Medicine*

The **National Academy of Sciences** is a private, nonprofit, self-perpetuating society of distinguished scholars engaged in scientific and engineering research, dedicated to the furtherance of science and technology and to their use for the general welfare. Upon the authority of the charter granted to it by the Congress in 1863, the Academy has a mandate that requires it to advise the federal government on scientific and technical matters. Dr. Ralph J. Cicerone is president of the National Academy of Sciences.

The **National Academy of Engineering** was established in 1964, under the charter of the National Academy of Sciences, as a parallel organization of outstanding engineers. It is autonomous in its administration and in the selection of its members, sharing with the National Academy of Sciences the responsibility for advising the federal government. The National Academy of Engineering also sponsors engineering programs aimed at meeting national needs, encourages education and research, and recognizes the superior achievements of engineers. Dr. Charles M. Vest is president of the National Academy of Engineering.

The **Institute of Medicine** was established in 1970 by the National Academy of Sciences to secure the services of eminent members of appropriate professions in the examination of policy matters pertaining to the health of the public. The Institute acts under the responsibility given to the National Academy of Sciences by its congressional charter to be an adviser to the federal government and, upon its own initiative, to identify issues of medical care, research, and education. Dr. Harvey V. Fineberg is president of the Institute of Medicine.

The **National Research Council** was organized by the National Academy of Sciences in 1916 to associate the broad community of science and technology with the Academy's purposes of furthering knowledge and advising the federal government. Functioning in accordance with general policies determined by the Academy, the Council has become the principal operating agency of both the National Academy of Sciences and the National Academy of Engineering in providing services to the government, the public, and the scientific and engineering communities. The Council is administered jointly by both Academies and the Institute of Medicine. Dr. Ralph J. Cicerone and Dr. Charles M. Vest are chair and vice chair, respectively, of the National Research Council.

**www.national-academies.org**

*v*

*Staff*

**RAYMOND WASSEL,** Project Director, Board on Environmental Studies and Toxicology
**STEVE MERRILL,** Director, Board on Science, Technology, and Economic Policy
**JAMES ZUCCHETTO,** Director, Board on Energy and Environmental Systems
**DAVID POLICANSKY,** Scholar
**KEEGAN SAWYER,** Associate Program Officer
**RUTH CROSSGROVE,** Senior Editor
**MIRSADA KARALIC-LONCAREVIC,** Manager, Technical Information Center
**RADIAH ROSE,** Editorial Projects Manager
**JOHN BROWN,** Program Associate
**PATRICK BAUR,** Research Assistant

*Sponsor*

**U.S. DEPARTMENT OF THE TREASURY**

# OTHER REPORTS OF THE BOARD ON
# ENVIRONMENTAL STUDIES AND TOXICOLOGY

Review of the Environmental Protection Agency's Draft IRIS Assessment of Tetrachloroethylene (2010)

Hidden Costs of Energy: Unpriced Consequences of Energy Production and Use (2009)

Contaminated Water Supplies at Camp Lejeune—Assessing Potential Health Effects (2009)

Review of the Federal Strategy for Nanotechnology-Related Environmental, Health, and Safety Research (2009)

Science and Decisions: Advancing Risk Assessment (2009)

Phthalates and Cumulative Risk Assessment: The Tasks Ahead (2008)

Estimating Mortality Risk Reduction and Economic Benefits from Controlling Ozone Air Pollution (2008)

Respiratory Diseases Research at NIOSH (2008)

Evaluating Research Efficiency in the U.S. Environmental Protection Agency (2008)

Hydrology, Ecology, and Fishes of the Klamath River Basin (2008)

Applications of Toxicogenomic Technologies to Predictive Toxicology and Risk Assessment (2007)

Models in Environmental Regulatory Decision Making (2007)

Toxicity Testing in the Twenty-first Century: A Vision and a Strategy (2007)

Sediment Dredging at Superfund Megasites: Assessing the Effectiveness (2007)

Environmental Impacts of Wind-Energy Projects (2007)

Scientific Review of the Proposed Risk Assessment Bulletin from the Office of Management and Budget (2007)

Assessing the Human Health Risks of Trichloroethylene: Key Scientific Issues (2006)

New Source Review for Stationary Sources of Air Pollution (2006)

Human Biomonitoring for Environmental Chemicals (2006)

Health Risks from Dioxin and Related Compounds: Evaluation of the EPA Reassessment (2006)

Fluoride in Drinking Water: A Scientific Review of EPA's Standards (2006)

State and Federal Standards for Mobile-Source Emissions (2006)

Superfund and Mining Megasites—Lessons from the Coeur d'Alene River Basin (2005)

Health Implications of Perchlorate Ingestion (2005)

Air Quality Management in the United States (2004)

Endangered and Threatened Species of the Platte River (2004)

Atlantic Salmon in Maine (2004)

Endangered and Threatened Fishes in the Klamath River Basin (2004)

Cumulative Environmental Effects of Alaska North Slope Oil and Gas Development (2003)

Estimating the Public Health Benefits of Proposed Air Pollution Regulations (2002)

Biosolids Applied to Land: Advancing Standards and Practices (2002)

The Airliner Cabin Environment and Health of Passengers and Crew (2002)

Arsenic in Drinking Water: 2001 Update (2001)

Evaluating Vehicle Emissions Inspection and Maintenance Programs (2001)

Compensating for Wetland Losses Under the Clean Water Act (2001)

A Risk-Management Strategy for PCB-Contaminated Sediments (2001)

Acute Exposure Guideline Levels for Selected Airborne Chemicals (seven volumes, 2000-2009)

Toxicological Effects of Methylmercury (2000)

Strengthening Science at the U.S. Environmental Protection Agency (2000)

Scientific Frontiers in Developmental Toxicology and Risk Assessment (2000)
Ecological Indicators for the Nation (2000)
Waste Incineration and Public Health (2000)
Hormonally Active Agents in the Environment (1999)
Research Priorities for Airborne Particulate Matter (four volumes, 1998-2004)
The National Research Council's Committee on Toxicology: The First 50 Years (1997)
Carcinogens and Anticarcinogens in the Human Diet (1996)
Upstream: Salmon and Society in the Pacific Northwest (1996)
Science and the Endangered Species Act (1995)
Wetlands: Characteristics and Boundaries (1995)
Biologic Markers (five volumes, 1989-1995)
Science and Judgment in Risk Assessment (1994)
Pesticides in the Diets of Infants and Children (1993)
Dolphins and the Tuna Industry (1992)
Science and the National Parks (1992)
Human Exposure Assessment for Airborne Pollutants (1991)
Rethinking the Ozone Problem in Urban and Regional Air Pollution (1991)
Decline of the Sea Turtles (1990)

*Copies of these reports may be ordered from the National Academies Press*
*(800) 624-6242 or (202) 334-3313*
*www.nap.edu*

# OTHER REPORTS OF THE BOARD ON ENERGY AND ENVIRONMENTAL SYSTEMS

Vision 21: Fossil Fuel Options for the Future (2000)

Renewable Power Pathways: A Review of the U.S. Department of Energy's Renewable Energy Programs (2000)

Letter Report on Recent Initiatives by the Office of Energy Efficiency & Renewable Energy and the Office of Power Technologies (2000)

Review of DOE's Office of Fossil Energy's Research Plan for Fine Particulates (1999)

Review of the Research Strategy for Biomass-Derived Transportation Fuels (1999)

Review of the Research Program of the Partnership for a New Generation of Vehicles, Fifth Report (1999)

Review of the Research Program of the Partnership for a New Generation of Vehicles, Fourth Report (1998)

Review of the R&D Plan for the U.S. Department of Energy's Office of Advanced Automotive Technologies (1998)

Effectiveness of the United States Advanced Battery Consortium as a Government-Industry Partnership (1998)

Review of the Research Program of the Partnership for a New Generation of Vehicles, Third Report (1997)

Application of Digital Instrumentation and Control Technology to Nuclear Power Plant Operations and Safety (Phase 1, 1995; Phase 2, 1997)

Review of the Research Program of the Partnership for a New Generation of Vehicles, Second Report (1996)

Decontamination & Decommissioning of Uranium Enrichment Facilities (1996)

Separations Technology and Transmutation Systems (1995)

Coal: Energy for the Future (1995)

Review of the Research Program of the Partnership for a New Generation of Vehicles, First Report (1994)

Review of the Strategic Plan of the U.S. Department of Energy's Office of Conservation and Renewable Energy (1993)

Nuclear Power: Technical and Institutional Options for the Future (1992)

Automotive Fuel Economy: How Far Should We Go? (1992)

The National Energy Modeling System (1992)

Potential Applications of Concentrated Solar Photons (1991)

Assessment of Research Needs for Wind Turbine Rotor Materials Technology (1991)

Alternative Applications of Atomic Vapor Laser Isotope Separation Technology (1991)

Fuels to Drive Our Future (1990)

Confronting Climate Change: Strategies for Energy Research and Development (1990)

Nuclear Engineering Education: Status and Prospects (1990)

University Research Reactors in the United States—Their Role and Value (1988)

*Copies of these reports may be ordered from the National Academies Press*
*(800) 624-6242 or (202) 334-3313*
*www.nap.edu*

# OTHER REPORTS OF THE BOARD ON SCIENCE, TECHNOLOGY, AND ECONOMIC POLICY

21st Century Innovation Systems for Japan and the United States: Lessons from a Decade of Change: Report of a Symposium (2009)

Innovative Flanders: Innovation Policies for the 21st Century: Report of a Symposium (2008)

Innovation in Global Industries: U.S. Firms Competing in a New World (Collected Studies) (2008)

India's Changing Innovation System: Achievements, Challenges, and Opportunities for Cooperation: Report of a Symposium (2007)

Innovation Policies for the 21st Century: Report of a Symposium

Committee on Comparative Innovation Policy: Best Practice for the 21st Century (2007)

Innovation Inducement Prizes at the National Science Foundation (2007)

Enhancing Productivity Growth in the Information Age: Measuring and Sustaining the New Economy (2007)

The Telecommunications Challenge: Changing Technologies and Evolving Policies— Report of a Symposium (2006)

Aeronautics Innovation: NASA's Challenges and Opportunities (2006)

Measuring and Sustaining the New Economy, Software, Growth, and the Future of the U.S Economy: Report of a Symposium (2006)

Reaping the Benefits of Genomic and Proteomic Research: Intellectual Property Rights, Innovation, and Public Health (2006)

Deconstructing the Computer: Report of a Symposium (2005)

Partnering Against Terrorism: Summary of a Workshop (2005)

Research and Development Data Needs: Proceedings of a Workshop (2005)

Productivity and Cyclicality in Semiconductors: Trends, Implications, and Questions: Report of a Symposium (2005)

A Patent System for the 21st Century (2004)

Patents in the Knowledge-Based Economy (2003)

Securing the Future: Regional and National Programs to Support the Semiconductor Industry (2003)

Government-Industry Partnerships for the Development of New Technologies (2002)

Partnerships for Solid-State Lighting: Report of a Workshop (2002)

Using Human Resource Data to Track Innovation: Summary of a Workshop (2002)

Medical Innovation in the Changing Healthcare Marketplace: Conference Summary (2002)

Measuring and Sustaining the New Economy: Report of a Workshop (2002)

Trends in Federal Support of Research and Graduate Education (2001)

The Advanced Technology Program: Assessing Outcomes (2001)

A Review of the New Initiatives at the NASA Ames Research Center: Summary of a Workshop (2001)

Capitalizing on New Needs and New Opportunities: Government-Industry Partnerships in Biotechnology and Information Technologies (2001)

Building a Workforce for the Information Economy (2001)

*Copies of these reports may be ordered from the National Academies Press*
*(800) 624-6242 or (202) 334-3313*
*www.nap.edu*

# Preface

The U.S. Congress directed the U.S. Department of the Treasury to arrange for a review by the National Academy of Sciences to define and evaluate the health, environmental, security, and infrastructural external costs and benefits associated with the production and consumption of energy—costs and benefits that are not or may not be fully incorporated into the market price of energy, into the federal tax or fee, or into other applicable revenue measures related to production and consumption of energy.

In response, the National Research Council established the Committee on Health, Environmental, and Other External Costs and Benefits of Energy Production and Consumption, which prepared this report. Biographic information on the committee members is presented in Appendix A.

In the course of preparing this report, the committee met six times. At two of the meetings, oral presentations were made by the following individuals at the invitation of the committee: Christopher Miller (staff for U.S. Senator Harry Reid); Mark Heil and John Worth (U.S. Department of the Treasury); Raymond Braitsch, Thomas Grahame, and Robert Marlay (U.S. Department of Energy); Robert Brenner and James Democker (U.S. Environmental Protection Agency); Arthur Rypinski (U.S. Department of Transportation); Nicholas Muller (Middlebury College); and Richard Tol (Economic and Social Research Institute, Dublin, Ireland). Interested members of the public at large were also given an opportunity to speak on these occasions. Subsequently, the committee held two teleconferences and one subgroup meeting to complete its deliberations.

In addition to the information from those presentations, the committee made use of peer-reviewed scientific literature, government agency reports, and databases.

This report has been reviewed in draft form by individuals chosen for their diverse perspectives and technical expertise in accordance with procedures approved by the National Research Council Report Review Committee. The purpose of this independent review is to provide candid and critical comments that will assist the institution in making its published report as sound as possible and to ensure that the report meets institutional standards for objectivity, evidence, and responsiveness to the study charge. The review comments and draft manuscript remain confidential to protect the integrity of the deliberative process. We wish to thank the following for their review of this report: David T. Allen, University of Texas, Austin; William F. Banholzer, the Dow Chemical Company; Eric J. Barron, National Center for Atmospheric Research; Donald Boesch, University of Maryland; Dallas Burtraw, Resources for the Future; Douglas M. Chapin, MPR Associates, Inc.; A. Myrick Freeman, III, professor emeritus, Bowdoin College; Charles H. Goodman, Southern Company Services, Inc. (retired); Dale W Jorgenson, Harvard University; Nathaniel Keohane, Environmental Defense Fund; Jonathan I. Levy, Harvard School of Public Health; Erik Lichtenberg, University of Maryland; Robert O. Mendelsohn, Yale University; Armistead Russell, Georgia Institute of Technology; Kumares C. Sinha, Purdue University; Kerry Smith, Arizona State University; Kirk R. Smith, University of California, Berkeley; Susan Tierney, Analysis Group; and Michael Walsh, Independent Consultant.

Although the reviewers listed above have provided many constructive comments and suggestions, they were not asked to endorse the conclusions or recommendations, nor did they see the final draft of the report before its release. The review of this report was overseen by Lawrence T. Papay, Science Applications International Corporation (retired) and Charles E. Phelps, University of Rochester. Appointed by the National Research Council, they were responsible for making certain that an independent examination of this report was carried out in accordance with institutional procedures and that all review comments were carefully considered. Responsibility for the final content of this report rests entirely with the author committee and the institution.

We wish to thank Eric Barron (National Center for Atmospheric Research) and Robert Stavins (Harvard University) for their service as members of the committee during the early stages of this study; they resigned from the committee for personal reasons.

Ronnie Brodsky (University of Maryland) and Paulina Jaramillo and Constantine Samaras (Carnegie Mellon University) helped with information gathering and literature reviews. Joseph Maher (Resources for the Future) assisted in data analysis and in developing report illustrations. The committee's work was assisted by staff of the National Research Council's Board on Environmental Studies and Toxicology (BEST); the Board on Energy and Environmental Systems (BEES); and the Board on Science, Technology, and Economic Policy (STEP). We wish to thank Raymond Wassel, project director, and James Reisa (director of BEST) Steve Merrill (director of STEP) and James Zucchetto (director of BEES). Scientific and technical information was provided by David Policansky, Keegan Sawyer, Patrick Baur, Alan Crane, Leah Nichols, Duncan Brown, and Mirsada Karalic-Loncarevic. Logistical support was provided by John Brown and Daniel Mullins. Radiah Rose managed the production of the report, Ruth Crossgrove was the editor, and Steve Marcus served as a contributing editor.

Jared Cohon, *Chair*
Committee on Health, Environmental,
    and Other External Costs and Benefits
    of Energy Production and Consumption

# Contents

### APPENDIXES

# Boxes, Figures, and Tables

## BOXES

## FIGURES

## TABLES

# HIDDEN COSTS OF
# ENERGY

## UNPRICED CONSEQUENCES OF ENERGY PRODUCTION AND USE

# Summary

Modern civilization is heavily dependent on energy from sources such as coal, petroleum, and natural gas. Yet, despite energy's many benefits, most of which are reflected in energy market prices, the production, distribution, and use of energy also cause negative effects. Beneficial or negative effects that are not reflected in energy market prices are termed "external effects" by economists. In the absence of government intervention, external effects associated with energy production and use are generally not taken into account in decision making.

When prices do not adequately reflect them, the monetary value assigned to benefits or adverse effects (referred to as damages) are "hidden" in the sense that government and other decision makers, such as electric utility managers, may not recognize the full costs of their actions. When market failures like this occur, there may be a case for government interventions in the form of regulations, taxes, fees, tradable permits, or other instruments that will motivate such recognition.

Recognizing the significance of the external effects of energy, Congress requested this study in the Energy Policy Act of 2005 and later directed the Department of the Treasury to fund it under the Consolidated Appropriations Act of 2008. The National Research Council committee formed to carry out the study was asked to define and evaluate key external costs and benefits—related to health, environment, security, and infrastructure—that are associated with the production, distribution, and use of energy but not reflected in market prices or fully addressed by current government policy. The committee was not asked, however, to recommend specific strategies for addressing such costs because policy judgments that transcend scientific and technological considerations—and exceed the committee's mandate—would necessarily be involved.

The committee studied energy technologies that constitute the largest portion of the U.S. energy system or that represent energy sources showing substantial increases (>20%) in consumption over the past several years. We evaluated each of these technologies over their entire life cycles—from fuel extraction to energy production, distribution, and use to disposal of waste products—and considered the external effects at each stage.

Estimating the damages associated with external effects was a multi-step process, with most steps entailing assumptions and their associated uncertainties. Our method, based on the "damage function approach," started with estimates of burdens (such as air-pollutant emissions and water-pollutant discharges). Using mathematical models, we then estimated these burdens' resultant ambient concentrations as well the ensuing exposures. The exposures were then associated with consequent effects, to which we attached monetary values in order to produce damage estimates. One of the ways economists assign monetary values to energy-related adverse effects is to study people's preferences for reducing those effects. The process of placing monetary values on these impacts is analogous to determining the price people are willing to pay for commercial products. We applied these methods to a year close to the present (2005) for which data were available and also to a future year (2030) to gauge the impacts of possible changes in technology.

A key requisite to applying our methods was determining which policy-relevant effects are truly external, as defined by economists. For example, increased food prices caused by the conversion of agricultural land from food to biofuel production, are *not* considered to represent an external cost, as they result from (presumably properly functioning) markets. Higher food prices may of course raise important social concerns and may thus be an issue for policy makers, but because they do not constitute an external cost they were not included in the study.

Based on the results of external-cost studies published in the 1990s, we focused especially on air pollution. In particular, we evaluated effects related to emissions of particulate matter (PM), sulfur dioxide ($SO_2$), and oxides of nitrogen ($NO_x$), which form criteria air pollutants.[1] We monetized effects of those pollutants on human health, grain crop and timber yields, building materials, recreation, and visibility of outdoor vistas. Health damages, which include premature mortality and morbidity (such as chronic bronchi-

---

[1]Criteria pollutants, also known as "common pollutants" are identified by the U.S. Environmental Protection Agency (EPA), pursuant to the Clean Air Act, as ambient pollutants that come from numerous and diverse sources and that are considered to be harmful to public health and the environment and to cause property damage.

tis and asthma), constituted the vast majority of monetized damages, with premature mortality being the single largest health-damage category.

Some external effects could only be discussed in qualitative terms in this report. Although we were able to quantify and then monetize a wide range of burdens and damages, many other external effects could not ultimately be monetized because of insufficient data or other reasons. In particular, the committee did not monetize impacts of criteria air pollutants on ecosystem services or nongrain agricultural crops, or effects attributable to emissions of hazardous air pollutants.[2] In any case, it is important to keep in mind that the individual estimates presented in this report, even when quantifiable, can have large uncertainties.

In addition to its external effects in the present, the use of fossil fuels for energy creates external effects in the future through its emissions of atmospheric greenhouse gases (GHGs)[3] that cause climate change, subsequently resulting in damages to ecosystems and society. This report estimates GHG emissions from a variety of energy uses, and then, based on previous studies, provides *ranges* of potential damages. The committee determined that attempting to estimate a single value for climate-change damages would have been inconsistent with the dynamic and unfolding insights into climate change itself and with the extremely large uncertainties associated with effects and range of damages. Because of these uncertainties and the long time frame for climate change, our report discusses climate-change damages separately from damages not related to climate change.

## OVERALL CONCLUSIONS AND IMPLICATIONS

### Electricity

Although the committee considered electricity produced from coal, natural gas, nuclear power, wind, solar energy, and biomass, it focused mainly on coal and natural gas—which together account for nearly 70% of the nation's electricity—and on monetizing effects related to the air pollution from these sources. From previous studies, it appeared that the electricity-*generation* activities accounted for the majority of such external effects, with other activities in the electricity cycle, such as mining and drilling, playing a lesser role.

---

[2]Hazardous air pollutants, also known as toxic air pollutants, are those pollutants that are known or suspected to cause cancer or other serious health effects, such as reproductive effects and birth defects, or adverse environmental effects.

[3]Greenhouse gases absorb heat from the earth's surface and lower atmosphere, resulting in much of the energy being radiated back toward the surface rather than into space. These gases include water vapor, $CO_2$, ozone, methane, and nitrous oxide.

*Coal*

Coal, a nonrenewable fossil fuel, accounts for nearly half of all electricity produced in the United States. We monetized effects associated with emissions from 406 coal-fired power plants, excluding Alaska and Hawaii, during 2005. These facilities represented 95% of the country's electricity from coal. Although coal-fired electricity generation from the 406 sources resulted in large amounts of pollution overall, a plant-by-plant breakdown showed that the bulk of the damages were from a relatively small number of them. In other words, specific comparisons showed that the source-and-effect landscape was more complicated than the averages would suggest.

*Damages Unrelated to Climate Change*   The aggregate damages associated with emissions of $SO_2$, $NO_x$, and PM from these coal-fired facilities in 2005 were approximately $62 billion, or $156 million on average per plant.[4] However, the differences among plants were wide—the 5th and 95th percentiles of the distribution were $8.7 million and $575 million, respectively. After ranking all the plants according to their damages, we found that the 50% of plants with the lowest damages together produced 25% of the net generation of electricity but accounted for only 12% of the damages. On the other hand, the 10% of plants with the highest damages, which also produced 25% of net generation, accounted for 43% of the damages. Figure S-1 shows the distribution of damages among coal-fired plants.

Some of the variation in damages among plants occurred because those that generated more electricity tended to produce greater damages; hence, we also reported damages per kilowatt hour (kWh) of electricity produced. If plants are weighted by the amount of electricity they generate, the mean damage is 3.2 cents per kWh. For the plants examined, variation in damages per kWh is primarily due to variation in pollution intensity (emissions per kWh) among plants, rather than variation in damages per ton of pollutant. Variations in emissions per kWh mainly reflected the sulfur content of the coal burned; the adoption, or not, of control technologies (such as scrubbers); and the vintage of the plant—newer plants were subject to more stringent pollution-control requirements. As a result, the distribution of damages per kWh was highly skewed: There were many coal-fired power plants with modest damages per kWh as well as a small number of plants with large damages. The 5th percentile of damages per kWh is less than half a cent, and the 95th percentile of damages is over 12 cents.[5]

The estimated air-pollution damages associated with electricity generation from coal in 2030 will depend on many factors. For example, damages

---

[4]Costs are reported in 2007 dollars.

[5]When damages per kWh are weighted by electricity generation, the 5th and 95th percentiles are 0.19 and 12 cents; the unweighted figures are .53 and 13.2 cents per kWh.

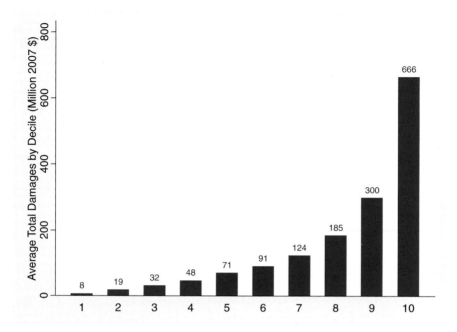

**FIGURE S-1** Distribution of aggregate damages among the 406 coal-fired power plants analyzed in this study. In computing this chart, plants were sorted from smallest to largest based on damages associated with each plant. The lowest decile (10% increment) represents the 40 plants with the smallest damages per plant (far left). The decile of plants that produced the most damages is on the far right. The figure on the top of each bar is the average damage across all plants of damages associated with sulfur dioxide, oxides of nitrogen, and particulate matter. Damages related to climate-change effects are not included.

per kWh are a function of the emissions intensity of electricity generation from coal (for example, pounds [lb] of $SO_2$ per megawatt hour [MWh]), which in turn depends on future regulation of power-plant emissions. Based on government estimates, net power generation from coal in 2030 is expected to be 20% higher on average than in 2005. Despite projected increases in damages per ton of pollutant resulting mainly from population and income growth—average damages per kWh from coal plants (weighted by electricity generation) are estimated to be 1.7 cents per kWh in 2030 as compared with 3.2 cents per kWh in 2005. This decrease derives from the assumption that $SO_2$ emissions per MWh will fall by 64% and that $NO_x$ and PM emissions per MWh will each fall by approximately 50%.

*Natural Gas*

An approach similar to that used for coal allowed the committee to estimate criteria-pollutant-related damages for 498 facilities in 2005 that generated electricity from natural gas in the contiguous 48 states. These facilities represented 71% of the country's electricity from natural gas. Again, as with coal, the overall averages masked some major differences among plants, which varied widely in terms of pollution generation.

*Damages Unrelated to Climate Change*   Damages from gas-fueled plants tend to be much lower than those from coal plants. The sample of 498 gas facilities produced $740 million in aggregate damages from emissions of $SO_2$, $NO_x$, and PM. Average annual damages per plant were $1.49 million, which reflected not only lower damages per kWh at gas plants but smaller plant sizes as well; net generation at the median coal plant was more than six times larger than that of the median gas facility. After sorting the gas plants according to damages, we found that the 50% with the lowest damages accounted for only 4% of aggregate damages. By contrast, the 10% of plants with the largest damages produced 65% of the air-pollution damages from all 498 plants (see Figure S-2). Each group of plants accounted for approximately one-quarter of the sample's net generation of electricity.

Mean damages per kWh were 0.16 cents when natural-gas-fired plants were weighted by the amount of electricity they generated. However, the distribution of damages per kWh had a large variance and was highly skewed. The 5th percentile of damages per kWh is less than 5/100 of a cent, and the 95th percentile of damages is about 1 cent.[6]

Although overall electricity production from natural gas in 2030 is predicted to increase by 9% from 2005 levels, the average pollution intensity for natural-gas facilities is expected to decrease, though not as dramatically as for coal plants. Pounds of $NO_x$ emitted per MWh are estimated to fall, on average, by 19%, and emissions of PM per MWh are estimated to fall by about 32%. The expected net effect of these changes is a decrease in the aggregate damages related to the 498 gas facilities from $740 million in 2005 to $650 million in 2030. Their average damage per kWh is expected to fall from 0.16 cents to 0.11 cents over that same period.

*Nuclear*

The 104 U.S. nuclear reactors currently account for almost 20% of the nation's electrical generation. Overall, other studies have found that

---

[6]When damages per kWh are weighted by electricity generation, the 5th and 95th percentiles are 0.001 and 0.55 cents; the unweighted figures are .0044 and 1.7 cents per kWh.

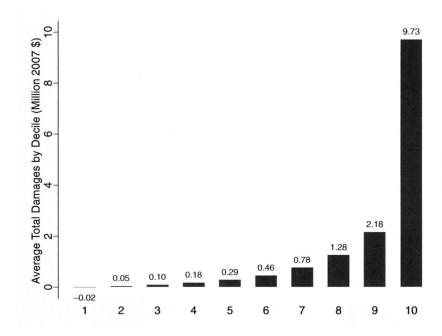

FIGURE S-2 Distribution of aggregate damages among the 498 natural-gas-fired power plants analyzed in this study. In computing this chart, plants were sorted from smallest to largest based on damages associated with each plant. The lowest decile (10% increment) represents the 50 plants with the smallest damages per plant (far left). The decile of plants that produced the most damages is on the far right. The figure on the top of each bar is the average damage across all plants of damages associated with sulfur dioxide, oxides of nitrogen, and particulate matter. Damages related to climate-change effects are not included.

damages associated with the normal operation of nuclear power plants (excluding the possibility of damages in the remote future from the disposal of spent fuel) are quite low compared with those of fossil-fuel-based power plants.[7]

However, the life cycle of nuclear power does pose some risks. If uranium mining activities contaminate ground or surface water, people could potentially be exposed to radon or other radionuclides through ingestion.

---

[7]The committee did not quantify damages associated with nuclear power. Such an analysis would have involved power-plant risk modeling and spent-fuel transportation modeling that would have required far greater resources and time than were available for this study.

Because the United States mines only about 5% of the world's uranium supply, such risks are mostly experienced in other countries.

Low-level nuclear waste is stored until it decays to background levels and currently does not pose an immediate environmental, health, or safety hazard. However, regarding spent nuclear fuel, development of full-cycle, closed-fuel processes that recycle waste and enhance security could further lower risks.

A permanent repository for spent fuel and other high-level nuclear wastes is perhaps the most contentious nuclear-energy issue, and considerably more study of the external cost of such a repository is warranted.

*Renewable Energy Sources*

Wind power currently provides just over 1% of U.S. electricity, but it has large growth potential. Because no fuel is involved in electricity generation, neither gases nor other contaminants are released during the operation of a wind turbine. Its effects do include potentially adverse visual and noise effects, and the killing of birds and bats. In most cases, wind-energy plants currently do not kill enough birds to cause population-level problems, except perhaps locally and mainly with respect to raptors. The tallies of bats killed and the population consequences of those deaths have not been quantified but could be significant. If the number of wind-energy facilities continues to grow as fast as it has recently, bat and perhaps bird deaths could become more significant.

Although the committee did not evaluate in detail the effects of solar and biomass generation of electricity, it has seen no evidence that they currently produce adverse effects comparable in aggregate to those of larger sources of electricity. However, as technology improves and penetration into the U.S. energy market grows, the external costs of these sources will need to be reevaluated.

*Greenhouse Gas Emissions and Electricity Generation*

Emissions of carbon dioxide ($CO_2$) from coal-fired power plants are the largest single source of GHGs in the United States. $CO_2$ emissions vary; their average is about 1 ton of $CO_2$ per MWh generated, having a 5th-to-95th-percentile range of 0.95-1.5 tons. The main factors affecting these differences are the technology used to generate the power and the age of the plant. Emissions of $CO_2$ from gas-fired power plants also are significant, having an average of about 0.5 ton of $CO_2$ per MWh generated and a 5th-to-95th-percentile range of 0.3-1.1 tons. Life-cycle $CO_2$ emissions from nuclear, wind, biomass, and solar appear so small as to be negligible compared with those from fossil fuels.

## Heating

The production of heat as an end use accounts for about 30% of U.S. primary energy demand, the vast majority of which derives from the combustion of natural gas or the application of electricity. External effects associated with heat production come from all sectors of the economy, including residential and commercial (largely for the heating of living or work spaces) and industrial (for manufacturing processes).

### *Damages Unrelated to Climate Change*

As with its combustion for electricity, combustion of natural gas for heat results in lower emissions than from coal, which is the main energy source for electricity generation. Therefore health and environmental damages related to obtaining heat directly from natural-gas combustion are much less than damages from the use of electricity for heat. Aggregate damages from the combustion of natural gas for direct heat are estimated to be about $1.4 billion per year, assuming that the magnitude of external effects resulting from heat production for industrial activities is comparable to that of residential and commercial uses.[8] The median estimated damages attributable to natural-gas combustion for heat in residential and commercial buildings are approximately 11 cents per thousand cubic feet. These damages do not vary much across regions when considered on a per-unit basis, although some counties have considerably higher external costs than others. In 2007, natural-gas use for heating in the industrial sector, excluding its employment as a process feedstock, was about 25% less than natural-gas use in the residential and commercial building sectors.

Damages associated with energy for heat in 2030 are likely to be about the same as those that exist today, assuming that the effects of additional sources to meet demand are offset by lower-emitting sources. *Reduction* in damages would only result from more significant changes—largely in the electricity-generating sector, as emissions from natural gas are relatively small and well controlled. However, the greatest potential for reducing damages associated with the use of energy for heat lies in greater attention to improving efficiency. Results from the recent National Research Council report *America's Energy Future: Technology and Transformation* suggest a possible improvement of energy efficiency in the buildings and industrial sectors by 25% or more between now and 2030. *Increased* damages would also be possible, however, if new domestic energy development resulted in higher emissions or if additional imports of liquefied natural gas, which

---

[8]Insufficient data were available to conduct a parallel analysis of industrial activities that generate useful heat as a side benefit.

would increase emissions from the production and international transport of the fuel, were needed.

## Greenhouse Gas Emissions

The combustion of a thousand cubic feet of gas generates about 120 lb (0.06 tons) of $CO_2$. Methane, the major component of natural gas, is a GHG itself and has a global-warming potential about 25 times that of $CO_2$. Methane enters the atmosphere through leakage, but the U.S. Energy Information Administration estimates that such leakage amounted to less than 3% of total U.S. $CO_2$-equivalent ($CO_2$-eq) emissions[9] (excluding water vapor) in 2007. Thus, in the near term, where domestic natural gas remains the dominant source for heating, the average emissions factor is likely to be about 140 lb $CO_2$-eq per thousand cubic feet (including upstream methane emissions); in the longer term—assuming increased levels of liquefied natural gas or shale gas as part of the mix—the emissions factor could be 150 lb $CO_2$-eq per thousand cubic feet.

# Transportation

Transportation, which today is almost completely reliant on petroleum, accounts for nearly 30% of U.S. energy consumption. The majority of transportation-related emissions come from fossil-fuel combustion—whether from petroleum consumed during conventional-vehicle operation, coal or natural gas used to produce electricity to power electric or hybrid vehicles, petroleum or natural gas consumed in cultivating biomass fields for ethanol, or electricity used during vehicle manufacture.

The committee focused on both the nonclimate-change damages and the GHG emissions associated with light-duty and heavy-duty on-road vehicles, as they account for more than 75% of transportation energy consumption in the United States. Although damages from nonroad vehicles (for example, aircraft, locomotives, and ships) are not insignificant, the committee emphasized the much larger highway component.

## Damages Unrelated to Climate Change

In 2005, the vehicle sector produced $56 billion in health and other nonclimate-change damages, with $36 billion from light-duty vehicles and $20 billion from heavy-duty vehicles. Across the range of light-duty technology and fuel combinations considered, damages expressed per vehicle

---

[9]$CO_2$-eq expresses the global-warming potential of a given stream of GHGs, such as methane, in terms of $CO_2$ quantities.

miles traveled (VMT) ranged from 1.2 cents to 1.7 cents (with a few combinations having higher damage estimates).[10]

The committee evaluated motor-vehicle damages over four life-cycle stages: (1) vehicle operation, which results in tailpipe emissions and evaporative emissions; (2) production of feedstock, including the extraction of the resource (oil for gasoline, biomass for ethanol, or fossil fuels for electricity) and its transportation to the refinery; (3) refining or conversion of the feedstock into usable fuel and its transportation to the dispenser; and (4) manufacturing and production of the vehicle. It is important that, in most cases, vehicle operation accounted for less than one-third of total damages; other components of the life cycle contributed the rest. Life-cycle stages 1, 2, and 3 were somewhat proportional to actual fuel use, while stage 4 (which is a significant source of life-cycle emissions that form criteria pollutants) was not.

The estimates of damage per VMT among different combinations of fuels and vehicle technologies were remarkably similar (see Figure S-3). Because these assessments were so close, it is essential to be cautious when interpreting small differences between combinations. The damage estimates for 2005 and 2030 also were very close, despite an expected rise in population. This result is attributable to the expected national implementation of the recently revised "corporate average fuel economy" (CAFE) standards, which require the new light-duty fleet to have an average fuel economy of 35.5 miles per gallon by 2016 (although an increase in VMT could offset this improvement somewhat).

Despite the general overall similarity, some fuel and technology combinations were associated with greater nonclimate damages than others. For example, corn ethanol, when used in E85 (fuel that is 85% ethanol and 15% gasoline), showed estimated damages per VMT similar to or slightly higher than those of gasoline, both for 2005 and 2030, because of the energy required to produce the biofuel feedstock and convert it to fuel. Yet cellulosic (nonfood biomass) ethanol made from herbaceous plants or corn stover had lower damages than most other options when used in E85. The reason for this contrast is that the feedstock chosen and growing practices used influence the overall damages from biomass-based fuels. We did not quantify water use and indirect land use for biofuels.[11]

Electric vehicles and grid-dependent hybrid vehicles showed somewhat

---

[10]The committee also estimated damages on a per-gallon basis, with a range of 23 to 38 cents per gallon (with gasoline vehicles at 29 cents per gallon). Interpretation of the results is complicated, however, by the fact that fuel and technology combinations with higher fuel efficiency appear to have markedly higher damages per gallon than those with lower efficiency solely due to the higher number of miles driven per gallon.

[11]Indirect land use refers to geographical changes occurring indirectly as a result of biofuels policy in the United States and the effects of such changes on GHG emissions.

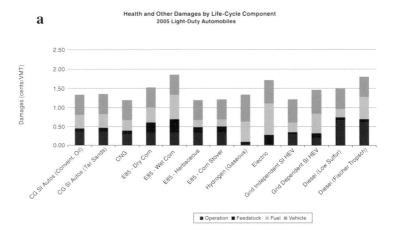

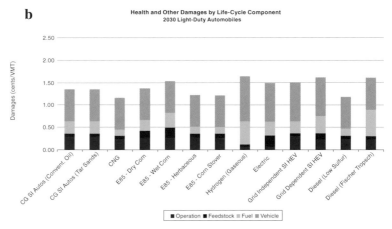

**FIGURE S-3** Health effects and other nonclimate damages are presented by life-cycle component for different combinations of fuels and light-duty automobiles in 2005 (*a*) and 2030 (*b*). Damages are expressed in cents per VMT (2007 U.S. dollars). Going from bottom to top of each bar, damages are shown for life-cycle stages as follows: vehicle operation, feedstock production, fuel refining or conversion, and vehicle manufacturing. Damages related to climate change are not included. ABBREVIATIONS: VMT, vehicle miles traveled; CG SI, conventional gasoline spark ignition; CNG, compressed natural gas; E85, 85% ethanol fuel; HEV, hybrid electric vehicle.

higher damages than many other technologies for both 2005 and 2030. Although operation of the vehicles produces few or no emissions, electricity production at present relies mainly on fossil fuels and, based on current emission control requirements, emissions from this stage of the life cycle are expected to still rely primarily on those fuels by 2030, albeit at significantly lower emission rates. In addition, battery and electric motor production—being energy- and material-intensive—added up to 20% to the damages from manufacturing.

Compressed natural gas had lower damages than other options, as the technology's operation and fuel produce very few emissions.

Although diesel had some of the highest damages in 2005, it is expected to have some of the lowest in 2030, assuming full implementation of the Tier 2 vehicle emission standards of the U.S. Environmental Protection Agency (EPA). This regulation, which requires the use of low-sulfur diesel, is expected to significantly reduce PM and $NO_x$ emissions as well. Heavy-duty vehicles have much higher damages per VMT than light-duty vehicles because they carry more cargo or people and, therefore, have lower fuel economies. However, between 2005 and 2030, these damages are expected to drop significantly, assuming the full implementation of the EPA Heavy-Duty Highway Vehicle Rule.

*Greenhouse Gas Emissions*

Most vehicle and fuel combinations had similar levels of GHG emissions in 2005 (see Figure S-4). Because vehicle operation is a substantial source of life-cycle GHGs, enforcement of the new CAFE standards will have a greater impact on lowering GHG emissions than on lowering life-cycle emissions of other pollutants. By 2030, with improvements among virtually all light-duty-vehicle types, the committee estimates that there will be even fewer differences in the GHG emissions of the various technologies than there were in 2005. However, in the absence of additional fuel-efficiency requirements, heavy-duty vehicle GHG emissions are expected to change little between 2005 and 2030, except from a slight increase in fuel economy in response to market conditions.

For both 2005 and 2030, vehicles using gasoline made from petroleum extracted from tar sands and diesel derived from Fischer-Tropsch fuels[12] have the highest life-cycle GHG emissions among all fuel and vehicle combinations considered. Vehicles using celluosic E85 from herbaceous feedstock or corn stover have some of the lowest GHG emissions because of the feedstock's ability to store $CO_2$ in the soil. Those using compressed natural gas also had comparatively low GHG emissions.

*Future Reductions*

Substantially reducing nonclimate damages related to transportation would require major technical breakthroughs, such as cost-effective con-

---

[12]The Fischer-Tropsch reaction converts a mixture of hydrogen and carbon monoxide—derived from coal, methane, or biomass—into liquid fuel. In its analysis, the committee considered only the use of methane for the production of Fischer-Tropsch diesel fuel.

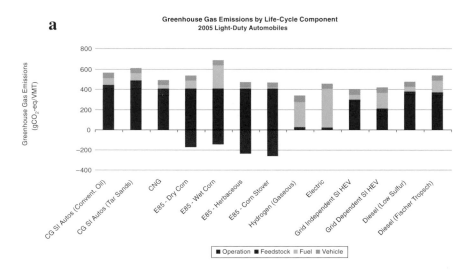

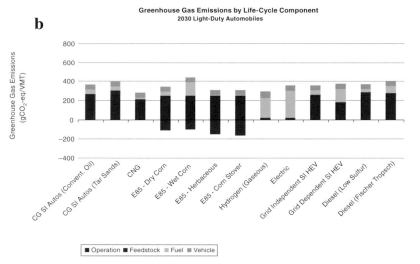

**FIGURE S-4** Greenhouse gas emissions (grams $CO_2$-eq)/VMT by life-cycle component for different combinations of fuels and light-duty automobiles in 2005 (*a*) and 2030 (*b*). Going from bottom to top of each bar, damages are shown for life-cycle stages as follows: vehicle operation, feedstock production, fuel refining or conversion, and vehicle manufacturing. One exception is ethanol fuels for which feedstock production exhibits negative values because of $CO_2$ uptake. The amount of $CO_2$ consumed should be subtracted from the positive value to arrive at a net value. AB-BREVIATIONS: g $CO_2$-eq, grams $CO_2$-equivalent; VMT, vehicle mile traveled; CG SI, conventional gasoline spark ignition; CNG, compressed natural gas; E85, 85% ethanol fuel; HEV, hybrid electric vehicle.

version of cellulosic biofuels, cost-effective carbon capture, and storage for coal-fired power plants, or a vast increase in renewable energy capacity or other forms of electricity generation with lower emissions.[13] Further enhancements in fuel economy will also help, especially for emissions from vehicle operations, although they are only about one-third of the total life-cycle picture and two other components are proportional to fuel use. In any case, better understanding of potential external costs at the earliest stage of vehicle research should help developers minimize those costs as the technology evolves.

### Estimating Climate-Change Damages

Energy production and use continue to be major sources of GHG emissions, principally $CO_2$ and methane. Damages from these emissions will result as their increased atmospheric concentrations affect climate, which in turn will affect such things as weather, freshwater supply, sea level, biodiversity, and human society and health.[14]

Estimating these damages is another matter, as the prediction of climate-change effects, which necessarily involves detailed modeling and analysis, is an intricate and uncertain process. It requires aggregation of potential effects and damages that could occur at different times (extending centuries into the future) and among different populations across the globe. Thus, rather than attempt such an undertaking itself, especially given the constraints on its time and resources, the committee focused its efforts on a review of existing integrated assessment models (IAMs) and the associated climate-change literature.

We reviewed IAMs in particular, which combine simplified global-climate models with economic models that are used to (1) estimate the economic impacts of climate change, and (2) identify emissions regimes that balance the economic impacts with the costs of reducing GHG emissions. Because IAM simulations usually report their results in terms of mean values, this approach does not adequately capture some possibilities of catastrophic outcomes. Although a number of the possible outcomes have been studied—such as release of methane from permafrost that could rapidly accelerate warming and collapse of the West Antarctic or Greenland ice sheets, which could raise sea level by several meters—the damages associated with these events and their probabilities are very poorly understood.

---

[13]The latter two changes are needed to reduce the life-cycle damages of grid-dependent vehicles.

[14]In response to a request from Congress, the National Research Council has launched America's Climate Choices, a suite of studies designed to inform and guide responses to climate change across the nation.

Some analysts nevertheless believe that the expected value of total damages may be more sensitive to the possibility of low-probability catastrophic events than to the most likely or best-estimate values.

In any case, IAMs are the best tools currently available. An important factor in using them (or virtually any other model that accounts for monetary impacts over time) is the "discount rate," which converts costs and benefits projected to occur in the future into amounts ("present values") that are compatible with present-day costs and benefits. Because the choice of a discount rate for the long periods associated with climate change is not well-established, the committee did not choose a particular discount rate for assessing the value of climate change's effects; instead, we considered a range of discount-rate values.

Under current best practice, estimates of global damages associated with a particular climate-change scenario at a particular future time are translated by researchers into an estimate of damages per ton of emissions (referred to as marginal damages) by evaluating the linkage between current GHG emissions and future climate-change effects. Marginal damages are usually expressed as the net present value of the damages expected to occur over many future years as the result of an additional ton of $CO_2$-eq emitted into the atmosphere. Estimating these marginal damages depends on the temperature increase in response to a unit increase in $CO_2$-eq emissions, the additional climate-related effects that result, the values of these future damages relative to the present, and how far into the future one looks. Because of uncertainties at each step of the analysis, a given set of possible future conditions may yield widely differing estimates of marginal damages.

Given the preliminary nature of the climate-damage literature, the committee found that only rough order-of-magnitude estimates of marginal damages were possible at this time. Depending on the extent of projected future damages and the discount rate used for weighting them, the range of estimates of marginal damages spanned two orders of magnitude, from about $1 to $100 per ton of $CO_2$-eq, based on current emissions. Approximately one order of magnitude in difference was attributed to discount-rate assumptions and another order of magnitude to assumptions about future damages from emissions used in the various IAMs. The damage estimates at the higher end of the range were associated only with emission paths without significant GHG controls. Estimates of the damages specifically to the United States would be a fraction of the levels in the range of estimates, because this country represents only about one-quarter of the world's economy, and the proportionate impacts it would suffer are generally thought to be lower than for the world as a whole.

## Comparing Climate and Nonclimate Damage Estimates

Comparing nonclimate damages to climate-related damages is extremely difficult. The two measures differ significantly in their time dimensions, spatial scales, varieties of impacts, and degrees of confidence with which they can be estimated. For 2005, determining which type of external effect caused higher damages depended on the energy technology being considered and the marginal damage value selected from the range of $1 to $100 per ton of $CO_2$-eq emitted. For example, coal-fired electricity plants were estimated to emit an average of about 1 ton of $CO_2$ per MWh (or 2 lb/ kWh). When multiplying that emission rate by an assumed marginal damage value of $30/ton $CO_2$-eq, climate-related damages equal 3 cents/kWh, comparable to the 3.2 cents/kWh estimated for nonclimate damages. It is important to keep in mind that the value of $30/ton $CO_2$-eq is provided for illustrative purposes and is not a recommendation of the committee.

**Natural Gas:** The climate-related damages were higher than the nonclimate damages from natural-gas-fired power plants, as well as from combustion of natural gas for producing heat, regardless of the marginal damage estimate. Because natural gas is characterized by low emissions that form criteria pollutants, the nonclimate damages were about an order of magnitude lower than the climate damages estimated by the models, if the marginal climate damage were assumed to be $30/ton $CO_2$-eq.

**Coal:** The climate-related damages from coal-fired power plants were estimated to be higher than the nonclimate damages when the assumed marginal climate damage was greater than $30/ton $CO_2$-eq. If the marginal climate damage was less than $30/ton $CO_2$-eq, the climate-related damages were lower than the nonclimate damages.

**Transportation:** As with coal, the transportation sector's climate-change damages were higher than the nonclimate damages only if the marginal damage for climate was higher than $30/ton $CO_2$-eq.

**Overall:** All of the model results available to the committee estimated that the climate-related damages per ton of $CO_2$-eq would be 50-80% worse in 2030 than in 2005. Even if annual GHG emissions were to remain steady between now and 2030, the damages per ton of $CO_2$-eq emissions would be substantially higher in 2030 than at present. As a result, the climate-related damages in that year from coal-fired power plants and transportation are likely to be greater than their nonclimate damages.

## Infrastructure Risks and Security

The committee also considered external effects and costs associated with disruptions in the electricity-transmission grid, energy facilities' vulnerability to accidents and possible attack, oil-supply disruptions, and other national security issues. We concluded as follows:

- The nation's electric grid is subject to periodic failures because of transmission congestion and the lack of adequate reserve capacity. These failures are considered an external effect, as individual consumers of electricity do not take into account the impact of their consumption on aggregate load. The associated and possibly significant damages of grid failure underscore the importance of carefully analyzing the costs and benefits of investing in a modernized grid—one that takes advantage of new smart technology and that is better able to handle intermittent renewable-power sources.
- The external costs of accidents at energy facilities are largely taken into account by their owners and, at least in the case of our nation's oil and gas transmission networks, are of negligible magnitude per barrel of oil or thousand cubic feet of gas shipped.
- Because the United States is such a large consumer of oil, policies to reduce domestic demand can also reduce the world oil price, thereby benefiting the nation through lower prices on the remaining oil it imports. Government action may thus be a desirable countervailing force to monopoly or cartel-producer power. However, the committee does not consider this influence of a large single buyer (known as monopsony power) to be a benefit that is external to the market price of oil. It was therefore deemed to be outside the scope of this report.
- Although sharp and unexpected increases in oil prices adversely affect the U.S. economy, the macroeconomic disruptions they cause do not fall into the category of external effects and damages. Estimates in the literature of the macroeconomic costs of disruptions and adjustments range from $2 to $8 per barrel.
- Dependence on imported oil has well-recognized implications for foreign policy, and although we find that some of the effects can be viewed as external costs, it is currently impossible to quantify them. For example, the role of the military in safeguarding foreign supplies of oil is often identified as a relevant factor. However, the energy-related reasons for a military presence in certain areas of the world cannot readily be disentangled from the nonenergy-related reasons. Moreover, much of the military cost is likely to be fixed in nature. For example, even a 20% reduction in oil consumption, we believe, would probably have little impact on the strategic positioning of U.S. military forces throughout the world.
- Nuclear waste raises important security issues and poses tough policy challenges. The extent to which associated external effects exist is hard to assess, and even when identified they are very difficult to quantify. Thus, although we do not present numerical values in this report, we recognize the importance of studying these issues further.

## Conclusion

In aggregate, the damage estimates presented in this report for various external effects are substantial. Just the damages from external effects the committee was able to quantify add up to more than $120 billion for the year 2005.[15] Although large uncertainties are associated with the committee's estimates, there is little doubt that this aggregate total substantially underestimates the damages, because it does not include many other kinds of damages that could not be quantified for reasons explained in the report, such as damages related to some pollutants, climate change, ecosystems, infrastructure, and security. In many cases, we have identified those omissions, within the chapters of this report, with the hope that they will be evaluated in future studies.

Even if complete, our various damage estimates would not automatically offer a guide to policy. From the perspective of economic efficiency, theory suggests that damages should not be reduced to zero but only to the point where the cost of reducing another ton of emissions (or other type of burden) equals the marginal damages avoided—that is, the degree to which a burden should be reduced depends on its current level and the cost of lowering it. The solution cannot be determined from the amount of damage alone. Economic efficiency, however, is only one of several potentially valid policy goals that need to be considered in managing pollutant emissions and other burdens. For example, even within the same location, there is compelling evidence that some members of the population are more vulnerable than others to a particular external effect.

Although not a comprehensive guide to policy, our analysis does indicate that regulatory actions can significantly affect energy-related damages. For example, the full implementation of the federal diesel-emission rules would result in a sizeable decrease in nonclimate damages from diesel vehicles between 2005 and 2030. Similarly, major initiatives to further reduce other emissions, improve energy efficiency, or shift to a cleaner electricity-generating mix (for example, renewables, natural gas, and nuclear) could substantially reduce the damages of external effects, including those from grid-dependent hybrid and electric vehicles.

It is thus our hope that this information will be useful to government policy makers, even in the earliest stages of research and development on energy technologies, as an understanding of their external effects and damages could help to minimize the technologies' adverse consequences.

---

[15]These are damages related principally to emissions of $NO_x$, $SO_2$, and PM relative to a baseline of zero emissions from energy-related sources for the effects considered in this study.

# 1

# Introduction

## GENESIS OF THE STUDY

Energy is essential to the functioning of society. From coal for electricity production to oil products for transportation to natural gas for space heating, every aspect of modern life depends on energy. Yet, as beneficial as energy is, its production, distribution, and consumption also have negative impacts especially on human health and the environment.

Most effects of energy are negative, but that does not imply that energy use has an overall negative impact on society. Quite the contrary; the benefits to society of U.S. energy systems are enormous. However, it was not the committee's task to estimate those benefits that are considered largely to be "internal" because they are reflected in energy prices or government policies.

The purpose of this study was to define and evaluate external effects of energy production, use, and consumption, which refer to costs and benefits not taken into account in making decisions (such as the siting of a power plant) or not reflected in market prices (for example, the price of gasoline at the pump). Under such conditions, the actions that follow might be sub-optimal—in the sense that the full social costs of the actions are not recognized—resulting in a loss of social welfare. When market failures like these occur, there is a case for government intervention in the form of regulation, taxes, fees, tradable permits, or other instruments that will cause economic agents to recognize the external effects in their decision making.

Before such public policies are pursued, the external effects of energy and their monetary values should be known. Thus, Congress directed the U.S. Department of the Treasury in the Energy Policy Act of 2005 (P.L. 109-58), Section 1352, to commission a study by the National Academy of Sciences that would "define and evaluate the health, environmental, security, and infrastructure external costs and benefits associated with the production and consumption of energy that are not or may not be fully incorporated into the market price of such energy, or into the Federal revenue measures related to that production or consumption." Funding for the study was later provided through the Consolidated Appropriations Act of 2008 (P.L. 110-161).

## STATEMENT OF TASK

In response to this mandate from Congress and the request from the Department of the Treasury, the National Research Council (NRC) established the Committee on Health, Environmental, and Other External Costs and Benefits of Energy Production and Consumption (see Appendix A). The Statement of Task (Box 1-1) was developed and served as the point of departure and guide for the committee's work. In the remainder of this chapter, we define key terms in the Statement of Task and explain the general procedures followed in executing the task.

This study is one of many related to energy that the NRC has recently undertaken. In the next section, we briefly discuss those NRC studies that have informed our work, especially the America's Energy Future (AEF) initiative, which is identified in the Statement of Task. We also briefly review previous studies on the external costs of energy.

Also in this chapter, we provide the definition of an externality—the focus and core concept of this study—and provide some examples.

The Statement of Task directed us to evaluate the externalities "associated with the production, distribution and consumption of energy from various selected sources." We explain how we selected the sources and the particular elements of the energy system on which we focused.

The approach that we took for identifying, quantifying, and evaluating externalities "in economic terms" is explained. A discussion of "appropriate metrics from each externality category" is included.

Although the committee was not asked to "recommend specific strategies for correcting observable externalities, because those choices will entail policy judgments"—a position with which we agree—it is important to understand and to keep in mind the policy contexts in which our results may be used.

The Statement of Task anticipated some of the methodological challenges of evaluating externalities. We discuss the particular difficulties of

**BOX 1-1**
**Statement of Task**

An NRC committee will define and evaluate key external costs and benefits—health, environmental, security, and infrastructure—associated with the production, distribution, and consumption of energy from various selected sources that are not or may not be fully incorporated into the market price of such energy, or into the federal tax or fee or other applicable revenue measures related to such production, distribution, or consumption. Although the committee will carry out its task from a U.S. perspective, it will consider broader geographic implications of externalities when warranted and feasible. The committee will not recommend specific strategies for internalizing observable externalities, because those choices would entail policy judgments that transcend scientific and technological considerations.

In carrying out its task, the committee will include the following activities:

Seek to build upon the results of the NRC initiative America's Energy Future: Technology, Opportunities, Risks, and Tradeoffs.

Identify key externalities to be assessed in the categories of human health, environment, security (including quality, abundance, and reliability of energy sources), and infrastructure (such as transportation and waste disposal systems not sufficiently taken into account by producers or consumers).

Consider externalities associated with producing, distributing, and consuming energy imported from foreign sources.

Define appropriate metrics for each externality category considered.

Identify state-of-the-science approaches for assessing external effects (actual or expected) and expressing their effects in economic terms.

Develop an approach for estimating externalities related to greenhouse gas emissions and climate change. Estimate externalities related to those changes.

Present qualitative and, to the extent practicable, quantitative estimates of externalities and associated uncertainties within a consistent framework that makes the discussion of externalities and uncertainties associated with energy production, distribution, and consumption more transparent.

When it is not feasible to assess specific externalities comprehensively, the committee will recommend assessment approaches and identify key information needs to inform future assessments.

dealing with space, time and uncertainty. The committee sought to build on the work of companion studies within the NRC, particularly the AEF and America's Climate Choices studies.

## RELATED STUDIES

### National Research Council Studies

With the National Academy of Sciences and National Academy of Engineering having identified energy as a high-priority topic, it has received attention from many NRC committees, both past and current. These studies were relevant to the work of this committee. We briefly discuss two studies here and cite them throughout our report where appropriate.

The AEF's effort at the NRC was concerned with future technology and its potential for reducing U.S. dependence on oil imports and lowering greenhouse gas emissions, while ensuring that affordable energy will be available to sustain economic growth. The AEF's task was to critically review recently completed major studies on the potential for change in energy technology and use; compare the studies' assumptions; analyze the currency and quality of the information used; and assess the relative states of maturity of technologies for potential deployment in the next decade. A secondary focus was on technologies with longer times to deployment.

A study committee and three panels produced an extensive analysis of energy technology options for consideration in an ongoing national dialogue. Collectively, they analyzed advanced coal technologies; nuclear power; renewable energy technologies (such as wind, solar photovoltaic, and geothermal); energy storage and infrastructure technologies; advanced transportation power-train technologies; technologies to improve energy efficiency in residential and commercial buildings, industry, and transportation; and the technical potential for reducing reliance on petroleum-based fuels for transportation. These topics were addressed for three time frames: present-2020, 2020-2035, and beyond 2035.

In response to a request from Congress concerning a related topic, the NRC also launched America's Climate Choices, a suite of studies designed to inform and guide responses to climate change across the nation (see NAS/NAE/NRC 2009a).

The results of the studies were intended to address the following key questions:

- What short-term actions could be taken to respond effectively to climate change?
- What promising long-term strategies, investments, and opportunities could be pursued to respond to climate change?
- What scientific and technological advances (for example, new observations, improved models, and research priorities) are needed to better understand and respond effectively to climate change?

- What are the major impediments—for example, practical, institutional, economic, ethical, and intergenerational—to responding effectively to climate change, and what can be done to overcome them?

The AEF and America's Climate Choices studies are important initiatives that have provided and will continue to provide valuable information on energy technology and policy options for the nation. Indeed, our own study has been informed by the AEF's analysis of future technology. However, neither of these efforts was designed to focus on the *monetary value* of energy's external effects, including climate change.

## Prior Externality Studies

The concept of externalities dates at least to the early twentieth century (Pigou 1920) and was discussed extensively in the post-war economics literature (Meade 1952; Scitovsky 1954; Mishan 1965; Arrow 1975). Interest in the externalities of energy production and use gathered momentum in the following decades. Hohmeyer (1988) was one of the more prominent studies during this period. He took a top-down approach in which he estimated the "toxicity weighted" emissions from electricity generation with fossil fuels and then multiplied this fraction by Wicke's (1986) estimates of total damages from pollution to various end points (such as those on health, forests, and animals). The most prominent study in the United States during this period (Ottinger et al. 1990) used estimates from previous studies that quantified the environmental costs from electric power generation. Results of Niemi et al. (1984, 1987) were among those used by Ottinger et al.; those studies focused on visibility and health effects of airborne particulate matter. Ottinger et al. followed a five-step procedure in using these studies to value environmental damages: emissions, dispersion, exposure, impacts, and damages.

Research in estimating the external costs of energy peaked in the early-to-mid-1990s when public utility commissions in the United States were interested in tilting electric utility investment choices toward sources with lower negative externalities, such as renewable energy. This policy was to be done with an "adder" equal to the marginal damages associated with each type of electricity generation. During this wave of interest, major studies were done by Oak Ridge National Laboratory and Resources for the Future (ORNL/RFF) for the U.S. Department of Energy, by Hagler Bailly for the New York State Energy Research and Development Authority (NYSERDA), by Research Triangle Institute for the State of Wisconsin (one of several states mounting these studies), and by several teams of European research organizations for the European Commission (EC). This latter study, called ExternE, worked in concert with the ORNL/RFF team to use similar protocols.

Around the same time that work began on the ORNL/RFF and ExternE studies, other studies were completed by the Union of Concerned Scientists (1992); Pearce et al. (1992) in their report to the U.K. Department of Trade and Industry; Triangle Economic Research (1995) for Minnesota; National Economic Research Associates (1993) in its study of Nevada; Regional Economic Research (1991) on California; and Consumer Energy Council of America Research Foundation (1993).

Later, the EC began a companion study on the external costs of transportation as well as other research efforts—the most recent being the New Energy Externalities Development for Sustainability (NEEDS) program—that further refined and developed methodologies extending those developed in ExternE as well as extending those to other energy technologies and study locations. Also in the transportation area, a series of studies were conducted by Greene et al. (1997) and Delucchi (2004), Parry et al. (2007), among several other studies performed since the mid-1990s.

The more notable differences between our committee's study and previous studies, particularly the major studies done in the early to mid-1990s, are in the different approaches to, and the extent to which, the studies addressed the following:

a.   Number of power plants—our study considered almost all coal and natural gas power plants in the country, whereas most other studies focused on a few sites or on plants within a state.

b.   Different power-generation options—our study considered fewer technologies than several of the previous larger studies; in particular, our study did not address the nuclear fuel cycle in the detail done in the ORNL/RFF and ExternE studies, which carried out extensive probabilistic risk modeling.

c.   The manner in which the dispersion of airborne pollutants and formation of secondary pollutants were modeled—our study used a reduced form approximation of these processes, whereas some previous models used more site-specific, detailed air dispersion and transformation models, albeit for a small number of power plant sites and regulatory scenarios. The early studies also had no or limited modeling and analysis of ozone and fine (2.5 microns or less in diameter) particulate matter formed from the chemical transformation of pollutants emitted by a power plant.

d.   Consideration of greenhouse gas emissions and their damages—like all previous externality studies, our study reviewed recent literature rather than undertaking new scientific research; our study reviewed more recent literature than most of the previous studies, although recent studies within the ExternE program used similar literature.

e.   Consideration of externalities associated with infrastructure and energy security—of the latter, ORNL/RFF and ExtenE focused only on oil security.

f.   The extent to which the entire life cycle of an energy technology (from feedstock through conversion through fuel distribution through energy service) was analyzed—the ways in which the different technology fuel cycles were analyzed—the ORNL/RFF, ExternE, and our study considered life-cycle impacts, whereas most other studies focused on electricity generation or use of vehicles in transportation and not on the upstream activities in the life cycle.

g.   The extent that externalities, other than those associated with electricity generation, were addressed in the same study (that is, transportation and energy used for heat).

With some exceptions, the "adders" studies of the early and mid-1990s took a place-based approach to damage estimation of energy. They would posit the construction of a new electricity-generating plant of a particular type at a given location. Each study considered a small number of alternative locations for each plant, generally from two to five. In those studies, the different results calculated for the different plants would reflect the influence of the specific location of the plant on the magnitude of the damages. In contrast, our study calculates the health-related and some of the environmental damages for most of the power plants in the United States and estimates the damages from each. In this respect, it is similar to the U.S. Environmental Protection Agency's regulatory impact analyses that use the BenMAP model and to the efforts of Muller and Mendelsohn (2007), which, although not studies of externalities per se, are comprehensive in their level of spatial resolution, such as addressing the specific location of all power-plant emissions.

The ORNL/RFF and ExternE studies included relatively detailed engineering descriptions of the technologies of the power plants, whereas our study and most other studies focused on estimates of emissions from power plants and not on the underlying technologies.

In estimating the health and environmental damages, the ORNL/RFF and ExternE studies used different detailed models to predict the dispersion of primary pollutants from the power plants and the atmospheric formation of secondary pollutants, specifically ozone and fine particulate matter. Studies of externalities associated with greenhouse gas emissions generally either focus exclusively on these emissions and the associated climate change, as exemplified by the authoritative reports of the Intergovernmental Panel on Climate Change, or focus on other pollutants. Our study, on the other hand, provides a range of quantitative estimates of the damages from climate change in monetary terms. The previous externality studies either do not include this important issue in their analysis or, as in the case of the ORNL/RFF and ExternE studies, draw on a more dated and limited scientific literature in this area.

Externalities associated with infrastructure and energy security are usually not addressed in other studies. Our report considers electric grid externalities, infrastructure vulnerability to attacks and accidents, and national security. The ORNL/RFF study provided estimates of damages from dependence on foreign oil, and other studies focused on this issue provided updated estimates. However, as discussed in our report, although damages might result from global dependence on oil in a cartel-dominated market, such damages are not considered externalities.

The ORNL/RFF and ExternE's consideration of damages from different parts of the life cycle, for example, coal mining, sets them apart from most other studies, which do not consider externalities on a life-cycle basis. Several studies have estimated life-cycle emissions of some fuels currently, or prospectively, used in ground transportation; these studies did not attempt to estimate the impacts and associated externalities of these emissions. Our study, on the other hand, takes an energy life-cycle approach somewhat similar to the ORNL/RFF and ExternE studies, but with more updated considerations and data.

Although many studies have addressed different aspects of the externalities from energy production, distribution, or use to varying degrees, they have focused on one type of externality (such as health effects), or one particular sector (usually electricity generation or transportation). Also, they generally focused on one part of the energy cycle. In contrast, our study has a relatively comprehensive scope that includes all types of externalities from energy life cycles of both electricity and transportation, as well as from production and use of energy for heat in residential, commercial, and industrial sectors.

## DEFINING AND MEASURING EXTERNALITIES

### Defining Externalities

External effects or "externalities" are important because failure to account for them can result in distortions in making decisions and in reductions in the welfare of some of society's members.

An externality, which can be positive or negative, is an activity of one agent (for example, an individual or an organization, such as a company) that affects the well-being of another agent and occurs outside the market mechanism. In the absence of government intervention, externalities associated with energy production and use are generally not taken into account in decision making. Box 1-2 provides definitions of the technical terms used in

**BOX 1-2**
**Definitions of Key Terms**

Much of the nomenclature for the key terms is taken from the damage function approach, which has been the standard approach to examine the costs and benefits of environmental regulations, required by OMB (NRC 2002a). This approach begins with some burden, say emissions, which ultimately has some physical effect; this effect is then monetized and termed damage. The monetary value of reductions in burdens is termed benefits (the opposite of damages).

Burdens: Externalities from economic activities are always by-products of those activities, some of which are useful and some of which cause health and environmental effects, for example. The by-products themselves are termed burdens. Emissions of air pollutants and discharges of pollutants into a river are examples of burdens.

Effects and Impacts: These burdens have a real effect in the environment, that is, they have a physical component that affects health, damages ecosystems or reduces visibility, for example. Sometimes, as with energy security, the physical component is not directly present. In any event, these physical effects are termed effects or impacts.

Damages: Damages are the monetary value of the physical effects, in its simplest form calculated by multiplying the quantity of physical effects of interest by a monetary value for that effect. This monetary value represents, ideally, the population average of the maximum willingness to pay for a unit improvement in this physical metric. That is, it reflects the preferences people have for reducing this physical effect, given their income and wealth. It is analogous to the price people are willing to pay for a product for sale in a market. Benefits are the opposite of damages.

this report that bear on externalities. Some examples of externalities using these terms are presented below. An additional illustration is presented in Appendix B.

**Example 1.** A coal-fired electricity-generating plant, which is in compliance with current environmental regulations, releases various pollutants to the atmosphere that adversely affect the health of residents. The pollution released by the plant is an example of a negative externality because it contributes to health problems for residents. The damage from this pollution is an additional cost of production to society (a "social cost"). If these social costs were not adequately taken into account in selecting the plant's site or the air pollution control technology that it uses, the true costs of the plant have not have been reflected in these decisions.

**Example 2.** Many thermal power plants use water for cooling; therefore, they emit heated effluent, which sometimes benefits anglers because fish in cold regions are attracted to the warmer water. Therefore, the fishing is better in the effluent plume. This is an example of a positive externality.

Total versus Average versus Marginal Damages: To be most helpful for policy, we want to estimate marginal damages and compare these to marginal costs to reduce these damages. The marginal damage is the damage that arises from the last unit of emissions or other type of burden. In many cases, marginal damages are constant over the relevant range of emissions. That is, the damage from the last unit is no different than the damage from the first unit of emissions. But, in some cases, say for pollutants that accumulate in the environment, marginal damages grow with more emissions.

In any event, for policy purposes, if the marginal damages from the last unit of emissions exceed the marginal costs from eliminating that unit, then it would benefit society to eliminate that last unit, since the damage prevented would exceed the cost of preventing that damage.

Total damages, in contrast, are the sum of marginal damages for all units of emissions. Average damage is the total damage divided by the number of units of emissions or other burdens in question. Average and marginal damages may equal one another under certain conditions, but in general they are different. In this report, sometimes we assume that they are equal because it is easier to calculate average rather than marginal damages and actual differences are expected to generally be within error margins.

Externalities: An externality, which can be positive or negative, is an activity of one agent (that is, an individual or an organization like a company) that affects the well-being of another agent and occurs outside the market mechanism. In the absence of government intervention, externalities associated with energy production are generally not taken into account in decision making.

The improved angling is a societal benefit that probably was not reflected in the utility's decisions about where to site the plant and effluent. This societal benefit does not take into account any other ecosystem changes, which might or might not be seen as beneficial.

**Example 3.** A company is building a new coal-fired power plant in a small community and hires a large number of construction workers. The increase in demand for construction workers drives up the local wage rate and adversely affects homeowners who wish to hire workers to remodel their homes. The price of their remodeling projects has gone up with the increase in wages. This is not an externality, however, because the activity of one agent (the company building the power plant) affects other agents (homeowners wishing to remodel their homes) through a market mechanism—the labor market. The company takes the increase in wages into account because it must also pay the higher wages to attract construction workers.

**Example 4.** Farmers respond to a demand for corn-based ethanol by diverting land from food production to fuel production. The reduction in the supply of feed corn and other grains drives up the prices of grains and meat, thereby making consumers worse off. This is not an externality since the activity of one agent (fuel buyers bidding up the price of corn for ethanol production) affects other agents (the food-buying public) through the markets for corn and other food products.

**Example 5.** Workers in high-risk occupations receive a higher wage than workers doing similar tasks in jobs with lower risks. This is not an externality because those bearing the risks are freely choosing within a market to accept this risk and are compensated for the risks they face through higher wages. Increased costs faced by a firm do not by themselves indicate whether the firm's activities are an externality. For example, electricity-generating plants participating in the Acid Rain Program of the U.S. Clean Air Act face higher costs because they must surrender valuable permits for each ton of sulfur dioxide ($SO_2$) emitted. This higher cost is the result of a government program to reduce the externality associated with acid rain. In the case of elevated costs to compensate for high-risk jobs, no government policy is involved in addressing the risks faced by employees due to the nature of the work they are offered.

Externalities matter because, when they are not accounted for, they can lead to a lower quality of life for at least some members of society. For example, suppose that the power plant in the first example has access to technology that, at a cost of $40/ton, can cut its emissions by 10 tons. Suppose further that the full cost of the effects that residents suffer (for example, health and psychological costs) is $50/ton. If the plant were to install the technology, total social welfare would increase—the additional cost to the plant would be $400, but the "savings" to the residents (that is, the reduction in adverse effects they suffer) would be $500. Human well-being would be increased by this change. However, if the externality had not been accounted for in the plant's decisions, aggregate well-being of all members of society would be lowered.

It is important to distinguish true negative externalities from unfortunate market signals (such as higher prices of food) that hurt some members of society but are not externalities. The reason for this distinction is that, in the case of a true externality, the possible well-being of society can be raised by accounting for it—the "pie" that represents the value of society's goods, services, and related intangibles is enlarged. If it is not a true externality, market intervention cannot alter the size of the pie but can only reallocate it.

Note the following additional points about externalities:

• The agents that produce externalities can be organizations or individuals. For instance, a restaurant diner who smokes an after-dinner cigarette (besides breaking the law in most states) creates an externality for others in the restaurant.

•   The activities that produce negative externalities usually also produce benefits for someone in society. The electricity that is produced by the coal-fired power plant provides value to the consumers who purchase that electricity. The externalities arise from the side-effects of that benefit-producing activity that are not reflected in its market price.

## How Externalities Are Characterized in This Study

Most of the externalities associated with the production and consumption of energy have been addressed, or corrected, to some degree, through public policies. Coal mining and oil and gas extraction are subject to federal, state, and local regulations that are intended to limit the environmental damages from fuel extraction. Air pollution emissions by power plants are regulated under the Clean Air Act, and tailpipe emissions from motor vehicles are regulated at the federal and state levels. Indeed, regulations designed to correct externalities have substantially reduced their magnitude over the past 30 years. What the committee evaluates in this study are the externalities that remain in 2005—and the externalities that are predicted to remain in 2030—after such regulations have been implemented.

To make clear what we evaluate in this study, consider Figure 1-1, which shows the marginal damage associated with $SO_2$ emissions and the marginal cost of emitting $SO_2$ for a hypothetical power plant in a particular year. According to the graph, the damage from emitting each additional ton of $SO_2$ is \$1,000. The cost of emitting another ton of $SO_2$ declines as more is emitted—equivalently, the marginal cost of reducing $SO_2$ emissions (moving from right to left on the horizontal axis) increases as less $SO_2$ is emitted. If Firm 1 (the hypothetical power plant) in 2005 is emitting E1 tons of $SO_2$, the external damages that we quantify equal the shaded area in Figure 1-1—that is, we quantify the total damages associated with the firm's current level of emissions.[1] If the firm were emitting E* of $SO_2$, the external damages that we evaluate in the study would correspond to the rectangle 0ABE*. We also express these damages per kilowatt-hour of electricity produced and quantify the marginal damage of each ton of $SO_2$ produced by the plant for a specific year.

It is important to note that, from the viewpoint of economic efficiency,

---

[1]We calculate aggregated damages by estimating the damages per ton of $SO_2$ (the horizontal line in Figure 1-1) and multiplying by the number of tons of $SO_2$ emitted in 2005, to which we add similar calculations for $NO_x$ and directly emitted PM. We therefore calculate the marginal damage per ton of $SO_2$ (and per ton of $NO_x$ and PM) at each plant, which are reported in Chapter 2.

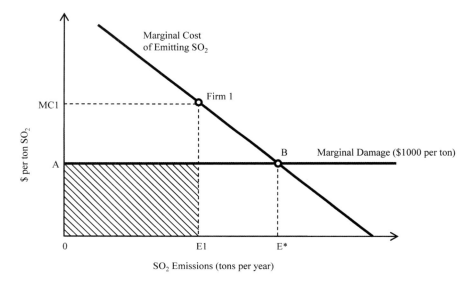

**FIGURE 1-1** Marginal damage associated with $SO_2$ emissions in a year (x-axis) and the marginal cost of emitting $SO_2$ in a year (y-axis) for a hypothetical power plant (Firm 1) emitting $SO_2$. E1 = amount of $SO_2$ (tons) emitted by Firm 1; E* = economically optimal level of $SO_2$ emissions; MC1 = marginal cost for Firm 1. Damages of emitting each additional ton of $SO_2$ are assumed to be constant.

the externalities that we characterize may be too large or too small. The economically optimal level of $SO_2$ emissions occurs at E* where the cost of reducing the last ton of $SO_2$ equals the corresponding reduction in damages. Even at the optimal point, damages occur—an externality remains, even though emissions are at an economically optimal level.[2]

What then is the significance of the externalities presented in this study? As is clear from the diagram, whether emissions should be reduced or increased from the viewpoint of economic efficiency depends on the current level of emissions and the cost of reducing them; it cannot be determined from the size of total damages alone. However, evaluating economic efficiency requires information about the costs of abating $SO_2$, which are outside the scope of this study. This does not mean that information about total damages is without value on its own: Plants with large total damages

---

[2]Some economists might say that there is no externality if emissions are at the optimal level (E* in Figure 1-1). However, the committee follows Baumol and Oates (1988) in saying that an externality exists, even though emissions are at an optimal level.

should likely be subject to a benefit-cost analysis of pollution control measures to see whether further pollution control is warranted.

Economic theory also suggests efficient methods for regulating the externalities associated with air pollution. For example, the first best solution to internalizing the damages in Figure 1-1 could be achieved by imposing a tax on $SO_2$ emissions equal to the marginal damage that the emissions impose—a tax equal to the height of the marginal damage curve. This solution could also be achieved by a pollution permit market in which firms traded rights to pollute denominated in damage terms (Roumasset and Smith 1990; Hung and Shaw 2005).[3] In either case, a firm would have an incentive to reduce its rate of $SO_2$ emissions to E*.[4] Information on the marginal damages associated with various pollutants, which we quantify for fossil-fueled power plants, is relevant to the efficient regulation of air pollution externalities.[5]

### Nonexternality Market Distortions and Impacts

In markets related to energy production and consumption, many other distortions occur that create opportunities for improvement of social welfare but that are not externalities. Because these distortions are outside the purview of this study, the committee touches only briefly on them here, largely to be clear that we do not attempt to identify and quantify the social costs associated with those distortions and to recognize that the magnitude of externalities can be directly affected by the presence of these distortions.

One form of market distortion that affects energy markets is the presence of market power—in the extreme, a single supplier of energy (monopoly) or a single buyer (monopsony). In such cases, a firm with market power can affect the price and quantity traded to its advantage and impose costs on others that exceed its gains. Cartels, such as OPEC, or large purchasers of oil, such as the United States, can exhibit market power.

Another form of market distortion that affects energy markets is the presence of taxes or subsidies ("tax breaks") that do not correct externali-

---

[3]An important caveat to these points is that no significant market distortions (for example, other externalities left unregulated or imperfectly competitive markets) are assumed for the rest of the economy.

[4]Imposing a tax equal to $SO_2$ damages per kilowatt-hour—that is, to the size of damages divided by electricity production—would not provide the same incentive. Indeed, it need bear no particular relationship to the marginal damages associated with $SO_2$.

[5]External damages per unit of output (for example, damages per kilowatt-hour or per mile) may also help to inform the choice among technologies (for example, whether a new power plant should be gas-fired or coal-fired). However, the choice among technologies should be based on the private as well as social costs.

ties but are imposed to raise revenue, provide support to an industry, or serve some other purpose. Of chief concern for this study is that these subsidies can affect the amount of an externality generated by an industry. For example, subsidies for oil exploration have encouraged the expansion of the petroleum industry, thereby increasing the magnitude of the externalities generated by the production and consumption of petroleum.

Information asymmetries and public goods are two additional cases of market failure that could affect energy markets. In this study, we quantify to the extent possible the noninternalized externalities conditional on the existing set of market regulations, taxes, subsidies, and market distortions from all sources.

In addition to market distortions that are not externalities, there are impacts of energy production and use that may be of public concern but that are not externalities. For example, as discussed earlier, increased corn prices due to the production of biofuels do not represent an externality as they result from the proper functioning of a market. This is not to say that higher corn prices are unimportant or that they should not be the subject of public concern and policy. However, there is no market failure to correct, nor is there an externality to internalize. Another important example is the distribution of social costs across space, time, or different population groups. Such distributional issues may be of great concern to policy makers, but they do not represent externalities.

In this study, we have restricted our attention to externalities, but we do point out some nonexternal impacts when relevant. For example, occupational injuries in coal mining and oil and natural gas extraction do not qualify as externalities according to our definition; however, there is interest in many quarters in documenting the magnitude of these impacts. Thus, we quantify (but do not monetize) them below.

## SELECTING ENERGY SOURCES AND USES FOR THIS STUDY

The committee's task was the evaluation of externalities "associated with the production, distribution and consumption of energy from various selected sources." Studying selected sources was necessary because it would have been infeasible to evaluate the entire energy system given the time and resources available to the committee. In selecting the sources for study, the committee was careful to include major elements of the energy system and the most significant externalities.

To create a basis for the committee's selection of energy sources and their end-use services, we started with the general framework shown in Figure 1-2. Energy sources depicted in the figure are fossil fuels (such as coal, petroleum, and natural gas), nuclear, and the various "renewables" (such as biomass, solar, and wind). Also shown are the various forms of

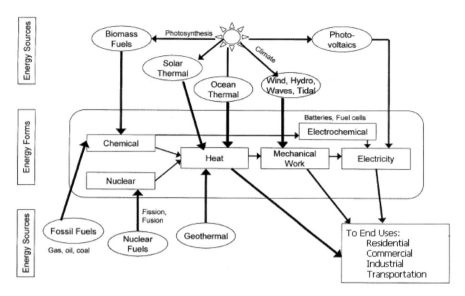

**FIGURE 1-2** Sources and forms of energy that provide the ability to do useful work. SOURCE: Tester et al. 2005. Reprinted with permission; copyright 2006, MIT Press.

energy that these sources represent or into which they are converted for use by industrial and residential users and in transportation. For example, coal is combusted in electric power plants to produce heat that is used to turn turbines (mechanical work) that turn generators that produce electricity that is transmitted to houses and factories.

Some energy sources, such as solar, produce heat directly without a combustion step, and others, such as wind, produce mechanical work directly. Furthermore, heat is, of course, used directly (in homes, for example) from, say, natural gas, without further conversion into mechanical work or electricity. Although they are not depicted in the diagram, it is worth keeping in mind that there are energy losses associated with the conversion, movement, and use of energy. In general, the more steps in the process, the less efficient the use of energy.

Figure 1-3 shows energy flows from primary energy sources to end uses in the United States in 2007. Total U.S. energy usage that year amounted to 101.5 quadrillion British thermal units (quads), so the numbers shown in the figure also approximately correspond to percentages. We note that electricity is an intermediate form of energy. From the perspective of the impacts of energy use in the United States, electric power is especially

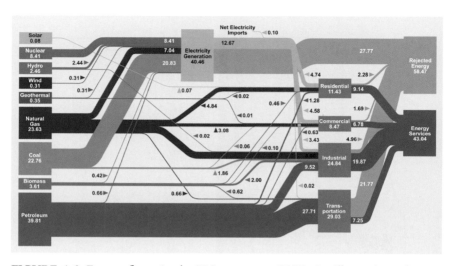

**FIGURE 1-3** Energy flows in the U.S. economy, 2007. An illustration of energy movement from primary sources (boxes on the left) to consumption by end-use sectors (residential, commercial, industrial, and transportation) (boxes on the right). SOURCE: Prepared by Lawrence Livermore National Laboratory and the Department of Energy (NAS/NAE/NRC 2009a, p. 17).

important, even though it is not an energy source or end use per se. The losses associated with energy transformation and transport are also shown as "rejected energy" or loss of useful energy. The amount of rejected energy (58.47 quads) in 2007 exceeded the amount of energy that provided energy services (43.04 quads).

The contribution percentage of each energy source to total U.S. energy consumption in 2007 is shown in Figure 1-4. Primary energy use and delivered energy use by sectors are shown in Figures 1-5 and 1-6, respectively. In Figure 1-5, the shaded regions in the end-use and electric-power bars show how much each source contributed to the total U.S. energy use of 101.5 quads. For example, petroleum use for transportation was approximately 28 quads in 2007. Figure 1-5 also shows that, of the 101.6 quads used in 2007, 40.6 quads were in the form of electricity.

In Figure 1-6, each end-use sector shows the total 2007 energy use, whether from electricity or from the "primary" energy sources in that sector. For example, the figure shows that industrial end users used 33.4 quads in 2007, 12.0 quads of which were in the form of electricity.

Figures 1-3, 1-4, 1-5, and 1-6 contain several notable facts about U.S. energy production and use. In addition to the overwhelming dependence of transportation on petroleum, noted above, we also see, for example, that

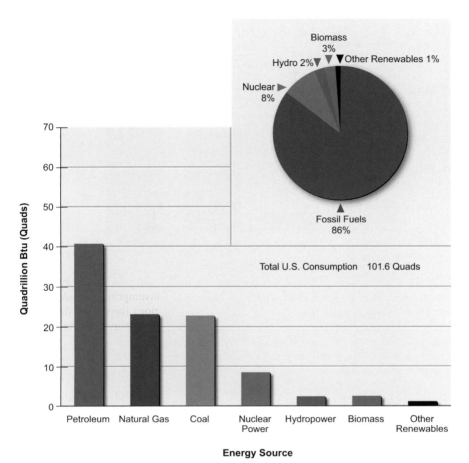

FIGURE 1-4 U.S. energy consumption by energy source in 2007. SOURCE: Data from EIA 2008a, Table 2.1a.

the vast majority of coal use in the United States is for electricity generation and that renewables and nuclear energy each represent relatively small fractions of total energy use.

Table 1-1 shows the sources and end uses selected for the analysis of external effects in this report. The energy sources are arrayed as rows. (Note that we have listed each of the renewables separately.) The end uses are shown as columns; here, for convenience, we have listed electricity production as an end use, and we have focused our attention in the industrial sector and commercial/residential sectors (buildings) on their use of energy for heating.

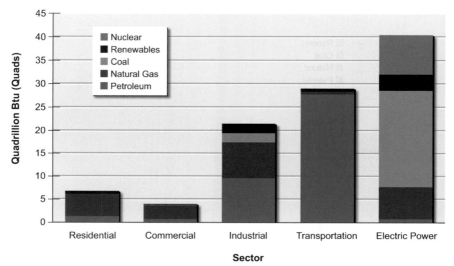

**FIGURE 1-5** U.S. consumption of energy by sector and fuel type in 2007. Natural gas is the major fuel type for the residential and commercial sectors. Petroleum and natural gas are the major fuel types for the industrial sector. Petroleum is by far the major fuel type for the transportation sector. For electric power, coal is the major fuel type, followed by natural gas and nuclear power. Energy consumed by the electric power sector is used to produce electricity consumed by the end-use sectors. SOURCE: EIA 2008b, in NAS/NAE/NRC 2009a, p. 22, Figure 1.8.

The committee discussed every cell in Table 1-1 at length and selected the shaded cells for evaluation. Given the available time and resources, we chose those cells that represented the largest portions of the U.S. energy system or those that represented energy sources with substantial increases (>20%) in consumption over the past 5 years (2002-2007). Thus, coal for electricity production and petroleum for transportation, which are the two largest single components of the U.S. energy system, were obvious choices. Natural gas is an important fuel, comparable to coal as a source. We included its use for electricity production and space heating in buildings. Wind, biomass, and solar were also included because of increased consumption in recent years.

The committee also included cells that represented sources and uses of particular current policy interest or that are expected to grow in significance (such as nuclear energy for electricity production). The base year for the committee's analyses was 2005 (or the nearest year for which data were available). We chose a future year to gain insight into the impact of future technologies on energy externalities. We chose 2030 for this purpose so as to be consistent with and leverage the work of the AEF.

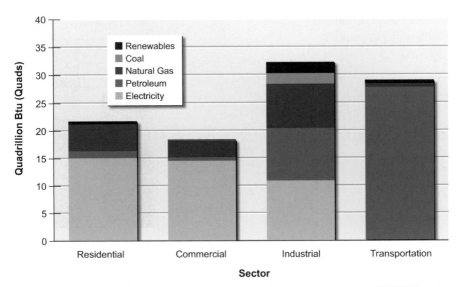

**FIGURE 1-6** U.S. delivered energy consumption by end-use sector in 2007. Electricity predominates in the residential and commercial sectors. Electricity, petroleum, and natural gas are the main forms of delivered energy for industry. Petroleum is by far the major fuel type for the transportation sector. SOURCE: EIA 2008b, in NAS/NAE/NRC 2009a, p. 22, Figure 1.9.

Identification of significant alternative energy technologies that are being considered for implementation before 2030 was guided by the AEF reports and other sources. Those that were evaluated are already in various stages of development and commercialization, so some information on potential externalities was available. Other technologies that are still in research and early demonstration phases may be available for commercialization by 2030, but the committee decided that credible evaluation of externalities would be premature until more was known about actual performance. The future technologies evaluated for the 2030 time-frame are as follows:

• Electricity Sector (Chapter 2): Nuclear power impacts are included in the present electricity-generation portfolio and will apply to continuing or expanding use of this source. Biomass, wind, and solar photovoltaics (PV) are already deployed to some extent in the present electricity production portfolio. In 2030, all of these technologies may be advanced and constitute a more important part of the electricity mix, so it is important to assess their externalities—both positive and negative. The AEF panel report on Electricity from Renewables (NAS/NAE/NRC 2009b) states that a "rea-

**TABLE 1-1** Committee Study Approach for Energy Sources and Consumption Sectors[a]

| Energy Source | Electricity Production (see Chapter 2) | Transportation (see Chapter 3) | Industry—Heat (see Chapter 4) | Buildings—Heat (see Chapter 4) |
|---|---|---|---|---|
| Oil | | MA | QE | QE |
| Coal | MA | | | |
| Natural gas/liquid | MA | MA | QE | MA |
| Uranium | QL | | | |
| Biomass | QL | MA | | |
| Hydropower | | | | |
| Geothermal | | | | |
| Wind | QL | | | |
| Solar power | QL | | | QE |
| Other fuels[b] | | MA | | |
| Electricity[c] | — | MA | QE | QE |

[a]The shaded cells indicate where the committee focused its consideration. MA = modeling analysis conducted by the committee; QL = quantitative information obtained from the literature; QE = qualitative evaluation.

[b]Other fuels includes hydrogen fuel cells and coal-based liquid fuels.

[c]Electricity is considered an intermediate energy source (generated from the combustion/use of coal, natural gas, uranium, and renewables). Electricity is included in this table because it is treated as a "whole" in Chapters 2 and 4, rather than by individual primary energy sources. For the transportation column, electricity also includes the manufacture and use of lithium batteries.

sonable target" for 2020 is to have nonhydropower renewables as 10% of the generation mix. Most of this is expected to come from wind, as solar PV technologies still face challenges of costs and technology development that limit likely deployment to a few percent on this time scale. Not evaluated were the following: hydropower (about 2.5% of the sector and unlikely to grow much); geothermal (about 0.3% of present use and unlikely to grow much until enhanced geothermal system technologies, for example, deep-heat mining, are improved); and fossil-fueled power plants with carbon capture and storage (CCS), which seem technically feasible but still are far from large-scale implementation and will require new infrastructure for carbon dioxide transportation to storage sites of proven integrity. The addition of significant amounts of variable and nondispatchable electricity sources (solar, wind) also presents challenges to the operability and reliability of the electric power grid, which has aging systems that need upgrading. These issues are considered in more detail in the AEF report *Electricity from Renewables: Status, Prospects, and Impediments* (NAS/NAE/NRC 2009b) and were not a major focus in this study, but we do discuss them in the context of "network externalities" in Chapter 6.

• Transportation Sector (Chapter 3): In the 2030 time frame, petroleum is still likely to be the major transportation fuel. There may be increased use of ethanol as a blend with gasoline or as a pure fuel. The analysis considers corn, cellulosic ethanol, cellulosic synthetic gasoline, and other feedstocks as biofuel options. Electric cars and plug-in hybrids are discussed and tied to externalities from the electricity sector analyses. Compressed natural gas is also an alternative fuel that may find use in both light- and heavy-duty vehicles. Some of the gasoline in use today comes from energy intensive production of Canadian oil sands; synthetic gasoline (or diesel) also can be made from coal, natural gas, or biomass. The AEF panel assessing possible options for producing alternative liquid transportation fuels concluded that CCS would likely be needed to reduce GHG emissions resulting from production of liquid fuels from coal (NAS/NAE/ NRC 2009c). CCS systems are not yet well enough evolved for any sort of a meaningful externalities analysis.[6] The committee evaluated hydrogen (made from fossil fuels or electrolysis of water) to some extent, as a possible conveyor of energy for transportation in the 2030 time frame, although its use will be limited because of the considerable need to develop storage and infrastructure technologies. Although aviation jet fuel is likely to remain petroleum based until alternative synthetic fuels are developed, there are some special externalities associated with aviation that will be addressed separately with regard to climate change (Chapter 5).

• Energy for Heat (Chapter 4): Consumption of energy for the production of heat in the buildings sector mostly involves natural gas, along with a little petroleum. The primary change for 2030 was assumed to be efficiency gains due, for example, to changes in building design and construction. Much of the information about potential efficiency gains was based on the AEF efficiency panel report, *Real Prospects for Energy Efficiency in the United States* (NAS/NAE/NRC 2009d). Because of its diversity and complexity, the industrial sector was discussed from a general perspective.

## FRAMEWORK FOR EVALUATING EXTERNAL EFFECTS

### The Life Cycle of Energy Use

When considering the external effects of using energy, people often focus on the end use—for example, the air pollution from burning gasoline

---

[6]To allow for widespread deployment of CCS technology starting around 2020, its technical and commercial viability will need to be demonstrated for a variety of 15-20 fossil-fuel-fired electricity-generating plants (NAS/NAE/NRC 2009a).

in an automobile. In fact, identifying energy's external effects requires a broader view that reflects energy's entire life cycle, from extraction of the energy source as it is found in nature through conversion, transportation, and transmission to its point of use and then to the ultimate fate of waste products from that use. Thus, the committee adopted life-cycle assessment (LCA) as its approach to identifying the external effects of energy.

There are two general types of LCA: process-analysis-based LCA and economic input-output (EIO) analysis-based LCA. Process-based LCA considers all inputs (including raw materials, energy, and water) and all outputs (including air emissions, water discharges, and noise of included processes associated with all life-cycle stages of a product or service). EIO-LCA helps address boundary-selection problems and data intensity by creating a consistent analytical framework for the economy of an area based on government-compiled input-output tables of commodity production and use coupled with material and energy use and emission and waste factors per monetary unit of economic output. It may not provide the level of detail of a process-based LCA. An important metric of the system studied is referred to as a functional unit (for example, vehicle miles traveled can be used as the functional unit for LCAs related to transportation). Because of the significant variability in assumptions, boundaries, and approaches, comparisons across different assessments must be done with caution (see discussions, especially in Appendix E, in NAS/NAE/NRC 2009b).

Figure 1-7 depicts the major elements of energy use from an LCA perspective. Taking the use of domestic petroleum for automobiles as an

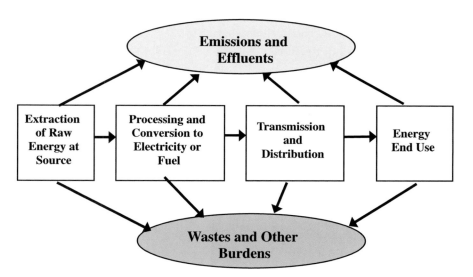

FIGURE 1-7 Life-cycle analysis for energy use.

example, the first box in Figure 1-7 would include the exploration and drilling for oil and sending it via pipeline, tanker, or truck to a refinery; the second box would include refining to produce gasoline; the third box would include transport of gasoline by pipeline and trucks to service stations; and the fourth box would include the burning of gasoline in automobiles. Each stage in the life cycle produces impacts, and all of them should be considered in estimating the external effects of energy use. Staying with the gas-powered auto as an example, the externalities associated with burning a gallon of gasoline are composed of the external effects of all the "upstream" activities necessary to produce and deliver that gallon of gas to the pump, as well as the direct impacts of burning the gasoline in the car. Further, on the basis of vehicle miles traveled, there are externalities associated with the car itself—the extraction and production of materials used in the manufacture of the car, the sales distribution network, the maintenance required during the life of the car, and the ultimate disposal of the car.

Energy efficiency and energy conservation are important aspects of overall energy policy, and often are considered as being equivalent to increasing the energy supply. Energy conservation refers to reducing energy use by reducing services that require energy; examples include lowering indoor temperatures in winter and raising them in summer to reduce the use of energy, and walking or riding a bicycle to reduce the use of gasoline. Energy efficiency refers to using less energy to achieve the same level of energy-based service; examples include using compact fluorescent light bulbs instead of incandescent ones to achieve the same ambient light level with less electricity and using more efficient internal combustion engines or hybrid systems to achieve the same transportation capacity with less fuel.

### Identifying and Quantifying Burdens, Impacts, and Damages

With the life-cycle stages identified, determining the externalities of energy required the committee to identify the burdens, impacts, and damages associated with each stage. The methodology followed what is termed the damage-function approach (Jolliet et al. 2004). The burdens come in many forms (for example, air emissions, liquid discharges, and solid wastes); they move through all media (air, water, and land); and they have a range of effects or impacts (on human health, natural ecosystems, and the built environment). All of these factors must be accounted for in producing a complete estimate of damages, although, as will be seen, for many energy uses some impacts and damages are far more important than others. Furthermore, some impacts are not externalities (see Box 1-2). The committee made judgments about what damages are likely to be externalities or are widely considered to be externalities (such as damages associated with energy security) and focused its work on those.

Table 1-2 shows impacts at each stage of the life cycle for generating electricity from coal. These impacts are all externalities, with the exception of occupational injuries among coal miners, which are viewed as a job characteristic that is traded in labor markets. The table is intended to be illustrative; Chapter 2 provides a more thorough discussion of the external effects of electricity generation.

Some impacts are direct, easy to understand, and often well-supported by data. The deaths and injuries suffered by coal miners, although not externalities, represent an example of this. Other impacts, such as the alterations in water availability or ecosystems due to climate change, are the result of very complicated physical, chemical, and biological processes about which there is great uncertainty.

In this study, given the constraints on time and resources, the committee relied on past work in the identification of impacts and their damages. We did not attempt to develop new methods for estimating impacts and damages, but we did identify areas where additional research would be particularly valuable. It is important to realize that estimating most of these impacts and damages is a several-step process based on many assumptions; this is true for even relatively well-understood impacts. Consider one of the

**TABLE 1-2** Illustrative Impacts of Producing Electricity from Coal

|  | Human Health | Ecosystems | Security and Infrastructure |
|---|---|---|---|
| Coal Mining | Coal miners' mortality and morbidity | Land disturbance river alteration, acid mine drainage | |
| Transportation of coal to power plants | Death and injury from accidents | Vegetation damage from air pollution | Load on transportation systems |
| Burning of coal | Mortality and morbidity from air pollution | Ecosystem effects from cooling Water discharges Ecological changes from climate change | Degradation of building materials Agricultural shifts and coastal community impacts due to climate change |
| Disposal of waste | Health effects of heavy metals in ash and other waste | Ecosystems effects of ash and other wastes | |
| Transmission of electricity | | Disturbance of ecosystems by utility towers and rights of way | Vulnerability of transmission system to attack or disaster |

boxes in Table 1-2: human health effects of air pollution from coal-fired power plants. Arriving at an estimate requires

1. Estimate *burdens*: air-pollutant emissions (which depend, among other things, on the particular coal that is used, the boiler technology, and the pollution-control technology).
2. Estimate the ambient concentrations of air pollutants (which depend, among other things, on the height of the exhaust stack, the location of the plant, and prevailing wind patterns).
3. Estimate the exposure of people to air pollutants (which depends, among other things, on where people live and how much time they spend outside).
4. Estimate *effects* or *impacts*: the health consequences of the exposures (which depend, among other things, on the age and health status of the exposed population and their life styles, for example, whether they smoke).
5. Estimate *damages*: the value of avoiding these impacts to society (which depends, among other things, on the incomes of individuals and the seriousness of the impact, as judged by individuals)

This emission → ambient concentration → exposure → impact → damage estimation process is generalizable; it underlies each box in Table 1-2 and the estimation of most of the damages considered in this study. Security of energy supplies and network externalities, both discussed in Chapter 6, do not fit as neatly into this framework as virtually everything else does.

A noted, many assumptions must be made. Many of the assumptions are specific to the case at hand, and we make clear in the context of each analysis what the key assumptions are. One cross-cutting issue is the distribution of impacts and damages, both spatially and within populations. The issue is discussed later in this chapter.

### Evaluating Impacts

Evaluating damages requires an estimation of the impacts—the tangible manifestations of the burdens of energy use. Thus, we have to express impacts in tangible terms and associate with each impact a metric or measure.

In Table 1-2, the impacts are all physical or biological. Describing some of them is fairly straightforward—for example, death or injury from transportation accidents can be associated with a metric. For example, a metric for the human health effects from coal transportation is the annual deaths from transportation accidents. Another metric is the number of injuries from transportation accidents. Both are meaningful, easily understood measures or indicators of the impact of interest, and data are available for quantifying them.

Other impacts are not nearly so straightforward. Virtually all of the eco-system impacts—no matter their cause—present a real challenge for evaluation. One could say that the overarching concern is ecosystem health, but this is not a useful metric: How would one define it or quantify it? We could use physical, chemical, or biological surrogates—for example, changes in water temperature, the concentrations of key chemical constituents in water, and biological productivity. Each of these is well-defined and measurable, but it is arguable whether they would adequately reflect changes in ecosys-tem health, individually or collectively (see EPASAB 2009).

A potential source of complexity in impact assessment is cumulative effects, which can be important but are often inadequately assessed. Our discussion here largely follows NRC (2003a).

Concern about cumulative effects started in a formal sense with the pas-sage of the National Environmental Policy Act (NEPA) of 1969, motivated by the environmental effects of multiple electric power plants. In 1978, the President's Council on Environmental Quality (CEQ) promulgated regula-tions for implementing NEPA (40 CFR Parts 1500-1508 [1978]), which defined a cumulative effect as "the incremental impact of the action when added to other past, present, and reasonably foreseeable future actions. . . . Cumulative impacts can result from individually minor but collectively significant actions taking place over a period of time." Thus, a single power plant might have only minor, or acceptable, impacts on, say, an estuary but the effects of multiple plants might be substantial and qualitatively differ-ent from the effects of only one plant. In other words, the effects can accu-mulate. Similarly, some of the effects of smokestack emissions from power plants can interact with those of automobile and industrial emissions; in addition, effects can accumulate over time. As NRC (2003a) pointed out, effects can be synergistic or antagonistic (in other words, they can be greater or smaller than the sum of their parts), and they can have thresholds.

Indeed, thresholds are an important complication in measuring im-pacts, especially for some environmental effects. The concern is not only with, say, increased water temperatures but also with temperatures above a certain level at which especially significant biological impacts occur, for example, massive fish kills.

Many threshold effects, like water temperatures, are relatively well-understood and predictable based on scientific knowledge. Others are more controversial due to the uncertainty of the underlying science and the eco-nomic and other consequences of dealing with the impact. For example, there has been a long standing disagreement over whether there is a nonzero level of exposure to radioactivity that is safe for humans. A more recent example relates to climate change and the potential for relatively rapid and perhaps irreversible shifts in certain subsystems of the earth, producing, for example, accelerated melting of ice sheets and caps.

Because the ultimate concern is human health and the environment (or more, generally, social welfare), these threshold effects are important. Water temperature can be measured reliably, but the temperature is not what people really care about; it is how life-forms respond to that temperature. If the two are proportional responses, then a physical measure, such as water temperature, can be a useful surrogate metric. When the proportionality breaks down, as it does when there is a sharp threshold, the surrogate loses its utility.

Beyond cumulative effects, thresholds, and other physical complexities, some impacts are very hard to quantify because they are highly subjective. Aesthetics is a good example of a factor whose quantification is highly uncertain. For example, harvesting trees in a forest affects the aesthetic appearance of the forest. One might consider the number of trees removed as a metric. Of course, the problem with such a metric is that it is not only the number of trees; it is also which trees.

Qualitative impacts present a special challenge to communicate, particularly to do so in a way that decision makers take them into account. The adage, "You can't manage what you can't measure," might be dangerous or, at least, limiting if the absence of quantification is taken to mean a value of zero (see the discussion of uncertainty at the end of this chapter).

Table 1-3 summarizes impact pathways and associated effects evaluated in an ExternE study on electricity and transportation.

## Damage Estimation: Monetizing Impacts

It is relatively straightforward to monetize goods that are routinely traded in markets, as the market prices give direct information about their monetary worth. Thus, if a family is willing to purchase salmon at the market price, then the value of the salmon must be at least as great as the purchase price, otherwise the family would not have been willing to give up the other items its members could have purchased with that money. The market price gives important information about how much this family is willing to trade off other items to have salmon for dinner. However, many of the externalities that we are interested in for this report do not trade in markets, so information on people's preferences is not readily available. Nonetheless, such "nonmarket" goods may have as much or more value to people as goods that do trade in markets (that is, if required, they would be willing to give up a lot to have them).

The main goal in monetizing the impacts of externalities is to place externalities on equal footing with other goods and services. When decision makers must decide, for example, whether to tax an externality, increase

**TABLE 1-3** Illustrative Impact Categories Pathways[a]

| Impact Category | Pollutant/ Burden | Effects |
|---|---|---|
| Human health: mortality | PM, $SO_2$ | Reduction in life expectancy |
| | Benzene, Benzo[a]pyrene 1,3-butadiene Diesel particles | Cancers |
| | Noise | Loss of amenity, impact on health |
| | Accident risk | Fatality risk from traffic and workplace accidents |
| Human health: morbidity | PM, $O_3$, $SO_2$ PM, $O_3$ PM, CO | Respiratory hospital admissions Restricted activity days Congestive heart failure |
| | Benzene, Benzo[a]pyrene 1,3-butadiene Diesel particles | Cancer risk (nonfatal) |
| | PM | Cerebrovascular hospital admissions Cases of chronic bronchitis Cases of chronic cough in children Cough in asthmatics Lower respiratory symptoms |
| | $O_3$ | Asthma attacks Symptom days |
| | Noise | Myocardial infarction Angina pectoris Hypertension Sleep disturbance |
| | Accident risk | Risk of injuries from traffic and workplace accidents |
| Building material | $SO_2$ Acid deposition | Ageing of galvanized steel, limestone, mortar, sandstone, paint, rendering, and zinc for utilitarian buildings |
| | Combustion particles | Soiling of buildings |
| Crops | $NO_x$, $SO_2$ | Yield change for wheat, barley, rye, oats, potato, sugar beet |
| | $O_3$ | Yield change for wheat, barley, rye, oats, potato, rice, tobacco, sunflower seed |
| | Acid deposition | Increased need for liming |

**TABLE 1-3** Continued

| Impact Category | Pollutant/ Burden | Effects |
|---|---|---|
| Global warming | $CO_2$, $CH_4$, $N_2O$, N, S | Worldwide effects on mortality, morbidity, coastal impacts, agriculture, energy demand, and economic impacts due to temperature change and sea-level rise |
| Amenity losses | Noise | Amenity losses due to noise exposure |
| Ecosystems | Acid deposition, nitrogen deposition | Acidity and eutrophication (avoidance costs for reducing areas where critical loads are exceeded) |

[a]ExternE developed this list to indicate the types of health and other impacts that were included in its investigations.

SOURCE: EC 2003, p. 3. Reprinted with permission; copyright 2003, European Communities.

public funding for education, or reduce support for antismoking programs, it is important for them to understand how their constituents view the value of these services compared with each other as well as with other uses of funds. The goal is simply to understand the trade-offs people are willing to make to get more of a good (or avoid doing without). This is a very useful concept for public policy decisions.

To monetize the impacts of externalities, the committee followed standard practice and defined the monetized value of an externality to an individual as the maximum amount that an individual would be willing to pay to obtain the good. This concept is called willingness to pay, often abbreviated WTP.[7]

There are both practical and conceptual arguments for attaching monetary values to impacts. One practical reason for monetization derives from the multiplicity of external effects and the difficulty of evaluating them in the context of national policy making. From just a partial list of metrics from external effects of coal-fired power plants, there are deaths from coal transportation, injuries from coal transportation, and water temperature increases from cooling-water discharges. The goal is to make sure that these

[7]There is an analogous concept of willingness to accept (WTA) that is defined as the minimum amount of money an individual would be willing to receive to give up a good that he or she owns. Discussions of the relationship between WTP and WTA and their technical counterparts, compensating and equivalent variation, can be found in graduate-level economic textbooks.

effects are factored into decisions related to producing electricity from coal. This requires a method for weighing these metrics against each other and, for policy making, against the costs of reducing the effects.

One class of methods is to use a numeraire to develop weights for each effect and to aggregate them. Indeed, in this report, the numeraire that we use is dollars, with the weights developed through the literature on WTP for reducing a given type of effect. Another class of methods is not to establish specific weights but to use multiattribute utility functions or, more generally, multiple criteria approaches (see, for example, Keeney and Raiffa [1993] and Cohon [2004]). Such an approach has been used in the past in the evaluation of federal water projects, but that approach dealt with three or four metrics at a time, not the dozens of metrics associated with the externalities of energy production and use. In contrast, using a monetary metric to the extent practicable, converting all the metrics into a single unit of dollars, and adding them up produces a single grand metric of all the external effects. This is a powerful result of great potential use to policy makers.

By estimating WTP, a monetary value can be placed on each external effect and then added up, thus producing a single dollar value for total external effects, which represents an estimate of the value that society places on those effects.

In this study, the committee did not have the time or resources to undertake new valuation studies and has therefore relied on existing studies of the monetary values of the externalities that we studied. One important example of a value taken from the literature is commonly referred to as the "value of a statistical life" (or VSL), which characterizes the rate at which people are willing to trade increased risk of death for other goods and services. By observing in many occupational and other settings how much people have been willing to pay to reduce the risk of death (or are paid in compensation for taking additional death risks) or by conducting surveys that ask people how much they are willing to pay to lower their death risks, estimates have been made for the VSL that are used in regulatory decision making around the world, including various agencies in the U.S. government. We used these values in our study, as explained in Chapters 2 through 5.

Using WTP as a monetary metric can make some people uncomfortable. There are some effects that are especially difficult to value (for example, ecological impacts) because there are many components of ecosystem services that people do not understand and for which people retain inherent cultural values. So at first blush, potential ecological impacts from climate change, such as lost polar bear habitat, seem to defy monetization. In practice, estimating the effect on polar bear habitat is very difficult because of the poorly understood relationship between changing greenhouse gas concentrations, climate effects, and, therefore, reduced habitat, but the fundamental economic valuation question remains conceptually straightforward: How much of other goods and services (such as education, housing, and health care) would people be willing to give up in exchange for preserving polar bear habitat? The conceptual answer will differ depending on whether there are many polar bears left in the wild or whether there are only a few. Whether a monetary value can be estimated that is accurate enough for use in policy decisions remains a challenge for many ecological services. The committee has used WTP to monetize external effects wherever possible, recognizing its limitations and controversies. Some effects are not monetized at all, and others are monetized with great uncertainty. Indeed, some effects cannot even be estimated, much less quantified, even though we know they exist.

The committee is especially aware that ecological impacts, including impacts on ecosystem services, have not been monetized in this report. Evaluating these impacts economically has a long and challenging history (for example, EPASAB 2009; NRC 2004a; Cropper 2000). Ecological effects that influence the production of economic goods, such as agricultural products, timber, fish, and recreational benefits, often have been monetized, although often incompletely. This report includes some aspects of agricultural production in its monetization of the damages from emissions from electricity generation that contribute to the formation of criteria air pollutants. However, changes in ecosystem services, such as nutrient cycling and provision of habitat, and more subtle changes in ecosystem functioning that can affect ecosystem performance have not generally been monetized, largely because it is difficult to quantify those changes at present (for example, Cropper 2000). Although the committee has described these impacts qualitatively, at least to some degree, they likely are significant monetarily and otherwise.

Despite these limits, the commmittee believes that using its results will improve federal policy making.

## Consideration of External Benefits

There are obviously considerable benefits to having energy. Most of these benefits are reflected in the prices paid for energy and are not *external* benefits. For the most part, external benefits are relatively few in number and small compared with the external damages that have been identified. For example, ORNL/RFF (1992-1998) identified the crop fertilization benefits of the nitrogen and sulfur from $NO_x$ and $SO_2$, respectively; the crop fertilization benefits of $CO_2$; and the recreational benefits of enhanced fishing opportunities in reservoirs formed from large hydro projects.

Of those, our study explicitly considered the crop fertilization benefits of $CO_2$; the results of the integrated assessment models considered in Chapter 5 account for the impacts. However, we did not explicitly consider other external benefits for the following reasons: ORNL/RFF (1992-1998) found the crop fertilization benefits of $NO_x$ and $SO_2$ to be small compared with the health-effect damages. We did not consider reservoir recreational benefits because we did not consider hydropower as important, for the purposes of this study, as the other technologies considered (refer to Chapter 2).

## THE POLICY CONTEXT FOR THIS STUDY

Externalities are important to analyze and understand because they provide an example of a situation where government involvement can potentially improve on the market outcome. Although the committee was not tasked to make recommendations for policy makers to address energy-related externalities, we did indicate how knowledge about the value of externalities can be used to improve market outcomes. This section relates the results of our study to existing policies that address externalities and discusses how the results of the study should and should not be used.

### The Nature of Externalities Evaluated in This Study

As noted earlier in this chapter, the committee evaluates the externalities associated with energy production and consumption that have not been corrected through existing policies—that is, the externalities remaining after policies have been implemented. Therefore, the study does not document the substantial progress that has been made in reducing the external damages associated with energy production and consumption over the past few decades. To illustrate, emissions from electric power plants that contribute to criteria pollutant formation are regulated by a variety of state and federal regulations. In particular, one of the goals of Title IV of the 1990 Clean Air Act Amendments was to reduce $SO_2$ emissions from coal-fired power plants by 50% from 1985 levels by the year 2010. Most of the reductions were already achieved by 2005, the year of this study. We quantify the damages associated with remaining $SO_2$ emissions from fossil-fueled power plants in 2005. A similar statement can be made regarding tailpipe emissions from motor vehicles. Emissions from cars per mile traveled have declined by 90% since the passage of the 1970 Clean Air Act as a result of various regulations. We calculate the remaining emissions from cars in 2005 and 2030.

We evaluate the damages associated with emissions in the years of 2005 and 2030, relative to zero emissions. For example, in the case of coal-fired power plants, we characterize the per plant aggregate damages associated with $SO_2$ emissions in 2005 compared with no $SO_2$ emissions. The same is true of the air-pollution damages associated with motor vehicles: We evaluate the per vehicle total damages from current emissions relative to zero emissions. This is not because emissions should be reduced to zero but because any other baseline would be arbitrary. The appropriate level of $SO_2$ emissions from power plants depends on the costs of reducing those emissions (see Figure 1-1), but estimating the appropriate level of emissions is beyond the scope of this study. The methods used to estimate air-pollution damages from fossil-fueled power plants and motor vehicles assume that the damages of each additional ton of pollution from a source are constant[8]—hence, we also compute the damages per ton of pollutant, which could be compared with control costs.

In the case of power plants, we provide estimates of the distribution of air-pollution damages across power plants. This is important for two reasons: First, the damages associated with a plant depend on where the plant is located, so damages vary spatially; second, total damages vary greatly across plants because of differences in plant size and pollution control. Variation in damages across plants is useful information from the perspective of pollution control. Plants with large total damages may warrant further air-pollution controls.

We also distinguish damages by the stage of the life cycle at which they are generated. Although it is possible to aggregate $NO_x$ damages associated with passenger transportation across all stages of the life cycle—oil exploration and extraction, oil refining, transportation of gasoline to the consumer, and consumption of gasoline by a car—regulations to limit $NO_x$ emissions will be targeted at different stages of the life cycle: Regulations to limit tailpipe emissions will differ from those to limit oil-refinery emissions. Similarly for damages associated with electricity generation, it is important for policy purposes to separate mining damages from those damages associated with electricity generation because policies to control each set of externalities will differ. Thus, although we present aggregate estimates of damages—per kilowatt hour or per mile traveled—they should be placed into proper context for policy.

## Policies to Correct Externalities

Policies to address or correct externalities include taxes, transferrable pollution permits, performance standards, and technology-based standards. Economic theory dictates that the most efficient policies for correcting

---

[8]This is a common assumption in the air-pollution literature. The concentration-response functions in the literature are essentially linear over the relevant range of ambient air pollution in the committee's study. Also, the emission-to-concentration relationship and unit costs of various health effects and other impacts are treated as constant. Unit costs are not necessarily constant across time and location.

externalities are those targeted at the externality itself—for example, a tax on $SO_2$ emissions rather than a tax on the electricity associated with those emissions or a tax on $NO_x$ emissions from motor vehicles rather than a tax on gasoline.[9] Taxing $SO_2$ emissions (or regulating them through a permit market or performance standard) provides an incentive to reduce $SO_2$ by using pollution-control equipment, by switching to low-sulfur coal, or by reducing the level of electricity produced. A tax on electricity generation does not provide the incentives to reduce $SO_2$ emissions per se. The same is true of a gasoline tax and $NO_x$ emissions.

For emissions related to criteria pollutants, the committee therefore notes that its estimates of externalities associated with emissions per kilowatt-hour of electricity produced or per gallon of gasoline should not be interpreted as recommendations for electricity or gasoline taxes equal to these monetized damages. Economically efficient methods of correcting emissions that contribute to criteria air pollutants include taxes on the emissions themselves or permit markets in which rights to pollute are denominated in terms of damages.[10] A similar statement can be made for $CO_2$ emissions. For fossil-fueled power plants, we provide estimates of the damages per ton for key emissions that contribute to criteria air pollutants, as a function of plant location. For $CO_2$ emissions, we provide ranges of estimates of marginal damages.

## Externalities and Technology Choice

A frequent use of estimates of the externalities associated with electric power generation and transportation is to inform technology choices when making public investment decisions. Should expansion of electricity-generating capacity take the form of coal, natural gas, nuclear power, or wind power? What technologies should be pursued as alternatives to gasoline-powered internal combustion engines for passenger vehicles? This study can help to inform such choices; however, it must be emphasized that we evaluate the externalities associated with various technologies independent of their costs. For example, an integrated gasification-combined cycle (IGCC) coal plant with carbon capture and storage is an extremely clean plant, but it is also an expensive one. Externalities are an important com-

---

[9]Policies that associate a price with the externality—for example, a tax or a permit market—are, in general, more efficient than policies that dictate the method of correcting the externality; for example, requiring coal-fired power plants to install flue gas desulfurization units (scrubbers).

[10]For example, if a power plant in a densely populated area creates more damages per ton of $SO_2$ emitted than a power plant in a remote area, the former plant would require more damage-denominated permits than the latter to emit a ton of $SO_2$.

ponent of the choice among various technologies but must be supplemented by estimates of private costs.

## SOME METHODOLOGICAL ISSUES: SPACE, TIME, AND UNCERTAINTY

Defining and evaluating externalities is unavoidably complicated by their spatial nature, by the fact that they manifest themselves over time, perhaps very far into the future, and by uncertainty. We discuss each of these issues in this section.

### Spatial Scales of Analysis

The external effects of energy, by their very nature, vary spatially. Some individuals and groups experience far greater effects from energy production and use than is reflected by the average amount—that is, than if the effects were evenly distributed—and others experience far less. In carrying out the committee's task, we focused on the spatial distribution of damages caused by coal-fired and gas-fired power plants wherever they were located and by transportation emissions in each of the U.S. counties in the 48 contiguous states. Note, however, that a lack of location data for stages upstream of the power plants and vehicle operations prevented us from estimating these kinds of damages in a spatially explicit manner.

#### Consideration of Effects on U.S. vs. Global Scales

Although the committee's task stipulated that the external costs and benefits of energy be analyzed from a U.S. perspective, we were also charged with consideration of broader, more global, implications when warranted and feasible. Some effects considered by the committee occur mainly in the United Sates, such as effects related to ozone-forming emissions from motor vehicles. However, other effects, such as those related to $CO_2$ emissions and climate change, will occur on a global scale. Likewise, for some of the security-related issues, or for transportation, which relies on energy production and distribution occurring outside the United States, ignoring the global consequences would result in substantial distortions. Moreover, as is apparent for climate change-related effects, some parts of the world are likely to suffer inherently different, and to some extent larger, burdens of these effects than the United States. In such situations, we have elected to characterize effects both in the United States and on a global scale, as consideration of them on different spatial scales might have an impact on policy choice. For practical reasons, we have provided sparing detail regarding differential impacts among non-U.S. regions.

*Consideration of Differential Effects on Local and Regional Scales*
*Within the United States*

Within different locations in the United States, many external costs and benefits related to energy are heterogeneously distributed as well—for reasons inherent to the nature of the economic activity or the geography or as a consequence of one or another policy choice. For example, one of the substantive health consequences of climate change in the United States is the impact on heat- and cold-related morbidity and mortality, for example, heat waves. These impacts are far stronger in northern cities with moderate climates within which temperatures fluctuate widely year to year. Because of differences in the extent of human physiologic adaptation to higher temperatures, more people die in heat waves in Chicago than, say, in Birmingham, Alabama, and rising average temperatures will accentuate that disparity further. Likewise, because of greater population density and prevailing winds, the distribution of harmful effects from emissions that form criteria air pollutants is highly nonhomogeneous. For example, populations in eastern seaboard counties bear more of the health-related external costs of this external impact of electricity production from fossil fuels than do populations in upwind areas and will continue to irrespective of any short-term policy choices. Thus, when aggregate damages are presented, the differential impact may be partly obscured.

For other impacts, such as the local—potentially devastating—effects of a power-plant disaster or disruption occurring in a distribution line (for example, an oil or gas line), local choices may be extremely important in determining "who pays." Often, siting of these types of facilities is partially determined by geographical factors, such as where production and utilization actually occur.

In some situations, aggregate damages may be juxtaposed against local damages, creating not only heterogeneity but also complex policy alternatives. For example, there is at least some evidence that centralized, rather than decentralized, management of spent nuclear fuel results in an inherently lower risk of adverse external consequences; yet arguably for the site or sites chosen for a centralized activity the local "costs" can be higher. Another similar consideration is that the damages of power-plant emissions vary by the population affected by the emissions.

*Differences in Susceptibility over Spatial Scales*

Even within the same locations, there is compelling evidence that some parts of the human population or that some species within an ecosystem are more vulnerable than others to a particular external effect. One of the factors responsible for differential effects is age; the very young and the very old are more susceptible to energy-related burdens, such as those imposed by heat stress, water constraints, or pollution. Likewise, the underlying health status of individuals or groups creates large disparities in effects. In highly developed societies, risks of the nutritional consequences of climate change or diarrheal illness are essentially nil, while these impacts dominate in societies with lower levels of overall health status. Conversely, air pollutants from electricity or transportation tend to affect, to a greater degree, individuals and societies with higher underlying rates of cardiovascular and chronic disease, which are more prevalent in richer societies. This same factor differentiates the consequence to an individual with chronic disease from his/her healthy partner, even living in the same house. These conditions may be confounded by disparities created by differential access to resources, for example, socioeconomic differences, within a nation or region. For example, during the last highly publicized heat wave—Chicago 2003—almost all of the excess deaths occurred among poor minorities without air-conditioning or ready access to health or social services. Once again, aggregate cost data would tend to mask, rather than emphasize, such differences.

## Temporal Issues

Some effects related to the production and use of energy may take years, decades, or longer to manifest themselves. For example, chronic health effects of air pollution attributable to fuel combustion are not the consequence of an exposure that occurred yesterday or a few weeks ago, but they are the cumulative result of conditions that develop over longer periods. As a more extreme example, health risks from the disposal of nuclear waste generated from electricity production may persist over millennia because of the long-lived nature of the radioactive waste. This persistence presents challenges in making judgments about the performance of a waste repository, the behavior of human society, and other key factors over a very long period.

One challenge is that it is very difficult to predict both the future physical effects and their monetary values because they depend on a host of uncertainties about how people in the future will live. A second challenge arises in comparing effects that are quantified in monetary values at different times (such as expenditures on control equipment now and fewer adverse health effects in the future). In making such evaluations, two factors should be considered. One is that many opportunities exist for investing resources now to yield future benefits. The future benefits of a proposed action should be compared with the future benefits that could be achieved by investing the same resources in other ways. The other factor is that the people affected may differ, especially if the delays are long enough that they are necessarily members of different generations.

It is conventional and appropriate to discount future values by a factor that depends on the distance into the future and the discount rate. (It is also necessary to account for future inflation, usually accomplished by valuing all consequences in "real"—that is, "constant" or "inflation-adjusted"—dollars.) In addressing the difficulties that arise in identifying the appropriate discount rate, two approaches are often used. The first is commonly referred to as "descriptive"; the second, as "prescriptive." The descriptive approach uses a discount rate that is similar to market interest rates, which are market prices that are determined by the interactions of individuals, firms, and other institutions seeking to borrow or save for various time periods. The prescriptive approach is often used for time periods of more than about 30 years, for which market interest rates rarely exist. This approach explicitly considers two factors: the rate at which future generations' utility should be valued relative to the current generation's utility (an ethical question), and the rate at which incremental resources will enhance the future generations' utility (a descriptive question) (see Chapter 5 for further discussion).

Estimates of the appropriate discount rate derived from the prescriptive approach are typically smaller than those derived from the descriptive approach. This divergence raises a number of ethical questions, such as whether individuals and governments currently are consuming too much and investing too little and how much individuals should sacrifice now to potentially benefit many future generations.

For valuation of climate-change effects (see Chapter 5), the discounted value referred to as the social cost of carbon is often used. It is the present-day value of the combined damages and benefits that will occur over many future years if an additional ton of greenhouse gas is emitted today. Estimating the discounted cost involves consideration of current greenhouse gas emissions' effects on climate over the next century or more, environmental and human welfare effects caused by climate change, how the effects may vary globally, the course of future economic development, the range and likelihood of economic and social effects arising from climate change, and the extent to which human society might adapt to climate change. Because the choice of a discount rate for such long periods involves great uncertainty, the committee does not recommend a particular discount rate for assessing the value of these effects.

## Model Selection and Evaluation

The committee made extensive use of computational models to evaluate available knowledge, compare alternative technologies, and provide a framework to assess damage. The committee recognizes that all models face inherent uncertainties because human and natural systems are more complex and heterogeneous than can be captured in a model. Moreover, the committee also recognizes that once a model is selected and applied, large uncertainties remain regarding input selection and choices of scale. In its selection and use of models, the committee relied on a report of the NRC's Committee on Regulatory Environmental Models (NRC 2007a), which recommended that models cannot be validated (declared true) but instead should be evaluated with regard to their suitability as tools to address a specific question. In following this approach, the committee first identified its specific questions, then identified the tools available, and finally made model selections.

The committee recognized that its analysis involved five key activities: (1) characterizing a range of technologies that provide electricity, transportation, and heating; (2) identifying the pollutant emissions (and other environmental hazards) attributable to each technology; (3) linking emissions (hazards) to exposures; (4) linking exposures to effects; and (5) translating effects into damages that can be monetized. Modeling was required for electricity production and heating—steps 3, 4, and 5—and for transportation impacts—steps 2, 3, 4, and 5. Therefore, the committee reviewed a number of models that could support this task and considered several alternatives, including several models to address the issues of model uncertainty. Ultimately, the committee determined that the use of a single model would make its results more transparent and open to evaluation than would trying to interpret results from several models. The committee selected the APEEP (Air Pollution Emission Experiments and Policy) model (see Chapter 2) for steps 3, 4, and 5 and the GREET (Greenhouse Gases, Regulated Emissions, and Energy Use in Transportation) model (see Chapter 3) for step 2 in transportation technologies. In making these choices, the committee did not consider these two models to be the only or even the best models for this task. Instead, the choice reflects the committee's recognition that these models were clearly appropriate for the task, were accessible to the committee, were transparent in their applications, and had received sufficient prior use and performance evaluation. To further evaluate the performance of these models for use in calculating external impacts, the committee carried out comparative evaluations where that was feasible.

### Intake Fraction and Other Tools for Model Evaluation

The committee sought other studies with comparable results to evaluate the consistency of its model approach with approaches used by others engaged in similar research. In making these evaluations, the concept of "intake fraction" was useful and transparent. It is defined by Bennett et al. (2002) as the integrated incremental intake of a pollutant summed over all exposed individuals and occurring over a given exposure time, released from a specified source or source class, per unit of pollutant emitted. Since that time, numerous studies have estimated intake fractions for various source categories (such as power plants, mobile sources, residential wood burning, indoor cleaning products, and aircraft) and pollutants (such as particulate matter and toxic air pollutants). Most important, use of the intake fraction approach has increasingly become a tool for model performance evaluation and model comparisons.

For source-receptor estimates from power plants, work by Nishioka et al. (2002) provided a model evaluation opportunity. To assess the health effects of increased pollution, Nishioka et al. (2002) modeled state-by-state exposures to fine particulate matter ($PM_{2.5}$) originating from power-plant combustion and used intake fraction as an intermediate output. The committee was able to compare its power-plant intake fraction obtained from APEEP with theirs and got consistent results. Moreover, Nishioka et al. (2002) multiplied their population-weighted exposures derived from intake fractions by exposure-response functions for premature mortality and selected morbidity outcomes, providing the committee with further opportunity to evaluate APEEP results.

In the transportation impact modeling, there were two studies that provide key evaluation opportunities. In an effort to better characterize the relationship between mobile-source emissions and subsequent $PM_{2.5}$ exposure, Greco et al. (2007) characterized $PM_{2.5}$ exposure magnitude and geographic distribution using the intake fraction. They modeled total U.S. population exposure to emissions of primary $PM_{2.5}$ as well as particle precursors $SO_2$ and $NO_x$ from each of 3,080 counties in the United States. Their mean $PM_{2.5}$ intake fraction was 1.6 per million with a range of 0.12 to 25 per million compared with 1.0 per million with a range of 0.04 to 33 per million obtained from APEEP. Greco et al. (2007) concluded that long-range dispersion models with coarse geographic resolution are appropriate for risk assessments of secondary $PM_{2.5}$ or primary $PM_{2.5}$ emitted from mobile sources in rural areas but that more-resolved dispersion models are warranted for primary $PM_{2.5}$ in urban areas because of the substantial contribution of near-source populations. One of the advantages of APEEP is better spatial resolution in urban counties, but it may still lack the necessary level of spatial detail, giving rise to some uncertainty about results.

Marshall et al. (2005) used three alternative methods to estimate intake fractions for vehicle emissions in U.S. urban areas. Their best estimate of the urban intake fraction for diesel particles was 4 per million, results that are consistent with the urban-county results in APEEP. However, the need for future efforts to provide exposure resolution below the county scale remains a priority.

*Addressing Uncertainty*

Assessment of uncertainty in model outputs is central to the proper use of model results for decision making. There are a number of uncertainties that arise in the calculation of damages from energy use. The committee elected to confront uncertainty using approaches recommended by the NRC's Committee on Regulatory Environmental Models (NRC 2007a). This committee considered the use of probabilistic (Monte Carlo) approaches to quantify all uncertainties to be problematic in many situations, especially when uncertainty analysis is used to reduce large-scale analyses of complex environmental and human health effects to a single probability distribution or when uncertainty is dominated by decision variables, as is the case for this current study. In this study, uncertainty is dominated by such factors as the selected value of a statistical life, which cannot easily be captured in a probability distribution. In situations where detailed probabilistic modeling is not appropriate, the models committee (NRC 2007a) recommended the use of scenario assessment and sensitivity analysis. The current committee chose to use this approach, and where feasible, it has used alternative scenarios and sensitivity analysis to characterize uncertainties.

## ORGANIZATION OF THE REPORT

The discussion in Chapter 2 focuses on the external effects and their valuations, resulting from electricity generation. Chapter 3 addresses externalities related to the production and use of transportation fuels. Chapter 4 discusses energy used to supply heat for industrial processes and to heat indoor spaces. Chapter 5 addresses effects attributable to climate change and their valuations. Chapter 6 discusses effects and valuations related to infrastructure and security. Chapter 7 presents overall conclusions from the committee's evaluations, including a comparison of climate and nonclimate damage estimates, and discusses factors to keep in mind when interpreting the results of the evaluations. Chapter 7 also recommends research to inform future consideration of various issues in this report.

# 2

# Energy for Electricity

## BACKGROUND

This chapter considers sources of energy used for the generation of electricity. The committee's analysis includes utilities, independent power producers, and commercial, and industrial sources. The generation data that we used are available at the Web site of the Energy Information Administration (EIA) (www.eia.doe.gov) of the U.S. Department of Energy, and are the official energy statistics from the U.S. government.

### The Current Mix of Electricity Sources

The total electricity generation[1] in the United States during 2008[2] was 4.11 million gigawatt hours (GWh), down very slightly from 2007. In terms of usage, the residential sector consumed the most electricity (36.6% of the total), followed by the commercial sector (36.3%). The industrial sector (26.9%), and transportation (0.2%) accounted for the rest.

The energy sources and the amount of electricity they contributed are given in Table 2-1.

The two largest classes of "other renewables" were wind, which produced 52,026 GWh or 1.3% of the 2008 electricity-generation total; and

---

[1]The amount of electricity used is less than the amount generated as a result of transmission losses. For 2007, EIA reported usage of 93.4% of the amount generated.

[2]We provide the latest data available here to establish the most recent context. Our analyses of power plant damages, however, were based on 2005 data, the latest for which full emissions information was available.

**TABLE 2-1** Net Electricity Generation by Energy

| Energy Source | Net Electricity Generation (GWh) | Percent of Total Net Generation |
|---|---|---|
| Coal | 2,000,000 | 48.5 |
| Petroleum liquids[a] | 31,200 | 0.8 |
| Petroleum coke | 14,200 | 0.4 |
| Natural gas | 877,000 | 21.3 |
| Other gases[b] | 11,600 | 0.3 |
| Nuclear | 806,000 | 19.6 |
| Hydroelectric | 248,000 | 6.0 |
| Other renewables[c] | 124,000 | 3.0 |

NOTE: Net electricity-generation numbers reported by the Energy Information Administration are rounded to three significant figures.

[a]Distillate fuel oil, residual fuel oil, jet fuel, kerosene, and waste oil.

[b]Blast furnace gas, propane gas, and other manufactured and waste gases derived from fossil fuels.

[c]Wind, solar thermal, solar photovoltaic (PV), geothermal, wood, black liquor, other wood waste, biogenic municipal solid waste, landfill gas, sludge waste, agricultural by-products, and other biomass.

SOURCE: Data from EIA 2008, 2009a.

wood and wood-derived energy sources (38,789 GWh, or 0.9%). Other renewable sources individually amounted to less than 0.5% each; the largest was other biomass, (16,099 GWh, or 0.4%. Generation from solar PV was approximately 600 GWh.

## Rationale for Choice of Fuel Sources to Analyze

This chapter provides detailed analyses of electricity generation from coal, natural gas, nuclear fission, wind, and solar. The first three sources were chosen because they together account for 88% of all electricity generated in the United States; moreover they feature prominently in current policy discussions about energy sources. Wind energy also is prominent in policy discussions concerning electricity, and it appears to have the largest potential among all renewable sources to provide additional electricity in the medium term according to current projections (see discussion later in this chapter). Solar energy for electricity (photovoltaics) also is discussed, although not in detail, because of recent legislative and public interest and because of the rapid increase in use over the past 10 years. For the above reasons, the committee concluded that analyzing the external costs and benefits associated with these sources would be of the greatest value to policy makers.

We mention biomass (briefly) because it is such a dispersed source of electricity (many very small generators). We did not focus on hydropower generation of electricity, even though its current contribution is far greater than that of all other renewable sources combined, because the potential use of hydropower to increase significantly is modest, and hydropower currently receives little attention in energy-policy discussions.

## Describing the Effects Caused by Life-Cycle Activities

In its analyses, the committee describes externalities—indeed, all *effects* caused by life-cycle activities—as being upstream or downstream. By "upstream," in the context of energy for electricity, the committee means effects that occur before electricity is generated at an electricity-generating unit (EGU) (such effects as EGU; steam turbine, wind turbine, and solar cell). For fossil-fuel and nuclear EGUs, the largest upstream effects are associated with obtaining and transporting fuel. They include effects of exploration, development, and extraction of geologic deposits of fuel or ore, refining and processing, and transportation of primary energy sources (for example, coal and natural gas). For solar, wind, and hydropower EGUs, the main upstream effects are associated with obtaining, fabricating, and transporting materials required for the EGU and with the construction of the EGU, including road building and other activities. Fossil-fuel and nuclear EGUs also have these effects, but they typically are smaller than those associated with the ongoing production and transportation of the primary energy sources. The committee's upstream limit for consideration of effects was exploration for fuel. Although effects even further upstream can occur, such as reactions to the announcement of a lease sale for oil, gas, or even the announcement of a proposed mine (for example, see NRC 2003a), those effects are generally unquantified. By "downstream" the committee means effects that are associated with generation of electricity and the subsequent transmission and distribution of electricity to end users. In other words, effects associated with the operation of an electricity-generating facility or with electricity transmission and distribution (that is, delivery to the end user) are considered downstream effects.

## General Approach Taken

The goal of this chapter is to describe and, when possible, to quantify the monetary value of the physical effects[3] (that is, the "damages") of electricity production. For electricity generation from nuclear fission, wind power, solar power, and biomass, our analysis summarizes effects reported

---

[3]The committee uses the term "physical effects" broadly, to include biological and human health effects, in order to distinguish them from monetary effects.

from previous studies, but does not monetize damages from externalities. For electricity generation from coal and natural gas we are able to quantify and monetize the externalities associated with local and global air pollution, both upstream and downstream. We express these externalities in costs per kWh of electricity generated and also in costs per ton of pollution generated.

As summarized in Chapter 1, this study is preceded by a large literature on the social cost of electricity. Two notable studies are those by Oak Ridge National Laboratory and Resources for the Future (ORNL-RFF) (1992-1998) and the ExternE project (EC 2003). The goal of each study was to estimate the life-cycle externalities associated with electricity production from various fuel types. Externalities were expressed in monetary terms per kWh to permit comparisons across fuel types. The social costs of electricity generation, together with the private costs of electricity generation, could thus be used to inform choices among fuel types when expanding or replacing generation capacity. Both studies conducted their analyses using representative plants in two geographic locations. Both studies were exhaustive in their descriptions of, and attempts to quantify, various categories of externalities throughout the fuel cycle.

In addition to literature on social costs of electricity, there have been studies on the environmental effects of electricity production. The National Research Council recently (2007b) reported on environmental effects of wind-energy projects, and the New York State Energy Research and Development Authority recently (NYSERDA 2009) reported on effects and risks to vertebrate wildlife in the northeastern United States from six types of electricity generation.[4] Both reports included assessments of all life-cycle stages, but did not quantify or monetize the effects.

This chapter builds on and extends these studies. We have attempted to describe externalities and other effects broadly, and to analyze them wherever possible. However, we have focused our efforts to monetize external costs for the categories of externalities that earlier studies found to be a significant component of damages. We extend the studies by measuring the externalities associated with local and global air pollution—a significant component of the costs of electricity generation—for individual coal-fired and gas-fired power plants in the United States. This allows us to characterize the diversity in the damages of electricity generation from fossil fuel across plants and to relate damages per kWh to the pollution intensity of the plant (that is, to pounds of sulfur dioxide [$SO_2$] or particulate matter [PM] emitted per kWh) and the location of the plant, which affects the size of the human and other populations exposed to pollution generated by the plant. We also express damages per ton of pollution emitted. While

---

[4]The six types were coal, oil, natural gas, nuclear, hydro, and wind.

a comparison of damages per kWh may (together with information about private costs) help inform the choice of fuel type, it is not particularly useful if the goal is to internalize the externalities associated with pollution emissions.[5] Economic theory suggests that the most economically efficient policy to address air-pollution externalities is a policy that targets the externality itself rather the output associated with it. We therefore present information on damages per ton of emissions from coal and natural gas plants that contribute to the concentrations of criteria pollutants.[6]

The core of our analysis of local air-pollution damages uses an integrated assessment model (the Air Pollution Emissions Experiments and Policy, or APEEP model) (Appendix C), which links emissions of $SO_2$, oxides of nitrogen ($NO_x$), $PM_{2.5}$, $PM_{10}$,[7] ammonia ($NH_3$), and volatile organic compounds (VOCs) to ambient levels of $SO_2$, $NO_x$, $PM_{2.5}$, $PM_{10}$, and ozone (see Box 2-1). The model calculates the damages associated with population exposures[8] to these pollutants in six categories: health, visibility, crop yields, timber yields, building materials and recreation. Health damages include premature mortality and morbidity (for example, chronic bronchitis, asthma, emergency hospital admissions for respiratory and cardiovascular disease), and are calculated using concentration-response functions employed in regulatory impact analyses by the U.S. Environmental Protection Agency (EPA). Damages to crops are limited to major field crops, and recreation damages are those associated with pollution damages to forests. A description of the concentration-response functions used in the model is in Appendix C, which also provides details on the choice of unit values used to monetize damages. Damages associated with carbon dioxide ($CO_2$) emissions are computed based on a review of the literature, and are described in Chapter 5. Not all impacts and externalities associated with electricity production have been quantified and monetized in this study. Table 2-2 summarizes which impacts are quantified, monetized, or qualitatively discussed within this chapter.

---

[5] An electricity tax equal to the marginal damage per kWh is a blunt instrument for internalizing the social costs of air pollution because it does not target the pollutants (for example, $SO_2$ or $PM_{2.5}$) that are the sources of the problem.

[6] As part of the U.S. Clean Air Act, the U.S. Environmental Protection Agency (EPA) establishes National Ambient Air Quality Standards PM, $SO_2$, $NO_x$, ozone, lead (Pb), and carbon monoxide (CO). These are referred to as criteria pollutants, which were established by the Clean Air Act as pollutants that are widespread, come from numerous and diverse sources, and are considered harmful to public health and the environment and cause property damage.

[7] $PM_{2.5}$ refers to particulate matter with an aerodynamic diameter less than or equal to 2.5 microns; $PM_{10}$ refers to particles less than or equal to 10 microns in diameter. Ultrafine particles—those less than 100 nanometers—were not treated as a separate category in this study.

[8] "Population exposure" is an aggregate figure derived from measurements or estimates of personal (individual) exposures that are extrapolated—based on statistical, physical, or physical-stochastic models—to a population (Kruize et al. 2003).

## BOX 2-1
## Airborne Particulate Matter

PM is a heterogeneous collection of solid and liquid particles that can be directly emitted from a source (primary pollutants) or can be formed in the atmosphere by interaction with other pollutants (secondary pollutants). Secondary PM can be formed by oxidation of $NO_x$ and $SO_x$ to form acids that can be neutralized by ammonia to form sulfates and nitrates. Organic PM may be chemically transformed by oxidants in the air to form secondary pollutants. Soot particles can be altered by adsorption of other pollutants on their surface.

PM is monitored for both mass and size. Ultrafine particles (less than 0.1 micron in aerodynamic diameter) can be emitted from combustion sources or can be formed by nucleation of atmospheric gases, such as sulfuric acid or organic compounds. Fine particles (less than 2.5 microns) are produced mainly by combustion of fossil fuels, either from stationary or mobile sources. Coarse particles (sometimes called $PM_{102.5}$) are mainly primary pollutants that may come from abrasive or crushing processes or the suspension of soil. PM larger than 10 microns is not of great concern for this report because they are not readily respirable and do not have a long half-life in the atmosphere.

Current research on PM is exploring the influence of particle composition (in addition to mass and size) on its toxicity, as recommended by the National Research Council (NRC 1998, 1999, 2001, 2004b). However, enough data are not yet available from this research to inform the estimation of damages in this report.

## Regulations

As noted in Chapter 1, the externalities examined in this study are those that have not been eliminated by regulation. Most stages of electricity production are subject to regulations at the federal, state, and local levels. Surface mining of coal, for example, is regulated under the 1977 Surface Mining and Control Act. Air-pollution emissions from electricity-generating facilities are regulated under the Clean Air Act. The U.S. Nuclear Regulatory Commission regulates and licenses nuclear power plants.

Relevant regulations for upstream and downstream activities related to electricity generation are varied and extensive. Their details are not necessarily of great import *for this study*, although they obviously are important for other reasons. For this study, though, the *existence* of regulations is of great importance, because in large part regulations are an attempt to reduce upstream and downstream damages from electricity generation, and they have substantially reduced these damages over time. We discuss only those damages that remain, with emphasis on those that can be quantified and monetized. Most of the committee's quantitative analyses of damages in

**TABLE 2-2** Energy for Electricity: Impacts and Externalities Discussed, Quantified, or Monetized

| | Energy Sources for Electricity | | | | | |
| Impact or Burden | Coal | Natural Gas | Nuclear | Wind | Biomass | Solar |
|---|---|---|---|---|---|---|
| *Upstream* | | | | | | |
| Air pollutant emissions ($SO_x$, $NO_x$, PM) | ✓ | ✓ | q | q | | q |
| $CO_2$-eq (carbon dioxide equivalent) emissions | ✓ | ✓ | q | q | | q |
| Metals, radionuclides, and other air pollutants | q | q | q | q | | q |
| Effluents | q | q | q | | | |
| Solid wastes | q | q | q | | | |
| Land cover/footprint | q | q | q | | q | q |
| Ecological effects | q | q | | | q | |
| Occupational and transport injuries | † | † | | | | |
| *Downstream* | | | | | | |
| Air pollutant emissions ($SO_x$, $NO_x$, PM) | $ | $ | ✓ | ✓ | | ✓ |
| $CO_2$-eq emissions | ✓ | ✓ | ✓ | ✓ | | ✓ |
| Metals, radionuclides, and other air pollutants | q | q | | ✓ | | q |
| Effluents | q | q | q | | | |
| Solid wastes | q | q | q | | | q |
| Land cover/footprint | q | q | q | q | | q |
| Ecological effects | q | | | †, q | | |

q = qualitative discussion.
✓ = emissions quantified.
† = impacts quantified.
$ = impacts monetized.

this chapter focus on emissions from electricity-generating facilities that are fired by coal or natural gas. Under the Clean Air Act, electric utilities are regulated at both the state and federal levels. The Clean Air Act requires states to formulate state implementation plans (SIPs) to pursue achievement of the National Ambient Air Quality Standards (NAAQS) (NRC 2004c). Under SIPs, electricity-generating units (EGUs) are assigned emissions limits for $SO_2$, $NO_x$, PM, and other pollutants, usually stated as performance standards (for example, maximum annual average tons of $SO_2$ that may be emitted per million British thermal units [MMBtu] of heat input). These performance standards vary widely across states. In addition, EGUs are subject under the Clean Air Act to "new source review," a series of regulations that pertain to newly constructed facilities and to modifications of existing facilities.[9] Coal-fired power plants built after 1970 are also subject to "new source performance standards" (NSPS), which impose strict limits on emissions that contribute to the formation of criteria air pollutants. For example, the 1978 NSPS for coal-fired power plants requires the installation of flue gas desulfurization units (scrubbers) on all new coal-fired EGUs.

Emissions of $SO_2$ and $NO_x$ are also regulated under various cap-and-trade programs. The goal of Title IV of the 1990 amendments to the Clean Air Act was to reduce $SO_2$ emissions from EGUs to 8.95 million tons by 2010. That goal has been achieved by issuing $SO_2$ permits (allowances) to EGUs equal to 1.2 pounds of $SO_2$ per MMBtu (based on 1985-1987 heat input) and allowing utilities to trade allowances, which may not violate the NAAQS. In 1998, EPA issued a call for SIPs to reduce emissions of $NO_x$. The rule provided the option for states to participate in a regional $NO_x$ Budget Trading Program. This program operated from 2003 to 2008, when it was replaced by a $NO_x$ ozone season trading program.

The net effect of the environmental regulations described above, as well as others, is that emissions per megawatt-hour (MWh) that contribute to criteria air pollution vary greatly among plants. Newer power plants have, on average, much lower emissions rates. As discussed later in this chapter, $SO_2$ (and $NO_x$) emissions per MWh are much lower for units installed after 1979 than for units installed before that date.

## ELECTRICITY PRODUCTION FROM COAL

### Current Status of Coal Production

Coal, a nonrenewable fossil fuel, accounts for approximately one-third of total U.S. energy production, and nearly half of all electricity produced. Coal is classified into four types based upon the relative mix of carbon, oxygen and hydrogen: lignite, sub-bituminous, bituminous, and anthracite (Table 2-3). The greater the carbon content, the greater the energy (heating) value of coal. Sub-bituminous and bituminous coal account for more than 90% of coal produced in the United States. Sub-bituminous coal has as much lower sulfur content but also as much lower energy content than bituminous coal. In electricity generation, replacing a ton of bituminous coal requires about 1.5 tons of sub-bituminous coal (NRC 2007c).

The United States has more than 1,600 coal-mining operations that pro-

---

[9]New source review applies to facilities in areas of pristine air quality where the goal is to prevent significant deterioration of air quality and also to facilities in areas that have not attained the NAAQS. Regulations governing each facility are determined on a case-by-case basis. See the regulatory overview in Chapter 2 of NRC 2006a.

**TABLE 2-3** Coal Classification by Type

| Type | Carbon Content (%) | Heating Value (Thousand Btu/lb) | U.S. Production (%) |
|------|--------------------|--------------------------------|---------------------|
| Lignite | 25-35 | 4.0-8.3 | 6.9 |
| Sub-bituminous | 35-45 | 8.3-13.0 | 46.3 |
| Bituminous | 45-86 | 11.0-15.0 | 46.9 |
| Anthracite | 86-97 | ~15.0 | <0.1 |

ABBREVIATION: Btu/lb = British thermal unit per pound.

SOURCE: EIA 2008a, Table 7.2; NEED 2008; EIA 2009b.

duced more than 1.18 billion short tons[10] in 2008. Major coal-producing regions are shown in Figure 2-1. The EIA estimates that 70% of coal production comes from surface mines, the majority of which are in Wyoming, Montana, West Virginia, Pennsylvania, and Kentucky. Large mining operations in the Powder River Basin (PRB) in Wyoming and Montana accounted for more than 50% of surface-mine coal production and 40% of nationwide coal production in 2007. Coal in the PRB is mainly sub-bituminous; coal in Appalachia is mainly bituminous (NRC 2007c). The top five coal-producing states in 2007 are listed in Table 2-4.

On average, more coal is produced in the United States than is consumed. The EIA estimates that nearly 95% of U.S.-mined coal is consumed domestically. In 2008, the United States exported 23.0, 7.0, and 6.4 million short tons to Canada, the Netherlands, and Brazil, respectively.

U.S. coal production is focused in a relatively small number of states, but coal is consumed throughout the country. As a result, coal is transported by all major surface transportation modes (Figure 2-2). Once mined, coal is typically transported to power plants, steel mills, and other commercial and industrial companies by rail. In 2007, approximately 70% of coal production was distributed by rail. The remaining 30% was transported by barge, tramway and pipelines, or truck.

Looking forward, it can be expected (barring shifts in current coal consumption trends) that western states will increase their production relative to other states (EIA 2008a). Table 2-5 below lists the ten states with the largest Estimated Recoverable Reserves (ERR). The ERR is derived by the Energy Information Administration (EIA) for each state by applying coal mine recovery and accessibility factors to the Demonstrated Reserve Base (NRC 2007c).

---

[10]A short ton is 2,000 pounds, or 907.2 kilograms.

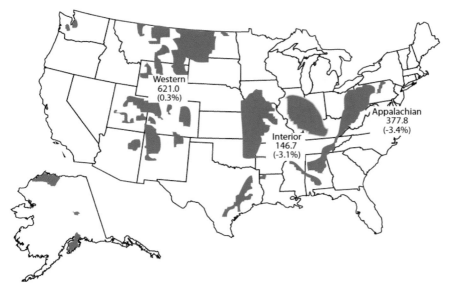

**FIGURE 2-1** Major coal-producing regions in the United States (million short tons and percent change from 2006). SOURCE: EIA 2009c, p. 2.

**TABLE 2-4** Five Leading Coal-Producing States, 2007, by Mine Type and Production (Thousand Short Tons)

| State | Number of Mines | Production |
|---|---|---|
| Wyoming | 20 | 453,568 |
| Underground | 1 | 2,822 |
| Surface | 19 | 450,746 |
| West Virginia | 282 | 153,480 |
| Underground | 168 | 84,853 |
| Surface | 114 | 68,627 |
| Kentucky | 417 | 115,280 |
| Underground | 201 | 69,217 |
| Surface | 216 | 46,064 |
| Pennsylvania | 264 | 65,048 |
| Underground | 50 | 53,544 |
| Surface | 214 | 11,504 |
| Montana | 6 | 43,390 |
| Underground | 1 | 47 |
| Surface | 5 | 43,343 |
| Total, Top Five States | 989 | 830,766 |
| Underground | 421 | 210,483 |
| Surface | 568 | 620,284 |
| Total, United States | 1,358 | 1,145,480 |

SOURCE: Adapted from EIA 2009c, p. 11, Table 1.

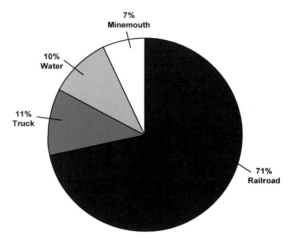

**FIGURE 2-2** Methods of U.S. coal transport. NOTE: Data exclude a small unknown component. SOURCE: EIA in AAR 2009.

## Brief History of Coal Production

Coal was the predominant source of U.S. energy from the late 19th century through the mid 20th century. Coal was used for electricity, space heating, industrial process heating for iron, steel, and other commodities, and fuel to power ship and train steam engines. During the latter 20th

**TABLE 2-5** Estimated Recoverable Reserves for the 10 States with the Largest Reserves by Mining Method for 2005 (million short tons)

| State | Underground Minable Coal | Surface Minable Coal | Total |
|---|---|---|---|
| Montana | 35,922 | 39,021 | 74,944 |
| Wyoming | 22,950 | 17,657 | 40,607 |
| Illinois | 27,927 | 10,073 | 38,000 |
| West Virginia | 15,576 | 2,382 | 17,958 |
| Kentucky | 7,411 | 7,483 | 14,894 |
| Pennsylvania | 10,710 | 1,044 | 11,754 |
| Ohio | 7,719 | 3,767 | 11,486 |
| Colorado | 6,015 | 3,747 | 9,762 |
| Texas | — | 9,534 | 9,534 |
| New Mexico | 2,801 | 4,188 | 6,988 |
| Total, Top 10 States | 137,031 | 98,896 | 235,927 |
| Total United States | 152,850 | 114,705 | 267,554 |

SOURCE: EIA 2006a. Adapted from NRC 2007c, p. 51, Table 3.2.

century, however, coal was rapidly replaced by petroleum and natural gas for fuel and space heating, respectively. Today, approximately 7% of coal is consumed to generate heat for a variety of industrial processes, including paper, concrete, and steel production.

## Upstream Impacts and Externalities of Electricity Production from Coal

### Injuries and Illnesses in Coal-Mining Operations

Although the gravity of occupational injuries and illnesses cannot be underestimated, the tradition in economics is to treat occupational injuries and deaths as job characteristics that are traded in labor markets rather than to treat them as externalities. In general, miners receive compensating wage differentials for the higher risks that they face on the job (Viscusi 1993).[11] In addition, some proportion of injuries and deaths are compensated after the fact through workmen's compensation, insurance, or court judgments. We also note that previous studies of the social cost of electricity (for example, ORNL-RFF 1994b) did not count occupational injuries and illnesses as externalities. However, occupational injuries are briefly discussed because they are an important societal concern related to energy production.

Coal-mining-related fatalities and nonfatal injuries have generally decreased over time, even though employee hours have not steadily declined (Figure 2-3). This is the result of increased regulation and safer mining technology. In 2008,[12] 29 fatal injuries (corresponding to 2 deaths per 10,000 workers) and 4,760 nonfatal injuries (an incidence rate of 3.83 per 100 workers) were reported.[13] This marked a 27% decrease from 2000 to 2007 in the incidence of both fatal and nonfatal injuries and, more dramatically, 35% and 54% decreases, respectively, in the incidence of fatal and nonfatal injuries from the previous decade. The majority of both fatal and nonfatal injuries occur in underground mines (67% in 2008), followed by strip mines (19%) and processing plants (8%).[14]

---

[11]It can be argued that wage differentials do not fully compensate for risk of death or injury because of the monopsony power on the part of employers or the lack of information on the part of workers. These are both examples of market imperfections but do not constitute externalities.

[12]All 2008 figures are preliminary.

[13]Injury data include all coal-operations incidents having occurred in mines, independent shops, processing plants, and offices. Contractors are included.

[14]Coal-mining disasters, defined by the U.S. Mine Safety and Health Administration as incidents resulting in five or more deaths, had decreased substantially in frequency and in number of fatalities since 1970. However, in 2006, a series of disasters resulted in the deaths of 19 miners. These events, particularly the January 2006 Sago Mine disaster, which resulted in the deaths of 12 miners, received nationwide attention and were the stimulus for the Mine Improvement and New Emergency Response (MINER) Act of 2006.

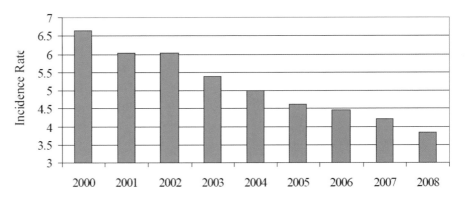

**FIGURE 2-3** Injuries in U.S. coal-mining operations from 2000 to 2008. SOURCE: Data from MSHA 2008, Table 08; MSHA 2009.

Most injuries in coal-mining operations result in workdays lost (WDL). In 2007, nonfatal injuries accounted for 220,284 WDL. Injuries classified as strain/sprain, cut or puncture, and fracture accounted for 76% of all injuries (31%, 24%, and 18%, respectively) but only 67% of nonfatal WDL and 6% of fatalities. Multiple injuries and bruises or contusions accounted for 79% and 12%, respectively, of fatalities, while accounting for only 3% and 11%, respectively, of total injuries. Coal-mining operations also reported a total of 159 occupational illnesses in 2007, 80 being disorders associated with repeated trauma and 40 being dust-related diseases of the lungs.

*Injuries and Fatalities in Coal Transport*

Coal transport introduces risks to the public and to employees of the transportation industry (primarily railroad, truck, and barge), which we describe below. As discussed above, occupational injuries and fatalities are not considered externalities. However, nonoccupational injuries and fatalities probably are externalities—that is, one could argue that the railroad operator might not take the full risk of death or injury to another person into consideration when choosing driving speed or safety equipment unless required to do so by law.

Domestic coal shipments represented 730 billion ton-miles in 2006, a 47% increase from 498 billion ton-miles in 1996. According to the Energy Information Administration, 71% of these U.S. coal shipments were delivered to their final domestic destinations by rail, followed by truck (11%) and barge (10%, mainly on inland waterways). Rail's share, along with the average length of haul for rail coal movements, has been increasing over the past 15 years (from 57% in 1990 to 71% in 2006). This is largely due to the growth of western coal. Waterborne traffic's share of coal shipments has been declining, while the share of coal shipped by truck has fluctuated. Trucks transport coal over short distances, thus accounting for a small proportion of coal ton-miles (less than 2% in 2002) but a more substantial amount of tonnage (12% that same year). The average distance traveled by truck per shipment of coal increased from 51 miles in 1997 to 88 miles in 2002.

Coal is by far the most significant commodity carried by rail. In 2007, coal transport accounted for almost 44% of tonnage, 24% of carloads, and 21% of gross revenue for U.S. Class I railroads as well as a significant portion of non-Class I railroad freight. The commodity dominates originated rail traffic in major coal-producing regions. For example, coal accounted for 79% of total rail tons originated in Kentucky, 95% in West Virginia, and 96% in Wyoming in 2006. Coal (not including coal coke) is also a significant commodity in waterborne commerce, accounting for approximately 9% of tonnage. Large trucking, by contrast, only owes 0.2% of vehicle miles traveled to coal transport. For these reasons, we focus on the externalities associated with the shipment of coal by rail.

Over the past several decades, rail transportation has seen considerable drops in accident/incident rates, thanks in part to numerous initiatives on grade crossings and trespasser prevention. In 2008, there were 571 freight rail fatalities and 4,867 nonfatal injuries, indicating a 9% decline in fatalities and 11% decline in nonfatal injuries since 2007, and, more notably, 48% and 76% declines, respectively, since 1990. Ninety-seven percent of fatalities occur among the public, while, in contrast, the majority of nonfatal injuries and illnesses are borne by employees.

To estimate fatal and nonfatal injuries attributable to coal transport via rail, we use revenue ton-miles[15] as a quantifiable proxy for risk of rail-associated injury. The reasoning for using revenue ton-miles as a proxy for risk of injury to railroad employees is that the number of employee hours, and hence the number of injuries, is more closely correlated with the revenue ton-miles measure than with train-miles or carloads. The reason for using revenue ton-miles as a proxy for risk of injury to the public is based on availability of information. A train-miles measure of coal transport would be the preferred metric for assessing risk to the public, but no such recent measure is available. We chose ton-miles of coal transport as the "next-best" measure for assessing risk to the public because it includes distance.

Our estimate of the number of fatal and nonfatal rail injuries attribut-

_____

[15]A revenue ton-mile is defined as the movement of one ton of revenue-generating commodity over the distance of 1 mile. It is calculated by multiplying tons moved by the number of miles involved.

able to shipping coal for electric power generation appears in Table 2-6. The estimate is computed by multiplying the total number of occupational and public injuries occurring on freight railroads[16] in 2007 by the proportion of ton-miles of commercial freight activity on domestic railroads accounted for by coal (43%).[17] This estimate is then multiplied by the percent of coal transported that is used for electric power generation (91%).

By analogy with coal mining, we assume that occupational deaths and injuries are not externalities. A key issue is whether deaths among the public constitute externalities. One can argue that they are externalities (most are people struck by a moving train); however, based on the magnitude of the resulting damages, we have not monetized them, and they are not included in our aggregated damages. Valuing the 241 lives lost in 2007 by using a value of a statistical life (VSL) of $6 million 2000 U.S. dollars (USD) (about $7.2 million 2007 USD) would result in damages less than $2 billion annually.

### Land-Use and Runoff Externalities from Surface and Underground Mines

This section describes, but does not quantify or monetize, environmental effects of coal mining. Over the past 58 years there has been a relative shift to surface mining and to coal from western states (Figures 2-2 and 2-4). Surface mining is used for shallow deposits. Techniques range from area strip mining more typical in the West to contour strip mining and mountaintop mining/valley fill (also known as "mountaintop removal") more typical in the East. Underground mining techniques range from drift mines and slope mines for deposits relatively near the surface to shaft mines for deposits deep underground.

Wyoming's Powder River Basin (PRB) has near-surface deposits of coal that are more than 100 feet thick, making surface mining easy and productive, and the coal is almost always shipped to market "raw" (that is, without processing). A single PRB surface mine can yield more than 90 million tons annually. In contrast, coal in Appalachia, whether from surface or underground mining, is generally produced at smaller, lower-yield mines, and the coal often is processed in order to lower ash and moisture content (NRC 2007c).

The negative environmental externalities of coal mines, both during operation and after closure, depend in part on the mining method:

---

[16]Counts of injury incidents for freight railroads include those occurring on Class I and switching freight railroads. While coal trains will be freight only, some freight railroads also operate passenger lines; to correct for this phenomenon, we remove passenger injuries and fatalities from the data.

[17]The most recent available statistics on ton-miles of coal transported via rail are for 2002 (DOT/DOC 2004).

**TABLE 2-6** Estimated Injuries, Illnesses, and Fatalities During Rail Transport of Coal for Electric Power, 2007

| | | Nonfatal Cases | | | | | Total Cases |
|---|---|---|---|---|---|---|---|
| | Fatalities | Injuries | Illnesses | NFDL | NDL | Total NF | |
| Employees on duty | 5 | 1,408 | 36 | 991 | 453 | 1,444 | 1,449 |
| Other (such as the public) | 241 | — | — | — | — | 698 | 939 |
| Total | 246 | — | — | — | — | 2,142 | 2,388 |

ABBREVIATIONS: NFDL = nonfatal days lost; NDL= no days lost; NF = nonfatal.

SOURCE: FRA 2008.

• *Underground mining.* In addition to its threats to human health and safety, underground mining can also have environmental externalities. Collapses or gradual subsidence above the mined void can affect surface and subsurface water flows. Mine fires can occur, especially in abandoned mines. The disposal of mine wastes, especially wastes resulting from coal processing, can present environmental problems (NRC 2002b, 2007c). As much as 50% of the material fed to a process for treating raw coal can result in waste, often in the form of slurry, which usually is pumped into an impoundment. Impoundments can give way, as in the October 2000 breakthrough of a 72-acre coal waste impoundment near Inez, Kentucky (NRC 2002b). Environmental problems also can be triggered by acid mine

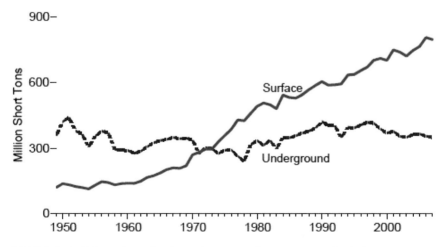

FIGURE 2-4 U.S. coal production 1949-2007, by mining method. SOURCE: EIA 2008a, p. 224, Figure 7.2.

drainage caused primarily by pyrite (FeS$_2$), which is found in coal, coal overburden, and mine waste piles (USGS 2009a).

- *Surface mining* (area and contour). Surface mining shares with underground mining the problem of mine waste disposal and acid mine drainage. It also poses the environmental challenge of reclaiming large tracts of land. The 1977 Surface Mining Control and Reclamation Act was intended to address surface-mining effects. It requires that sites be returned to their prior condition or to a condition that supports "higher and better uses."
- *Mountaintop mining/valley fill* (MTM/VF). MTM/VF is a type of surface mining used on steep terrain. Since its inception in the 1970s, this mining method has become widespread in Appalachia. Mountaintop mining often generates a large volume of rock, or "excess spoil," that cannot be returned to its original locations and typically is placed in adjacent valleys. MTM/VF shares the negative externalities of other types of surface mining (see above) and has other externalities as well.

A Final Programmatic Environmental Impact Statement (FPEIS) on MTM/VF was released in October 2005 to consider developing agency policies regarding the adverse environmental effects of MTM/VF. Prepared by the U.S. Army Corps of Engineers, EPA, the U.S. Department of Interior's Office of Surface Mining and Fish and Wildlife Service, and the West Virginia Department of Environmental Protection, the FPEIS focused on approximately 12 million acres encompassing most of eastern Kentucky, southern West Virginia, and western Virginia as well as scattered areas of eastern Tennessee. About 6.8% of the study area (816,000 acres) has been or may be affected by recent and future (1992-2012) mountaintop mining (EPA 2002, 2005a).

The study area is largely forested and contains about 59,000 miles of streams, most of which are considered headwater streams. The FPEIS comments that "headwater streams are generally important ecologically" and that "the study area is valuable because of its rich plant life and because it is suitable habitat for diverse populations of migratory songbirds, mammals, and amphibians" (EPA 2005a, p. 3).

The EPA Region 3 Web site on MTM/VF and the FPEIS note that valley fills generally are stable, but "based on studies of over 1,200 stream segments affected by mountaintop mining and valley fills, the following environmental issues were noted:

- An increase of minerals in the water—zinc, sodium, selenium, and sulfate levels may increase and negatively impact fish and macroinvertebrates leading to less diverse and more pollutant-tolerant species.
- Streams in watersheds below valley fills tend to have greater base flow.

- Streams are sometimes covered up.
- Wetlands are, at times inadvertently and other times intentionally, created; these wetlands provide some aquatic functions, but are generally not of high quality.
- Forests may become fragmented (broken into sections).
- The regrowth of trees and woody plants on regraded land may be slowed due to compacted soils.
- Grassland birds are more common on reclaimed mine lands as are snakes; amphibians such as salamanders, are less likely. . . .
- Cumulative environmental costs have not been identified . . . (EPA 2009a).

The Web site also notes that there may be social, economic, and heritage issues with MTM/VF. Similarly, a USGS study of the Kanawha Basin (Paybins et al. 2000) shows significant degradation in the biotic communities of this mid-Atlantic river basin as a result of coal-mining operations, and other USGS studies show similar effects elsewhere (see USGS 2009b).

A possible benefit of coal mining can be the roads, utilities, and other infrastructure that accompany a mining operation. With proper planning, especially integration of the mine decommissioning and closure plan with local master plans, this infrastructure can be used for other economic enterprises following mine closure (NRC 2007c).

## Upstream Emissions of Greenhouse Gases and Other Pollutants

The upstream life cycle of power generation from coal includes many relevant activities such as construction of infrastructure and power plants (see, for example, Pacca and Horvath 2002), but the most significant, from a perspective of greenhouse gas (GHG) emissions and criteria-pollutant-forming emissions, are surface and underground mining and transportation of coal. Mining and transport are fuel- and energy-intensive, requiring combustion of fossil fuels for cutting, moving, and preparing the coal from the mine and delivering it to power plants and other industrial facilities. Beyond emissions from engines, there are also significant emissions of methane, a GHG that exists within coal seams and is released as the seams are cut to extract the coal. As methane is a much more potent GHG than $CO_2$, methane emissions are a significant concern.

In surface mining, the overburden (layers of rock and earth above the coal) is broken and removed to get to the underlying coal. The breaking and removal of both overburden and coal, and its movement from mine to transportation network is done with enormous machinery and engines that operate mostly by burning liquid fuels that release GHG emissions and criteria-pollutant-forming emissions. Underground mining uses similar technologies, but shafts need to be drilled down to the seam depth, and the subsurface coal cutting and moving equipment is generally less energy efficient due to its smaller size since it has to fit beneath the surface.

Prior studies have assessed the relative contribution of air emissions from mining and transport of coal in the life cycle of coal-fired power generation (Jaramillo et al. 2007, Spath et al. 1999, ORNL/RFF 1992-1998). While not negligible, these studies found that upstream activities lead to relatively small life-cycle air emissions because of the dominance of GHG emissions and criteria-pollutant-forming emissions on site at coal-fired power plants. For example, Jaramillo et al. (2007) report that the mid-point GHG emission factors for coal combustion (at the power plant) and the entire coal life cycle are 2,100 lb $CO_2$ equivalent (eq)/MWh and 2,270 lb $CO_2$-eq/MWh, respectively.

## Downstream Externalities of Electricity Production from Coal

### Analysis of Current Air-Pollution Damages from Coal-Fired Power Plants

The air-pollution emissions from fossil-fueled power plants constitute a significant portion of the downstream damages associated with electric power generation. In this section, we quantify the impacts on human health, visibility, agriculture, and other sectors associated with coal-fired power-plant emissions contributing to criteria pollutant formation. The effects of those emissions on ambient air quality are modeled using the APEEP model (Muller and Mendelsohn 2006) and are calculated for each of 406 coal-fired power plants for the year 2005. We use the APEEP model to calculate the damages associated with emitting a ton of each of four pollutants ($SO_2$, $NO_x$, $PM_{2.5}$, and $PM_{10}$) at each power plant. Damages per ton are multiplied by the tons of each of the four pollutants emitted by the plant in 2005. This produces an estimate of aggregate damages associated with criteria-pollutant-forming emissions from each plant. Damages are also expressed per kWh.

### Choice of Modeling Platform

Calculating the damages associated with air-pollution emissions involves three steps: (1) translating changes in emissions into changes in ambient air quality; (2) using concentration-response functions to calculate health impacts, environmental impacts, and others; and (3) valuing those impacts. This section describes the choices the committee made along each of these dimensions and discusses their strengths and limitations.

*Approach to Air-Quality Modeling*

There are two general approaches one can take to air-quality modeling: process-based modeling and reduced-form modeling. A process-based model captures the complexities of environmental processes by including exhaustively detailed representations of each mechanism in the atmosphere. Process-based models attempt to reflect the natural processes that govern the relationship between emissions and concentrations. The models are often applied to simulations with very fine spatial and temporal scales. The Community Multiscale Air Quality (CMAQ) model is widely considered the state of the science in process-based air-quality modeling (Byun and Schere 2006).

Despite these advantages, there are downsides to process models. Because of their exhaustive embodiment of a multitude of atmospheric processes, such models are time-intensive and expensive to operate. The implied cost of running process models limits the number of times researchers can run these models for a particular application. This constraint forces policy analyses using these models to make other compromises. For example, process models cannot be used to conduct large numbers of experiments. As a result, national applications of CMAQ and other process models feature a relatively small number of modeling runs in which many sources have their emissions modified at once. This approach may be appropriate for simulating a national or regional policy, but the simulation design is fundamentally unable to isolate the impact of emissions from individual sources over a large modeling domain. If that is the objective of the research, which is the objective in this study, then a simpler, reduced-form air-quality model.[18]

The reduced-form modeling approach depicts the environment with a simple representation that mimics the overall behavior of the entire system. Reduced-form models do not include all the complex relationships of the process-based models. Their advantages are that they are relatively fast, inexpensive to operate, and easy to interpret. The most critical drawback of reduced-form models is that they may omit or misrepresent a key element in the environmental process. The model used in this analysis, APEEP, uses a source-receptor matrix with county-level sources and receptors that are derived from a Gaussian air-quality model. The cells of the matrix, which are generated by the Gaussian model, represent estimates of the concentrations of a given pollutant (per unit of emission). The cells were systematically adjusted to implicitly represent the spatial effects of the dispersion and

---

[18]Both approaches are valid. The use of CMAQ in regulatory impact analysis considers a limited number of scenarios in which emissions from many sources are simultaneously reduced as a result of the contemplated regulation. In contrast, we wish to consider separately the impacts of emissions from each power plant.

transformation processes embodied in the CMAQ model. An alternative approach to develop a reduced-form model is to fit a "response surface" to CMAQ output, which has been used by EPA. The latter is a purely statistical approach.

APEEP has been carefully calibrated to CMAQ to reflect the relationships between emissions and concentrations that CMAQ estimates. However, APEEP has some drawbacks: It cannot effectively represent episodic events because of the use of annual and seasonal average meteorologic data. Although its use of county-level resolution is quite fine-grained for a national study, a preferred approach would be grid-cell-level resolution for large western counties.

Our choice of air-quality modeling approach in this study is motivated by the desire to model the impact of emissions from individual power plants. Power plants vary greatly in the amount of pollution they emit and, by virtue of their location, in the impact of the pollution on human health and on ecosystems. Exploring the heterogeneity of pollution impacts across space is important from a policy perspective because it provides regulators with a means to set priorities for emissions abatement by identifying the relative damage caused by emissions from different sources. To explore these effects, many model runs must be conducted. Reduced-form models are the optimal modeling choice in such a context.

## Choice of Concentration-Response Functions

In analyses of air-pollution damages and the benefits of reducing them (for example, the Benefits and Costs of the Clean Air Act, 1990-2010 [EPA 1999]), impacts on human health constitute the vast majority of monetized damages, with premature mortality constituting the single largest damage category. The concentration-response functions for human health end points (including premature mortality, chronic bronchitis, and hospital admissions) used in APEEP are listed in Table C-1 of Appendix C. They are the same concentration-response functions as those used in the EPA regulatory impact analyses; therefore, those functions have been vetted by the EPA Clean Air Science Advisory Committee. In particular, the impact of PM on premature mortality is calculated using the relationship between $PM_{2.5}$ and all-cause mortality in Pope et al. (2002).[19] The concentration-response functions used to calculate impacts on agriculture, forestry, and

---

[19]We have chosen not to calculate the quality-adjusted life years (QALYs) or disability-adjusted life years (DALYs) associated with power-plant emissions. The goal of this study is to monetize damages. A recent Institute of Medicine study (IOM 2006) recommended that QALYs and DALYs not be monetized.

other sectors are listed in Appendix C and further described in Muller and Mendelsohn (2006).

One limitation of the APEEP model as used in this analysis is its limited treatment of ecosystem damages. For example, the model does not measure the impacts of acid rain associated with $NO_x$ and $SO_2$ emissions either on tree canopy or on fish populations. It also fails to capture eutrophication of fresh-water ecosystems from nitrogen deposition.

*Valuation*

As in most analyses of damages associated with criteria-pollutant-forming emissions, health damages figure prominently in aggregate monetized damages—especially premature mortality associated with $PM_{2.5}$. The value of monetized damages is particularly sensitive to the VSL used to monetize cases of premature mortality. The value that we use for our central case analysis is $6 million 2000 USD. This value is supported by recent meta-analyses of the literature on the VSL as well as by values used in EPA regulatory impact analyses. In their 2003 meta-analysis, Viscusi and Aldy (2003) reported a mean value of $6.7 million (2000 USD), and Kochi et al. (2006) reported a value of $5.4 million based on an empirical Bayes estimator. These values are in line with values used in recent EPA regulatory impact analyses: The Clean Air Interstate Rule (CAIR) regulatory impact analysis (RIA) uses a value of $5.4 million (1999 USD), and the EPA National Center for Environmental Economics recommends using a $7.4 million VSL (2006 USD). (This amount is equivalent to $6.3 million in 2000 USD.) A $6 million VSL (2000 USD) is also used by other researchers (for example, Levy et al. 2009) who recently examined the health impacts of power-plant emissions.

We applied the same VSL to persons of all ages. Although there is some evidence that willingness to pay for changes in mortality risks varies with age, the EPA Environmental Economics Advisory Committee of the Science Advisory Board judged in 2007 that the literature on this issue was not sufficiently mature to determine exactly how the VSL varies with age. The practice of valuing lives lost by multiplying the number of life years lost by the value of a statistical life year (VSLY) was also rejected. The empirical evidence on the impact of age on the VSL does support the use of the VSLY approach, which assumes that the VSL is proportional to remaining life expectancy (EPASAB 2007).

In calculating the value of premature mortality, we treated the lives lost due to changes in $PM_{2.5}$ concentrations as occurring in the same year as the change in the concentrations. EPA (1999, Appendix D) assumed that the impact of a reduction in $PM_{2.5}$ concentrations was spread over 5 years, with 25% of the change in deaths occurring in same year as the change in concentrations, 25% occurring the next year, and one-sixth occurring in each of the following 3 years. At a 3% discount rate, the present discounted value of damages using EPA's lag structure would be 95% of the mortality damages that we calculated. At a discount rate of 7%, the damages would be 89% of the mortality damages that we calculated. However, selecting a particular lag structure is associated with great uncertainty. In its review of the NAAQS for PM, EPA indicated that it is difficult to assess the time between the occurrence of a cause and its purported effect based on the studies it reviewed of PM exposures, given that airborne PM concentrations are generally correlated over time in any given area. For all-cause mortality and cardiovascular mortality, EPA observed that the greatest effect size is generally reported for the 0-day lag and 1-day lag. The effect generally tapered off for longer lag periods (EPA 2005b).

### Treatment of Uncertainty

The version of APEEP used in our analysis does not provide error bounds that reflect either statistical uncertainty in the concentration-response functions used in the model or in the range of VSL estimates in the literature. The relationship between emissions and ambient air quality is likewise treated as certain, as is the case in regulatory impact analyses of air-quality regulations. Due to the importance of the VSL in determining the size of air-pollution damages, we used a value of $2 million (2000 USD) as a sensitivity analysis. The likely impact of using alternative concentration-response functions (for example, Dockery et al. 1993) is discussed below.

*Methodology* The APEEP model calculates the damages associated with emitting an additional ton of each of six pollutants ($SO_2$, $NO_x$, $PM_{2.5}$, $PM_{10}$, $NH_3$, and VOCs) as a function of the county in which the pollutant is emitted and the effective stack height of the emissions. The categories of damages covered by APEEP and reflected in our estimates include premature mortality associated with $PM_{2.5}$, cases of chronic bronchitis and respiratory and cardiovascular hospital admissions associated with $PM_{2.5}$ and $PM_{10}$, changes in crop and timber yields associated with ozone, damage to building materials from $SO_2$, impairments to visibility associated with $PM_{2.5}$ and recreation damages associated with ozone-related changes in forest canopy. As described in more detail in Appendix C, APEEP calculates the impact of a ton of emissions of each pollutant on ambient air quality, and the effect of the change in ambient air quality on population-weighted exposures to PM, ozone, $SO_2$, and $NO_x$. The impact of changes in exposure on health, crop yields, visibility, and other categories of damages is estimated using concentration-response functions from the literature. Damages are monetized using unit values from the literature. (Appendix C lists the concentration-response functions used in the analysis and the unit values used to monetize damages.)

We calculated damages associated with each power plant by multiplying the damages per ton of each pollutant by the number of tons of each pollutant emitted by the plant in 2005, implying that we calculated the damages associated with 2005 emission levels compared with zero emissions. In practice, installing additional pollution control devices (or switching to low-sulfur coal) could reduce emissions very close to zero at most plants. We could have calculated damages relative to some estimate of the lowest emissions levels achievable by using existing control technologies; however, a zero baseline is more transparent. This approach implies that the damages calculated at each plant are an upper bound to the benefits from additional pollution controls.[20]

*Results*   The monetized damages associated with emissions of $SO_2$, $NO_x$, $PM_{2.5}$, and $PM_{10}$ in 2005 are calculated for each of 406 coal-fired electricity-generating facilities by combining damages per ton from APEEP with emissions data from the 2005 National Emissions Inventory (NEI).[21] Estimates of the damages associated with a ton of each of four kinds of emissions ($SO_2$, $NO_x$, $PM_{2.5}$, and $PM_{10}$) that form criteria air pollutants are obtained from APEEP as a function of the county in which the pollutant is emitted and the effective stack height of the emissions. These are combined with data on emissions of these pollutants, by stack, from the 2005 NEI.[22] This allows us to calculate the monetized damages associated with each pollutant at the plant level. Data from the Energy Information Administration on net generation of electricity from coal were used to compute monetized damages per kWh.

---

[20]The installation of some pollution-control devices may lower the efficiency with which the plant operates, but this effect is likely to be small. It should be emphasized that lowering emissions is not equivalent to closing the plant. Net generation of electricity, and hence the benefits of the electricity generated by the plant, would remain essentially unchanged if damages were reduced.

[21]APEEP calculates damages associated with ammonia ($NH_3$) and volatile organic chemicals (VOCs). These pollutants were dropped from our analysis due to missing emissions data for a significant fraction of plants. Damages from ammonia were recorded for 310 out of our sample of 406 coal plants. When the damage per kWh estimates were recalculated to include the impacts of ammonia ($PM_{10}$-related visibility reduction and morbidity, as well as $PM_{2.5}$-related mortality), these components were found to be small, accounting for less than 1% of damage per kWh in all but 19 plants. The latter group contained significant outliers, for which ammonia-related impacts accounted for as much as 14% of these facilities' adjusted damages per kWh. Consequently, the ammonia-inclusive damages per kWh are generally very close to the original estimates in the report.

[22]Specifically, we obtained emissions data for each stack at each plant associated with coal-fired generation and used information on meteorological conditions and exit velocity to approximate the effective height of the stack.

Damages from the criteria-pollutant-forming emissions were calculated, as described above, for each of 406 plants that generated electricity from coal in 2005.[23] Table 2-7 and Figure 2-5 present the distribution of monetized damages across plants. (In Table 2-7 all plants are weighted equally, hence the mean figures are arithmetic means of damages across all plants.) As Table 2-7 makes clear, most damages come from $SO_2$ (85%), followed by $NO_x$ (7%), $PM_{2.5}$ (6%) and $PM_{10}$ (2%). This reflects the size of $SO_2$ and $NO_x$ emissions from coal-fired power plants and the damages associated with fine particles formed from $SO_2$ and $NO_x$.[24] Directly emitted $PM_{2.5}$ has very high damages per ton (see Table 2-8), but very little $PM_{2.5}$ is emitted directly by power plants; most is formed from chemical transformations in the atmosphere.

Table 2-8 shows how the damages per ton of pollutant vary across plants, again weighting all plants equally. Variation in damages per ton reflects differences in the size of the populations (human and other) exposed to pollution from each plant, as well as differences in effective stack heights across plants. The assumption implicit in our calculations—that the damage per ton of pollutant emitted is independent of the number of tons emitted at the plant—is consistent with the epidemiological literature and with the calculation of air-pollution damages by EPA and other agencies.[25]

Damages from the criteria-pollutant-forming emissions in 2005 averaged $156 million per plant, but the range of damages across plants was wide—the 5th and 95th percentiles of the distribution are $8.7 and $575 million dollars, respectively (2007 USD). As Figure 2-5 shows, the distribution is highly skewed. After ranking all the plants according to their damages, we found that the most damaging 10% of plants produced 43% of aggregate air-pollution damages from all plants, and the least damaging 50% of the plants produce less than 12% of aggregate damages.[26] Where are the plants with the highest damages located? The map in Figure 2-6 shows the size of damages created by each of the 406 plants, by plant location. Plants with large damages are concentrated to the east of the Mississippi, along the Ohio River Valley, in the Middle Atlantic and the South.

Some of the variation in damages across plants occurs because plants that generate more electricity tend to produce greater aggregate damages;

---

[23]Each of our plants is classified as SIC 4911. Together they accounted for 94.6% of electricity generated from coal and sold to the grid (EIA 2009d, Table 1.1).

[24]Approximately 99% of the damages associated with $SO_2$ come from secondary particle formation, that is, the transformation of $SO_2$ into $PM_{10}$ and $PM_{2.5}$.

[25]The concentration-response functions in the air pollution literature are approximately linear in ambient concentrations. The unit values assigned to health and other endpoints are likewise assumed to remain constant over the relevant ranges of the endpoints.

[26]Each set of plants—the most damaging 10% and the least damaging 50%—account for approximately one quarter of electricity generated by the 406 plants.

**TABLE 2-7** Distribution of Criteria-Air-Pollutant Damages Associated with Emissions from 406 Coal-Fired Power Plants in 2005 (2007 U.S. Dollars)

|  | Mean | Standard Deviation | 5th Percentile | 25th Percentile | 50th Percentile | 75th Percentile | 95th Percentile |
|---|---|---|---|---|---|---|---|
| $SO_2$ | 1.6E + 08 | 1.9E + 08 | 4.3E + 06 | 2.4E + 07 | 6.5E + 07 | 1.6E + 08 | 5.2E + 08 |
| $NO_x$ | 1.1E + 07 | 1.1E + 07 | 7.5E + 05 | 3.1E + 06 | 7.2E + 06 | 1.6E + 07 | 3.0E + 07 |
| $PM_{2.5}$ | 9.0E + 06 | 1.3E + 07 | 2.3E + 05 | 1.3E + 06 | 4.0E + 06 | 1.0E + 07 | 3.6E + 07 |
| $PM_{10}$ | 5.2E + 05 | 6.9E + 05 | 1.8E + 04 | 9.8E + 04 | 2.6E + 05 | 6.2E + 05 | 1.9E + 06 |
| Total | 1.6E + 08 | 2.0E + 08 | 8.7E + 06 | 3.4E + 07 | 8.1E + 07 | 1.8E + 08 | 5.8E + 08 |

NOTE: All plants are weighted equally, rather than by the fraction of electricity they produce.

ABBREVIATIONS: $SO_2$ = sulfur dioxide; $NO_x$ = oxides of nitrogen; PM = particulate matter.

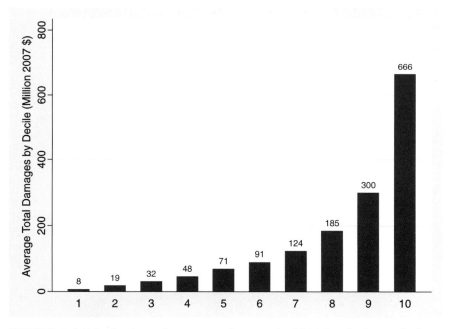

**FIGURE 2-5** Distribution of aggregate damages in 2005 by decile: coal plants (U.S. dollars, 2007). NOTE: In computing this graph, power plants were sorted from smallest to largest based on aggregate damages. The lowest decile represents the 40 plants with the smallest aggregate damages. The figure on the top of each bar is the average, across all plants, of damages associated with $SO_2$, $NO_x$, $PM_{2.5}$, and $PM_{10}$.

**TABLE 2-8** Distribution of Criteria-Air-Pollutant Damages per Ton of Emissions from Coal-Fired Power Plants (2007 U.S. Dollars)

|            | Mean  | Standard Deviation | 5th Percentile | 25th Percentile | 50th Percentile | 75th Percentile | 95th Percentile |
|------------|-------|--------------------|----------------|-----------------|-----------------|-----------------|-----------------|
| $SO_2$     | 5,800 | 2,600              | 1,800          | 3,700           | 5,800           | 6,900           | 11,000          |
| $NO_x$     | 1,600 | 780                | 680            | 980             | 1,300           | 1,800           | 2,800           |
| $PM_{2.5}$ | 9,500 | 8,300              | 2,600          | 4,700           | 7,100           | 10,000          | 26,000          |
| $PM_{10}$  | 460   | 380                | 140            | 240             | 340             | 490             | 1,300           |

NOTE: All plants are weighted equally, rather than by the fraction of electricity they produce.

ABBREVIATIONS: $SO_2$ = sulfur dioxide; $NO_x$ = oxides of nitrogen; PM = particulate matter.

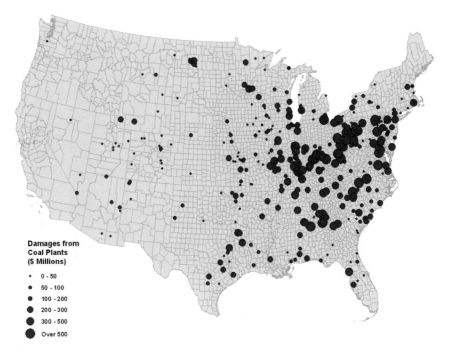

**FIGURE 2-6** Air-pollution damages from coal generation for 406 plants, 2005 (U.S. dollars, 2007). Damages related to climate-change effects are not included.

hence we also report damages per kWh of electricity produced.[27] Table 2-9 and Figures 2-7 and 2-8 show damages per kWh for all four pollutants. Mean damages per kWh (2007 USD) from four criteria-pollutant-forming emissions are 4.4 cents per kWh if all plants are weighted equally and 3.2 cents per kWh if plants are weighted by the electricity they generate. The lower figure reflects the fact that larger plants are often less damaging per kWh.[28] What is equally important as mean damages is the distribution of damages across plants. As Table 2-9 indicates the 95th percentile of the distribution—damages of 12 cents per kWh—is more than an order of magnitude larger than the 5th percentile. The distribution of damages per kWh (Figure 2-7) is very skewed: There are many coal-fired power plants with low damages per kWh as well as a small number of plants with high damages. Using generation-weighted figures, the damages per kWh from the least damaging 5% of plants were very small: 94% lower than the average coal-fired plant and almost as low as the average damage per kWh at natural gas power plants (0.16 cents). Figure 2-8 maps damages per kWh for each power plant. As in the case of aggregate damages, the plants with lowest damages per kWh are in the West. Plants with the largest damages per kWh are concentrated in the Northeast and the Midwest.

What explains variation in damages per kWh across plants? Damages per kWh associated with a criteria air pollutant (for example, $SO_2$) are the product of emissions per kWh and the damage per ton of pollutant emitted. For the 406 plants examined, variation in damages per kWh is primarily due to variation in pollution intensity (emissions per kWh) across plants, rather than variation in damages per ton of pollutant, which varies with plant location. In the case of $SO_2$, emissions per kWh reflect the sulfur content of the coal burned, adoption of control technologies (for example, scrubbers), as well as the vintage of the plant. Pounds of $SO_2$ emitted per MWh (see Tables 2-10 and 2-11) vary greatly across plants, and this variation explains approximately 83% of the variation in damages attributed to $SO_2$ emissions per kWh. As Table 2-11 indicates, pounds of $SO_2$ and $NO_x$ emitted per MWh vary significantly with plant vintage, reflecting the fact that newer plants are subject to more stringent pollution controls. Variation in damages per ton of $SO_2$ emitted (see Table 2-8) accounts for only

---

[27]It is, however, the case that less than half of the variation in damages is explained by variation in the amount of electricity generated. A regression of damages on net generation yields an $R^2 = 0.32$; the $R^2$ is 0.48 when the logarithms of the variables are used.

[28]The correlation coefficient between damages per kWh and net generation is = −0.26, significant at <0.01 level of significance.

**TABLE 2-9** Distribution of Criteria-Air-Pollutant Damages per Kilowatt-Hour Associated with Emissions from 406 Coal-Fired Power Plants in 2005 (2007 Cents)

|  | Mean | Standard Deviation | 5th Percentile | 25th Percentile | 50th Percentile | 75th Percentile | 95th Percentile |
|---|---|---|---|---|---|---|---|
| $SO_2$ | 3.8 | 4.1 | 0.24 | 1.0 | 2.5 | 5.2 | 11.9 |
| $NO_x$ | 0.34 | 0.38 | 0.073 | 0.16 | 0.23 | 0.36 | 0.91 |
| $PM_{2.5}$ | 0.30 | 0.44 | 0.019 | 0.053 | 0.13 | 0.38 | 1.1 |
| $PM_{10}$ | 0.017 | 0.023 | 0.001 | 0.004 | 0.008 | 0.023 | 0.060 |
| Total (equally weighted) | 4.4 | 4.4 | 0.53 | 1.4 | 2.9 | 6.0 | 13.2 |
| Total (weighted by net generation) | 3.2 | 4.3 | 0.19 | 0.71 | 1.8 | 4.0 | 12.0 |

NOTE: In the first five rows of the table, all plants are weighted equally; that is, the average damage per kWh is 4.4 cents, taking an arithmetic average of the damage per kWh across all 406 plants. In the last row of the table, the damage per kWh is weighted by the electricity generated by each plant to produce a weighted damage per kWh.

ABBREVIATIONS: $SO_2$ = sulfur dioxide; $NO_x$ = oxides of nitrogen = PM, particulate matter.

24% of the variation in damages per kWh.[29] A ton of pollution emitted by plants located closer to population centers does more damage than the same ton emitted in a sparsely populated area; however, while plant location is important, coal plants are not located in counties with the highest damages per ton of $SO_2$ in the United States.

To summarize, the aggregate damages associated with criteria-pollutant-forming emissions from coal-fired electricity generation in 2005 were approximately $62 billion (USD 2007), or 3.2 cents per kWh (weighting each plant by the fraction of electricity it produces); however, damages per plant

---

[29]A regression of $SO_2$-related damages per kWh on pounds of $SO_2$ emitted per kWh produces an $R^2$ of 0.83. Regressing $SO_2$-related damages per kWh on damages per ton of $SO_2$ emitted produces an $R^2$ of 0.24. Even so, this last result does not elucidate the substantial heterogeneity in marginal damages that arises purely because of location. To more clearly highlight the role of geography, we took the $SO_2$ emission intensity of a national tall-stack coal-fired integrated gasification combined-cycle (IGCC) plant (0.043 tons/kWh) (NETL 2007) and applied this value to the marginal damages in the year 2030 estimated by APEEP for 485 counties in which there are currently coal-fired electricity-generating facilities. (The use of the APEEP model to generate marginal damages for 2030 is discussed later in this chapter.) The coefficient of variation of the resulting estimates is 0.38.

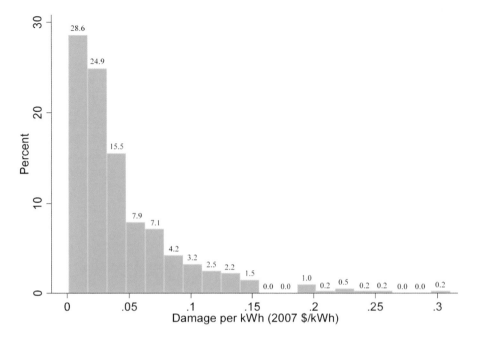

**FIGURE 2-7** Distribution of air-pollution damages per kWh for 406 coal plants, 2005 (U.S. dollars, 2007). NOTE: All plants are weighted equally rather than by the electricity they produce.

varied widely. The lowest-damage 50% of plants, which accounted for 25% of net generation, produced 12% of damages, and the highest-damage 10% of plants, which also accounted for 25% of net generation, produced 43% of the damages. Although damages are larger for plants that produce more electricity, less than half of the variation in damages across plants is explained by differences in net generation.

Damages per kWh also varied widely across plants: from approximately half a cent (5th percentile) to over 13 cents per kWh (95th percentile). (These are unweighted figures.) Most of the variation in damages per kWh can be explained by variation in emissions intensity across plants. In the case of $SO_2$, which accounts for 85% of the damages associated with $SO_2$, $NO_x$ and PM, over 80% of the variation in $SO_2$ damages per kWh is explained by variation in pounds of $SO_2$ emitted per kWh. Damages per ton of $SO_2$ emitted, which vary with plant location, are less important in explaining variation in $SO_2$-related damages per kWh. (They are, by themselves capable of explaining only 24% of the variation in damages per kWh.)

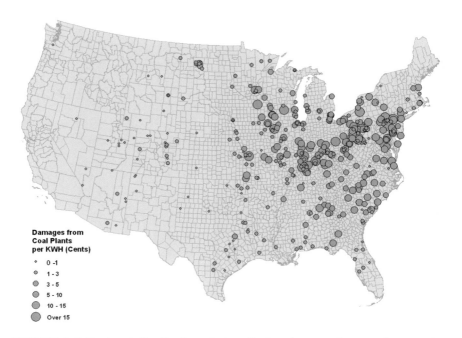

FIGURE 2-8 Regional distribution of air-pollution damages from coal generation per kWh in 2005 (U.S. dollars, 2007). Damages related to climate change are not included.

Of the 14 categories of criteria-air-pollutant damages included in AP-EEP, 6 relate to human health and the remainder to physical impacts (materials damage, ozone damage to crops and forests, the cost of foregone recreation due to $SO_2$, $NO_x$, ozone, and VOCs, and the cost of reduced visibility due to airborne particulate matter).

*Sensitivity Analysis and Comparison with the Literature*

The results of any analysis of the damages associated with air-pollution emissions depend critically on (1) the size of the emissions reduction analyzed; (2) the air-quality model used to translate emissions into ambient air quality; (3) the choice of concentration-response function for premature mortality and (4) the VSL used to monetize premature mortality. Premature mortality constitutes 94% of the damages reported above. When a VSL of $2 million is used (Mrozek and Taylor 2002), premature mortality constitutes 85% of total damages, and the weighted-average cost per kWh falls to 1.2 cents. If we had chosen to use Dockery et al. (1993) as the

**TABLE 2-10** NO$_x$ and SO$_2$ Emissions (2002) from Coal-Fired Electricity Generation by Age of Power Plant

*a.* 2002 NO$_x$ Emissions and Share of Generation of Coal-Fired Capacity by Vintage

| Power Plant Established | Avg. NO$_x$ Emission Rate (lb/MWh) | % Total NO$_x$ Emitted | % of Coal-Fired Electricity Generation | % of NO$_x$ Emitted per % of Electricity Generated[a] | % of Coal-Fired Electricity Capacity |
|---|---|---|---|---|---|
| Pre-1950 | 5.51 | 0.65 | 0.50 | 1.31 | 0.92 |
| 1950-1959 | 5.07 | 15.11 | 12.56 | 1.20 | 14.32 |
| 1960-1969 | 4.56 | 21.27 | 19.65 | 1.08 | 20.51 |
| 1970-1979 | 4.28 | 39.31 | 38.76 | 1.01 | 38.13 |
| 1980-1989 | 3.53 | 21.74 | 25.97 | 0.84 | 23.84 |
| Post-1990 | 3.15 | 1.92 | 2.56 | 0.75 | 2.27 |

*b.* 2002 SO$_2$ Emissions and Performance of Coal-Fired Capacity by Vintage

| Power Plant Established | Avg. SO$_2$ Emission Rate (lb/MWh) | % of Total SO$_2$ Emitted | % of Coal-Fired Electricity Generation | % of SO$_2$ Emitted per % of Electricity Generated[a] | Average Capacity Factor (%)[b] | Average Heat Rate (Btu/kWh generated) |
|---|---|---|---|---|---|---|
| Pre-1950 | 20.58 | 1.02 | 0.50 | 2.04 | 36.35 | 12,549 |
| 1950-1959 | 15.78 | 19.64 | 12.56 | 1.56 | 58.93 | 10,668 |
| 1960-1969 | 13.92 | 27.12 | 19.65 | 1.38 | 64.37 | 10,150 |
| 1970-1979 | 9.31 | 35.75 | 38.76 | 0.92 | 68.29 | 10,270 |
| 1980-1989 | 6.02 | 15.49 | 25.97 | 0.60 | 73.17 | 10,401 |
| Post-1990 | 3.88 | 0.98 | 2.56 | 0.38 | 75.80 | 9,982 |

*c.* 2002 NO$_x$ Emissions and Share of Generation of Coal-Fired Capacity by NSPS[c]

| NSPS Status According to EIA 767 | Avg. NO$_x$ Emission Rate (lb/MWh) | % Total NO$_x$ Emitted | % of Coal-Fired Electricity Generation | % of NO$_x$ Emitted per % of Electricity Generated[a] | % of Coal-Fired Electricity Capacity |
|---|---|---|---|---|---|
| Unknown | 2.93 | 0.16 | 0.23 | 0.69 | 0.27 |
| Not Affected by NSPS | 4.67 | 65.90 | 59.51 | 1.11 | 62.62 |
| Subject to Aug. 1971 Standards (D) | 3.57 | 26.73 | 31.58 | 0.85 | 29.56 |
| Subject to Sept. 1978 Standards (Da) | 3.50 | 7.21 | 8.68 | 0.83 | 7.56 |

*continued*

96

**TABLE 2-10** Continued

*d.* 2002 SO$_2$ Emissions and Performance of Coal-Fired Capacity by NSPS[c]

| | Average SO$_2$ Emission Rate (lb/MWh) | % of Total SO$_2$ Emitted | % of Coal-Fired Electricity Generation | % of SO$_2$ Emitted per % of Electricity Generated[a] | Average Capacity Factor[b] (%) | Average Heat Rate (Btu/kWh Generated) |
|---|---|---|---|---|---|---|
| **Unknown | 4.56 | 0.10 | 0.23 | 0.45 | 56.58 | 11,247 |
| Not Covered by NSPS | 12.93 | 76.25 | 59.51 | 1.28 | 63.85 | 10,250 |
| Subject to Aug. 1971 Standards (D) | 6.66 | 20.86 | 31.58 | 0.66 | 71.79 | 10,519 |
| Subject to Sept. 1978 Standards (Da) | 3.23 | 2.78 | 8.68 | 0.32 | 77.17 | 10,185 |

NOTES: All quantities, including percentages of emissions and generation capacity, are calculated with reference only to coal-fired generating units. Percentages (taking account of rounding) add to 100% because other types of generating capacity are not considered. These tables and the associated dataset were constructed by David Evans of Resources for the Future. Data used to make these tables come from three sources: emission data are from EPA's CEM system database; generation and capacity data are from EIA's 767 dataset; and information on vintage of generating units is from EIA's Form 860 dataset.

[a]If the generators of a particular vintage (or in a particular NSPS category) emitted a particular pollutant in proportion to its share of total electricity generation, the value would be 1.

[b]Capacity factor of units that operated that are strictly associated with boilers in CEM system database.

[c]The Subpart D standards apply to fossil-fuel-fired steam boilers for which construction began after August 17, 1971. The Subpart Da standards affect those boilers that began construction after September 18, 1978. For boilers not covered by NSPSs construction began before August 17, 1971. A new NSPS for NO$_x$ was promulgated in 1998, but no new coal-fired generating facilities have been permitted since this new standard was issued.

ABBREVIATIONS: SO$_2$ = sulfur dioxide; NO$_x$ = oxides of nitrogen; lb/MWh = pound per megawatt-hour; Btu/kWh = British thermal unit per kilowatt-hour; CEM = continuous emission monitoring; NSPS, new source performance standards.

SOURCE: EIA 2004a,b; EPA 2004a. As presented in NRC 2006a.

TABLE 2-11 Distribution of Pounds of Criteria-Pollutant-Forming Emissions per Megawatt-Hour by Coal-Fired Power Plants, 2005

| | Mean | Standard Deviation | 5th Percentile | 25th Percentile | 50th Percentile | 75th Percentile | 95th Percentile |
|---|---|---|---|---|---|---|---|
| $SO_2$ | 12 | 11 | 1.5 | 5.4 | 8.9 | 16 | 33 |
| $NO_x$ | 4.1 | 2.3 | 1.3 | 2.6 | 3.7 | 4.9 | 9.0 |
| $PM_{2.5}$ | 0.59 | 0.58 | 0.092 | 0.20 | 0.35 | 0.81 | 1.8 |
| $PM_{10}$ | 0.72 | 0.67 | 0.12 | 0.28 | 0.48 | 0.94 | 2.1 |

ABBREVIATIONS: $SO_2$ = sulfur dioxide = $NO_x$, oxides of nitrogen; PM = particulate matter.

concentration-response function for premature mortality instead of Pope et al. (2002), our damages would have been approximately three times as large as what is reported above.

How do our estimates of damages compare with the literature? Levy et al. (2009) estimated the criteria-air-pollutant damages associated with individual coal-fired power plants using a methodology similar to what is used here; however, their estimates of damages are much higher, ranging from $0.02 to $1.57 per kWh, with a median estimate of 14 cents per kWh (1999 USD).[30] Converting the results of Levy et al. to 2007 USD, their median estimate is almost 6 times as high as our median estimate of 2.9 cents per kWh (Table 2-9).[31] It is, however, possible to reconcile the two sets of estimates. Two notable differences are that Levy et al.'s estimates are based on emissions data for 1999 rather than 2005 and their estimates depend on a concentration-response function for premature mortality based on Schwartz et al. (2008) rather than Pope et al. (2002).[32] Emissions of $NO_x$ from coal-fired power plants were approximately 50% higher in 1999 than in 2005; emissions of $SO_2$ were approximately one-third higher. The concentration-response function in Schwartz et al. (2008) yields about three times more deaths associated with a microgram of $PM_{2.5}$ than those estimated using Pope et al. (2002)—the concentration-response function used in APEEP. These differences lead to much higher estimates of mortality associated with $PM_{2.5}$, and over 90% of the damages associated with air emissions in our study come from $PM_{2.5}$ mortality. Levy et al. (2009) also performed uncertainty propagation involving asymmetric triangular distributions, which would contribute modest upward bias to the median

---

[30]The mean value of a statistical life used in Levy et al. (2009) was identical to ours—$6 million USD. They reported monetary values in 1999 USD.

[31]The figures in Levy et al. (2009) were unweighted by electricity production.

[32]The concentration-response function for premature mortality in APEEP is the all-cause mortality function in Pope et al. (2002).

damage estimates. In short, if Levy et al. (2009) had used the same mortality concentration-response function and the same emissions as APEEP, and had not done uncertainty propagation, the results would have been nearly identical to ours.

Estimates of the benefits of reducing $SO_2$ and $NO_x$ emissions under the Clean Air Interstate Rule (CAIR) (EPA 2005b) are also higher than ours because of differences in air-quality modeling. The regulatory impacts analysis of CAIR examined the benefits of reducing emissions of $SO_2$ and $NO_x$ at power plants in 28 states in the eastern United States. The analysis predicted that in 2015 a reduction in $SO_2$ emissions of approximately 4 million tons and a reduction in $NO_x$ emissions of approximately 1.5 million tons would reduce premature mortality by 17,000 deaths. Our analysis, in contrast, estimates that in those states a reduction in $SO_2$ and $NO_x$ that is approximately twice as large would result in 10,000 fewer deaths in 2005. This result is due to differences in air-quality modeling: The use of CMAQ in the CAIR regulatory impact analysis (EPA 2005b) leads to an estimate of 1.15 µg/m$^3$ reduction in population-weighted $PM_{2.5}$ exposure, a much larger effect than is predicted by APEEP.[33] A study evaluating the performance of the version of CMAQ used in the CAIR study (version 4.3) found that it overestimated sulfate PM concentrations at sample locations in the eastern United States by 9% in one sample of largely rural sites (the Interagency Monitoring of Protected Visual Environments) and by 6% in another sample of largely urban sites (Speciated Trends Network) (EPA 2005c). This estimation bias was higher in the summer months, when sulfate concentrations are higher—14%. However, the estimation bias still does not fully account for the difference between the CMAQ and APEEP predictions.

Air-quality modeling results from APEEP agree well with other studies that use Gaussian plume models to model dispersion of pollutants from power plants (Nishioka et al. 2002; Levy et al. 2009), but concentrations of $PM_{2.5}$ from power plants are lower in APEEP than in CMAQ (EPA 2005c; Fann et al. 2009).[34] One of the advantages of APEEP is better spatial resolution in urban counties, but it may still lack the necessary level of spatial detail in urban areas, giving rise to some uncertainty about results.

In contrast to Levy et al. (2009), Muller et al. (2009) report estimates of criteria air-pollutant damages from coal-fired power plants that are slightly lower than those presented here (mean damages of approximately 2 cents

---

[33]The CAIR regulatory impact analysis uses the same concentration-response function as APEEP (all-cause mortality from Pope et al. (2002)) and a slightly lower VSL ($5.5 million 1999 USD). The U.S. population in 2015 is predicted to be about 9% higher than in 2005.

[34]Fann et al. (2009), using the Response Surface Model based on CMAQ, found damages per ton of $SO_2$ from power plants of $15,000 in Atlanta and $18,000 in Chicago. The 95th percentile of damages in our study is $11,000.

per kWh, on the basis of 2007 USD), using a value of a statistical life year (VSLY) approach.

## Downstream $CO_2$ Emissions of Electricity Generation from Coal

The emissions of $CO_2$ from coal-fired power are the largest single source of GHG emissions in the United States. The heat rate (energy of coal needed to generate 1 kWh of electricity) varies widely among coal-fired plants; thus the $CO_2$ emissions vary (with an average of about 1 ton of $CO_2$ per MWh of power generated [the 5th-95th percentile range is 0.95-1.5 tons]). The main factors affecting differences in the $CO_2$ generated are the technology used to generate the power and the age of the plant. The effect of $CO_2$ and other GHG emissions on global warming are discussed in Chapter 5.

## Externalities Associated with Heavy-Metal Emissions of Electricity Generation from Coal

Heavy metals are toxic both to the environment and to public health. The combustion of coal to produce electricity results in emissions of heavy metals, depending on the source of the coal, the conditions of combustion, and the cleanup technologies used. Among the heavy metals found in coal-combustion wastes are antimony (Sb), arsenic (As), beryllium (Be), cadmium (Cd), chromium (Cr), cobalt (Co), copper (Cu), lead (Pb), manganese (Mn), mercury (Hg), molybdenum (Mo), nickel (Ni), selenium (Se), silver (Ag), thallium (Tl), vanadium (V), and zinc (Zn). To determine the risks for human health and for the environment associated with particular heavy metals, one must consider both the toxicity of the metal and the potential for exposure to the metal.

Information on the toxicity of individual metals and their various metallic species can be found in the Integrated Risk Information System (IRIS) database at the EPA Web site (epa.gov/IRIS). Highly toxic metals for humans and the environment include Hg, As, Cd, Pb, and Se. Major routes of exposure are through air emissions and through leaching of contaminants from landfills or surface impoundments of wastes.

Trace metals, including heavy metals, have been classified according to how they partition among waste streams from coal combustion (EPA 1995):

Class 1. Elements that are approximately equally concentrated in the fly ash and bottom ash or that show little or no small particle enrichment (that do not contain many small particles). Examples include manganese, beryllium, cobalt, and chromium.

Class 2. Elements that are enriched in fly ash relative to bottom
ash, or show increasing enrichment with decreasing particle size.
Examples include arsenic, cadmium, lead, and antimony.
Class 3. Elements emitted in the gas phase (primarily mercury and in
some cases, selenium).

The main concern for human health is the risk associated with metals that end up in small, respirable particles or in the gas phase. Some of
the most toxic heavy metals (arsenic, lead, cadmium) are enriched in the
smaller particles. Particle control technologies will have limited impact on
the emissions of mercury, which is emitted as a gas. Metals are deposited
from the atmosphere and enter the food chain, where they can affect humans who eat contaminated organisms, mainly fish, as described in more
detail below.

Mercury from coal-fired power plants has been the subject of regulatory attention for some time. In March 2005, the EPA issued the Clean
Air Mercury Rule (CAMR) to establish emissions limits and a voluntary
cap-and-trade system for mercury from electricity-generating units (EGUs).
Concurrently, it "delisted" EGUs as a source of hazardous air pollutants
that would be regulated according to the strict requirements of Section 112
of the Clean Air Act as amended. In February 2008, the D.C. Circuit Court
vacated both CAMR and the delisting. In February 2009, the EPA withdrew
its appeal of this vacatur; instead, it is developing standards for EGU emissions of hazardous air pollutants, including mercury, under Section 112.
(A companion rule—CAIR, which was promulgated in May 2005—targets
EGU emissions of $SO_2$ and $NO_x$ that cross state boundaries. In December
2008, the D.C. Circuit Court decided to remand rather than vacate CAIR,
leaving the rule in place while EPA addresses concerns raised in a July 2008
D.C. Circuit Court decision.) This and additional information are at EPA's
Web site (EPA 2009b).

EPA recently developed a draft, site-based, probabilistic (Monte Carlo)
risk assessment of onsite coal combustion waste disposal practices at coal-
fired power plants across the United States (RTI 2007). The risk assessment
includes a screening step to determine if the toxicity of the contaminant
and the known routes of exposure constitute a risk of excess lifetime cancer greater than 1 in $10^5$ or a hazard quotient for noncancer end points
greater than 1. These risk assessments include those for trace metals, including heavy metals, and should be published soon. The metals exceeding
the human health risk criteria described above at the 90th percentile for
cancer included arsenic and for noncancer end points included boron, molybdenum, selenium, and cadmium. For ecological receptors, exceedances
were found for lead, boron, arsenic selenium, and cadmium at the 90th
percentile. A limitation of the risk assessments is that while they take into

account exposure from leachates of landfills and impoundments, they do not appear to take into account emissions into the air nor do they consider speciation of metals.

Unlike most of the other heavy metals, the dominant human exposure pathway for mercury is dietary. Mercury is emitted atmospherically from burning coal in elemental, particle-bound, and reactive forms that are deposited locally, regionally, and globally. After deposition, Hg enters water bodies where it is converted to methylmercury by microbes in the water column and sediment. Methylmercury bioaccumulates in aquatic species, reaching its highest concentration in high trophic-level fish such as shark, swordfish, and tuna; it also is found in many freshwater species. Consumption of fish is the major source of human exposure. Prenatal exposure to methylmercury is associated with subtle cognitive deficits and adult exposure may increase risk of fatal heart attack (Salonen et al. 1995; NRC 2000). Because of the complex pathway that mercury follows from its emission by power plants to its ingestion by people, affected by meteorological, chemical, physical, biological, and behavioral factors, it is difficult to estimate ecological and human health effects, which include impairment of cognitive function due to mercury exposure. Estimating monetary damages is even more difficult because of the lack of information on willingness to pay for reducing the risk of subtle cognitive effects from mercury exposure.

## Coal Combustion By-Products

By-products of burning coal to generate electricity include fly ash, bottom ash, flue gas desulfurization (FGD) materials, and fluidized bed combustion (FBC) residues (OSMRE 2009). In 2007, approximately 131 million tons of coal combustion by-products (CCBs) were produced in the United States (ACAA 2008a).[35] Of this total, about 56 million tons were reused. CCBs and their reuse by type of CCB in 2007 are summarized in Table 2-12. As shown in Figure 2-9 the tonnage of CCBs produced annually has increased more than fourfold since 1966. Reuse of CCBs also has increased but has not kept pace.

CCBs can contain traces of naturally occurring radioactive materials (regarding NORMs, see USGS 1997), as well as mercury, arsenic, lead, and other toxic materials. While CCBs have not been made subject to hazardous waste regulations under Subtitle C of the Resource Conservation and Recovery Act (RCRA), a 2006 NRC report noted that "CCRs [coal combustion residues] often contain a mixture of metals and other constituents

---

[35]The osmre.gov Web site distinguishes between CCBs and CCPs (coal-combustion products). The latter are "beneficially used" and are thus a subset of CCBs; however, this nomenclature has not been universally adopted. The more generic term "CCB" is used in this text.

**TABLE 2-12** 2007 Coal Combustion Product (CCP) Production and Use Survey Results

| CCP Categories | Fly Ash | Bottom Ash | Bolier Stag | FGD Gypsum | FGD Material Web Scrubbers |
|---|---|---|---|---|---|
| Total CCPs produced by category | 71,700,000 | 18,100,000 | 2,072,695 | 12,300,000 | 16,600,000 |
| Total CCPs used by category | 31,626,037 | 7,303,538 | 1,663,980 | 9,228,271 | 810,080 |
| Concrete/concrete products/grout | 13,704,744 | 665,756 | 0 | 118,406 | 0 |
| Blended cement/raw feed for clinker | 3,635,881 | 608,533 | 6,888 | 656,885 | 0 |
| Flowable fill | 112,244 | 0 | 0 | 0 | 0 |
| Structural fills/embankments | 7,724,741 | 2,570,163 | 158,767 | 0 | 97,610 |
| Road base/sub-base | 377,411 | 802,067 | 20 | 0 | 0 |
| Soil modification/stabilization | 856,673 | 314,362 | 169 | 0 | 0 |
| Mineral filler in asphalt | 17,223 | 21,771 | 63,729 | 0 | 0 |
| Snow and ice control | 0 | 736,979 | 44,367 | 0 | 0 |
| Blasting grit/roofing granules | 0 | 71,903 | 1,377,658 | 0 | 0 |
| Mining applications | 1,306,044 | 165,183 | 0 | 0 | 299,793 |
| Gypsum panel products | 0 | 0 | 0 | 8,254,849 | 0 |
| Waste stabilization/solidification | 2,680,328 | 7,056 | 0 | 0 | 10,378 |
| Agriculture | 49,662 | 2,546 | 0 | 115,304 | 9,236 |
| Aggregate | 135,331 | 806,645 | 450 | 70,947 | 0 |
| Miscellaneous/other | 1,025,724 | 530,574 | 11,932 | 11,880 | 393,063 |
| Totals by CCP type/application | 31,626,037 | 7,303,538 | 1,663,980 | 9,228,271 | 810,080 |
| Category use to production rate (%) | 44.11% | 40.35% | 80.28% | 75.03% | 4.88% |
| Supplemental: Cenospheres sold (pounds) | 12,659,597 | | | | |

SOURCE: ACAA 2008a. Reprinted with permission; copyright 2008, American Coal Ash Association.

| FGD Material Dry Scrubbers | FGD Other | FBC Ash (not including ARIPPA FBC Ash data) | CCP Production/ Utilization Totals | FBC Ash combined with ARIPPA FBC Ash production | CCP Production/ Utilization Totals (including ARIPPA FBC Ash data) |
|---|---|---|---|---|---|
| 1,812,511 | 2,449,731 | 1,273,061 | 126,307,998 | 6,092,756 | 131,127,693 |
| 150,365 | 113,298 | 323,741 | 51,219,310 | 5,143,436 | 56,039,005 |
| 21,266 | 0 | 5,518 | 14,515,690 | 5,518 | 14,515,690 |
| 0 | 81,801 | 0 | 4,989,988 | 0 | 4,989,988 |
| 12,417 | 2,735 | 0 | 127,406 | 0 | 127,406 |
| 555 | 0 | 46,282 | 10,598,118 | 46,282 | 10,598,118 |
| 0 | 0 | 0 | 1,179,509 | 0 | 1,179,509 |
| 154 | 429 | 199,441 | 1,371,228 | 199,441 | 1,371,228 |
| 0 | 0 | 0 | 102,723 | 0 | 102,723 |
| 0 | 0 | 0 | 781,346 | 0 | 781,346 |
| 0 | 0 | 0 | 1,449,561 | 0 | 1,449,561 |
| 111,195 | 0 | 0 | 1,882,215 | 4,819,695 | 6,701,910 |
| 0 | 0 | 0 | 8,254,849 | 0 | 8,254,849 |
| 1,416 | 28,333 | 72,500 | 2,800,031 | 72,500 | 2,800,031 |
| 3,352 | 0 | 0 | 180,100 | 0 | 180,100 |
| 0 | 0 | 0 | 1,013,373 | 0 | 1,013,373 |
| 0 | 0 | 0 | 1,973,173 | 0 | 1,973,173 |
| 150,365 | 113,298 | 323,741 | 51,219,310 | 5,143,436 | 56,039,005 |
| 8.30% | 4.62% | 25.43% | 40.55% | 84.42% | 42.74% |

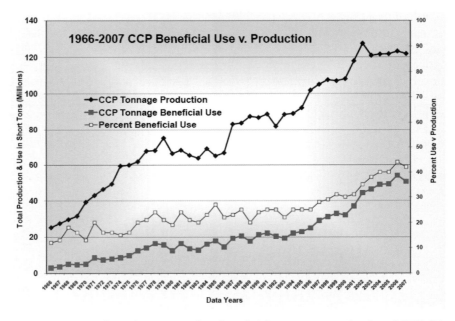

FIGURE 2-9 Coal combustion product beneficial use versus production. SOURCE: ACAA 2008b. Reprinted with permission; copyright 2008, American Coal Ash Association.

in sufficient quantities that they may pose public health and environmental concerns, if improperly managed. . . . Risks to human health and ecosystems may occur when CCR-derived contaminants enter drinking water supplies, surface water bodies, or biota" (NRC 2006b, p. 3). In addition, while inhalation of dust from CCBs is primarily a worker safety issue, precautions are needed to protect the public from CCB dust if it becomes airborne (EPA 2009c).

Under RCRA, states may regulate CCBs as a solid waste, a special waste, or, on a case-by-case basis, as a hazardous waste; they may do so by statute, generic or specific regulations, policy, or guidance (Archer 2000). States vary widely in the extent to which they regulate CCBs. Unlike disposal of other solid wastes such as household wastes, no uniform practices have been required by federal regulation (Buckley and Pflughoeft-Hassett 2007).

If only because of the quantities of fly ash produced annually (71 million tons in 2007, of which 31 million tons were directed to reuse), fly ash storage and disposal are of particular concern. With the spill in December

2008 of more than 1 billion gallons of fly ash sludge from a retention pond at the Tennessee Valley Authority's coal-fired plant in Kingston, Tennessee, fly ash became a matter of national attention. Fly ash usually is stored in ponds or landfills on or near to their power plant sites, which typically are located on waterways because of the plant's need to use and release water. Storing fly ash dry in landfills is considered safer, but even then, fly ash landfills often do not have the liners, leachate collection systems, and caps required under RCRA Subtitle D regulations for municipal solid waste landfills (EPA 2008a).

EPA has identified 431 slurried CCB impoundments through a national survey. Of the impoundments identified, 49 have been given a "high-hazard" rating by EPA (2009d).

## Externalities from Coal in 2030

### Technology in 2030

It is impossible to consider the future of coal-fired generation without considering the prospect of carbon capture and storage (CCS). CCS is a technology where the $CO_2$ emissions are first separated from the stack emissions, then collected and typically sent for offsite storage via small pipelines. Most current discussions about this nascent technology relates to where the carbon would be stored, with the most prominent discussions suggesting storage in underground geological sites such as aquifers or depleted gas fields, as well as in oil fields via enhanced oil recovery (EOR). $CO_2$ could also be liquefied, with potential storage in oceans. While beyond the scope of this chapter, there are significant risks due to accidental release of sequestered carbon.

The most common coal-fired technology being discussed for the future is IGCC (integrated gasification combined cycle), in which coal is first gasified before being used to generate electricity. IGCC plants are not only the most obvious next step in coal technology, but are more compatible with carbon capture systems. CCS is expected to be able to divert 80-90% of the $CO_2$ generated at these power plants. However, an IGCC/CCS system has an energy penalty in that more energy is needed to run the system, and thus more coal is required per kWh of electricity generated.

The current dominant technology, pulverized coal (PC), is compatible with CCS, but is generally more costly. As PC will remain the dominant technology in the "fleet" of power plants for several decades, and PC plants are being used decades past their original design lifetimes, the need for considering CCS for PC plants is inevitable. It is likely that PC technology will also have CCS and, depending on incentives and motivations, could be the dominant source of sequestered carbon. In general, IPCC estimates the cost per kWh of electricity from IGCC to be less than PC, including CCS systems (Table 2-13 [IPCC 2005, Table 8.3a]).

There are few IGCC projects in the world as of 2009, and relatively few CCS demonstration projects, especially for geological sequestration other than EOR. If IGCC and CCS technology is to be incorporated into the electricity sector, then ramp-up of siting, design, and construction of these plants needs to begin immediately for it to have any significant impact on air emissions within 20 years.

A relevant scenario is that in a future with 80-90% capture of $CO_2$ from coal-fired power, the upstream air emissions from mining and transportation will become much more significant, and possibly the largest

**TABLE 2-13** IPCC Range of Aggregate Costs for $CO_2$ Capture, Transport, and Geological Storage

| | Pulverized Coal Power Plant | Natural Gas Combined Cycle Power Plant | Integrated Coal Gasification Combined Cycle Power Plant |
|---|---|---|---|
| Cost of electricity without CCS [carbon capture and storage] (US$ MWh$^{-1}$) | 43-52 | 31-50 | 41-61 |
| *Power plant with capture* | | | |
| Increased fuel requirement (%) | 24-40 | 11-22 | 14-25 |
| $CO_2$ captured (kg MWh$^{-1}$) | 820-970 | 360-410 | 670-940 |
| $CO_2$ avoided (kg MWh$^{-1}$) | 620-700 | 300-320 | 590-730 |
| % $CO_2$ avoided | 81-88 | 83-88 | 81-91 |
| *Power plant with capture and geologic storage[a]* | | | |
| Cost of electricity (US$ MWh$^{-1}$) | 63-99 | 43-77 | 55-91 |
| Electricity cost increase (US$ MWh$^{-1}$) | 19-47 | 12-29 | 10-32 |
| % increase | 43-91 | 37-85 | 21-78 |
| Mitigation cost (US$/tCO$_2$ avoided) | 30-71 | 38-91 | 14-53 |
| Mitigation cost (US$/tC avoided) | 110-260 | 140-330 | 51-200 |
| *Power plant with capture and enhanced oil recovery[b]* | | | |
| Cost of electricity (US$ MWh$^{-1}$) | 49-81 | 37-70 | 40-75 |
| Electricity cost increase (US$ MWh$^{-1}$) | 5-29 | 6-22 | (−5)-19 |
| % increase | 12-57 | 19-63 | (−10)-46 |
| Mitigation cost (US$/tCO$_2$ avoided) | 9-44 | 19-68 | (−7)-31 |
| Mitigation cost (US$/tC avoided) | 31-160 | 71-250 | (−25)-120 |

[a]Transport costs range from 0-5 US$/tCO$_2$. Geological storage cost (including monitoring) range from 0.6-8.3 (US$/tCO$_2$).

[b]Transport costs range from 0-5 US$/tCO$_2$ stored. Costs for geological storage including EOR range from −10 to −16 US$/tCO$_2$ stored.

SOURCE: IPCC 2005, Table 8.3a, p 347. http://www.ipcc.ch/pdf/special-reports/srccs/srccs_chapter8.pdf. Reprinted with permission; copyright 2005, Intergovernmental Panel on Climate Change.

single source of emissions in the coal power life cycle. Further, if the EIA's long-term scenarios related to electricity mix hold true (that is, still 50% coal in 2030) [EIA 2009e], then significantly more coal will be mined, and these upstream externalities, while still relatively small on a per-kWh basis, will probably grow in magnitude in the local areas where coal is mined and where unit trains of coal deliveries pass through.

### Air-Pollution Damages from Coal-Fired Power Plants in 2030

The air-pollution damages associated with electricity generation from coal in 2030 depend on many factors. Aggregate damages depend on the growth in electricity demand and the extent to which coal is used to satisfy this demand, as opposed to other fuels. Damages per kWh are a function of the emissions intensity of electricity generation from coal (for example, pounds of $SO_2$ per MWh), which depends on future regulations governing power plant emissions. The damages per ton of $SO_2$ and $NO_x$ depend on the location of coal-fired power plants and on the size of the populations affected by them.

To give a sense of how damages in 2030 might compare with estimates for the year 2005, we use EIA forecasts of electricity production from coal and of $SO_2$ and $NO_x$ emissions, together with estimates of damages per ton of pollutant emitted in 2030 from APEEP. The assumptions underlying our analysis are outlined below. Because of the greater uncertainties associated with the 2030 analysis, we focus on estimates of aggregate damages from coal-fired power generation, rather than presenting a detailed distribution of damages, as in the section above.

*Methodology*   The 2030 thermal power-plant analysis relied on EIA's Annual Energy Outlook 2009 projections (EIA 2009f, Table 72-100) for the growth of net generation and emissions of $SO_2$ and $NO_x$. On average, net generation from coal-fired power plants is estimated to be 20% higher in 2030 than in 2005. Estimates are available by type of generator, fuel type, and North American Electric Reliability Corporation (NERC) region generation. EIA does not project changes in $PM_{2.5}$ and $PM_{10}$ emissions. These were imputed as the average of the projected changes in these two species. These regional trends were used to construct multipliers for 2005 net generation by plant and emissions by stack. We applied each regional multiplier to all the plants with a given fuel in that region of the country. We assumed that coal plants in 2030 will be sited in the same locations as current plants.

Our 2030 results therefore embody all the regulatory and technological assumptions made by EIA. We deliberately took this analytical tack because our charge precluded us from considering policies to remedy externalities.

Thus, we did not attempt to substitute our own judgments about future regulatory developments in place of EIA's projections, which are widely used and generally regarded as authoritative.

We used EIA estimates of $SO_2$ and $NO_x$ from electricity generation in 2030, together with estimates of electricity generation by fuel type and emission intensities by fuel type in 2005 to estimate the percentage reduction in tons of $SO_2$ and $NO_x$ per MWh at coal plants. On average, pounds of $SO_2$ per MWh are assumed to decrease from 10.1 lb (weighted by electricity generation) in 2005 to 3.65 lb in 2030. The corresponding figures for $NO_x$ are 3.42 lb/MWh in 2005 (weighted by electricity generation) and 1.90 lb in 2030.[36] Estimates of 2030 emissions intensities together with forecasts of net generation produce estimates of emissions of $SO_2$, $NO_x$, $PM_{2.5}$, and $PM_{10}$ at the location of each plant in 2030.

APEEP was used to generate estimates of damages per ton of pollutant by county and effective stack height, in 2030. These estimates assume that the meteorological conditions and other assumptions used in modeling the impact of a change in emissions on ambient air quality are the same in 2030 as in 2005, and that emissions are emitted at the same effective stack heights at each plant as in 2005. The same concentration-response functions used in the 2005 analysis are used to translate changes in ambient concentrations into cases of premature mortality and morbidity in 2030; however, the U.S. population will have changed, according to forecasts from the U.S. Census Bureau. An increase in population size was reflected in the 2030 analysis, but the age structure of the population was not changed. The VSL is assumed to increase with income growth. Using an elasticity of the VSL with respect to income of 0.50 (Viscusi and Aldy 2003) and assumptions in EPA's national Energy Modeling System (NEMS) about growth in per capita income, the VSL is 27% higher in 2030 (in 2000 USD) than in 2005, as are the unit values applied to other health end points. The combined effect of increases in population and increases in the VSL and other health values is to increase damages per ton of pollution, on average, by over 50% compared with 2005 values. The percentage change, however, varies considerably by pollutant and county. In the counties in which coal plants are currently located—where we assume they will be located in 2030—the mean increase in damage per ton of pollutant emitted is 36% for $SO_2$ and 32% for $NO_x$.

*Results*  Damages from $NO_x$, $SO_2$, $PM_{10}$, and $PM_{2.5}$ were calculated, as described above, for each of 406 plants that generated electricity from coal

---

[36]The corresponding figures for $PM_{2.5}$ are 0.215 lb/MWh (2030) versus 0.491 lb/MWh (in 2005). For $PM_{10}$ the emissions intensities are 0.263 lb/MWh (2030) versus 0.594 lb/MWh (in 2005).

in 2005. In spite of the fact that net generation is 20% higher in 2030 than in 2005, monetized air-pollution damages (in 2007 USD) are approximately $38 billion—about 40% lower than in 2005. Damages per kWh (weighted by electricity generation) are 1.7 cents per kWh, compared with 3.2 cents per kWh in 2005. The fall in damages per kWh is explained by the assumption that pounds of $SO_2$ per MWh will fall by 64% and that $NO_x$ and PM emissions per MWh will fall by approximately 50%. This counteracts the increase in damages per ton.

For future technologies at coal-fired plants, such as IGCC with CCS, criteria-pollutant-forming emissions per kWh are expected to be significantly lower than emissions per kWh from a typical plant in 2030 (NETL 2007). Plants using future technologies would also be expected to have damages at the lower end of current distributions. On the other hand, damages that would be attributable to providing the expected infrastructure (for example, pipelines and geological sites) for long-term geological sequestration of $CO_2$ are much more uncertain.

## ELECTRICITY PRODUCTION FROM NATURAL GAS

### History and Current Status of Natural Gas Production

Natural gas, a non-renewable energy source that consists primarily of methane, is consumed in the United States for heat, fuel, and electricity. During the mid 20th century, natural gas was predominantly used for residential and commercial space heating, as well as for industrial process heating. Since then, natural gas has taken an increasing share in production of electricity. In 2008, approximately 30% of produced natural gas was used to produce electricity.

U.S. natural gas production matched domestic consumption until the early 1970s. Natural gas productivity (volume of natural gas extracted per well) peaked in 1971 with 119,251 wells producing on average 435,000 cubic feet per day (Figure 2-10). Total annual domestic production reached 22.6 trillion cubic feet in 1973, after which it began to decline (Figure 2-11). By 2007, the United States had 452,768 producing gas wells, nearly four times as many as in 1971, indicating that the mean productivity per well had declined substantially. However, preliminary data from EIA suggest that gross withdrawals in 2008 of natural gas were the highest recorded, exceeding 26 trillion cubic feet; marketed production was 21.4 trillion cubic feet. Currently, more than 75% of domestic NG production comes from Texas, Wyoming, Oklahoma, New Mexico, Louisiana, and the federal offshore Gulf of Mexico.

The United States has increased its reliance on natural gas imports to keep pace with consumption, which was 23.0 trillion cubic feet in 2007

**Natural Gas Well Average Productivity**

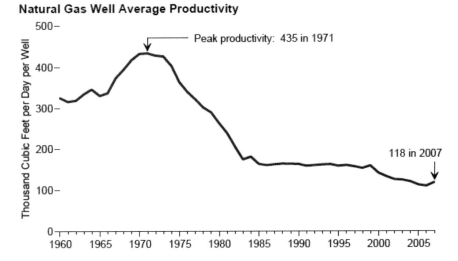

FIGURE 2-10 U.S. natural gas well average productivity. SOURCE: EIA 2008a, p. 188, Figure 6.4.

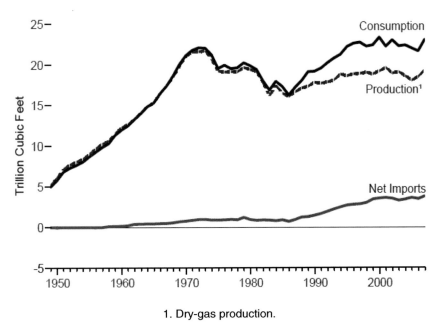

1. Dry-gas production.

FIGURE 2-11 Natural gas production, consumption, and imports in the United States. SOURCE: EIA 2008a, p. 182, Figure 6.1.

and 23.2 trillion cubic feet in 2008. Imports have increased since 1970. In the past few years (2003-2008), gross imports have averaged around 4 trillion cubic feet annually. Exports have increased from about 0.7 trillion cubic feet in 2003 to just over 1 trillion cubic feet in 2008. More than 90% of imported NG is transported by pipeline from Canada and Mexico. The United States also imports liquid natural gas (LNG) by ocean tankers from Trinidad, Egypt, Norway, Nigeria, and Qatar.

Natural gas is gathered and transmitted from producing fields and storage sites by pipeline. The United States has more than 300,000 miles of inter- and intrastate NG transmission pipelines. Domestic and imported NG is stored underground in natural geologic spaces. The United States had 400 storage sites (depleted fields, aquifers, and salt caverns) with greater than 8,400,000 million cubic feet of storage in 2007.

## Upstream Externalities of Electricity Production from Natural Gas

*Natural Gas Exploration and Drilling*

*Exploration and Development*   Exploratory activities to locate natural gas reservoirs are similar to those for oil. Exploratory drilling for natural gas uses the same rotary equipment and methods for development and production drilling, and it produces wastes mostly in the form of pollutants in water, primarily from the use of drilling fluids. Drilling also produces drill cuttings and mud. Exploration and development of natural gas occurs onshore and offshore, with potentially different types and levels of pollution. Initial exploration often uses seismic operations—the use of artificial shock waves directed into the earth to assess geologic strata based on reflection of the energy—both onshore and offshore. On land, transportation of the equipment can damage terrestrial ecosystems, especially in roadless areas (NRC 2003a). Offshore seismic exploration can adversely affect fish and marine mammals, especially if explosives are used (NRC 2003a).

For onshore drilling, significant proven reserves in the United States are along the Gulf Coast and in the Rocky Mountain region. Although rotary drilling is generally for exploration and development, cable-tool drilling can be utilized for shallow, low-pressure gas reservoirs. The amount of land required for a typical gas field of approximately 120 wells ranges from 420 to 640 acres depending on the size of the natural gas reservoir (on average 3.5 to 5.33 acres per gas well). This is a smaller area than is required for oil wells, which require approximately 40 acres per well. The primary waste products from gas well exploration and development are oils, heavy metals, and dissolved solids contained in the drilling mud or produced water. Specifically, the waste products are oil and grease, suspended solids, phenol, arsenic, chromium, cadmium, lead, and barium. These drilling wastes do not change significantly from region to region.

Drilling operations potentially create significant amounts of air pollution. Large diesel engines typically power the drilling equipment and emit significant quantities of PM, sulfur oxides, and oxides of nitrogen. These emissions can be substantial during drilling of deep wells requiring large power outputs or in large fields where multiple drilling operations occur simultaneously. Other sources of air pollution include organic compounds that may volatilize from reserve and other holding pits used as waste repositories during drilling operations, although the volume of these compounds is insignificant compared with diesel engine emissions. Oil and gas wells abandoned at the end of their productive life may cause environmental damage to the surrounding land surfaces and underground freshwater aquifers.

A considerable amount of natural gas exploration and development is located offshore on platforms, primarily in the Gulf of Mexico. For offshore drilling operations, drilling rigs may either be stationary or mobile. For drilling in waters up to 300 feet deep and marsh areas, mobile drilling rigs are mounted on barges and rest on the bottom. In water deeper than 300 feet, drilling rigs are mounted on floating or semi-submersible vessels with special equipped hulls that support the drill rig above the water level. To transport drill rigs to marsh areas, canals are dredged to the drill sites to float the rigs into place.

The wastewater from offshore platforms includes production wastes, deck drainage, and sanitary and domestic wastes. It can contain oils, toxic metals, and organic chemicals. Significant pollutants in produced waters include oil and grease, arsenic, cadmium, copper, cyanide, lead, mercury, nickel, silver, zinc, and organic carbon. Spilled oil and grease can adhere to fish and destroy algae and plankton, thereby altering the aquatic food chain. Additionally, damage is likely to occur to the plumage and coats of water animals and fowl. Lead, zinc, and nickel are toxic to fish even in low concentrations. However, offshore drilling rigs attract fish and can reduce fuel costs for recreational fishermen—an economic benefit.

*Extraction*  Natural gas is extracted by using either the existing pressure of the gas reservoir or by using pumps. Gas wells produce not only dry gas but also can produce varying quantities of light hydrocarbon liquid condensates and salt water. The resulting produced water (also known as "formation water" or "brine water") includes all waters and particulate matter associated with the gas producing formation. Produced water is the primary waste from offshore platforms. It can contain oils, toxic metals, salts, and organic compounds, which can cause environmental damage. For both onshore and offshore extraction, the type of technology used to treat produced water depends on state or local regulations as well as cost-effectiveness. Additionally, air emissions can include hydrogen sulfide that can be as high as 6% by volume in sour gas (that is, natural gas that contains hydrogen sulfide).

Natural gas is also produced using enhanced gas recovery extraction (EGR) methods. The primary technologies used for EGR are fracturing and directional drilling. Fracturing involves the use of either chemical explosives or water under pressure. Adverse impacts from the use of advanced hydraulic fracturing include air emissions and noise from the pressurized injection process. Preparing the well casing can cause leaks to groundwater or the surface. Water forced into gas-bearing shale can cause contamination or disruption of nearby wells. The use of chemical explosive fracturing has environmental impacts that are similar to those of advanced hydraulic fracturing. When the wells are constructed, noise, air emissions, soil erosion, and aesthetic deterioration may occur. There is also the danger of gas leaks or explosions from pipelines or storage tanks.

Directional or slant drilling (drilling that is not vertical) for recovery of natural gas can result in air emissions and soil erosion during the preparation of the drilling site. Drilling and production activities result in noise and risk of explosions. EGR processes have a considerably greater potential for causing air-quality degradation than do conventional recovery technologies. In both conventional and EGR processes, air-quality impacts result from emissions associated with production and injection pumps and fugitive emissions from wellheads and handling and storage facilities. Additionally, EGR technologies produce emissions from the combustion engines of compressors and from steam boilers in steam flood operations.

*Other Impacts*   The "footprint" for locations for natural gas exploration, development, and extraction is smaller than that for similar oil wells. While the impacts may not be as great for natural gas field operations, there are a number of additional impacts for both land and offshore activities that should be mentioned, in addition to those described above, that can have significant, although difficult to quantify, impacts.

For land-based operations, seismic measurements are a problem due to noise, aesthetics, and land use impacts, although most of these are temporary (for example, NRC 2003a). For the longer term, there are potential impacts related to habitat destruction. Wastewaters from all aspects of operation must be treated or they can cause significant degradation to the surface waters.

Offshore operations have different impacts in some cases. First, there is the overall impact of land degradation along the Gulf Coast. For both land-based, but nearshore operations, and for offshore operations, there is significant deterioration of onshore land, leading to salt water encroachment, land subsidence, and loss of land to the sea. The offshore operations can also have an impact on the surrounding ecosystem. Despite previous comments on benefits to recreational fishermen, natural gas platforms can have deleterious effects on larger ecosystems and can impact commercial fishing operations.

In a life-cycle analysis performed by Dones et al. (2005), it is estimated that approximately 25% of $CO_2$ emissions come from the processes discussed above, as treated as total production emissions (exploration, field production, purification). Other values include 10% of the methane, 50% of the nonmethane volatile organic hydrocarbons, 40% of the particulate matter, 20% of the nitrogen oxides, and 80% of sulfur dioxide emissions for the total fuel cycle.

## Occupational Injuries Associated with Oil and Gas Extraction and Transport

*Fatal and Nonfatal Injuries in Oil and Natural Gas Extraction*[37]   As in the case of mining, we assume that fatal and nonfatal occupational injuries do not constitute externalities, but we briefly discuss them because of their societal importance. In 2007, oil and gas industry fatalities accounted for almost two-thirds of fatal work injuries in mining. Unlike in coal mining, the number of fatalities in oil and gas extraction has been increasing, reaching in 2006 levels seen only decades ago (Figure 2-12). The incidence of fatalities, approximately 3 per 10,000 workers, is also higher than in coal mining (2 per 10,000 workers). The number of reported injuries has also increased (Figure 2-13).

*Fatal and Nonfatal Injuries in Transportation of Natural Gas*   In 2003, U.S. pipelines moved 590 billion total ton-miles of crude oil and petroleum products, and 278 billion ton-miles of natural gas (Dennis 2005). This includes gathering pipelines, which carry products from production fields; transmission pipelines, which transport products to terminals and refineries; and distribution pipelines, which carry products to final market and consumption points. Electric power plants receive 98% of their natural gas from direct mainline pipeline deliveries; 2% is provided by local distribution companies. In 2007, natural gas transport incurred 2 fatalities and 7 injuries. Natural gas distribution caused 8 fatalities and 35 injuries. Although the number of fatalities from natural gas pipeline activity has fluctuated, averaging 12 annually from 2000 to 2007, related injuries have

---

[37]It is difficult to separate injuries associated with oil extraction from injuries associated with natural gas extraction.

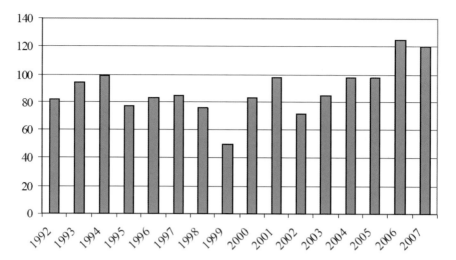

**FIGURE 2-12** U.S. fatalities in oil and gas extraction from 1992 to 2007. SOURCE: BLS 2009a.

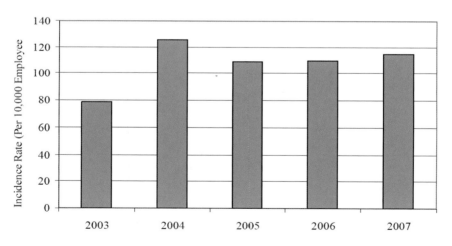

**FIGURE 2-13** Injuries and illnesses in U.S. oil and natural gas extraction operations. SOURCE: BLS 2009b.

steadily decreased over time (BTS 2009, Table 2-46). The majority of fatal and nonfatal injuries during natural gas transport are occupational and therefore are not treated as externalities.

### Upstream GHG Emissions and Other Pollutants

The upstream life cycle of power generation from natural gas includes many relevant activities such as construction of the infrastructure and power plants, but the most significant from a perspective related to GHG emissions and criteria-pollutant-forming emissions are the extraction and transportation of gas. These activities are generally fuel- and energy-intensive, requiring combustion of fossil fuels for drilling and removing the gas from underground and delivering to the power plant. Beyond emissions from engines, there are also significant GHG emissions of methane, which is from fugitive emissions of natural gas.

Of increasing relevance is the use of liquefied natural gas (LNG) to generate power. Over the past decade, a global market has begun for the extraction of gas for export via liquefying it, shipping it by tanker (similar to petroleum), and regasification. Each of these stages increases the energy use and air emissions (related to criteria pollutants and GHG) associated with the life cycle of the power generated.

Transportation of natural gas in the United States occurs via pipelines. While pipelines are a very cost- and energy-efficient transportation mode, they use significant amounts of fuels and electricity to move the gas from well to power plant. In addition, pipelines leak natural gas as methane into the air. As noted above, the transportation of LNG involves ocean tankers.

The prior studies mentioned above for coal also assessed the relative contribution of the upstream life cycle of gas-fired power generation for domestically sourced NG (Jaramillo et al. 2007, Meier et al. 2005, Spath and Mann 2000, ORNL/RFF 1992-1998). As was the case for coal, these studies found that upstream activities lead to relatively small life-cycle effects because of the dominance of criteria-pollutant-forming emissions and GHG emissions from gas-fired power plants (although the percentage share of upstream emissions in the life cycle are higher). For example, Jaramillo et al. (2007) reports that the mid-point GHG emission factors for domestic natural gas combustion (at the power plant) and the entire natural gas life cycle are 1,100 lb $CO_2$-eq/MWh and 1,250 lb $CO_2$-eq/MWh, respectively. Thus in this study we have focused on quantifying the air emissions associated with the burning of gas at power plants. This assumption would need to be revisited in a future scenario that had order-of-magnitude increases in the amount of LNG consumed for power generation (and its higher per unit emissions), but it is not considered in this study.

## Downstream Externalities of Electricity Production from Natural Gas

### Analysis of Current Air-Pollution Damages from Gas-Fired Power Plants

The air-pollution emissions from gas-fueled power plants constitute a significant portion of the downstream damages associated with electricity generation. In this section, we quantify the impacts of criteria-pollutant-forming emissions from gas-fired power plants on human health, visibility, agriculture, and other sectors, using the methods outlined in the section on coal. The effects of emissions on ambient air quality are calculated for each of 498 facilities that used gas to generate electricity in 2005. These facilities, which include electric utilities, independent power producers and combined heat and power facilities, each generated at least 80% of their electricity from gas and had installed capacity of at least 5 MW. Together they accounted for 71% of electricity generation from natural gas in 2005.[38]

Damages related to emissions of $NO_x$, $SO_2$, $PM_{10}$, and $PM_{2.5}$ were calculated for each of the 498 plants described above. Table 2-14 presents the distribution of monetized damages across the 498 natural-gas-fired power plants. (All plants are weighted equally in the table; hence the mean figures are arithmetic means of damages across all plants.) Most damages are related to directly emitted $PM_{2.5}$ (56%), followed by $NO_x$ (37%), $SO_2$ (4%), and $PM_{10}$ (3%), unlike coal plants where most damages (85%) are related to $SO_2$ emissions. Damages, however, are much lower than for coal plants. Average annual damages per plant are $1.49 million, which reflects both lower damages per kWh at natural-gas-fired power plants, but also smaller plants: Net generation at the median coal plant is more that 6 times as large as at the median gas facility.[39]

Some of the variation in damages across plants reflects differences in net generation; hence, we also report damages per kWh of electricity produced.[40] Table 2-15 presents the distribution of air-pollution damages per kWh. (All plants are weighted equally in the first five rows of the table; in the last row, plants are weighted by the fraction of electricity they produce.) Mean damages per kWh from the criteria-pollutant-forming emissions are 0.43 cents per kWh if all plants are weighted equally and 0.16 cents per kWh if plants are weighted by the fraction of electricity they generate. Dam-

---

[38]Emissions data in the National Emissions Inventory are reported at the stack level. When generating units powered by different fuels use the same stack, an attempt is made to apportion emissions by fuel type. To reduce errors in emissions data we analyze gas plants that use no coal and generate 80% of more of their electricity from natural gas.

[39]Median annual net generation is 3.01 billion kWh for coal plants and 0.469 billion kWh for gas plants.

[40]It is, however, the case that less than 40% of the variation in damages is explained by variation in the amount of electricity generated. A regression of damages on net generation yields an $R^2 = 0.09$; the $R^2$ is 0.37 when the logarithms of the variables are used.

**TABLE 2-14** Distribution of Criteria-Pollutant Damages Associated with Emissions from 498 Gas-Fired Power Plants in 2005 (2007 U.S. Dollars)

|  | Mean | Standard Deviation | 5th Percentile | 25th Percentile | 50th Percentile | 75th Percentile | 95th Percentile |
|---|---|---|---|---|---|---|---|
| $SO_2$ | 6.40E+04 | 2.58E+05 | 1.80E+02 | 1.96E+03 | 1.02E+04 | 2.92E+04 | 2.23E+05 |
| $NO_x$ | 5.49E+05 | 1.25E+06 | 4.86E+03 | 4.32E+04 | 1.43E+05 | 4.74E+05 | 2.37E+06 |
| $PM_{2.5}$ | 8.31E+05 | 3.23E+06 | 4.70E+02 | 1.50E+04 | 1.04E+05 | 4.12E+05 | 3.17E+06 |
| $PM_{10}$ | 4.47E+04 | 1.75E+05 | 4.07E+01 | 9.72E+02 | 5.44E+03 | 2.22E+04 | 1.62E+05 |
| Total | 1.49E+06 | 4.10E+06 | 1.02E+04 | 1.02E+05 | 3.57E+05 | 1.28E+06 | 5.50E+06 |

NOTE: All plants are weighted equally.

ABBREVIATIONS: $SO_2$ = sulfur dioxide; $NO_x$ = oxides of nitrogen; PM = particulate matter.

**TABLE 2-15** Distribution of Criteria-Pollutant Damages per Kilowatt-Hour Associated with Emissions from 498 Gas-Fired Power Plants in 2005 (Cents based on 2007 U.S. Dollars)

|  | Mean | Standard Deviation | 5th Percentile | 25th Percentile | 50th Percentile | 75th Percentile | 95th Percentile |
|---|---|---|---|---|---|---|---|
| $SO_2$ | 0.018 | 0.067 | 0.00013 | 0.00089 | 0.0022 | 0.006 | 0.075 |
| $NO_x$ | 0.23 | 0.74 | 0.0014 | 0.013 | 0.038 | 0.16 | 1.0 |
| $PM_{2.5}$ | 0.17 | 0.56 | 0.00029 | 0.007428 | 0.026 | 0.08 | 0.75 |
| $PM_{10}$ | 0.009 | 0.029 | 0.00003 | 0.00043 | 0.0014 | 0.0042 | 0.036 |
| Total (unweighted) | 0.43 | 1.2 | 0.0044 | 0.041 | 0.11 | 0.31 | 1.7 |
| Total (weighted by net generation) | 0.16 | 0.42 | 0.001 | 0.01 | 0.036 | 0.13 | 0.55 |

NOTE: In the first five rows of the table, all plants are weighted equally; that is, the average damage per kWh is 0.43 cents, taking an arithmetic average of the damage per kWh across all 498 plants. In the last row of the table, the damage per kWh is weighted by the fraction of electricity generated by each plant to produce a weighted damage per kWh.

ABBREVIATIONS: $SO_2$ = sulfur dioxide; $NO_x$ = oxides of nitrogen; PM = particulate matter.

ages per kWh are, on average, an order of magnitude lower—0.16 cents per kWh for natural gas compared with 3.2 cents per kWh for coal.[41] The lower figure reflects the fact that larger plants are often cleaner.[42] It should,

---

[41]Both figures weight damages per kWh at each plant by electricity generated by the plant.

[42]The correlation coefficient between damages per kWh and net generation is −0.18. It is −0.49 between the logarithms of the variables.

however, be emphasized that the distribution of damages per kWh has a high variance and is very skewed: Although, on average, damages from natural-gas-fired plants are an order of magnitude lower than damages from coal-fired power plants, there are some gas facilities with damages per kWh as large as coal plants.

As Figure 2-14 shows, the distribution of damages across plants is highly skewed. After sorting the plants according to damages, we found that the 10% of plants with highest damages produce 65% of the air-pollution damages from all 498 plants, and the lowest emitting 50% of plants within the lowest damages account for only 4% of aggregate damages. Each group of plants accounts for approximately one-quarter of sample electricity generation. The map in Figure 2-15 shows that the natural gas plants that produce the largest damages are located in the Northeast (along the Eastern seaboard), Texas, California, and Florida.

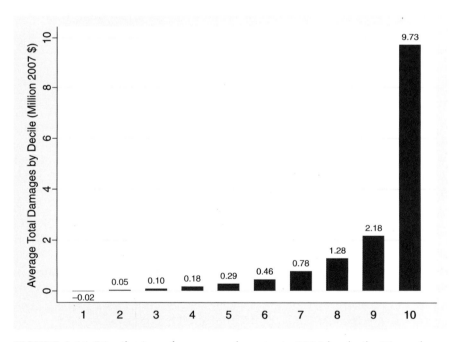

**FIGURE 2-14** Distribution of aggregate damages in 2005 by decile: Natural-gas-fired plants. NOTE: In computing this graph plants were sorted from smallest to largest based on aggregate damages. The lowest decile represents the 50 plants with the smallest aggregate damages. The figure on the top of each bar is the average across all plants of damages associated with $SO_2$, $NO_x$, $PM_{2.5}$, and $PM_{10}$. Damages related to climate change are not included.

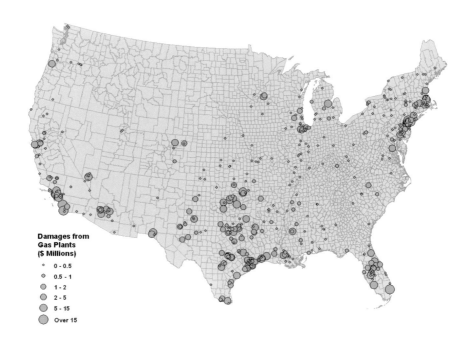

**Damages from Gas Plants ($ Millions)**

- 0 - 0.5
- 0.5 - 1
- 1 - 2
- 2 - 5
- 5 - 15
- Over 15

**FIGURE 2-15** Criteria-air-pollutant damages from gas generation for 498 plants, 2005 (U.S. dollars, 2007). Damages related to climate change are not included.

Table 2-16 shows amounts of pollutants emitted and Figures 2-16 and 2-17 show damages per kWh. Figure 2-17, which maps damages per kWh for the natural-gas-fired power plants in our sample, shows where these facilities are located. As in the case of coal-fired power plants, variation in damages per kWh across natural gas plants is explained both by variation in emissions of pollution per kWh and also by variation in damages per ton pollutant. In the case of $PM_{2.5}$, variation in pollution intensity and variation in damages per ton of $PM_{2.5}$ explain equal amounts of the variation in $PM_{2.5}$ damages per kWh.[43] In contrast to coal plants, natural gas plants are located in areas of high marginal damages per ton of $PM_{2.5}$ (Table 2-17). However, variation in damages per ton of $NO_x$ accounts for only 5% of the variation in $NO_x$ damages per kWh, while variation in pounds of $NO_x$

---

[43]Regressing $PM_{2.5}$-related damages per kWh on pounds of $PM_{2.5}$ emitted per kWh produces an $R^2$ of 0.26. Regressing $PM_{2.5}$-related damages per kWh on damages per ton of $PM_{2.5}$ also produces an $R^2$ of 0.26.

**TABLE 2-16** Distribution of Pounds of Criteria-Pollutant-Forming Emissions per Megawatt-Hour by Gas-Fired Power Plants, 2005

|  | Mean | Standard Deviation | 5th Percentile | 25th Percentile | 50th Percentile | 75th Percentile | 95th Percentile |
|---|---|---|---|---|---|---|---|
| $SO_2$ | 0.045 | 0.20 | 0.00069 | 0.0044 | 0.0065 | 0.012 | 0.15 |
| $NO_x$ | 2.3 | 9.0 | 0.052 | 0.17 | 0.48 | 1.7 | 5.5 |
| $PM_{2.5}$ | 0.11 | 0.39 | 0.00057 | 0.016 | 0.045 | 0.091 | 0.28 |
| $PM_{10}$ | 0.12 | 0.39 | 0.00092 | 0.018 | 0.050 | 0.094 | 0.32 |

NOTE: All plants are weighted equally, rather than by the electricity they produce.

ABBREVIATIONS: $SO_2$ = sulfur dioxide; $NO_x$ = oxides of nitrogen; PM = particulate matter.

emitted per MWh accounts for 75% of the variation in $NO_x$ damages per kWh.

To summarize, the aggregate damages associated with criteria-pollutant-forming emissions from the facilities in our sample in 2005, which generated 71% of the electricity from natural gas, were approximately $0.74 billion, or 0.16 cents per kWh (2007 USD); however, damages per plant varied

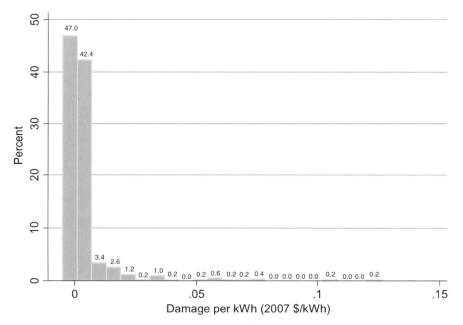

**FIGURE 2-16** Distribution of criteria-air-pollutant damages per kWh of emissions for 498 natural-gas-fired power plants, 2005. Damages related to climate change are not included.

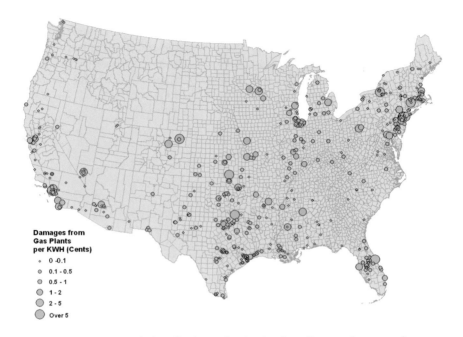

**FIGURE 2-17** Regional distribution of criteria-air-pollutant damages from gas generation per kWh (U.S. dollars, 2007). Damages related to climate change are not included.

widely. The 50% of plants with the lowest damages per plant, which accounted for 23% of net generation, produced 4% of the damages, and the 10% of plants with the highest damages per plant, which accounted for 24% of net generation, produced 65% of the damages. Although damages are

**TABLE 2-17** Distribution of Damages per Ton of Criteria-Pollutant-Forming Emissions by Gas-Fired Power Plants (2007 U.S. Dollars)

|         | Mean   | Standard Deviation | 5th Percentile | 25th Percentile | 50th Percentile | 75th Percentile | 95th Percentile |
|---------|--------|--------------------|----------------|-----------------|-----------------|-----------------|-----------------|
| $SO_2$  | 13,000 | 29,000             | 1,800          | 3,100           | 5,600           | 9,800           | 44,000          |
| $NO_x$  | 2,200  | 2,000              | 460            | 990             | 1,700           | 2,800           | 4,900           |
| $PM_{2.5}$ | 32,000 | 59,000          | 2,600          | 6,900           | 12,000          | 26,000          | 160,000         |
| $PM_{10}$ | 1,700  | 3,400            | 170            | 330             | 630             | 1,300           | 7,800           |

NOTE: All plants are weighted equally, rather than by the fraction of electricity they produce.

ABBREVIATIONS: $SO_2$ = sulfur dioxide; $NO_x$ = oxides of nitrogen; PM = particulate matter.

larger for plants that produce more electricity, less than 40% of the variation in damages across plants is explained by differences in net generation.

Damages per kWh also varied widely across plants: from about one-thousandth of a cent (5th percentile) to 0.55 cents per kWh (95th percentile). (These are weighted figures.) Most of the variation in $NO_x$ damages per kWh can be explained by variation in emissions intensity across plants; however, for $PM_{2.5}$, which constitutes over half of the monetized air-pollution damages, variation in damages per ton of $PM_{2.5}$ are as important in explaining variation in $PM_{2.5}$ damages per kWh as differences in $PM_{2.5}$ emissions intensity.

### Downstream $CO_2$ Emissions from Electricity Production from Natural Gas

The emissions of $CO_2$ from gas-fired power plants are significant. As the heat rate (energy of coal needed to generate 1 kWh of electricity) varies widely among coal-fired plants, so does it vary among gas-fired plants (with an average of about 0.5 ton of $CO_2$ per MWh of power generated (the 5th-95th percentile range is 0.3 to 1.1 tons per MWh).

## Externalities from Natural Gas in 2030

### Technology in 2030

In comparison to coal, less drastic technological change for central-station power generation by natural gas is expected. However, natural-gas powered fuel cells could become mainstream and generate significant amounts of electricity (such technology exists but is not currently at power-station scale).

Additionally, more natural gas could become available through discovery or more-aggressive development of existing sources. While domestic production has been relatively flat for years, new deposits such as the Marcellus Shale in the eastern United States hint at increasing domestic production. The prospect of this gas, however, is balanced against deeper drilling and more complicated extraction, which would increase the life-cycle energy use and associated emissions of using the resource.

Liquefied natural gas (LNG) is becoming an increasingly likely source of global natural gas-fired power. LNG has significant additional life-cycle stages compared with natural gas, which leads to additional energy use and air emissions. Synthetic natural gas (SNG) from coal is also a possible pathway. LNG and SNG both have substantially higher upstream emissions than natural gas, which would need to be taken into account in assessing their effects for future natural-gas-fired power.

Natural-gas-fired power plants have been discussed as candidates for CCS technology in the future. This combination is generally estimated to have smaller incremental costs (only about 1-2 cents per kWh) than for coal, but it captures less $CO_2$ per kWh than coal. Thus from a cost-effectiveness (and related damage avoidance) perspective, coal-fired plants will continue to be a more desirable target for CCS in the future.

A beneficial feature of natural gas power plants is their ability to quickly increase or decrease power output as needed. Thus they can fill in power demand for intermittent renewables such as wind and solar when other fast-ramping sources such as hydropower are not available. However, today's gas turbines are not designed to be ramped up and down continuously, and emit more GHG emissions and criteria-pollutant-forming emissions while ramping up and down (Katzenstein and Apt 2009). If a large percentage of renewables is installed by 2030, and natural gas is relied on for fill-in power, then considerable design improvements will be needed for those natural gas plants.

*Downstream Air-Pollution Damages from Gas-Fired Power Plants in 2030*

Our analysis of the criteria air-pollution damages associated with electricity generation from natural gas in 2030 follows the analysis for coal-fired electricity generation described earlier in the chapter. Specifically we ask how damages at the locations of the 498 facilities examined for 2005 would change if electricity generation were to increase at the rate predicted by the EIA and if emission intensities were to decline at rates consistent with EIA projections of emissions of $SO_2$ and $NO_x$ from fossil fuel. These assumptions are combined with estimates of damages per ton of the criteria-pollutant-forming emissions estimated from APEEP.

EIA projections of electricity generation from natural gas were used to estimate net generation in 2030. On average, electricity production from natural gas is predicted to increase by 9% from 2005 levels; hence we assumed that generation at each facility increases by this percentage. Reductions in pollution intensity for natural gas facilities are not as dramatic as for coal plants: pounds of $NO_x$ emitted per kWh are estimated to fall, on average, by 19%; emissions of $PM_{2.5}$ and $PM_{10}$ per MWh are each estimated to fall by about 32%.[44] Damages per ton of pollutant will, of course, rise, as described in section on coal-fired electricity.

The net effect of these changes is to decrease the projected aggregate damages generated by the 498 gas facilities from $0.74 billion (2007 USD)

---

[44]Emissions of $SO_2$ per MWh are estimated to fall by about 51%, but little $SO_2$ is emitted by gas-fired power plants.

in 2005 to $0.65 billion in 2030. Average damage per kWh from gas generation falls to 0.11 cents (2007 USD) from 0.16 cents in 2005.

## ELECTRICITY PRODUCTION FROM NUCLEAR POWER

### Current Status of Nuclear Power Production

In 2009, according to EIA, 104 commercial nuclear generating units are fully licensed to operate by the U.S. Nuclear Regulatory Commission. Their locations are shown in Figure 2-18. In addition, 14 nuclear power reactors are undergoing decommissioning, as shown in Figure 2-19 and listed in Table 2-18.

Of the 104 reactors in operation, 69 are pressurized light-water reactors (PWRs), totaling 65,100 net megawatts (electric[45]); and 35 units are boiling water reactors (BWRs), totaling 32,300 net megawatts (electric). Other reactor technologies exist or are being developed (see discussion later in this section on new developments in nuclear technology), but as of February 2009 none of these technologies operated commercially in the United States.

There has been no recent construction of nuclear generating plants in the United States. Nuclear generating capacity has been expanded by upgrading or adding capacity at existing power plants; the most recent reactor, Watts Bar No. 1, in Tennessee, was connected to the grid in February 1996.

### Brief History of Nuclear Power

Electricity from nuclear fission was first generated in the United States on December 20, 1951, by the U.S. Atomic Energy Commission's Experimental Breeder Reactor (DOE 2006a). The first commercial electricity-generating nuclear power plant at Shippingsport, Pennsylvania, reached its designed power-production level in 1957. It was shut down in 1982, when decommissioning began.

The growth of nuclear-powered electricity was rapid in the 1960s, though it slowed in the 1970s. In 1986, Ohio's Perry plant became the 100th U.S. commercial nuclear power reactor in operation. By 1991, the United States had 111 nuclear power units.[46] The highest number reached

---

[45]The total power capacity of a thermal power plant is greater than its electric power capacity because it is less than 100% efficient in converting heat into electricity. The output of interest here is electric power.

[46]A "unit" refers to a single nuclear power generating reactor. A nuclear installation can consist of more than one unit.

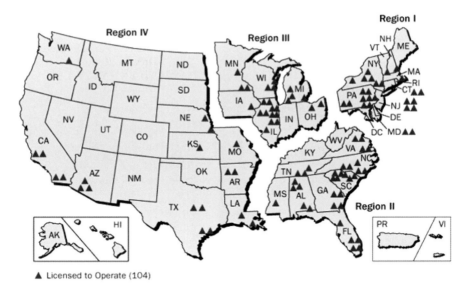

**FIGURE 2-18** Locations of operating nuclear power reactors in the United States. SOURCE: U.S. NRC 2008a.

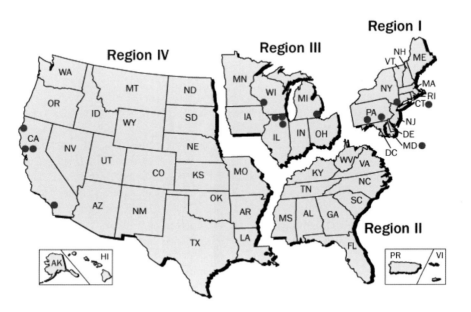

**FIGURE 2-19** Locations of nuclear power reactor sites undergoing decommissioning in the United States. SOURCE: U.S. NRC 2008b.

**TABLE 2-18** U.S. Nuclear Power Reactors Undergoing Decommissioning

|          | Name                            | Location          |
|----------|---------------------------------|-------------------|
| 1        | Dresden—Unit 1                  | Dresden, IL       |
| 2        | Fermi—Unit 1                    | Newport, MI       |
| 3        | Humboldt Bay                    | Eureka, CA        |
| 4        | Indian Point—Unit 1             | Buchanan, NY      |
| 5        | LaCrosse boiling water reactor  | Genoa, WI         |
| 6        | Millstone—Unit 1                | Waterford, CT     |
| 7        | Nuclear Ship Savannah           | Baltimore, MD     |
| 8        | Peach Bottom—Unit 1             | Delta, PA         |
| 9        | Rancho Seco                     | Herald, CA        |
| 10       | San Onofre—Unit 1               | San Clemente, CA  |
| 11       | Three Mile Island—Unit 2        | Middletown, PA    |
| 12       | Vallecitos boiling water reactor| Sunol, CA         |
| 13 and 14| Zion—Units 1 & 2                | Warrenville, IL   |

SOURCE: U.S. NRC 2008b.

was 112 in 1990, then constituting one-fourth of the world's nuclear power units; they provided almost 20% of the electricity produced in the United States (DOE 2006a, EIA 2008a). By 1998, the number of operating units was 104, as it remained in 2008 (EIA 2008a). Net electricity generation grew from 1.7 GWh in 1961 to 38.1 GWh in 1971, 272.7 GWh in 1981, 612.6 GWh in 1991, 768.8 GWh in 2001, and 806.5 GWh in 2007. The nuclear share of total electricity production reached 19.5% in 1988, and it has since ranged between 17.8% and 20.6% (EIA 2008a).

### Upstream Externalities

*Uranium Mining*

Canada and Australia currently account for 44% of global uranium production, with 18 other countries—notably, Kazakhstan, Niger, Russian, Namibia, and Uzbekistan—for supplying the remainder (IAEA 2008). Reduction in uranium stockpiles for weapons has contributed to an abundance of uranium on the market. The United States currently accounts for 5% of global production, much of the U.S. share coming from Wyoming.

Uranium is produced from open-pit (surface) mining, underground mining, or in situ leaching (ISL) techniques. Surface and underground mining for uranium is similar to mining for coal (described earlier). The ISL technique requires drilling several wells and pumping in a solution to leach the uranium out of the surrounding rock. The uranium-bearing solution is then pumped out of the wells and treated on-site to produce yellowcake (uranium ore). In Wyoming at present, all uranium production occurs at in situ facilities in the Powder River Basin (Paydirt 1999). Other states where ISL facilities could be located include Nebraska, South Dakota, and New Mexico (U.S. NRC 2008c). In 2006, uranium mining and milling in the United States produced 4,692,000 lbs of $U_3O_8$. Of this total, nearly 91% was produced at ISL facilities; with the remainder coming from underground mining (EIA 2008c, Table 2).

With uranium mining in general, radiological exposure can occur in three main ways: through inhalation of radioactive dust particles or of radon gas, ingestion of radionuclides in food or water, and direct irradiation of the body. For surface mine workers, exposure to radon exposure is generally less important than direct irradiation or dust inhalation; however, exposure to radon can be important for underground miners, although occupational radiological exposure is not an externality (see discussion and explanation in Chapter 1). For members of the public, the most significant pathways from an operating mine are radon and other radionuclide ingestion following surface water transport. From a rehabilitated mine, the pathways most significant over the long term are likely to be groundwater as well as surface water transport and bioaccumulation in animals and plants located at the mine site or associated water bodies (Australian Government 2009).

The draft *Generic Environmental Impact Statement for In Situ Leach Uranium Milling Facilities* (GEIS) released by the Nuclear Regulatory Commission in July 2008 assessed the impacts of four phases of ISL—construction, operation, aquifer restoration activities, and decommissioning—on land use, transportation, geology and soils, surface water and groundwater, terrestrial and aquatic ecology, air quality, noise, historical and cultural resources, visual and scenic resources, socioeconomic characteristics, public and occupational health and safety, and waste management. Impacts were qualitatively evaluated according to whether they were small ("not detectable or so minor that they will neither destabilize nor alter noticeably important attributes of the resource"), moderate ("sufficient to alter the resource noticeably, but not to destabilize, important attributes"), or large ("clearly noticeable and . . . sufficient to destabilize important attributes") (U.S. NRC 2008c, p. 4.1-1).

According to the draft GEIS, there is the potential for large impacts on historical and cultural resources, depending on local conditions; on localized ecological resources, especially on a few rare and endangered species, depending on site-specific habitat; and on groundwater. The possibility of groundwater impacts due to leaks and spills, excursions, and deep-well injection of processing waste historically has been an area of particular concern with ISL. The draft GEIS notes that the magnitude of groundwater impacts will depend on factors such as contamination during construction activities, which could be mitigated by best management practices; failure of well seals or other operational conditions, which could be detected by monitoring and testing; and the potential for impacts on deep aquifers from deep-well injection of processing wastes, which would depend on the state's permitting process.

Adverse environmental and human health effects can occur from legacy (discontinued) uranium mining and milling sites as well as from some current operations, especially in developing countries and in the former Soviet Union (Waggitt 2007). In the United States, a 1978 law—the Uranium Mill Tailings Remediation Control Act (UMTRCA)—as amended in 1983—provides for the remediation by the U.S. Department of Energy of 26 legacy uranium production facilities. U.S. laws do not classify uranium mining overburden as a radioactive waste, so its placement in radioactive waste disposal facilities is not required; however, EPA has the authority under various legal statues to protect the public and the environment from exposure to the hazardous and toxic characteristics of conventional (open-pit and underground) uranium mining wastes (EPA 2009e). Nevertheless, concern remains about some about the negative effects of both past and current mining practices (see, for example, WISE-Uranium 2009). A law passed by the Navajo Nation Council in 2005 banned uranium mining and milling altogether on sites within Navajo territory (SRIC 2009).

## Uranium Conversion and Enrichment

The only uranium conversion facility in the United States is at Metropolis, IL. This facility produces about 14,000 metric tons[47] of uranium per year. The process converts uranium oxide (yellowcake) into uranium hexafluoride, which is a gas. At the end of the conversion process, the amount of U-235 in the gas is about 0.7%. In order to enrich the material to that needed for reactor fuel to between 3 and 4%, the material is sent to a gaseous diffusion facility. Currently, the only facility in the United States is at Paducah, KY.

Although this facility is expected to be replaced by other centrifuge facilities being constructed at Piketon, OH, and Eunice, NM, the Paducah facility will remain in operation for several more years. The electricity intensity assumed for such a facility (Dones et al. 2005) is about 2600 kWh/separative work unit (SWU). When the Piketon facility is completed, the electricity use drops to approximately 40 kWh/SWU. The Piketon facility is due to begin operation in 2011; uncertainties about financing made the likelihood of meeting that deadline uncertain (Mufson 2009). The Eunice facility is scheduled to begin production even sooner, but, as of this writing, neither facility is in production. For that reason, it is reasonable to utilize

---

[47]A metric ton, sometimes written tonne, is 1,000 kilograms, or 2,205 pounds.

the analyses by Dones et al. that were based on life-cycle assessments for pressurized water reactor facilities. While these analyses cover the entire life cycle, the majority of the atmospheric emissions come from power plants producing electricity that is needed for part of the enrichment process using centrifuge technology. Thus, the estimated emissions values (all as g/MWh) are as follows:

$SO_2$: 22.5
$PM_{2.5}$: 5.4
$NO_x$: 33.9
Nonmethane volatile organic hydrocarbons: 7.7

### Upstream Emissions of GHG Emissions and Other Pollutants

It is often mentioned that nuclear power produces no air-pollutant emissions. Although that is generally true for the generation of nuclear power, the upstream part of the life cycle of nuclear power generation includes the mining, milling, and processing of uranium; transportation of the nuclear fuel; and construction of facilities, all of which entail criteria-pollutant-forming emissions and GHG air emissions. In short, the non-generation impacts dominate (Dones et al. 2005, Weisser 2007).

Koch (2000) estimated the $CO_2$, $SO_2$, $NO_x$, and PM emissions of nuclear power to be 1-2 orders of magnitude less than those of coal-fired power. Sovacool (2008) summarized a range of studies on the life-cycle GHG emissions of nuclear power and estimated that the mean was about 66 g $CO_2$-eq/kWh. Sovacool also noted that the "frontend" of the fuel cycle (including mining and milling uranium ore, conversion, and enrichment) represented 38% of the total emissions. NAS/NAE/NRC (2009a) cited and agreed with the conclusion reached by Fthenakis and Kim (2007) that life-cycle $CO_2$ emissions for nuclear plants, assuming that the current U.S. nuclear fuel cycle is maintained, could range from 16 to 55 g $CO_2$-eq/kWh. For comparison, coal plants without CCS produce an average of 1,000 g $CO_2$/kWh.

### Downstream Externalities

### Damages from Routine Plant Operations and Estimated Accident Damages

The main downstream burdens from operations of nuclear power plants are related to radioactive waste, discussed in some detail below. Other routine burdens are related to the release of heated cooling water. There are also land-use and ecological effects associated with nuclear plants, which are similar to those experienced at other thermal power plants (for example, see Box 2-2).

There have not been significant damages associated with release of radioactive materials from an operating nuclear power plant in the United States, but a few such accidents have occurred elsewhere. Although the potential for such accidents is a public concern—and also a concern to industry and regulatory bodies—the committee has not attempted to monetize or even quantify such potential. Previous studies, such as ORNL/RFF (1992-1998) and ExternE (EC 1995a), estimated the risk of accidents using detailed fault-tree models and found the risks and associated externalities to be small (as summarized later in this chapter). The committee did not undertake a modeling effort because such an analysis would have involved power-plant risk modeling and spent-fuel transportation modeling that would have required far greater resources and time than were available for this study. Also, apparently there were no developments since the earlier studies that would have led to any appreciable increase in the estimated probabilities of a reactor accident (a decrease in the estimate would be more likely).

Nuclear power plants routinely generate not only electricity but also radioactive wastes, including low-level radioactive waste (LLRW); "greater than Class C" (GTCC) wastes; and high-level radioactive waste (HLRW), mainly from spent nuclear fuel.

---

**BOX 2-2**
**Entrainment and Impingement of Aquatic Organisms by Thermal Power Plants**

Entrainment and impingement of fish and other aquatic organisms in intake structures of thermal power plants has received much attention. Impingement occurs when organisms are trapped by the force of the intake of water at intake screens; entrainment occurs at power plants with once-through cooling systems when the organisms—usually eggs, larvae, and juveniles—are carried with the water through the plant's heat exchanger and returned to the water body with the discharged water. Mortality from impingement and entrainment can approach 100%. Despite many studies, the population effects of impingement and entrainment usually are not well-known (Heimbuch et al. 2007). It appears that the most likely conditions for serious ecological impacts occur when there are many power plants in an area or when a power plant is sited in an area with a localized population of an organism that could be threatened with serious population consequences. These impacts, which are common to all thermal plants with once-through cooling systems, have not been quantified or monetized. Sovacool (2009a) has more broadly reviewed water-related impacts of thermal power plants and the effects of those plants on water resources.

Nuclear power plants are a significant source of LLRW; their LLRW may include anything from clothing and rags to ion-exchange resins, filters, tank residues, and irradiated reactor components. LLRW is either stored for decay to background levels before being disposed of as conventional nonradioactive waste (a practice possible only with slightly contaminated materials), or it is disposed of in near-surface engineered landfills. An interstate compact system for the disposal of commercially-generated LLRW was established through the 1980 Low-Level Radioactive Waste Policy Act (LLRWPA) as amended in 1985. Intended to spur the development of regional LLRW disposal sites, the process mandated by the act largely has failed. As of early 2009, there were only three LLRW disposal sites in the nation: one in Barnwell, SC, which is licensed to take Classes A, B, and C LLRW but as of July 2008 was restricted to take only waste generated in the Atlantic Compact states (South Carolina, Connecticut, and New Jersey); one in Richland, WA, which takes Classes A, B, and C waste from the nine states in the Northwest and Rocky Mountain compacts; and one in Clive, Utah, which accepts waste from all states but is licensed for Class A waste only. (Class A waste, which has the lowest concentration of long-lived radionuclides, requires fewer protective measures.) In 2005, approximately 4 million cubic feet of LLRW was shipped for disposal (U.S. NRC 2009). Nuclear power plants have the means to safely store LLRW on-site, including storage for decay to background levels if the waste is only slightly contaminated. Limited access to LLRW disposal sites—especially for Classes B and C waste—is an inconvenience for nuclear power plants, particularly those that are due for rehabilitation, up-rating, or decommissioning, but it is not likely to be an immediate environmental, health, and safety hazard.

The GTCC wastes from nuclear power plants come mainly from highly irradiated reactor components. Under the LLRWPA as amended, the federal government is responsible for all commercially generated GTCC waste (as well as GTCC-like waste generated by federal activities). In 2007, the DOE initiated a scoping process for a draft environmental impact statement to assess the environmental, social, and economic impacts of one or more facilities for GTCC and GTCC-like waste disposal. Disposal methods being considered include enhanced near-surface disposal, intermediate depth borehole disposal, and disposal at a geologic repository (GTCC LLRW EIS 2009).

According to the 1982 Nuclear Waste Policy Act (NWPA), the federal government is required to develop one or more geologic repositories to store HLRW generated by commercial activities and federal defense activities. The DOE is responsible for developing the site, the NRC for licensing it, and the EPA for setting radiation protection standards for humans and the environment. The NWPA was amended in 1987 to designate Yucca Mountain in Nevada as the only candidate for a geologic repository in the United States. After years of investigation and analysis by DOE,

Yucca Mountain was found suitable in 2002 by Energy Secretary Spencer Abraham and President George W. Bush. Nevada Governor Kenny Guinn vetoed the decision, but the veto was overturned by Congress in July 2002. An application for a license is before the Nuclear Regulatory Commission. The future of Yucca Mountain as a repository is unclear, however, because President Barack Obama's budget for FY 2010 significantly reduced funding for the program, and the Obama administration has generally voiced skepticism about it. DOE is studying alternate strategies for dealing with nuclear wastes that do not involve a repository at Yucca Mountain. With the disposal of HLRW being arguably the most contentious issue concerning nuclear energy, a detailed assessment of the externalities associated with its disposal would be a high priority for future study. Such a study would be extremely complex, given the considerable uncertainties, long timeframe, and severe impacts under certain scenarios.

As of 2002, about 45,000 tons of spent fuel from nuclear power plants were in storage—virtually all on-site. Most of the spent fuel rod assemblies are stored in water pools; less than 5% are stored in dry casks (U.S. NRC 2002). Unlike wet storage, dry cask storage is almost totally passive: It is simpler and uses few human or mechanical support systems. However, it is not suitable until the nuclear rod assemblies have been out of the reactor for a few years, allowing the heat generated by radioactive decay to decline. The NWPA limits the amount of waste to be stored at the geologic repository to 70,000 metric tons of heavy metals, of which 90% (63,000 metric tons) could be attributable to commercial spent nuclear fuel. However, one analysis suggests that Yucca Mountain would be technically capable of storing at least four and possibly nine times that amount (EPRI 2007).

Transportation of radioactive waste is jointly regulated by the U.S. NRC and the U.S. Department of Transportation (DOT). The U.S. NRC sets requirements for packaging radioactive materials; the DOT regulates shipments while they are in transit. For shipping spent fuel, casks or containers that shield and contain the radioactivity and dissipate the heat are required. Many shipments of spent fuel have been made, typically between different reactors of a utility, in order to share storage space. Lacking a geologic repository or its centralized storage equivalent, very little HLRW has been transported for long distances. Low-level waste has been transported long distances without significant incident for decades.

### Reprocessing Nuclear Fuel

Since 1977, there has been a moratorium in the United States on the reprocessing of spent nuclear fuel. In limited recycling processes that are commercially available in France, Japan, and the United Kingdom, uranium and plutonium are separated from spent nuclear fuel for eventual reuse as fuel, and the remaining transuranics, along with the fission products, are converted to vitrified waste for storage (Finck 2005; DOE 2006b). This process reduces the volume of waste to be stored by a factor of 4 but creates a separated pure plutonium product, which could present a proliferation and security risk.

Recently, research has been conducted in France, Japan, and the United States to develop a full-recycle, closed-fuel process to make more efficient use of the nuclear fuel and to avoid large storage problems (Finck 2005; DOE 2006b). This recycling process makes use of advanced separation techniques that can separate out (1) long-lived fission products, such as technetium and iodine, for immobilization and eventual disposal as high-level waste; (2) short-lived fission products, such as cesium and strontium, which can be prepared for decay storage until they meet the requirements for disposal as low-level waste; and (3) transuranic elements, including plutonium, neptunium, americium, and curium, which can be fabricated into fuel for advanced fast reactors (DOE 2006b).

The reprocessing and recycling of spent nuclear fuel through advanced separation techniques and fast reactors increases the efficiency of fuel use and decreases the need for high-level radioactive waste disposal capacity. The DOE has stated that reprocessing offers the opportunity for significant cost reduction (Finck 2005); others, however, have argued that it would be more expensive than current "once-through" practices (von Hippel 2001). It also has been argued that no reprocessing technique is as proliferation-resistant as not reprocessing spent fuel at all and leaving the plutonium mixed with highly radioactive fission products (von Hippel 2001).

## Estimates of Aggregate Damages from Nuclear Power Plants

We present here the results of two previous, studies of damages from nuclear power plants, by ExternE (EC 1995b) for France, and by Oak Ridge National Laboratory and Resources for the Future (ORNL/RFF 1995) for two sites in the United States. They are comprehensive and well documented; the range of values they produced and the reasons for the differences are informative.

ExternE (EC 1995b) estimated that the cost of damages for all stages of the nuclear fuel cycle, including reprocessing and accidents, was about 2.5 ECU mils (mECU) per kWh if no discount rate was applied. The ECU, the predecessor of the euro, was worth an average of 0.77 USD in 1995; thus the estimate was about 1.9 mils/kWh, equivalent to about 2.5 mils/kWh in 2007 USD. This is about 10% of the damage estimate in this study for criteria-pollutant-forming emissions from coal. When 3% and 10% discount rates were applied, the damage cost declined to 0.1 and 0.05 mECU/kWh, respectively. The large sensitivity to discount rate results from the adoption by ExternE of a time horizon of 100,000 years for estimating the total collective radiation dose due to release of radionuclides, and the ExternE estimate suggests that over a short or medium term, the aggregate damages for nuclear power are at least 3 orders of magnitude less than the air-pollutant damages alone from coal.

The contemporaneous study by Oak Ridge National Laboratory and Resources for the Future (ORNL/RFF 1995) produced estimates of the aggregate costs of nuclear operations of an average of 0.25 mils/kWh for two sites, including accidents; and 0.2 mils/kWh without accidents. These numbers are one order of magnitude less than the ExternE values with zero discount rate, and twice and four times as large, respectively, as the ExternE values with 3% and 10% discount rates. However, the U.S. (ORNL/RFF) assessment did not include reprocessing, as that did not then (and does not now) exist in the United States; the conversion, enrichment, fuel-fabrication, and low-level waste disposal stages were considered separately in that study. The ExternE assessment also expanded the physical boundaries to 1,000 km (regional) and to global dimensions. In addition, the technologies and sites in the two assessments are different. When these factors are taken into account, the results of the two studies are directly comparable, and the estimated damage costs of nuclear power remain significantly lower than those for coal.

These results depend in part on the estimated probability of accidents and their probable consequences, and those values are a function of many factors, including reactor design, training and motivation of personnel, population density and distribution, emergency response, and so on. Additional information and experience would likely help to refine those estimates (EC 2005).

## New Developments in Nuclear Technology

Nuclear power has the potential to produce large amounts of dependable electricity without emitting $CO_2$. In recent years, nuclear reactors have produced about 20% of U.S. electricity, but this contribution will drop unless new capacity is added. This section considers both updated versions of today's light-water reactors (LWRs), and possible advanced reactors for the future.

### Updated Light-Water Reactors

The current generation of nuclear reactors continues to function reliably, but considerable research has been conducted in recent years to improving designs. New reactors are expected to be simpler, easier to operate, and generally more resilient than current designs, and several utilities are planning on constructing them.

*Next-Generation Reactors*

There are plans in the United States to build several evolutionary light-water reactors (LWRs). America's Energy Future estimates that 5 to 9 such reactors could be built by 2020 (NAS/NAE/NRC 2009a). If they are built on time and within budget, perhaps additional similar reactors could follow. Nonetheless, new LWRs will be very expensive. It is important to examine alternative approaches that might have advantages and cost less.

DOE's Generation IV Program (DOE 2009a) includes research on five reactor concepts. Only one, the very-high-temperature reactor (VHTR) is receiving significant funding, about $70 million requested for FY 2009. It is a helium gas-cooled, graphite-moderated reactor which is an updated redesign of the experimental high-temperature gas reactors (for example, Fort St. Vrain). The technology has significant advantages, including a low probability of a major radioactive release and the amount of heat that it produces. The VHTR is expected to operate above 1000 degrees °C (1800 °F) and could be used for industrial process heat and hydrogen production as well as electricity.

DOE also has a related Nuclear Hydrogen Initiative (NHI), which focuses on thermochemical splitting of water molecules. Such processes, using the VHTR as the energy source, are projected to be significantly more efficient than electrolysis.

According to a 2008 National Research Council review of DOE's Nuclear Energy programs (NRC 2008a), both Gen IV and NHI are well-designed, and funding should be kept at levels according to progress towards milestones. The result, if successful, could lead to operating reactors before 2030, but probably only a few. Major decisions have yet to be made, including the basic core design.

DOE also is supporting work on a reactor that is intended to consume long-lived components of waste LWR fuel. The reactor could also produce power, but the primary goal is to reduce the nuclear waste disposal problem from tens of thousands to hundreds of years. Given the level of R&D required, and uncertainties in economics and the licensing path, such a reactor is unlikely to be operating by 2030.

## ELECTRICITY PRODUCTION FROM WIND

### Current Status of Wind Energy

It is difficult to keep current with respect to the status of wind energy in the United States because it is increasing so rapidly. By the end of 2008, the total installed capacity[48] in the United States was 25,170 MW (25.17

[48]Installed capacity, also called nameplate capacity, is the maximum rated electricity output in MW.

GW), up from 16,824 MW at the end for 2007 (AWEA 2009). For the 12 months ending November 30, 2008, 44,689 GWh of electricity were generated by wind-powered turbines out of a total of 4,118,000 GWh, or 1.1% (EIA 2009a, Table 1.1a).

The American Wind Energy Association (AWEA) lists 6,922 active turbines as of September 30, 2008, ranging in nameplate capacity from less than 1 KW to 3,000KW (3 MW).[49] The number of wind turbines per project ranges from 1 to more than one thousand. The largest project in terms of nameplate capacity is 766 MW. The smallest are many 50 KW installations consisting of single turbines. It is not possible to characterize an "average" wind plant in any meaningful way, but it is common for modern plants to have a nameplate capacity of between 40 and 300 MW and to consist of turbines ranging in individual capacity from about 1.5 to 3 MW. The earliest utility-scale projects were commissioned in California in the early 1980s; a few of those turbines, most on the order of tens of KW to about 100 KW, still are producing electricity.

### Brief History of Wind Energy

The first utility-scale wind-energy plants in the United States began operation in 1981, with a total installed capacity of less than 10MW. The increase was rapid at first, reaching 1.2 GW by 1986, but then slowed, with total capacity of only 1.8 GW in 1998. Then a period of rapid increase began again; capacity reached 4.3 GW by 2001, 6.6 GW by 2003, 9 GW by 2005, and more than 25 GW by the end of 2008. Much of the increase is fueled by federal production-tax credits (PTCs), which have been sporadic. The current federal PTC extends through 2009 as of December 2008. State-mandated renewable-energy portfolios, which require the state's energy use to be based on renewable sources (mainly wind) by target dates, also have affected the penetration of wind-generated electricity, as do general economic conditions.

### Future Considerations for Wind Energy

As indicated above, with the passage of time, the most-obvious change in wind-energy plants has been the reduction in total number of turbines and increase in the size (both physical size and nameplate capacity) of the individual turbines. Even some early plants had total nameplate capacities of from 40 to 80 MW, but projects exceeding 100 MW became common only in the late 1990s. These changes are largely technology-driven, result-

---

[49]Because the wind does not blow all the time (it is intermittent), the actual generation capacity of a wind turbine is only about 30% of the "nameplate capacity."

ing in larger turbines, and it seems likely that individual turbines of 5 MW will be commercially deployed in the United States soon.

A July 2008 report of the U.S. Department of Energy assessed the possibility of providing 20% of the nation's electricity from wind by 2030. The report noted that

> The 20% Wind Scenario in 2030 would require improved turbine technology to generate wind power, significant changes in transmission systems to deliver it through the electric grid, and large expanded markets to purchase and use it. In turn, these essential changes in the power generation and delivery process would involve supporting changes and capabilities in manufacturing, policy development, and environmental regulation (DOE 2008a, p. 4).

The report also noted (p. 57) that a 20% Wind Scenario would require a substantial development of offshore technology as well as improvements to land-based technology.

As of mid-2009, all U.S. wind-energy plants were on land. A number of offshore projects had been proposed, but none had been permitted. The Cape Wind project proposed for Nantucket Sound in Massachusetts had advanced the furthest: In January 2009, the Minerals Management Service (MMS) of the U.S. Department of the Interior (DOI) released a final environmental impact statement (FEIS) for the project, which was proposed to have 130 3.6 MW turbines located 4.7 miles offshore. Because the project was to be sited in federal waters, a lease with the federal government ws required. (The 2005 Energy Policy Act amended the Outer Continental Shelf (OCS) Lands Act to give the U.S. Department of the Interior authority to issue leases, easements, or rights-of-way for activities supporting renewable energy production; DOI delegated this authority to one of its bureaus, the MMS.) The FEIS identified most impacts as negligible or minor. (For a summary of the impacts cited in the FEIS, see MMS 2009, Table E-1.)

In April 2009, the DOI finalized its framework for renewable energy production on the OCS. In May 2009, the Energy Facilities Siting Board of the Commonwealth of Massachusetts granted a Certificate of Environmental Impact and Public Interest for the Cape Wind project, combining nine state and local permits required into one "super permit." A Record of Decision from MMS on the Cape Wind application for construction, operation, and eventual decommissioning is expected shortly.

### Upstream Impacts of Wind Energy

As noted at the beginning of this chapter, upstream effects of wind-energy generation of electricity differ substantially from those of fossil-fuel and nuclear plants in that there is no production, refinement, and transportation of fuel. As a result, the effects described below comprise all the upstream effects. Other kinds of EGUs also have such upstream effects, but because of their requirements for fuel, these effects are only a very small part of the total.

## Materials and Transportation

Metal components make up nearly 90% by weight and more than one-third by value of a modern wind turbine. For a 150 MW project, transportation requirements have been as much as 689 truckloads, 140 railcars, and 8 ships to the United States (Ozment and Tremwell 2007). Raw materials used include copper, iron (steel), rare earths for permanent magnets in rotors. The metal parts can be cast, forged, or machined. Turbine rotors are made of composites, balsa wood, carbon fiber, and fiberglass. Blades can approach 50 meters in length (and the nacelle of a turbine can be 70-90 meters above the ground). The mining of metals, fabrication and transportation of parts, and the assembly of the components have impacts that have been qualitatively described elsewhere in this chapter.

## On-site and Downstream Impacts of Wind Energy

### Ecological Effects

Assessment of the ecological effects of generating electricity from wind has focused primarily on deaths of flying animals caused by interactions with turbines. Bird deaths attributable directly to wind generation of electricity probably are less than 100,000 per year in the United States (for example, NRC 2007b; Sovacool 2009b). The only bird deaths considered to potentially reflect a population-level problem currently are of raptors, occurring mainly in older installations in California (NRC 2007b). Total anthropogenic bird deaths probably exceed 100 million per year in the United States,[50] and could be as high as 1 billion (NRC 2007b).

Bat deaths caused by wind turbines, especially in the eastern United States, have been higher than expected (NRC 2007b, Arnett et al. 2008), although they are extremely difficult to quantify, because bats are small

---

[50]Estimating the number of anthropogenic bird deaths is difficult, but the largest sources of mortality include birds' flying into buildings, flying into transmission lines, collisions with vehicles, exposures to toxic chemicals, and predation by domestic cats; this last factor alone could cause more than 100 million bird deaths per year (NRC 2007b and references therein).

and hard to find (Kunz et al. 2007). To date, no member of any bat species listed under the Endangered Species Act has been reported killed by a wind turbine. Bat populations of many species have been declining in the eastern United States, and because so little is known about the demography of bats, and because it is so difficult to quantify bat deaths, it is possible that the number of bats killed by wind turbines is a significant population-level threat to some species in some locations. The concern is intensified to the degree that the number of turbines continues to increase (NRC 2007b).

Although the primary focus of ecological effects of wind has been on deaths of flying animals, wind-generated electricity also can have wider ecosystem and habitat effects. Land-use changes to accommodate wind-energy installations are similar in kind to those for many other kinds of electricity-generating plants, including the need for roads and rights-of-way for transmission lines. The overall footprint of a wind-energy plant tends to be larger than for others, but the intensity of land-use change can be lower, because in many cases, the land between the turbines is not affected. On forested ridge lines in the eastern United States, the forest generally is cleared, or at least cut back, throughout the installation's footprint (NRC 2007b).

Most studies, including the NRC's 2007b report, have not identified significant ecological impacts other than those described above. However, the total installed wind-energy capacity when most recent reports were published was less than 12 GW, as compared with the more than 25 GW at the end of 2008. The rapid recent and projected future growth of wind-powered electricity generation in the United States means that ecological assessments probably will need to be repeated.

### Aesthetic and Visual Effects

There have been few quantitative studies of aesthetic and visual impacts, although there are well-established methods for assessing them quantitatively (NRC 2007b).

### Noise, Flicker, Radar Interference, Other

Adverse effects caused by noise—annoyance, sleep disturbance, and discomfort—have been documented and may be locally significant. Electromagnetic interference with television and radio broadcasting and radar also has been documented (NRC 2007b). Flicker effects have not been documented in the United States. All the above effects appear to be relatively small compared with effects related to other energy technologies considered by the committee.

## Assessment of Externalities from Wind Energy

The life-cycle damages associated with wind energy have not been fully quantified, but for those effects for which the committee has information, it is safe to say that, in aggregate for 2009, potential damages associated with wind turbines are small compared with those associated with coal and natural gas as electricity sources. Criteria-pollutant forming emissions and GHG emissions are much smaller per kWh than for coal or natural gas, and wind power produces far less electricity than do coal and natural gas, and so the aggregate emissions are very much smaller. Aggregate land-use effects considered over the entire life cycle are not significantly larger at present than those for other generation types, especially if one considers that in some cases former land uses can continue between wind turbines.

Anthropogenic causes of bird deaths include collisions with power lines, implying contributions from all sources of electric-power generation and use. Collisions with power lines likely account for the deaths of more than 130 million birds each year, dwarfing the estimated number of bird deaths caused by direct collisions with wind turbines (20,000 to 37,000 in 2005) (NRC 2007b and references therein). We do not have enough information to reliably compare the death rates of birds across all electricity-generation sources per kWh,[51] but if wind power ever provides 20% of U.S. electricity supply, as some scenarios suggest it will, then its significance as a cause of bird deaths would increase.

Damages associated with bat deaths are difficult to analyze. Bat deaths appear to be largely, if not uniquely, associated with wind generation of electricity, but no good estimate of the numbers of bats killed is available (NRC 2007b). In addition, the lack of understanding of the demography and ecology of bats makes it difficult to assess the importance of bat deaths. It appears likely to this committee that societal damages associated with the killing of bats by wind turbines are currently small by comparison with the aggregate damages associated with electricity generation by coal, natural gas, and the sum of all other sources. We agree with the NRC (2007b) that better information is needed, especially in light of the probable future increase in the number and density of wind turbines.

---

[51]Such a comparison was attempted by Sovacool (2009b). He concluded that wind energy killed 0.3 birds per kWh, nuclear power killed 0.4 birds per kWh, and fossil-fuel powered electricity killed 5.2 birds per kWh. Most of the fossil-fuel-related bird deaths were attributed to future climate change, and thus they represented a projection rather than an actual estimate of current bird deaths; he estimated the nonclimate-related avian mortality rate at 0.2 bird deaths per kWh.

## ELECTRICITY PRODUCTION FROM SOLAR POWER

### Background and Current Status

Solar power, a renewable energy source, refers to the capture and conversion of solar radiation (that is, sunlight) into electricity or heat. The use of solar power to generate electricity most commonly involves photovoltaic (PV) modules, or "solar panels," that are installed in large solar power plants ("solar farms") or on the walls or roofs of buildings. Other methods also exist to use heat generated by solar collectors or by other technology to generate electricity from steam turbines. The passive use of solar power for heating is discussed in Chapter 4. Concentrating solar power (CSP) systems use optics to concentrate direct incident solar radiation, which is converted into thermal energy that can be used to generate electricity. CSP-system use in the United States is limited, primarily to sites in the Southwest, which have abundant direct solar radiation.

PV- and CSP-system electricity generation by the electricity sector combined to supply 500 GWh in 2006 and 600 GWh in 2007, which constitute about 0.01% of the total U.S. electricity generation. EIA data indicate that the compounded annual growth rate in net U.S. generation from solar was 1.5% from 1997 to 2007 (NAS/NAE/NRC 2009b). However, this estimate does not account for the growth in residential and other small PV installations, which are applications that have displayed the largest growth rate for solar electricity.

U.S. solar panel and module imports increased from 45,313 peak KW in 2002 to 280,475 peak KW in 2007. EIA estimates that 90% of end use for domestic PV shipments is grid-interactive electricity production. Approximately 3% is remote electricity production. The remaining 7% is distributed among uses that include communications, consumer goods, transportation, water pumping, health, and others.

### Upstream and Downstream Impacts of Photovoltaic Energy

PV installations have two main parts: the solar panels and the balance of system (BOS) components. Generating electricity from PV modules, which produce direct-current (DC) electricity, requires a BOS to convert the DC power into the more commonly used alternating-current (AC) electricity. As such, upstream life-cycle activities involve mining of materials required for both solar panels and BOS components, panel and BOS manufacturing and construction, and finally the PV system installation.

Solar panels are made of semi-conducting materials similar to those used in the electronics industry. "Solar grade" silicon, derived from quartz sand, is the most commonly used material to make solar panels. However, emerging thin-film technology, which allow use of solar panels as roof tiles and other building features, can be made of a variety of materials, including amorphous silicon, gallium arsenide (GaAs), cadmium-telluride (CdTe), and copper indium gallium selinide (CIGS). With the exception of silicon and arsenic (for arsenide), the metals required for thin-film technologies are rare, and their use may depend on foreign imports. Materials for both CdTe and CIGS can be obtained from waste streams of zinc and copper smelting (USGS 2008).

Manufacturing these panels is a very high-technology, material- and energy-intensive process. A number of the metals for thin-film PV technology are toxic (for example, arsenic and cadmium), thus raise environmental and public health concerns about metal emissions during the extraction, material upgrading, and manufacturing activities associated with PV systems. The intense energy requirements for upstream PV activities are another concern. Various studies have considered the relevant life-cycle flows of materials, energy, and resources for PV systems. Most studies have focused on the life cycle of solar photovoltaic (PV) systems, specifically on crystalline silicon systems, and on energy and GHG emissions. Fewer studies have considered life-cycle material and substance use, or emerging thin-film technologies like cadmium-tellurium (CdTe) PV panels (Fthenakis and Alsema 2006).

Unlike other energy-generation technologies, for which the underlying technology has not changed significantly over 30 years, the manufacture of PV panels has undergone significant efficiency improvements and material shifts over that time (for example, the cost/watt decreased from $6 to $2 from 1990 to 2005). Studies in Europe that focused on previous generation technology estimated that producing solar power had 30% higher health impacts than natural gas, and GHG emissions of 180 g/kWh—an order of magnitude higher than nuclear (EC 2003). Follow-on studies, including CdTe systems, showed lower but nonzero life-cycle health impacts from PV of about 0.1-0.2 cents per kWh, primarily caused by GHG, lead, and particulate matter emissions (Fthenakis and Alsema 2006). The life-cycle GHG emissions are estimated to be 20-60 g/kWh, comparable to those of nuclear power (Fthenakis and Kim 2007), while $NO_x$ and $SO_2$ emissions are estimated at 40-180 and 50-450 mg/kWh respectively, far less than other generation methods (Fthenakis et al. 2008). Fthenakis and colleagues (2008) also evaluated heavy metal emissions (that is, Ar, Cd, Cr, Pb, Hg, and Ni), and found that that they are greatly reduced in comparison to emissions from fossil fuels, even with PV technology that makes direct use of the emitted compounds.

Generally excluded from LCA studies are transport considerations of raw materials to panel manufacturers in the United States. Transport considerations are important depending on type of PV system. For example, the United States has very little or no domestic production of arsenic, gallium, or indium, and must rely upon imports for these materials (USGS 2008). Because of intense energy requirements for upstream activities, research has begun to evaluate the "energy payback"—the amount of time a PV system must operate in order to recover the energy used to produce a PV system (DOE 2004).

Downstream life-cycle activities include electricity generation, storage, and disposal or recycling of worn-out panels. As with wind power, the production of electricity with PV systems does not emit air pollutants, including GHGs. Externalities associated with downstream PV activities may arise due to intermittency, that is, the need for grid electricity when sunlight is not available. Chapter 6 further discusses grid interruptions associated with renewable energy sources. Other externalities may arise from the disposal of worn-out PV systems. Worn-out solar panels have potential to create large amount of waste, a concern exacerbated by the potential for toxic chemicals in solar panels to leach into soil and water. Many components of solar panels can be recycled, but the United States currently does not have or require a solar PV recycling system.

To capture enough solar energy to produce large amounts of electricity requires a certain amount of land. Much of the United States receives enough solar energy to produce around 1 kWh per square meter of PV panel per day in the summer, less in winter, but more if the panel is tracked to follow the sun. The economic and other values of the land that would be needed to capture enough solar energy to provide substantial amounts of electricity would depend on a host of factors, including the land's location, ownership, and proximity to population centers, and other potential uses for the land. However, other factors also could affect solar-powered electricity at such a scale.

### Future Considerations for Solar Energy

While solar PV and CSP are still developing technologies, they will be an increasing, but still small, part of electricity generation through 2020. Although solar power represents a very small fraction the U.S electricity generation, the energy potential of solar power is enormous. A 2009 NRC report, *Electricity from Renewable Resources: Status, Prospects, and Impediments*, notes that current domestic solar power potential is 13.9 TWh, more than 3,000 fold greater than current electricity demand (NAS/NAE/NRC 2009b, p. 4).

If solar energy for electricity were to become a significant part of the

U.S. energy mix, more attention would need to be paid to damages resulting from the manufacture, recycling, and disposal of equipment. Land-use issues also would probably be a concern.

## ELECTRICITY PRODUCTION FROM BIOMASS

The nature of electricity generated from biomass feedstock is difficult to quantify and, for its externalities, even more difficult to obtain reasonable numbers. This is because the production and utilization of biomass for electricity production is inherently localized, resource-specific, and small scale. In addition, the term "biomass" can refer to a variety of feedstocks. The following discussion addresses issues associated with biomass use for electricity generation; because different feedstocks often are used for ethanol production for transportation fuel, the issues associated with them are somewhat different as well (see Chapter 3).

### Feedstock Production

Feedstock comes from forestry and agricultural residues and from harvesting of forest and agricultural products. Some electricity generation uses either industrial biomass residues or municipal solid waste.

In the case of energy crops, land could be used for other activities. For agricultural residues, farming practices and the viability of the land for farming could be affected. In some cases, changes in land use can increase carbon emissions. Other uses can enhance terrestrial carbon sequestration.

Sufficient water is needed to raise crops, forest products, and their residues. Non-point-source runoff can impact surrounding surface-water systems. Use of pesticides can affect water quality through non-point-source runoff. Energy use can have impacts through life cycles for growing biomass feedstock and the related harvesting of crops or agricultural residues.

Use of fertilizers, particularly petroleum-based, constitutes an additional life-cycle issue, since much fertilizer is produced using natural gas. Additionally, there could be an increase in GHG emissions from energy use in the treatment of the fields and emissions of nitrous oxide from the fertilizers.

Labor and related societal issues are related to changes in farming and forestry practices and in harvesting residues. Ecological effects, primarily destruction of habitat, mainly involve taking marginal lands for energy crops and forest products.

Most impacts from the use of municipal solid waste as a feedstock for electricity are expected to be positive, since the need for landfilling waste and the related potential for runoff to surface waters from landfills is minimized. However, concerns remain about atmospheric emissions from conversion facilities and land use (siting).

Emissions from the combustion of biomass can include polychlorinated biphenyl compounds, although the focus of recent analysis has been primarily on enclosed systems, such as cook stoves (Gullet et al. 2003). Although damages from biomass-generated electricity on a per-kWh basis might equal or even exceed those from other sources in some cases, the committee has not provided detailed analyses because this technology probably will have only limited market penetration in 2030.

### Transportation

Similar to the harvesting of biomass feedstock, transportation of feedstocks has localized impacts. Many facilities use biomass as a feedstock, derived from processes and residues generated on site. Where energy crops or biomass residues are collected away from the location of the power plant, the cost of transportation limits how far from the power plant these low-energy-density feedstocks can be obtained. The impacts associated with transportation are similar to standard transportation impacts associated with vehicle miles driven in terms of air quality impacts, energy penalties, and accidents.

### Power Generation

In 2008, not quite 40,000 GWh were generated from wood and wood waste, about 0.9% of the total (see Table 2-1 and associated text). Biomass accounted for about 16,000 GWh (0.3%).

The National Electric Energy Data System indicates that in 2003 there was less than 1.6 GW capacity of biomass-fired power plants in the United States (EPA 2004b). This is a small amount compared with overall generating capacity.

Many of the issues facing biomass combustors are similar to issues faced by larger-scale fossil-fuel generation, although they typically are more localized, because the generators are small, which may limit the control technologies placed on the system. In addition, many of these systems have been in operation since 1937, and therefore presumably "grandfathered" in on some environmental rules.

Air quality is a local issue, particularly for particulate matter from smaller, older combustors. Facility health and safety are important for older facilities.

Siting issues, such as aesthetics, are significant for newer facilities, such as those utilizing municipal solid waste. Citizens can be concerned about aesthetics and possible odors from atmospheric emissions.

For potential new technologies such as biogasifiers and use of liquid fuels derived from bio-oils, other environmental issues are unlikely to be a large factor, but there could be a public perception that these facilities will use feedstock from land that has been clear cut for energy crops, such as tropical oils.

## TRANSMISSION AND DISTRIBUTION OF ELECTRICITY

Here, we briefly discuss effects and damages associated with electricity transmission lines. Chapter 6 provides a discussion of security issues associated with interruptions or intermittencies in transmission/transport and distribution systems for electricity and for fuels such coal, oil, and natural gas.

Perceptions exist that high-voltage power lines and substations pose health risks (for example, of childhood leukemias and adult cancers, as well as acute effects) through their emission of extremely low frequency (ELF) electromagnetic radiation, but despite many studies, adverse health effects of transmission lines have not been conclusively established. The World Health Organization recently assessed this issue in detail (WHO 2007), and WHO's International Agency for Cancer Research addressed it further in 2008 (IARC 2008). The reports conclude that the evidence on some impacts of ELF on human health is inconclusive, including childhood leukemias; and that on other aspects the information leads to the conclusion that there are no adverse effects. The IARC report further concludes that if there are any excess cancer cases the number is very small, and that more than 99% of people are not exposed to enough ELF radiation from transmission lines for there to be a possibility of their suffering increased incidence of cancer.

Transmission lines also have raised concerns—as have various electricity-generating facilities—about loss of property values along and near them due to visual impairment and perceived or actual health risks, as well as possible land-use effects. The loss of property values is not an externality, being instead a market-mediated reflection of real or perceived physical damages. However, the visual impairment or any health risks associated with transmission lines are an externality.

Some renewable sources of energy, especially wind and solar, often need to be sited far from end users, thus requiring more new transmission lines than some other sources would need. For these reasons, proposals for new transmission lines often have been controversial, and managing the need for transmission lines and building new ones is thus a significant policy issue. However, because externalities associated with them appear to be very small by comparison with other aspects of electricity generation, the committee has not considered them in detail.

## SUMMARY

This chapter has examined information on burdens, effects, and damages associated with electricity generation from coal, natural gas, nuclear power, wind energy, solar energy, and biomass. In the case of fossil fuel and nuclear power, this discussion includes consideration of the exploration, extraction, and processing of fuel, and the transportation of fuel to generating facilities (upstream externalities) as well as electricity generation and distribution (downstream externalities).[52] Some burdens and effects have been discussed in qualitative terms, and others have been quantified and, when possible, monetized.

Our main goal is to examine the uncompensated external costs (and benefits) associated with electricity production. Many external costs have been reduced through regulation: For example, the criteria-air-pollutant damages associated with electricity generation from fossil fuel have been substantially reduced by federal and state regulations over the past 30 years. We examine only those damages that remain. Occupational injuries and deaths are of importance to society, but they do not constitute external costs associated with coal mining and oil and gas production. We therefore do not monetize them and do not add them to external costs, such as the health costs associated with air-pollution emissions.

There are at least two reasons for examining the externalities associated with electricity generation. One is to inform the choice among fuel types when increasing electricity production or replacing existing plants. This is typically done by comparing the external cost per kWh of electricity generation across fuel types. Another reason for examining externalities is to help identify situations where additional regulation may be warranted to reduce the external costs produced by current electricity generation. Identifying sources with large aggregate air-pollutant damages can help identify facilities where further analysis of the costs and benefits of reducing emissions is warranted. This chapter helps to inform both issues.

### Electricity from Coal and Natural Gas

In the case of electricity generation from coal and natural gas, we have described the upstream externalities associated with fuel extraction and processing and have quantified the air-pollution damages associated with

---

[52]We have not conducted a fully comprehensive life-cycle analysis of the external costs of electricity generation. In particular, we have estimated the external costs associated with power plant construction. Those costs probably are small compared with all other life-cycle costs, because thermal power plants often last more than 50 years, so when annualized, the costs are small over the plant's life span.

electricity generation at 406 coal-fired and 498 gas-fired power plants in 2005. This is based on emissions data from the 2005 National Emissions Inventory and estimates of damages per ton of pollutant from the APEEP model. Damage estimates are based on emissions of $SO_2$, $NO_x$, $PM_{2.5}$ and $PM_{10}$ and include impacts on human health, visibility, agriculture and other sectors. The average damage associated with these emissions per kWh at coal plants, weighting plants by the electricity they generate, is 3.2 cents per kWh (2007 USD), using a value of a statistical life (VSL) of $6 million (2000 USD).[53] The corresponding figure for gas facilities is 0.16 cents per kWh (2007 USD). However, the distribution of damages per kWh is wide for each set of plants, reflecting variation in the emissions intensity of plants and in their location. As a result, the coal plants with the lowest damages per kWh are cleaner than the natural gas plants with the highest damages per kWh. Specifically, the 9% of natural gas plants with the highest damages per kWh exceed the damages per kWh for the 10% of coal plants with the lowest damages.

The aggregate damages associated with emissions of $SO_2$, $NO_x$, $PM_{2.5}$, and $PM_{10}$ from coal generation in 2005 were approximately $62 billion (2007 USD), or $156 million per plant on average. The 50% of plants with the lowest damages per plant, which accounted for 25% of net generation, produced 12% of damages, and the 10% of plants with the highest damages per plant, which also accounted for 25% of net generation, produced 43% of the damages. The situation for gas is similar, although damages per plant are lower: the 10% of natural gas facilities in our sample with the highest damages per plant produce 65% of the air-pollution damages associated with the 498 facilities that we examined.

What are criteria air-pollution damages from coal and natural gas plants likely to be in 2030? To examine damages in 2030 we increase electricity generation at the plants analyzed in 2005 by amount consistent with EIA forecasts of electricity production from coal and natural gas. This implies, on average, a 20% increase in electricity produced from coal and a 9% increase in electricity produced from natural gas. We also assume that the emissions intensity of plants will fall in a manner consistent with EIA estimates of total emissions from fossil fuel plants. The APEEP model was used to estimate damages per ton from $SO_2$, $NO_x$, $PM_{2.5}$, and $PM_{10}$ in 2030. In spite of increases in damages per ton of pollutant, due to population and income growth, average damages per kWh (weighted by electricity generation) at coal plants are 1.7 cents per kWh (electricity-weighted),

---

[53]Premature mortality constitutes over 94% of total damages. When a VSL of $2 million is used, premature mortality constitutes 85% of total damages and the cost per kWh (electricity-weighted) falls to 1.2 cents.

compared with 3.2 cents per kWh in 2005 (also electricity-weighted). The fall in damages per kWh is explained by the assumption that pounds of $SO_2$ per MWh will fall by 64% and that $NO_x$ and PM emissions per MWh will fall by approximately 50%. Average damage per kWh from gas generation falls to 0.11 cents (2007 USD) from 0.16 cents in 2005 (weighting plants by net generation).

### Electricity from Nuclear Power

The committee did not quantify damages associated with nuclear power because the analysis would have involved power-plant risk modeling and spent-fuel transportation modeling that would have taken far greater resources and time than were available for this study. Notwithstanding that this modeling was not undertaken, previous studies suggest that the monetized value of these risks are small (ORNL/RFF 1992-1998; EC 1995b). The upstream damages result largely from uranium mining, most of which occurs outside the United States. With uranium mining in general, radiological exposure can occur through inhalation of radioactive dust particles or radon gas, ingestion of radionuclides in food or water, and direct irradiation from outside the body. For surface mine workers, exposure to radon exposure is generally less important than direct irradiation or dust inhalation; however, exposure to radon can be important for underground miners. If radiological exposure is taken into account in the miners' wages, it would not be considered an externality. For members of the public, the most significant pathways from an operating mine are radon or other radionuclide ingestion following surface water transport; from a rehabilitated mine, the more significant pathways over the long term are likely to be groundwater as well as surface water transport and bioaccumulation in animals and plants located at the mine site or on associated water bodies. Upstream impacts also include air emissions, including GHG emissions, but they are one or two orders of magnitude smaller than the emissions from coal-fired plants.

Downstream burdens are largely confined to the release of heated water used for cooling—such releases occur at any type of thermal plant—and the production of low-level radioactive wastes (LLRW) and high-level radioactive wastes (HLRW) from spent fuel; release of highly radioactive materials has not occurred on a large scale in the United States (but obviously has occurred elsewhere). Either LLRW is stored for decay to background levels and then disposed of as non-radioactive waste (a practice possible with slightly contaminated materials) or it is disposed of in near-surface landfills designed for radioactive wastes.

For spent nuclear fuel that is not reprocessed and recycled, HLRW is usually stored at the plant site. No agreement has been reached on a geologic repository for HLRW in the United States, and therefore little HLRW is transported for long distances. LLRW has been transported for decades without serious incident. The issue of having a permanent repository is perhaps the most contentious nuclear-energy issue, and considerably more study on the externalities of such a repository is warranted.

## Electricity from Wind

Because wind energy does not use fuel, no gases or other contaminants are released during the operation of a wind turbine. Upstream effects are related to the mining, processing, fabrication, and transportation of raw materials and parts; those parts are normally transported to the wind-energy plant's site for final assembly. The committee concludes that these life-cycle damages are small compared with the life-cycle damages from coal and natural gas. Downstream effects of wind energy include visual and noise effects, the same kinds of land-use effects that accompany the construction of any electricity-generating plant and transportation of electricity, and the killing of birds and bats that collide with the turbines.

Far more birds—by at least three orders of magnitude—are estimated to be killed by collisions with transmission lines, which are associated with all forms of electricity generation, than by collision with wind turbines. Therefore, although the detailed attribution of transmission-line-caused bird deaths by electricity source would be difficult, the committee concludes that bird deaths caused by wind-powered electricity generation are small compared with deaths from all other sources.

Wind-energy installations often have larger footprints than nuclear or coal plants, but the land use within the footprint often is less intensive than within the smaller footprints of thermal plants. In most cases, wind-energy plants do not currently kill enough birds to cause population-level problems except perhaps locally, mainly affecting raptors. The numbers of bats killed and the population consequences of those deaths have not been quantified, but could be significant. If wind-powered energy generation continues to grow as fast as it has recently, bat and perhaps bird deaths could become more important.

The committee has not quantified any effects of solar or biomass generation of electricity, but has not seen evidence that, at current generation capacity, there are effects that are comparable to those from larger sources of electricity generation. However as technology and penetration into the U.S. energy market improves, the externalities from these sources will need to be reevaluated.

## Research Recommendations

Many of the significant externalities associated with electricity generation can be estimated quantitatively, but there are several important areas where additional research is needed:

- Although it appears that upstream and downstream (pre- and post-generation) activities are generally responsible for a smaller portion of the life-cycle externalities than the generation activities themselves for some sources, it would be useful to perform a systematic estimation and compilation of the externalities from these other activities, comparable in completeness to the externality estimates for the generation part of the life cycle. In this compilation, damages from activities that are locally or regionally significant (for example, the storage and disposal of coal combustion by-products, in situ leaching techniques for uranium mining) need to be taken into account.
- The "reduced-form" modeling of pollutant dispersion and transformation is a key aspect in estimating externalities from airborne emissions, which constitute most of the estimated externalities for fossil-fuel-fired power plants. These models should continue to be improved and tested and compared with the results of more complex models, such as CMAQ.
- The health effects associated with toxic air pollutants, including specific components of PM, from electricity generation should be quantified and monetized. Because of the importance of VSL in determining the size of air-pollution damages, further exploration is needed of how willingness to pay varies with mortality-risk changes and with such population characteristics as age and health status.
- For fossil-fuel options, the ecological and socioeconomic impacts of coal mining, for example, of mountaintop removal and valley fill, are a major type of impact in need of further research in to quantify their damages.
- For nuclear power, the most significant challenges in estimating externalities are appropriately estimating and valuing risks when the probabilities of accidents and of radionuclide migration (for example, at a high-level waste repository) are very low but the consequences potentially extreme, and whether the cost to utilities of meeting their regulatory requirements fully reflects these externalities.
- The analysis of risks associated with nuclear power in the RFF/ORNL study should be updated to reflect advances in technology and science.
- For wind technologies, the major issues are in quantifying bird, and especially bat deaths; disturbances to both the local animal populations and landscape; and valuing them in terms comparable to economic damages.

• For solar, an important need is a life-cycle analysis of the upstream activities that quantifies the possible releases of toxic materials and their damages; another is a better understanding of the externalities that would accompany dedicating tracts of land to solar panels.

• For transmission lines needed in a transition to a national grid system, better estimates are needed of both the magnitude and the spatial distribution of negative and positive externalities that would accompany this transition.

# 3

# Energy for Transportation

## BACKGROUND

### The Current Mix of Energy Sources for Transportation

According to the U.S. Energy Information Agency, approximately 28% of all energy used in the United States is currently in the transportation sector (NAS/NAE/NRC 2009d). Of that used, approximately 96% is in the form of petroleum, 2.6% is natural gas, and less than 1% is biomass, electricity, or other fuels. Overall, transportation is responsible for approximately 70% of all U.S. petroleum consumption.

In its recent report, the National Research Council (NRC) Committee on America's Energy Future reports that, as of 2003, the transportation sector used approximately 28.4 quadrillion British thermal units (quads) of energy, of which more than 75% was expended in highway transportation, 17% in nonhighway transportation (for example, air, rail, and pipeline), and 8% in other off-highway use (for example, agriculture and construction) (NAS/NAE/NRC 2009d).[1] Figure 3-1 from its report illustrates that, of the highway sector, cars account for 43% of highway energy use (approximately, 34% of all transportation energy use), light trucks for 32% (approximately 26% of the total), and medium and heavy trucks for 24% (approximately 19% of the total).

Of the fuels consumed, AEF reports that gasoline accounted for approximately 62% of the energy used (measured in British thermal units)

---

[1]The *Transportation Energy Data Book* (Davis et al. 2009) indicates that highway transportation expended 80% of the energy used by the entire transportation sector in 2007.

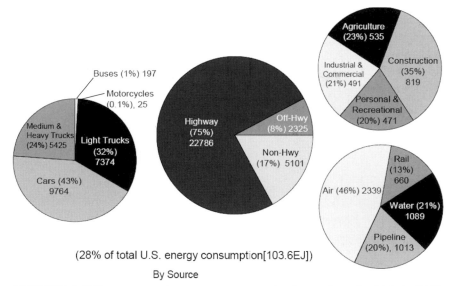

FIGURE 3-1 U.S. transportation energy consumption by mode and vehicle in 2003. SOURCE: U.S. Department of Energy's *Transportation Energy Data Book* (Bodek 2006) in NAS/NAE/NRC (2009d). Reprinted with permission; copyright 2006, Massachusetts Institute of Technology.

(EIA 2006b), and diesel (primarily in medium- and heavy-duty vehicles) accounted for approximately 17% of energy used.

### Regulation of Transportation Air Quality Emissions

The past four decades have seen a substantial national effort to regulate the emissions from transportation, starting with light-duty vehicles in the 1970s, and moving to heavy-duty on-road vehicle, and most recently to a range of other transportation sources, including construction and agricultural equipment, locomotives, boats, and ships (NRC 2004c). These efforts have been driven in part by even stricter standards adopted by California, which have in turn been adopted by a number of states. The result has been substantial reductions in emissions and ambient levels of a number of pollutants, even as vehicle miles have increased. For example, there have been substantial reductions of ambient levels of carbon monoxide (CO), in most cases to levels below[2] the current National Ambient Air Quality Standards (NRC 2003b).

---

[2] As of July 31, 2009, Clark County, Nevada is the only U.S. county in nonattainment for carbon monoxide (see EPA 2009f).

Starting in the late 1980s in the states and in 1990 on the national level, a number of rules have been aimed at changing the formulation of fuels to reduce a variety of emissions (for example, benzene and other volatile organic compounds [VOCs]) and to facilitate the introduction of new emission-control technologies (for example, ultra-low-sulfur diesel fuel) (NRC 2004c). Substantial requirements have also been enacted this decade that require enhanced use of biofuels (more details provided later in chapter).

### Improving Vehicle Efficiency

In addition to regulation to reduce emissions in the transportation sector, the United States has seen substantial efforts, beginning in the 1970s and renewed recently, to improve vehicle efficiency (NRC 2002c). The recent AEF efficiency panel report (NAS/NAE/NRC 2009d) assessed the opportunities for reducing energy consumption in the transportation sector through advances in efficiency.

That report notes that energy usage in transportation has grown rapidly in the United States over the past decades except for brief pauses during economic recessions in 1974, 1979-1982, 1990-1991, and 2001. The present economic decline, along with the 2008 spike in petroleum prices, is also likely to slow the demand for transportation fuels. Globally, the major drivers for energy efficiency are the price of fuel (influenced by taxes), regulations, personal choice, and the personal environmental values movement. In Europe, where high fuel and vehicle taxes raise owner costs and where diesel fuel is taxed less than gasoline, new-vehicle fuel economy is approaching 40 miles per gallon (mpg). In 1999, Japan instituted a fuel economy program to encourage vehicle efficiency per mile traveled, and its present new-vehicle fuel economy is similar to Europe's. In 2006, Japan revised its fuel economy standard to 47 mpg by 2015 (Ann et al. 2007).

In the United States, technological efficiency improvements are available at fairly modest costs. With present market structures, vehicle drivetrain efficiency has been improving at a rate of about 1% per year. However, rather than reducing their fuel expenses as a result of these improvements, most U.S. consumers have opted to purchase larger vehicles with more acceleration and accessories that consume even more energy. So in spite of technological improvements in the efficiency of vehicle components, the fuel demand has continued to rise, and the U.S. light-duty vehicle fleet now has an average new-vehicle fuel efficiency of about 25 mpg.

Recently, California adopted so-called GHG emission standards that would require substantial reductions in GHG emissions, primarily through enhancements in fuel economy, by 2016; 13 additional states indicated that they would adopt the standards once the U.S. Environmental Protection

Agency (EPA) approved a waiver of the Clean Air Act to allow the standards to move forward. Although EPA had originally rejected California's application for a waiver, in January 2009 EPA began a formal process to reconsider the waiver, and in May 2009, after detailed discussions among California, EPA, and auto makers, President Obama announced an approval of the waiver and a new unified approach to both federal corporate average fuel efficiency (CAFE) and GHG emissions standards that will result in a national standard comparable to the California standards. This action is expected to result in the achievement of the former 35.5 miles per gallon CAFE goal by 2016, several years sooner than originally envisioned.

A wide variety of technologies are available to improve fuel economy, in particular those to improve drive-train efficiency, vehicle aerodynamics, rolling resistance, and weight reduction (NRC 2008b). Many of these will be widely deployed by 2020, but further gains will be possible. Diesel engines and hybrid electric vehicles (HEVs), such as the Toyota Prius, are currently available and can reduce fuel consumption by more than 25% relative to today's gasoline vehicles. A shift to these technologies, coupled with other improvements, could result in a new-vehicle fleet with substantially improved fuel efficiency.

## APPROACH TO ANALYZING EFFECTS AND EXTERNALITIES OF TRANSPORTATION ENERGY USE

### Rationale for the Selection of Vehicle Fuels and Technologies

In considering its task, the committee recognized that it could not estimate quantitative externalities for every possible energy use in the transportation sphere. Therefore, the committee attempted to place transportation energy uses in order of importance on the basis of two key factors: (1) the degree to which a current transportation energy use is a significant part of energy use, and (2) the degree to which an emerging fuel and technology is likely to become a significant part of transportation energy use in the future. In applying these criteria and assessing the degree to which the data would support quantitative analysis, the committee focused on two key areas:

- A quantitative analysis of current and 2030 energy use, emissions, and externalities for highway transportation for both petroleum-based fuels and conventional biofuels (for example, corn ethanol) using the GREET (Greenhouse Gases, Regulated Emissions, and Energy Use in Transportation) model for primary analysis tied to the APEEP (Air Pollution Emission Experiments and Policy) model to estimate physical effects and monetary damages. This analysis applies to more than 75% of all current U.S. energy use in the transportation sector.

- A qualitative and quantitative synthesis of what is currently known on several other key fuels and technologies, including emerging biofuels (for example, corn stover and grasses); hybrid, plug-in hybrid, and electric vehicles; and other fuels (natural gas and hydrogen fuel cells).

## Transportation Life-Cycle Analysis

Our goal is to develop and apply an LCA framework that can provide more detailed quantitative assessments of the comparative health and environmental benefits, risks, and costs of existing fossil fuels (petroleum), as well as future mixes of transportation technologies and fuels. To meet this goal, we build on state-of-the-art life-cycle-impact-assessment (LCIA) methods that have been developed for evaluating and allocating the health, resource, and environmental impacts of industrial, agricultural, and energy technology systems (Guinée and Heijungs 1993; Horvath et al. 1995; Hoffstetter 1998; IAEA 1999; Hertwich et al. 2001; Bare et al. 2002; EC 2008). This effort and its resulting framework provide quantitative estimates of impacts that can be considered "external" in the context of Chapter 1.

One can take either a top-down or bottom-up approach when allocating health and environmental costs to transportation technologies. The top-down approach considers morbidity and mortality statistics for a specific population, such as the inhabitants of a country or of a large urban region, and attempts to allocate these impacts to a specific source, such as transportation emissions or power-plant emissions. The bottom-up approach provides a list of hazard sources (such as pollutant releases) and tracks these hazards from the source to exposure and damage. Top-down assessments for air pollution have been carried out for many regions, making it possible to provide a disease-burden estimate for air pollution. However, allocation to specific energy systems cannot be resolved because the top-down approach lacks the spatial and temporal resolution needed to track impacts to specific technologies. In contrast, the impact pathway assessment used in the ExternE study (EC 2003, p. 3) and the more recent analysis by Hill et al. (2009) of air-emission impacts from transportation fuels both used a bottom-up approach in which environmental benefits and costs are estimated by following the pathway from source emissions through pollutant-level changes in air, soil and water to health and environmental impacts.

The life cycle of effects associated with using energy for transportation includes upstream effects, such as extracting and processing the fuels, building the infrastructure needed to use transportation systems (for example, roads), building the infrastructure needed to deliver energy for vehicles (for example, pipelines and tankers), and manufacturing the vehicles. The life cycle of effects also includes the use of energy in vehicles, such as effects associated with emissions from vehicle tailpipes.

With respect to the categories of interest in this study, the committee summarized some of the key pathways by which energy sources for transportation lead to impacts. In general, most of the emissions occur as a result of burning fossil fuels in the life cycle of transportation fuels. Such energy use occurs across the supply chain, including fuel use for drilling oil wells or farming biomass fields, to transporting feedstocks and fuels to and from refineries, the refining process, transporting fuel to and from consumers, and the use of the fuels by consumers.

The movement of feedstocks and fuels in the supply chain of transportation fuels is different from that of electricity. Petroleum and petroleum products (for example, gasoline or diesel fuel) are generally transported by pipeline or truck; whereas coal, the primary energy source for electricity, is predominantly transported by rail. A significant share of the petroleum used to make fuels is from foreign sources (where it is extracted and delivered to the U.S. market via ocean tanker).

Various studies have been conducted of externalities of energy use in transportation. Before the phrase "life cycle" became popular, studies of this scope in the energy domain were referred to as "fuel-cycle" studies. The term fuel cycle was intended to represent the entire cycle of effects associated with using fuels. Today, such studies are often called "well-to-wheel" analyses because their scope goes from the oil well to powering the wheels of the car. In general, these terms all refer to the holistic study of impacts from extraction through combustion of the fuel for transportation. Other scopes exist too, for example, "well to tank," which involves all steps needed to get a fuel to the vehicle, but not using the fuel.

Prior studies around the world have assessed the relative contribution of environmental burdens from producing and using fuels for transportation (for example, Delucchi 1993, MacLean and Lave 2003a,b, Ogden et al. 2004, Brinkman et al. 2005, EC-JRC 2008, Ruether et al. 2005). Different from the study of environmental burdens related to electricity, those studies presented a mixed view of the relative importance of upstream-emissions versus in-use vehicle emissions. In prior studies, for petroleum-based fuels, the largest amount of emissions generally occurred when burning fossil fuels in vehicles while driving them, and upstream emissions were relatively modest (although they did not, in general, include vehicle manufacturing in those upstream effects).

## Scope of the Analysis

Because this study is about externalities associated with *energy* production, distribution, and use, this chapter considers the externalities from transportation technologies that use different forms of energy and fuels. The externalities of transportation per se are not within the scope of the study. Thus, the committee generally does not consider vehicle safety issues and traffic accidents, damage to road pavement from heavy trucks, or traffic congestion. These are not related to energy options. We consider them only to the extent that there are significant damages from the transport of fuels. For instance, Chapter 2 considers rail accidents associated with the transport of coal, but not all rail accidents. Similarly, our study considers oil tanker accidents, but not all transportation accidents.

The committee's goal was to estimate the external damages, in dollars per additional mile traveled, of different types of vehicle-fuel technologies, both current (2005) and future (2030). To do this properly, the committee recognized that it would be necessary to keep track of each type of pollutant and its source location and other factors that would vary spatially and over time. We also wanted to track the life-cycle stage of the damage and the end point category (for example, mortality and morbidity).

To obtain the estimates of emissions per vehicle miles traveled (VMT) by vehicle-fuel technology and life-cycle stage, the committee relied primarily on the GREET model. Sponsored by the U.S. Department of Energy's Office of Energy Efficiency and Renewable Energy (EERE), Argonne National Laboratory developed a full-life-cycle model called GREET. It allows researchers and analysts to evaluate various vehicle and fuel combinations on a full fuel-cycle and vehicle-cycle basis. The GREET model and analyses using the model have been published in a large number of peer-reviewed journals. The model has been widely used by Argonne, and other organizations have used GREET for their evaluation of advanced vehicle technologies and new transportation fuels. GREET users include government agencies, the auto industry, the energy industry, research institutions, universities, and public interest groups. GREET users are in North America, Europe, and Asia.[3]

GREET includes more than 100 fuel production pathways and more than 70 vehicle and fuel systems. Fuels include conventional and oil-sands-based petroleum fuel, natural gas, coal-based liquid fuels; biofuels derived from soybeans, corn, sugarcane, and cellulosic biomass; and grid-independent hybrids, grid-dependent hybrids, and all electric and hydrogen fuel cells. Unfortunately, although GREET covers light-duty autos and two types of light-duty trucks,[4] it does not contain information on heavy-duty

---

[3]A comparison of GREET 1.8b and Mobile6.2 emission factors for gasoline vehicles reveals that the latter are generally higher. See Appendix F for details.

[4]Class 1 trucks are under 6,000 lb gross vehicle weight rating (GVWR) and less than 3,750 lb loaded vehicle weight (LVW); class 2 trucks have the same GVWR and greater than 3,750 LVW.

trucks, which represent almost the entire U.S. fleet diesel fuel consumption, which is sizable compared with the consumption of all transportation fuels. Accordingly, the committee made separate estimates of direct emissions from heavy-duty trucks based on EPA's Mobile6.2 model and then used GREET to calculate the upstream emissions for the given fuel cycle. The committee decided in the interest of time, given their relatively smaller overall contribution, to omit rail, sea, and air transport and off-road vehicles from consideration in the modeling analysis of emissions from transportation energy use (that is, less than 25% of total transportation energy use).

Table 3-1 provides the complete list of vehicle-fuel technologies that the committee modeled with GREET[5] and the heavy-duty vehicles modeled by the committee outside of GREET.

To address technology improvements over time, GREET simulates fuel-production pathways and vehicle systems over a period from 1990 to 2020 in 5-year intervals. The results for any given year reflect GREET's estimates from 5 years before, so as to reflect the average fleet on the road in the year being analyzed. Thus, the committee, which was interested in external damages for 2005 (the base year for our analysis), used the 2000 GREET results for 2005. For its 2030 estimates, the committee used the 2020 results (that is, those vehicles on the road in 2020) with one major adjustment, replacing the default vehicle fuel efficiency for light-duty autos in GREET with the 35.5 mpg, which will be required by 2016 under the recently announced new efficiency and GHG emission standards. For heavy-duty diesels (HDDs), the committee captured emission improvements expected as dirtier trucks are retired from 2021 to 2030 and are replaced by HDDs meeting the 2007 and 2010 tailpipe standards. This approach will probably overestimate emissions in those years if emissions continue to fall with efficiency improvements (as GREET assumes until 2020).

For a given vehicle and fuel system, GREET separately calculates the following:

•      Consumption of total energy (energy in nonrenewable and renewable sources), fossil fuels (petroleum, natural gas, and coal together), petroleum, coal, and natural gas.

•      Emissions of carbon dioxide ($CO_2$)-equivalent GHGs—primarily $CO_2$, methane ($CH_4$), and nitrous oxide ($N_2O$). (The committee recognizes the potential importance of other climate-change agents, such as black carbon and ozone. Although our estimates of damages unrelated to climate change included particulate matter and ozone, it was not feasible to obtain climate-change-related estimates through GREET.)

---

[5]The committee used Version 1.8b for estimating fuel-related emissions and Version 2.7a for estimating vehicle manufacturing emissions.

**TABLE 3-1** Vehicle-Fuel Technologies in the Committee's Analysis

| Light-Duty Autos and Class 1 and 2 Trucks | Heavy-Duty Vehicles |
|---|---|
| RFG SI autos (conventional oil) | HDGV2B |
| RFG SI autos (tar sands) | HDGV3 |
| CG SI autos (conventional oil) | HDDV2B |
| CG SI autos (tar sands) | HDDV3 |
| RFG SIDI autos (conventional oil) | HDDV4 |
| RFG SIDI autos (tar sands) | HDDV5 |
| CNG | HDDV6 |
| E85—dry corn | HDDV7 |
| E85—wet corn | HDDV8A |
| E85—herbaceous | HDDV8B |
| E85—corn stover | |
| E10—dry corn | |
| E10—wet corn | |
| E10—herbaceous | |
| E10—corn stover | |
| Electric | |
| Hydrogen (gaseous) | |
| Grid-independent SI HEV | |
| Grid-dependent SI HEV | |
| Diesel (low sulfur) | |
| Diesel (Fischer Tropsch) | |
| Diesel (soy BD20) | |

NOTES: The modeling analysis included 33 vehicle-fuel technologies (23 light-duty vehicle fuels and 10 heavy-duty vehicle fuels). BD20 = 20% biodiesel blend; CG = conventional gas; CNG = compressed natural gas; E10 = 10% ethanol blend; E85 = 85% ethanol blend; HEV = hybrid electric vehicle; HDDV = heavy-duty diesel vehicle; RFG = reformulate gasoline; SI = spark ignition; SIDI = spark ignition, direct injection.

- Emissions of six substances that form criteria air pollutants: VOCs, CO, nitrogen oxides ($NO_x$), particulate matter smaller than 10 microns ($PM_{10}$), particulate matter smaller than 2.5 microns ($PM_{2.5}$), and sulfur oxides ($SO_x$).

GREET represents "well-to-wheel" life-cycle emissions in four stages: feedstock, fuel, vehicle manufacturing, and operations. For gasoline vehicles, these stages translate to the following:

- Feedstock: Extraction of oil and its transportation to the refinery.
- Fuel: Refining of the oil and its transportation to the pump.
- Vehicle: All emissions associated with production of the vehicle, which accounts for all life-cycle stages because it involves energy use.
- Operations: Tailpipe and evaporative emissions.

For other types of vehicles, the stages are analogous. For grid-dependent hybrids, a more complicated example, several energy types are involved. First is the gasoline life cycle for that portion of driving that uses gasoline. Second is the electricity life cycle. In this case, the feedstock emissions are those involving such activities as extraction of coal and natural gas that are weighted to reflect a default mix of electricity-generating technologies. (The committee used a national electricity-generation mix of fuel types taken from the national energy modeling system (NEMS) model for estimating 2030 electricity emissions.) The fuel emissions are those from the power sector's smokestacks. Emission estimates for vehicle manufacturing are adjusted to reflect the differences between energy and materials requirements for hybrids vs. conventional vehicles, say, regarding battery manufacturing.

The GREET model is fully assumption-driven but comes with a series of default values representing various assumptions. The committee set these values primarily at their default values but tested alternative values when it appeared warranted. See Appendix D for details on settings chosen by the committee.

The level of spatial detail in GREET is limited to whether the emissions are from urban or rural use. This choice appears to be primarily related to considerations of how direct grams per mile emissions from vehicles are dependent on vehicle speeds, which, in turn, are different in an urban vs. rural setting. To estimate damages, however, particularly by air pollutants, a finer degree of spatial detail is necessary.

The committee's strategy was to define U.S. counties in the 48 contiguous states as either urban or rural and then assign urban or rural emission factors to counties. This approach probably works well for direct vehicle emissions, since every county has vehicle emissions. However, decisions had to be made on where to locate sources of upstream emissions, such as refineries for petroleum and ethanol.

In general, such sources (except for emissions from electricity produced for electric vehicles and grid-dependent vehicles) were assumed to be located in every county, although some adjustments were made for oil refineries, ethanol production, and vehicle manufacturing). The committee located refineries by petroleum administration for defense districts (PADD), calculated damages per unit emissions by PADD from the APEEP model, assigned counties to PADDs, and from there assigned the PADD-specific unit damages to each county. Clearly, these assumptions simplify a complex situation where fuels can be imported as well as domestically produced. But the purpose of the analysis is to examine damages from sources in the United States. Thus, one should interpret the GREET results as what the damages would be *if* the county featured all the stages of the life cycle, for example, a refinery (see Appendix D for details).

Once GREET produces estimates of the emissions per mile associated with various vehicles and fuel types, this information (with the exception of emissions associated with vehicle operation and electricity production for electric vehicles and grid-dependent hybrids) was paired with results from the APEEP model, which provides estimates of the physical health and other non-GHG effects and monetary damages per ton of emissions that form criteria air pollutants.[6] For electric and grid-dependent hybrid vehicles, a similar approach was used to estimate damages for the feedstock and vehicle manufacturing components of the life cycle; however, the allocation of electric-utility-related damages to the operations and electricity production components of the life cycle were better approximated by applying a GREET-generated kWh/VMT and applying that to the estimated average national damages per kWh from the electricity analysis presented in Chapter 2 (details of this approach can be found in Appendix D).

Damages are estimated for mortality, morbidity and "other," which includes recreational damages related to visibility and crop damage related to ozone. These estimates are delivered for individual U.S. counties and for four stack heights, including tall stacks (appropriate for modeling source-receptor relationships [SRRs] associated with electric utility emissions), medium stacks (appropriate for modeling SRRs for industrial emissions), low stacks (appropriate for modeling SRRs for commercial emissions), and ground level (appropriate for modeling mobile-source SRRs). Thus, one can think of there being four matrices of physical and dollar per ton estimates, one matrix for each stack height, with each matrix covering counties and effects and damages. Because we have life-cycle emissions information, emissions per mile estimates at various stages of the life cycle were paired with the appropriate stack-height estimates.

## Presentation of Results

Results are provided by light-duty autos, two classes of light-duty trucks and eight classes of heavy-duty diesel trucks, covering 2005 and 2030, for all the vehicle-fuel technologies, all the pollutants, and all the life-cycle stages, as well as for alternative assumptions about the value of statistical life (VSL). All damages are expressed in dollar (2007 USD) per VMT terms, unless specified otherwise. With damages estimated at the county level for the 48 contiguous U.S. states, a distribution of damages over all counties was obtained. Thus, for all life-cycle stages, the 5th and 95th percentile range and median county damages are presented for each

---

[6]A more detailed description of the APEEP model is given in Chapter 2 and Appendix D. In estimating monetary damages, APEEP uses a value of a statistical life of $6 million/year (in 2007 dollars), as discussed further in Chapter 2.

pollutant, type of effect, year (2005 and 2030), and vehicle-fuel technology combination. For the operation stage, damage estimates are averaged over all the counties, both unweighted and weighted by population. The latter is more realistic as more-populated counties are doing more damage.

The committee also made estimates of these health and other non-GHG damages on a per gallon basis, although interpretation of these estimates is complicated by the fact that those fuel and technology combinations with higher inherent fuel efficiency would appear to have markedly higher damages per gallon than those with lower efficiency solely because of the higher number of miles driven per gallon. Also, GHG-related life-cycle emissions per mile are presented in this chapter, but damages are not discussed here (that occurs in Chapter 5). Information on energy use per mile was also calculated.

Finally, the committee did attempt to estimate aggregate annual damages for light-duty vehicles and heavy-duty vehicles in 2005—by multiplying per mile damages for each of the fuel and technology combinations in use in 2005 by the best estimates available of total VMT. Estimates for light-duty vehicles are somewhat conservative because, given the limitations on separating VMT among light-duty autos and light-duty trucks, we estimated aggregate damages using the per VMT damages we estimated for autos only. Similar estimates could not be made for 2030 given the substantial uncertainty in what fuels and what technology market shares will be at that time.

## PRODUCTION AND USE OF PETROLEUM-BASED FUELS

### Current Status and Brief History of Petroleum

Crude oil, a nonrenewable energy source, comprises the largest fraction of energy consumed in the United States (Figures 1-3 and 1-4 in Chapter 1). In 2007, the United States consumed 7.5 billion barrels of crude oil and petroleum products, of which nearly 70% was used by the transportation sector. U.S. consumption declined briefly in 1973 because of the Arab OPEC oil embargo (Figure 3-2). In response to the embargo, the U.S. government created the Strategic Petroleum Reserve (SPR). As of 2007, the SPR holds 697 million barrels of crude oil. Once the Arab OPEC embargo was lifted, U.S. consumption dramatically increased until rising oil prices in early 1980s caused a steep decline in consumption. Since the mid-1980s, U.S. oil consumption has steadily risen. In 2007, motor gasoline consumption reached a record high of 9.29 million barrels per day (390 million gallons/day).

Since the mid-20th century, U.S oil consumption has exceeded domestic oil production, thus nearly 60% of crude oil and petroleum products are

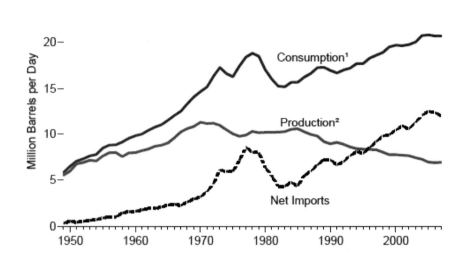

1. Petroleum products supplies is used as an approximation for consumption
2. Crude oil and natural gas plant liquids production

**FIGURE 3-2** Overview of petroleum consumption, production, and imports from 1949 to 2007. SOURCE: EIA 2008a, p. 124, Figure 5.1.

imported. In 2007, 71% of net crude-oil imports came from five countries: Nigeria (11%), Venezuela (12%), Mexico (14%), Saudi Arabia (15%), and Canada (19%). Domestic and imported crude oil are transported to U.S. refineries primarily by pipeline, barge, and ocean tankers (EIA 2008d).

The United States currently has 150 operable oil refineries capable of processing 17.6 million barrels of crude oil per day. Refineries are located in urban and rural areas across the United States. A map of current refineries is provided in Figure 3-3.

Approximately one barrel of crude oil produces 44 gallons of finished petroleum products, including jet fuel, diesel, and gasoline (Figure 3-4). More than 40% of crude oil is refined to finished motor gasoline (Figure 3-5).

### U.S. Vehicle Fleet

The U.S. Department of Transportation maintains an online report entitled "National Transportation Statistics." The report is updated quarterly and includes data beginning in 1960. Table 3-2 provides a summary of the most recent transportation statistics.

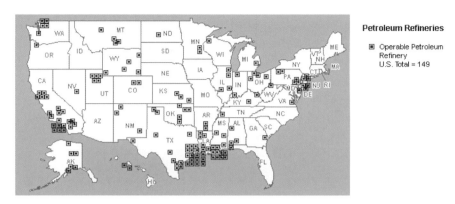

**Petroleum Refineries**

☐ Operable Petroleum
   Refinery
   U.S. Total = 149

FIGURE 3-3 Location of U.S. oil refineries. Texas and Alaska each account for large shares of U.S. crude-oil production, but the federal offshore areas in the Gulf of Mexico and California produce roughly a one-fourth share of the U.S. total, which surpasses the individual shares of Texas and Alaska. SOURCE: EIA 2009g.

## Technology and Fuel Pathways

Hydrocarbon fuels (gasoline, diesel fuel, and their potential substitutes) have a complex web of production and transport processes that include resource extraction, transport, refining storage, transfers, and combustion. Therefore, to understand externalities, one has to develop a map of the life

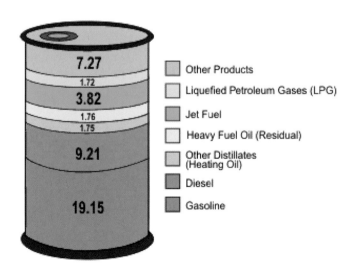

☐ Other Products

☐ Liquefied Petroleum Gases (LPG)

☐ Jet Fuel

☐ Heavy Fuel Oil (Residual)

☐ Other Distillates
   (Heating Oil)

☐ Diesel

☐ Gasoline

FIGURE 3-4 Products made from one barrel of crude oil (gallons). One barrel of crude oil is approximately equal to 45 gallons. SOURCE: EIA 2009h

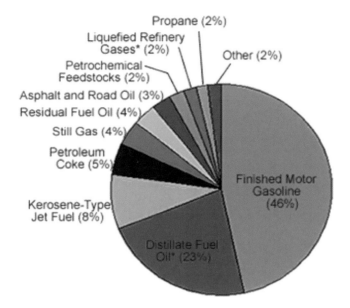

* Distillate Fuel Oil includes heating oil and diesel fuel. Liquid Refinery Gases include ethane/ethylene, propylene, butane/butylene, and isobutane/isobutylene.

**FIGURE 3-5** U.S. refinery and blender net production of refined petroleum products in 2007 (total = 6.57 billion barrels). SOURCE: EIA 2008d.

cycle of fuel. Different populations are affected at different stages of the life cycle. Despite the complexity, there are a few components of the fuel life cycle that tend to dominate with respect to overall health and environmental damage associated with the full life cycle of a transportation fuel. Figure 3-6 illustrates how the committee conceived the different stages of the fuel life cycle in several key phases: extraction and transport of petroleum feedstock; production and transport of refined product; transport, retail storage, and distribution; fuel use; and waste generation and management to carry out a life-cycle impact assessment. The potential effects of each of these phases are described briefly below.

In general, each phase of the cycle can contribute to deleterious effects from components of the hydrocarbon mixture itself; from activities and materials associated with a particular phase in the fuel cycle (for example, road development for oil production); and from generated wastes or by-products that pollute air, water, and soil or that contribute to climate-change effects.

**TABLE 3-2** Number of U.S. Aircraft, Vehicles, Vessels, and Other Conveyances

| | 1960 | 1990 | 2000 | 2006 |
|---|---|---|---|---|
| *Air* | | | | |
| Air carrier[a] | 2,135 | 6,083 | 8,055 | U |
| General aviation[b] (active fleet) | 76,549 | 198,000 | 217,533 | 221,943 |
| *Highway, total* | 74,431,800 | 193,057,376 | 225,821,241 | 250,851,833 |
| *(registered vehicles)* | | | | |
| Passenger car | 61,671,390 | 133,700,496 | 133,621,420 | 135,399,945 |
| Motorcycle | 574,032 | 4,259,462 | 4,346,068 | 6,686,147 |
| Other 2-axle 4-tire vehicle | N | 48,274,555 | 79,084,979 | 99,124,775 |
| Truck, single-unit 2-axle 6-tire or more | N | 4,486,981 | 5,926,030 | 6,649,337 |
| Truck, combination[c] | 11,914,249 | 1,708,895 | 2,096,619 | 2,169,670 |
| Bus | 272,129 | 626,987 | 746,125 | 821,959 |
| *Transit[d]* | | | | |
| Motor bus | 49,600 | 58,714 | 75,013 | (P) 83,080 |
| Light rail cars | 2,856 | 910 | 1,327 | (P) 1,801 |
| Heavy rail cars | 9,010 | 10,567 | 10,311 | (P) 11,052 |
| Trolley bus | 3,826 | 610 | 652 | (P) 609 |
| Commuter rail cars and locomotives | N | 4,982 | 5,498 | (P) 6,403 |
| Demand response | N | 16,471 | 33,080 | (P) 43,509 |
| Other[e] | N | 1,197 | 5,208 | (P) 8,741 |
| *Rail* | | | | |
| Class I, freight cars | 1,658,292 | 658,902 | 560,154 | 475,415 |
| Class I, locomotive | 29,031 | 18,835 | 20,028 | 23,732 |
| Nonclass I, freight cars | 32,104 | 103,527 | 132,448 | 120,688 |
| Car companies and shippers freight cars | 275,090 | 449,832 | 688,194 | 750,404 |
| Amtrak, passenger train car | N | 1,863 | 1,894 | 1,191 |
| Amtrak, locomotive | N | 318 | 378 | 319 |
| *Water* | | | | |
| Nonself-propelled vessels[f] | 16,777 | 31,209 | 33,152 | 32,211 |
| Self-propelled vessels[g] | 6,543 | 8,236 | 8,202 | 8,898 |
| Oceangoing steam and motor ships (1,000 gross tons and over)[h] | 2,914 | 635 | 461 | 286 |
| Recreational boats[i] | 2,450,484 | 10,996,253 | 12,782,143 | 12,746,126 |

NOTE: N = data do not exist; R = data are revised; U = data are not available.

[a]Air carrier aircraft are those carrying passengers or cargo for hire under 14 CFR 121 and 14 CFR 135. Beginning in 1990, the number of aircraft is the monthly average of the number of aircraft reported in use for the last 3 months of the year.

[b]1991-1994 are data revised to reflect changes in adjustment for nonresponse bias with 1996 telephone survey factors; 1995-1997 data may not be comparable to 1994 and earlier years because of changes in methodology. Includes air taxi aircraft.

[c]In 1960, this category includes all trucks and other 2-axle 4-tire vehicles.

[d]Prior to 1984, transit excludes most rural and smaller systems funded via Sections 18 and

*continued*

**TABLE 3-2** Continued

16(b)(2), Urban Mass Transportation Act of 1964, as amended. Also prior to 1984, includes total vehicles owned and leased.

*eOther includes aerial tramway, automated guideway transit, cablecar, ferry boat, inclined plane, monorail, and vanpool.

*fNonself-propelled vessels include dry-cargo barges, tank barges, and railroad-car floats.

*gSelf-propelled vessels include dry-cargo and passenger, offshore supply vessels, railroad-car ferries, tankers, and towboats.

*hBeginning in 2006, vessels are reported if they are greater than 10,000 deadweight tons, and prior to 2006, boats of greater than 1000 deadweight tons were reported.

*iRecreational vessels include those required to be numbered in accordance with Chapter 123 of Title 46 U.S.C.

SOURCE: BTS 2009, Table 1-11.

This section describes pollutant releases and other stressors that can lead to effects described above. It does not attempt to quantify effects and does not attempt to assess the efficacy of the various approaches used to manage risks of those effects.

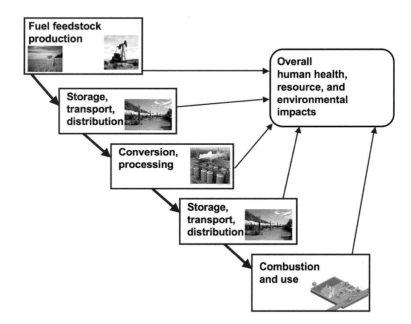

**FIGURE 3-6** Conceptual stages of fuel life cycle. SOURCE: Adapted from Energy Biosciences Institute, University of California.

Each phase of the petroleum cycle involves the use of electricity. Because Chapter 2 discusses life-cycle effects associated with electricity production, they are not included here.

## Extracting Crude Oil

*Conventional Oil Reserve*   The major oil producing areas in the United States are in the Gulf of Mexico region (onshore and offshore), California, and Alaska. As of 2007, there were about 500,000 active oil wells in the United States (onshore and offshore) (EIA 2008a, p. 127, Table 5.2). Much of U.S. extraction activities take place near sensitive coastal and estuarine habits.

When a potential oil reservoir is discovered, exploratory drilling is conducted to confirm the presence of oil. For onshore drilling, the land is cleared and leveled to construct a drill platform and install ancillary equipment. Depending on the location, roads, air strips, and buildings may also be constructed. Offshore, floating barges, semi-submersible vessels, or specially designed floating oil rigs are used to support exploratory drilling (API 2009). Inland water and wetland drilling and transportation can have significant effects on wetlands and estuarine habitats, requiring additional techniques to reduce disruption of those ecologic habitats.

Land is excavated to form a reserve pit where wastes from drilling are placed. Drilling wastes from offshore operations can cause a rapid build-up of a debris layer on the ocean floor that can degrade benthic communities. Drilling wastes may contain trace amounts of mercury, cadmium, arsenic, and hydrocarbons.

Drilling operations also produce combustion-related emissions, such as exhaust from diesel engines and turbines that power the drilling equipment. Hydrogen sulfide may be released.

When the presence of oil is confirmed, oil wells are constructed to extract the crude oil. Initially, oil may rise by "natural lift." Over time, mechanical pumps or injection methods, using steam, for example, are needed to bring the oil to the surface. Storage tanks, pipelines, and processing plants are also built.

Crude oil is prepared for shipment to storage facilities and then to off-site refineries. Natural gas can be separated from the oil at the well site and processed for sale, or the gas can be flared as a waste (usually at onshore operations), releasing CO, $NO_x$, and possibly sulfur dioxide ($SO_2$) if the gas is sour. Triethylene glycol is commonly used as a desiccant to remove water from the gas.

Wastewater generated at the production facility may contain organic compounds (for example, benzene and naphthalene), inorganics (for example, lead and arsenic), and radionuclides. VOCs may be emitted via leaks from the production process equipment. Emissions also occur from combustion of fuel to operate machinery.

Oil spills may come from storage tanks, during transfers, or from pipes, valves, joints, or gauges. For onshore spills, concern is for surface-water contamination via runoff and for seepage into groundwater. Effects of offshore spills can vary substantially, depending on factors such as coastal proximity and degree of turbulence. Accidents known as well blowouts can result in large releases of contaminated water, oil, methane, or other fluids. The mixture can be spread in a wide area around the rig, possibly leaching through the soil to a freshwater aquifer or running off into nearby surface waters. The blowout may also result in a well fire.

*Nonconventional Oil Reserve: Oil Sand*   Oil sands (also called tar sands or bituminous sands) contain a viscous oil referred to as bitumen that serves as a nonconventional source of synthetic crude oil. Oil sands can be extracted by surface mining using methods similar to those used for coal. The sands are transported to an extraction plant, where bitumen is separated from the sands using hot water and agitation. Once separated, the bitumen is upgraded to synthetic crude oil, which can then be refined into fuels. Approximately 2 tons of tar sands generates one barrel of synthetic crude oil In situ extraction is generally used for deep oil-sand deposits. Heat is applied underground, and bitumen is pumped to the surface for subsequent refining.

Currently, there is no production of synthetic crude oil from tar sands in the United States. The largest commercial oils sands industry is located in Alberta, Canada. Oil sands contribute more than 40% of total crude-oil production in Canada. Approximately 20% of crude oil imported into the United States is from Canada.

Impacts of oil-sand extraction and processing generally arise from degradation of ecological habitats, water and consumption, and waste (tailings) disposal. Canada's National Energy Board reports surface and in situ mining operations require 2-4.5 barrels of water to produce one barrel of synthetic crude oil (NEB 2006). Tar-sand extraction and upgrading also requires a high level of energy input. Natural gas is used to heat steam and generate electricity required for in situ recovery, as well as to upgrade bitumen. The government of Alberta reports that oil-sand production is responsible for 5% of Canada's GHG emissions (Alberta 2008).

*Nonconventional Oil Reserve: Oil Shale*   Oil shale is a sedimentary rock that contains kerogen, a solid bituminous material that can be processed to create synthetic crude oil. The United States contains the world's largest deposit of oil shale. The Green River Formation of Colorado, Utah, and Wyoming contains an estimated 800 billion barrels of recoverable oil (BLM 2008). Although technological methods exist to extract crude oil from oil shale, commercial extraction and processing is not economically or environmentally viable in the United States.

Potential impacts from oil-shale extraction arise from changes in land use, habitat disturbance, mining waste production, water consumption, and energy consumption. Oil-shale development is expected to consume between two and five barrels of water per barrel of oil produced (Bartis et al. 2005).

From among this wide range of potential impacts at different stages of resource extraction, the committee was constrained—by the limitations of the GREET model and the scarcity of available national databases on many ecosystem impacts and other impacts—to quantify only those impacts that result directly or indirectly from energy use and the air-quality emissions produced during these operations.

## Refining Crude Oil

Refineries separate conventional and synthetic crude oil into different petroleum products that can be used as fuels, lubricants, chemical feedstocks, and other oil-based products. Fuels make up the vast majority of the output (see Figure 3-5). Pollutants generated during crude-oil refining typically include VOCs, CO, $SO_x$, $NO_x$, particulates, ammonia ($NH_3$), hydrogen sulfide ($H_2S$), metals, spent acids, and numerous toxic organic compounds. Emissions occur throughout refineries and arise from the thousands of potential sources, such as valves, pumps, tanks, pressure relief valves, and flanges. Emissions also originate from the loading and unloading of materials (such as VOCs released during charging of tanks and loading of barges), as well as from wastewater-treatment processes (such as aeration and holding ponds).

Relatively large volumes of wastewater are generated by the petroleum refining industry, including contaminated surface water runoff and process water. Accidental releases of liquid hydrocarbons have the potential to contaminate large volumes of groundwater and surface water, possibly posing a substantial risk to human health and the environment.

Storage tanks are used throughout the refining process to store crude oil, intermediate products, finished products, and other materials. The tanks are a considerable source of VOC emissions. Hazardous and nonhazardous wastes are generated from many of the refining processes, petroleum handling operations, as well as wastewater treatment.

## Transporting and Distributing Crude Oil and Refined Products

Oil imported to the United States from outside North America is transported predominantly by ocean tanker. Imports from Canada flow through several pipelines that connect with the U.S. pipeline infrastructure in Illinois, Oklahoma, Wyoming, and Washington. Crude oil is transported from production operations to refineries by tankers, barges, rail tank cars, tank trucks, and pipelines. Refined petroleum products are conveyed to fuel marketing terminals and petrochemical industries by these same modes. From the fuel marketing terminals, the fuels are delivered by tank trucks to service stations, other commercial facilities, and local bulk storage plants. The final destination for gasoline is usually a motor-vehicle gasoline tank.

The United States has an extensive oil pipeline network used to transport oil from wells and ocean tankers to refineries. There are about 30,000-40,000 gathering pipelines and 55,0000 trunk pipelines to transport oil in the United States (API/AOPL 2007). Pipelines also carry refined petroleum products from oil refineries to bulk terminal storage sites. There is an estimated 95,000 miles of pipelines carrying refined petroleum products (API/AOPL 2007). Airports often have dedicated pipelines to carry fuel directly to them (API/AOPL 2007).

Transport and distribution of oil is a source of air pollution. Each of the transport and distribution activities is a potential source of evaporation loss. Transport of crude oil and refined petroleum products also present risks of oil leaks, spills, and large scale accidents (for example, the 1989 Exxon Valdez oil spill). Environmental releases of crude oil or refined petroleum products can pollute terrestrial and aquatic habitats as well as drinking water. The NRC report *Oil in the Sea III: Inputs, Fates, and Effects* (NRC 2003c) assessed data gathered between 1990 and 1999 and estimated that 9,100 tons of petroleum are released in North American waters as a result of transportation of crude oil and refined products. Pipeline leaks and other accidents related to petroleum fuel are discussed in Chapter 6.

### Storing Refined Products

Crude oil and refined petroleum products are stored in large volumes throughout the fuel cycle. In 2008, more than 338 million barrels of crude oil and refined products were held in storage at refineries. Bulk terminal storage facilities held more than 320 million barrels of refined petroleum products, including distillate fuel oils (diesel fuel), gasoline, and aviation fuels. Finished gasoline and diesel fuel are also stored in underground storage tanks (USTs) at gasoline stations. EPA regulates more than 623,000 USTs at approximately 235,000 locations (EPA 2009g).

The primary concern surrounding storage tanks is the potential for leaks, spills, and explosions. Similar to pipelines, crude oil and refined petroleum products leaking from storage tanks can accumulate into soils and seep into surface and groundwater, contaminating terrestrial and aquatic habitats as well as drinking-water resources. Since 1988, there have been about 479,800 confirmed releases from USTs and about 377,000 completed clean-ups (EPA 2008b). Storage tanks are also a source of evaporative emissions.

## Using Fuel for Light-Duty and Heavy-Duty Transportation

The category of on-road or highway mobile sources includes vehicles used on roads for transportation of passengers or freight. On-road vehicles are further divided in such categories as light-duty vehicles, light-duty trucks, heavy-duty vehicles, and motorcycles. The vehicles may be fueled with gasoline, diesel fuel, or alternative fuels, such as alcohol or natural gas. Nonroad sources include vehicles, aircraft, marine vessels, and locomotives, and other vehicles and equipment used for construction, agriculture, and recreation.

Cars, trucks, and buses consumed about 80% of the transportation energy used in the United States in 2007. Portions used by other transport modes are air (9%), water (5.2%), pipeline (3%), and rail (2.4%) (Davis et al. 2009).

NAS/NAE/NRC (2009d) indicates that incremental improvements in vehicle technology could reduce the fuel consumption of gasoline internal-combustion-engine vehicles by up to 35% over the next 25 years. Diesel-fueled trucks are expected to continue dominating the freight transportation sector for at least the next 25 years. The report estimates 10-20% reductions in fuel use by heavy-duty and medium-duty vehicles by 2020, resulting mostly from technological and design improvements. Advances in jet engine and aircraft technology have the potential to improve the efficiency of new aircraft (for passenger and freight) by up to 35% over the next two decades. The AEF report indicates that it is feasible to reduce energy consumption in marine shipping by 20-30% by 2020 through a combination of technological innovations (such as improved hull design) and systems improvements (such as speed reduction). Technological improvements could reduce $CO_2$ emissions by 5-30% in new vessels and 4-20% in existing ones.

Combustion of petroleum-based fuels by motor vehicles results in exhaust emissions that include VOCs, $NO_x$, particulate matter, CO, and $CO_2$. Evaporative emissions from the onboard reservoir of unburned fuel can occur while the vehicle is in use or when the engine is turned off.

Vehicle emissions include a class of pollutants referred to as air toxics. These include known carcinogens, such as benzene, and probable human carcinogens, such as formaldehyde and diesel particulate matter. EPA estimates that mobile sources of air toxics account for about half of all cancers attributed to outdoor sources of air toxics. Some toxic compounds occur naturally in petroleum and become more concentrated when petroleum is refined. Others are not present in fuel but are formed as by-products in the vehicle exhaust or formed from reactions of vehicle emissions in the atmosphere.

Lead emissions occur from piston-engine aircraft that use a commonly available aviation gas 100LL (100 octane low lead). Lead is added to 100LL in the form of tetraethyl lead to improve engine performance. Lead is not added to jet fuel that is used in commercial aircraft, military aircraft, or other turbine-engine-powered aircraft (EPA 2008c).

## Modeled Estimates of Life-Cycle Emissions and Damages from Petroleum Use in Light-Duty and Heavy-Duty Highway Transportation

The committee selected VMT as the primary unit for characterizing external damages for highway transportation. Rather than a gallon of fuel, which is difficult to compare because of large variations in energy content, or a joule of delivered energy, which depends strongly on vehicle efficiency, the VMT best characterizes the type of service associated with transportation vehicles. The use of VMT as the functional unit for comparison makes it possible to address the life-cycle impacts of fuel and vehicle technology combinations, which was a key goal for comparing current and future damages for transportation options. There is also the option of using person-VMT, but this option requires assumptions about vehicle passenger loads that confuse the goal.

Modeling damages from the life-cycle emissions attributable to petroleum requires characterization of emission factors for both the life cycle of the fuel and the production and operation of the vehicle. Both GHG emissions expressed as $CO_2$-equivalent and local air-pollution emissions are included. For air-pollution emissions, not only the magnitude of the emissions (per VMT) but also the geographic distribution of the emissions is important. The committee modeled the monetized damages associated with pollutant emissions using the APEEP model. For GHG emissions, for which damages do not depend on the geographic location of release, only the life-cycle $CO_2$-equivalent emissions for the petroleum and vehicle life cycle are reported. Damages for $CO_2$-equivalent emissions are discussed in Chapter 5.

### Emissions Characterization

Emissions characterization included life-cycle emissions for light-duty gasoline- and diesel-fueled vehicles and for heavy-duty diesel vehicles for 2005 and 2030 fuel and vehicle technology combinations. Life-cycle emissions for gasoline- and diesel-fueled light-duty vehicles are obtained primarily from GREET (Argonne National Laboratory 2009) and include emissions of GHGs, VOCs, $NO_x$, $SO_x$ direct $PM_{2.5}$, secondarily formed

$PM_{2.5}$ (from VOCs, $NO_x$, and $SO_x$ emissions) and secondarily formed ozone (from VOC and $NO_x$ emissions). The committee carried out its own analysis to obtain (1) life-cycle $NH_3$ emissions related to $PM_{2.5}$ formation for 2005 and 2030, (2) emissions from gasoline- and diesel-fueled heavy-duty vehicles for 2005 and 2030 (which are not covered in GREET), and (3) estimated emissions for those substances covered in GREET for the year 2030 based on using the most current 2020 data in GREET, further updated to incorporate the expect 35.5 mpg required fuel efficiency after 2016 (see discussion in Scope of the Analysis above and in Appendix D on how this analysis was accomplished). For each pollutant-vehicle mix, emissions per VMT include emissions from (1) feedstock production, (2) fuel production, (3) vehicle operation, and (4) vehicle production (except heavy-duty vehicles). To assess health and other monetized damages, emissions from vehicle operation are allocated to U.S. counties based on the estimated fraction of aggregate U.S. VMT that occur within that county. Emissions for other stages are allocated to regions based on the geographic distribution of the economic activity associated with each specific life stage, for example, the distribution for refineries.

*Results*

Table 3-3 contains a summary of the results from the GREET-APEEP modeling effort related to gasoline and diesel fuels in light-duty autos. Calculations were also carried out for light-duty trucks, but these did not vary significantly from the results for light-duty autos. Each row of Table 3-3 contains the range- and population-adjusted mean for health damages in 2005 and 2030 reported on a VMT basis. There is also a column showing the health costs per gasoline gallon equivalent (gge). It can be seen from this table that year 2005 health impacts do not vary significantly among the fuel-vehicle technology options. Only compression ignition, direct injection using Fischer-Tropsch diesel shows a significant difference from other options, largely due to the more-intense energy use needed to process that type of fuel. (In its analysis, the committee considered only the use of methane for the production of Fischer-Tropsch diesel fuel.) Although damages from 2005 to 2030 would be expected to increase due to population growth, the increase is largely offset in these analyses by the substantial increase in fuel economy to 35.5 mpg by 2016.

Table 3-4 provides a summary of the modeling results from the GREET-APEEP modeling effort related to gasoline and diesel fuels in heavy-duty vehicles. Each row of Table 3-4 lists the range- and population-adjusted mean for health and other non-GHG damages on a VMT basis in 2005 and 2030. A column shows the health and other non-GHG damages per gasoline gallon equivalent for light-duty vehicles. Within the heavy-duty class, larger vehicles have a greater impact per VMT, as is expected. A

**TABLE 3-3** Health and Other Non-GHG Damages from a Series of Gasoline and Diesel Fuels Used in Light-Duty Automobiles[a]

|  | 2005 | | | 2030 | |
|---|---|---|---|---|---|
|  | 5th and 95th Percentile Range[b] (Cents/VMT) | Population-adjusted Mean (Cents/VMT) | Population-adjusted Mean (Cents/gge)[c] | 5th and 95th Percentile Range[b] (Cents/VMT) | Population-adjusted Mean (Cents/VMT) |
| Conventional gasoline (SI, petroleum) | 0.34-5.07 | 1.34 | 29.02 | 0.43-4.87 | 1.35 |
| Conventional gasoline (SI, tar sands) | 0.35-5.36 | 1.35 | 29.26 | 0.45-4.99 | 1.36 |
| Reformulated gasoline (SI, petroleum) | 0.35-5.12 | 1.38 | 29.83 | 0.45-4.87 | 1.35 |
| Reformulated gasoline (SI, tar sands) | 0.35-5.40 | 1.39 | 30.07 | 0.45-4.99 | 1.36 |
| Reformulated gasoline (SIDI, petroleum) | 0.33-4.89 | 1.32 | 32.68 | 0.45-4.96 | 1.37 |
| Reformulated gasoline (SIDI, tar sands) | 0.33-5.14 | 1.33 | 32.92 | 0.45-5.09 | 1.38 |
| CIDI using low-sulfur diesel | 0.30-7.57 | 1.49 | 38.65 | 0.40-4.22 | 1.19 |
| CIDI using Fischer-Tropsch diesel | 0.41-7.77 | 1.80 | 46.65 | 0.58-5.48 | 1.61 |

[a]Costs are in 2007 USD.

[b]From the distribution of results for all counties in the 48 contiguous states in the United States.

[c]Cents/gallon of gasoline equivalent, calculated by multiplying average miles per gallon by per VMT damages. This calculation will therefore show highest damages for the most fuel-efficient vehicles. Costs are in 2007 USD.

ABBREVIATIONS: GHG = greenhouse gas; VMT = vehicle miles traveled; gge = gasoline gallon equivalent; SI = spark ignition; SIDI = spark ignition, direct injection; CIDI = compression ignition, direct injection.

**TABLE 3-4** Health and Other Damages Not Related to Climate Change from a Series of Gasoline and Diesel Fuels Used in Heavy-Duty Vehicles[a]

| | 2005 | | | 2030 | |
|---|---|---|---|---|---|
| | 5th and 95th Percentile Range[b] (Cents/VMT) | Population-adjusted Mean (Cents/VMT) | Population-adjusted Mean (Cents/gge) | 5th and 95th Percentile Range (Cents/VMT) | Population-adjusted Mean (Cents/VMT) |
| HDGV2B Heavy-duty gasoline vehicles class 2B | 1.01-31.89 | 6.14 | 61.39 | 0.36-11.43 | 1.87 |
| HDGV3 Heavy-duty gasoline vehicles class 3 | 1.15-38.82 | 7.23 | 66.47 | 0.41-13.86 | 2.41 |
| HDDV2B Heavy-duty diesel vehicles class 2B | 0.46-18.79 | 3.23 | 41.34 | 0.24-8.63 | 1.23 |
| HDDV3 Heavy-duty diesel vehicles class 3 | 0.51-20.76 | 3.58 | 41.50 | 0.27-9.87 | 1.39 |
| HDDV4 Heavy-duty diesel vehicles class 4 | 0.20-22.83 | 3.90 | 39.40 | 0.29-10.26 | 1.53 |
| HDDV5 Heavy-duty diesel vehicles class 5 | 0.68-31.87 | 5.29 | 51.87 | 0.33-13.47 | 1.76 |
| HDDV6 Heavy-duty diesel vehicles class 6 | 0.88-38.38 | 6.49 | 56.48 | 0.38-15.92 | 1.97 |
| HDDV7 Heavy-duty diesel vehicles class 7 | 1.08-47.53 | 8.01 | 60.08 | 0.45-15.92 | 2.39 |
| HDDV8A Heavy-duty diesel vehicles class 8A | −0.50-56.61 | 9.47 | 61.52 | 0.47-16.77 | 2.53 |
| HDDV8B Heavy-duty diesel vehicles class 8B | −2.20-62.65 | 10.41 | 64.53 | 0.49-16.94 | 2.63 |

[a]Costs are in 2007 USD.

[b]From the distribution of results for all counties in the 48 contiguous states in the United States.

ABBREVIATIONS: VMT = vehicle miles traveled; gge = gasoline gallon equivalent.

significant decrease in impacts is also seen from 2005 to 2030 in spite of rising populations. The decrease is attributable to lower particulate matter and $SO_2$ emissions from heavy-duty vehicles in 2030 relative to 2005. Negative cost estimates represent conditions for which $NO_x$ emissions from vehicles would contribute to a decrease in ambient ozone concentration, when particulate matter emissions are reduced. For the pollutants considered and for these few cases, the negative results reflect benefits within this analytical framework.

Table 3-5 shows how emissions of $CO_2$-equivalent vary among different fuel types, among different vehicle types, and between the years 2005 and 2030 on a VMT basis. Although there is a significant difference between $CO_2$-equivalent emissions from light-duty vehicles and those from heavy-duty vehicles, there is not a significant difference among light-duty

**TABLE 3-5** Carbon Dioxide Equivalent ($CO_2$-eq) Emissions of GHGs from a Series of Gasoline and Diesel Fuels

| Fuel and Vehicle Combination | $CO_2$-eq 2005 g/VMT | $CO_2$-eq 2030 g/VMT |
|---|---|---|
| RFG SI autos (conventional oil) | 552 | 365 |
| RFG SI autos (tar sands) | 599 | 399 |
| CG SI autos (conventional oil) | 564 | 365 |
| CG SI autos (tar sands) | 611 | 399 |
| RFG SIDI autos (conventional oil) | 487 | 366 |
| RFG SIDI autos (tar sands) | 527 | 399 |
| Diesel (low sulfur) | 476 | 372 |
| Diesel (Fischer Tropsch) | 537 | 401 |
| HDGV2B heavy-duty gasoline vehicles class 2B | 1,095 | 1,080 |
| HDGV3 heavy-duty gasoline vehicles class 3 | 1,187 | 1,165 |
| HDDV2B heavy-duty diesel vehicles class 2B | 969 | 957 |
| HDDV3 heavy-duty diesel vehicles class 3 | 1,071 | 1,064 |
| HDDV4 heavy-duty diesel vehicles class 4 | 1,224 | 1,216 |
| HDDV5 heavy-duty diesel vehicles class 5 | 1,262 | 1,255 |
| HDDV6 heavy-duty diesel vehicles class 6 | 1,433 | 1,424 |
| HDDV7 heavy-duty diesel vehicles class 7 | 1,650 | 1,647 |
| HDDV8A heavy-duty diesel vehicles class 8A | 1,903 | 1,882 |
| HDDV8B heavy-duty diesel vehicles class 8B | 2,007 | 1,969 |

NOTE: 2030 estimates assume 35.5 mpg for all light-duty vehicles.

ABBREVIATIONS: GHGs = greenhouse gases; VMT = vehicle miles traveled; RFG = reformulated gasoline; SI = spark ignition; SIDI = spark ignition, direct injection; CG = conventional gas.

vehicles in $CO_2$-equivalent emissions, with the exception that the vehicles fueled with petroleum derived from oil shale had notably higher life-cycle emissions.

Regarding GHG emissions, there is no significant reduction in $CO_2$-equivalent releases per VMT between 2005 and 2030.

## PRODUCTION AND USE OF BIOFUELS

### History and Current Status

It has long been known that alcohols, which are produced from the fermentation of sugars, can be used as a fuel in internal combustion engines. Serious and recent interest in the production of biofuels for transportation was spurred by the oil embargo and petroleum supply disruptions that occurred in the 1970s. This interest in producing biofuels from biomass was of interest because biomass could be grown domestically and could serve as a possible substitution for petroleum. Also, if the crops growing the biomass feedstock were managed properly, it could serve as a renewable fuel—that is, each year, or on some appropriate crops rotation bases, fuels could be continually produced. In recent years, the potential benefit of biofuels to reduce the amount of GHGs per unit energy content of fuel compared with petroleum and other fossil-fuel-based sources of transportation fuels has become another important factor in developing production and vehicle technologies for the use of biofuels.

Ethanol produced from corn is currently the largest and most economically viable biofuel being produced in the United States (biodiesel from soy is the second largest). Ethanol's production has grown over the years stimulated by federal subsidies and rose to a level of about 8 billion gallons per year in 2008. Corn is the primary feedstock in the United States and is converted to ethanol through dry-milling or wet-milling production processes (NRC 2008c). One bushel of corn produces about 2.8 gallons of ethanol. In Brazil, sugar cane is the primary crop used, and an extensive ethanol industry has evolved, producing about 4.5 billion gallons per year to fuel vehicles that can use mixtures of gasoline and ethanol.

### Regulations and Technologies Current and Anticipated in 2030

The Energy Independence and Security Act (EISA) of 2007 stipulates that 36 billion gallons per year of biofuels should be produced and used by 2022, with 21 billion gallons per year produced from cellulose-based technologies beyond an expected corn-based ethanol target for 2015 of 15 billion gallons per year. Both the legislation and energy analysts see cellulosic-based biofuels as the most important long-term feedstock for producing ethanol, with the underlying assumption that in the long term, breakthroughs and bioengineering of organisms might lead to processes that convert cellulose through more advanced production technologies to produce other fuels, such as gasoline, biobutanol, or possibly hydrogen. Whether these targets stipulated in EISA will be realized depend on how quickly the technology for production of biofuels evolves, the cost of such fuels, federal policies, and the economics of the fuel market.

*Biofuel Supply*

The *Liquid Transportation Fuels from Coal and Biomass: Technological Status, Costs, and Environmental Impacts* (NAS/NAE/NRC 2009c) report from the America's Energy Future (AEF) study provides summaries of the likely technologies and growth in the use of coal and biomass liquid transportation fuels until 2020. They identify the primary technologies for converting cellulosic feedstocks (biochemical and thermochemical), discussing many of the technological challenges associated with each.

Table 3-6 lists the main feedstocks that the AEF panel discussed, the time frame in which these feedstocks are expected to be technologically and economically viable, the region of the United States in which they are most likely to be of significant magnitude and a qualitative list of the externalities that may be associated with the production of these feedstocks.

Of the feedstocks identified in Table 3-6, only corn-grain ethanol is in production at a scale that can be viewed as significant and technologically mature. Thus, only reasonable speculations can be made about the other feedstocks and their market location and associated set and magnitude of externalities. Overall, the recently completed analysis of the prospects for these sources by the AEF study estimated that approximately 420 million tons of a variety of such fuels could be produced using technologies available today, and 550 million tons could be produced using technologies expected by 2020 (NAS/NAE/NRC 2009c, Table S-1). It is difficult to accurately project what that will mean in terms of actual fuel produced; indeed, the rapidity with which the corn ethanol market has grown and the volatility of prices mean that neither the technology nor the set of externalities generated by its presence are particularly well understood or appraised.

Given the uncertainties associated with these feedstocks, the committee has identified three that are among the most likely to be relied upon and for which some data are available from which we can produce educated guesses concerning the likely externalities associated with them. The feedstock we focus on for further analysis include the following: corn grain, corn stover, and a perennial grass to produce transportation fuels. These feedstocks represent the current technology (corn grain), a likely mid-term technology (corn stover) and a likely long-term, so-called "second generation" technology (perennial grasses).

**TABLE 3-6** Feedstocks Identified in AEF Report and Partial List of Their Externalities

| Feedstock | Time Frame | Likely Location | Potential Externalities |
|---|---|---|---|
| Corn, grain | Current | Corn Belt | Water quality (nutrients, sediment, pesticides), wildlife habitat, GHG |
| Corn stover | [a] | Corn Belt | Soil erosion and water quality, carbon sequestration in soils, GHG |
| Traditional hay crops (alfalfa and clover) | [a] | Pacific Northwest, Great Plains | Wildlife habitat |
| Perennial grasses Switchgrass Miscanthus Diverse mixes | [a] | Existing CRP land (spread throughout the U.S.), marginal lands, existing crop land | Water quantity, water quality (nutrients, sediment, pesticides), wildlife habitat, GHG |
| Woody biomass (hybrid polar and willow, forest industry residues, fuel treatment residues, forest product residues, and urban wood residues) | [a] | [a] | Forest fires |
| Animal manure | [a] | [a] | Water quality (positive externality if diverted from excess agricultural application or storage spills) |
| Waste paper and paperboard | [a] | [a] | |
| Municipal solid wastes | [a] | [a] | |

[a]An analysis of the potential for these fuels can be found in NAS/NAE/NRC 2009c.

Another biofuel under consideration and in some use is the so-called biodiesel, that is, fuels derived from biomass that can replace diesel fuels for use in diesel engines. Typically, biodiesel refers to fuels produced from crops that contain oils, such as soy beans, which can be converted quite efficiently with well-known processes into diesel fuel. The NRC (2008c) estimates that because of limitations on soy bean production, only about 1.5 billion gallons per year of soy-based diesel fuel could be produced without significant impacts on the food and agricultural markets. Demand for biodiesel greater than that would probably have to be satisfied with imports. Other investigators are pursuing research and development on the production of biodiesel through the growth of algae in algal farms, but there is disagreement on how far from commercial readiness this technology might be. Its potential role probably lies in the 2020-2035 time frame and beyond.

### Fuel Cycle and Externalities

The upstream production externalities of feedstock effects will be location-specific because different feedstocks will be economically viable in different locations (for example, corn stover and switchgrass in the Corn Belt region, miscanthus in warm climates, and trees and forestry in the southeast). In addition, the externalities associated with any given feedstock are also likely to vary by specific field and watershed within a region (such as depending on climate, land-use history, soils, slope of the land, and proximity to water bodies) and can be attenuated by farming practices (such as the use of conservation tillage, nutrient management of both fertilizer and manure applications, and placement of buffers or wetlands).

Finally, transportation of feedstocks to processing facilities is expected to remain expensive even after technological improvements so that numerous, small processing facilities located throughout the region is a likely configuration of the industry. Therefore, externalities associated with production and transportation of the feedstock and liquid fuels will be both site-specific and widespread (the AEF reports that "hundreds of conversion plants, and associated fuel transportation and delivery infrastructure" (NAS/NAE/NRC 2009c, p. 5) will be needed. The AEF report also calls for watershed-specific studies to address the suite of externalities and technological challenges associated with alternative feedstocks.

In characterizing the externalities associated with liquid transportation fuels from biomass, the externalities generated at each of the following stages need to be considered:

1. Production of the feedstock (farm or forest externalities).
2. Transportation of the feedstock to the processing facility.
3. Processing of the feedstock into liquid fuels.
4. Transportation of the fuel to distribution endpoints.
5. Downstream effects of using the fuel.

There may be different external effects and different magnitudes of externalities along each of those steps associated with each type of feedstock. A complete externality accounting would need to include those occurring at each step. The externalities listed in Table 3-6 are those associated primarily with the first step, the production of the feedstock.

The technology associated with transforming alternative feedstocks into fuel is developing for the cellulosic feedstocks and include biochemical and thermochemical conversion processes. In both cases, a large quantity of water is required for processing that, in water-constrained areas, will probably constitute an externality (measured in terms of increased water scarcity via quicker drawdown of reservoirs and increased pumping costs). Water and air emissions will probably be externalities as well.

## The Sources of Externalities in Production of Feedstocks

The three feedstocks that the committee targeted for analysis all require land for their growth and production. Eventually, all three feedstocks as well as a mixture of others may be used to produce biofuels and may compete with each other for land and profitability or may be located in different regions of the country. Briefly, the externalities investigated and the way in which feedstock can generate the externality are described next.

*GHG Emissions* The production and harvesting of corn generates GHG emissions in a number of ways, including the use of fuel for tillage, planting, applying inputs (nutrients and pesticides), harvesting, and shipping of the product. By tilling the soil in preparation for planting, carbon that is stored in soil is released into the atmosphere. Farmers that practice conservation tillage (one of many forms of reduced or no tillage) generally increase the carbon stored in the soil (carbon is sequestered), but this tillage practice is not profitable for all farmers and depends on the characteristics of the land, climate, and crop grown. Currently, regardless of tillage practice, corn stover is left to decompose and rebuild carbon and other nutrients in the soil.

If corn stover were to be used for ethanol production, it would be removed from the soil and therefore not left to decompose and rebuild the soil. Agronomists and others debate about how much stover can be removed to maintain soil productivity, but there is no reason to believe that soil carbon storage does not decline immediately as stover is removed (although the magnitude could be quite small). Thus, on any given field, biofuel production using stover can be expected to have the same GHG emission consequences associated with planting and harvesting corn as just described, with additional losses of carbon sequestration. There may be additional fuel usage needs for the stover to be harvested, and almost certainly there will be high fuel needs for the transportation of stover from the field to the processing facility.

Switchgrass or other perennial grass will not need annual planting or tilling once established, and, hence, should have lower fuel usage and corresponding GHG emissions than corn production. The degree to which switchgrass or other perennial will be fertilized is unclear. Large-scale commercial production of switchgrass is not currently viable, so the amount of inputs that farmers will use to maximize their profitability of growing this crop is unclear. Heggenstaller et al. (2009) provides estimates of fertilization that maximizes profitability.

A number of studies have looked at the life-cycle emissions of GHGs associated with ethanol produced from various feedstocks. Delucchi (2006) provides such an analysis and a review of earlier analyses. Estimates by NAS/NAE/NRC (2009c) of well-to-wheels $CO_2$-equivalent emissions in tonnes per barrel gasoline equivalent[7] are the following: for petroleum-based gasoline, 0.40; for corn ethanol, 0.22; for biochemical cellulosic ethanol, –0.02; for thermochemical coal and biomass conversion, 0.5, and with carbon capture and storage (CCS), –0.19; and for thermochemical biomass conversion, –0.12, and with CCS, –0.95. Thus, there could be significant GHG benefits for biomass fuels as well as a reduced dependence on imports, although the projections by NAS/NAE/NRC (2009c) indicated that biofuels could replace only a proportion of what is consumed in the transportation sector. The impacts on reducing $CO_2$ emissions could be significant, especially if CCS technology is developed between now and 2020, becomes ready for commercial deployment, and can be coupled to some of the technologies, such as gasification-based systems.

The committee used estimates from GREET to generate estimates of the GHG emissions from alternative feedstocks.

*Water Quality and Soil Erosion*    Corn is a heavily fertilized crop with large water demands. The major water-quality issues related to corn production include the runoff of nitrogen, phosphorus, and sediment. The amount of nutrients and sediment that leave a field can vary greatly depending on the slope of the land, climate (particularly heavy rains), field tiling, cropping history, soil type, tillage practice, and a variety of other factors. Thus, to undertake a careful analysis of the water-quality consequences of corn production, one must know where the additional corn that would be used for feedstocks would be produced as well as such information as how the field is managed and whether any conservation practices are in place. To further complicate the issue, the amount of pollutant that enters a waterway and how far it moves within a waterway depend on a variety of geologic and hydrologic factors.

Additional corn to produce biofuels can come from producing more corn on land that is already in corn production. This can occur by the use

---

[7]One tonne is equal to 2,200 pounds.

of more inputs or by changing rotation practices, for example, by moving from a 2-year corn–soybean rotation to the continuous planting of corn. Although continuous corn planting has lower yields than rotated corn, if price differentials are high enough, it will be profitable for farmers to grow corn more often in their rotations. The second way that additional corn can be produced is to grow it on land that was previously not in agricultural production or that was used for a lower valued crop. A major potential source of such land is land that has been placed in the Conservation Reserve Program (CRP), a federal program that pays farmers to have idle land. About 5% of agricultural land nationwide is enrolled in the CRP; this land tends to be of lower value and higher environmental sensitivity than average. The water-quality effects of additional corn production will depend on how and where the additional corn is generated; that in turn will depend on the profitability of corn production for biofuel usage. There is evidence that the higher corn prices experienced in the past 2 years or so has resulted in increased conversion of CRP to working land (Secchi et al. 2009).

If corn stover is used to produce biofuels, the same set of water-quality externalities described above will apply, but will be magnified for two reasons. First, when stover is left on the land, it acts to reduce soil erosion and helps to retain nutrients on the land (especially phosphorus). Second, when stover can be sold for biofuels, the overall profitability of corn production will rise (because the ears can still be sold for biofuels or feed), thus making corn more profitable and increased production more likely.

The water-quality effects associated with perennial grasses will also depend on the location in which they are grown, but their perennial nature and lower input use (although again without evidence of how these crops will be commercially grown, this is difficult to gauge) should translate into lower water-quality impacts than corn-production impacts.

*Wildlife Habitat and Biodiversity*   The effect on wildlife and biodiversity from using more land for corn or perennial grass production will depend on how the land was used prior to production (for example, whether a different row crop was planted, left idle in CRP, or used as pasture). Perennial grasses are more likely than corn to be suitable habitat for more wildlife, but may be less suitable than the land use prior to biofuels production.

A number of other externalities related to biofuels production and industry expansion should be noted. First, a significant expansion of the industry will require a major increase in production facilities that will generate externalities associated with the building and maintaining of these facilities. Depending upon the technology used to convert feedstocks to ethanol, there may be solid waste or other pollution externalities associated with the ongoing production of ethanol in these facilities. There also are potentially significant concerns about water consumption and ethanol production. For an extensive discussion, see the recent National Research Council report *Water Implications of Biofuels Production in the United States* (NRC 2008d).

## Indirect Land Use and Externalities

The role of "indirect land use"—changes occurring indirectly as a result of biofuels policy in the United States and the effect of such changes on GHGs—has been a major source of discussion since a paper by Searchinger and colleagues was published in *Science* in 2008. The argument put forth by Searchinger et al.(2008) is that when demand for corn or farmland in general increases, crop prices increase, making it profitable for farmers to increase their acreage. If this increased acreage comes from plowing up land that has not been in agricultural production and is particularly environmentally sensitive (for example, rainforests in Brazil and pristine ecosystems in the United States), GHG emissions could increase (for example, burning rainforests would release large amounts of carbon) and have other detrimental environmental concerns. The loss of Brazillian rainforests due to these market pressures is particularly cogent, but the issue of increased GHG emissions applies to a variety of land-use changes as long as the land that is brought into production to grow biofuel feedstocks results in lower carbon storage.

Under the requirements of EISA, EPA recently released its revised Renewable Fuels Standards (RFS2). As mandated, EPA performed its life-cycle computation of GHG contributions of corn ethanol, two types of biodiesel, and three cellulosic ethanol feedstocks (sugarcane, switchgrass, and corn stover) using indirect land-use effects as a component of the GHG contribution. In recognition of the uncertainty associated with measuring indirect land-use effects, EPA presented its emission estimates both with and without the indirect effects (see EPA [2009h] for a summary of its life-cycle analysis). The effect of indirect land uses can be quite large; in some cases, EPA's analysis suggests that significant positive gains in GHG of a fuel relative to gasoline could be largely offset by indirect land use changes. The state of California has also adopted the approach of including indirect land-use effects in its fuel standards.

The committee's task in this report is to identify and monetize the externalities associated with energy production and consumption. We discussed whether these externalities should include both the direct and the indirect land-use effects and chose to report only the direct land-use effects (as captured in GREET). In doing so, we by no means dismiss the potential importance of indirect land-use effects in policy design, but we do not wish to treat externalities associated with the production of biofuels any differently than the externalities associated with the production of other fuels.

Why did we come to this choice? First, there is an important distinction between the externalities associated with the direct use of land to grow crops for biofuels and the externalities associated with the indirect effects. The indirect effects are induced by price changes and are associated with the production of a second product. To avoid double counting, it is important that the externalities associated with the indirect effects be associated with the second product and not have both assigned to the first product.

For example, when a crop is planted and grown to produce a gallon of ethanol, there are externalities (such as GHGs and changes in water quality) associated with its production; these externalities of course are appropriately counted against the production of that gallon of ethanol. These externalities include the direct land-use effects. In contrast, the indirect effects occur from a market response due to some price changes. When the price of a biofuel crop increases due to a policy that promotes biofuels, farmers elsewhere will find it profitable to plant that crop, which will then be used to produce a second product (perhaps another gallon of biofuels or a food product). This "indirect effect" will generate externalities, but these externalities should be associated with the second product, not the first.

In the specific context of the biofuels land-use debate, the lost carbon and ecosystem services from indirect land-use changes are appropriately viewed as an externality from growing crops elsewhere, say Brazil, not from production of biofuels in the United States. Or if these indirect land uses occur within the United States, they would already be counted as the direct land-use effects of growing biofuels for carbon in that second location. Thus, when estimating the externalities associated with U.S. biofuel production, analysts shouuld include the externalities associated with direct land-use changes to produce the feedstocks, but not the market-induced indirect effects.

The second reason we do not attempt to incorporate indirect land-use effects is that if we were to do so, for consistency we would need to include all market-induced changes in externalities that could be linked to any other energy source. For example, an increase in the price of electricity generated by an expanded electric-vehicle requirement could result in more people using wood-burning stoves in lieu of electric heaters, more usage of gas-powered lawn mowers, and earlier turning out of lights in the evening. The first two changes would increase the negative externalities of smog, GHGs, and noise, whereas the third would reduce light pollution. The accounting of the indirect-effects argument would be to add all of the effects of these externalities on to electric vehicles. These are just a few of the externalities that could be induced by price changes.

The fact that there are two separate externalities associated with production at two locations is not merely an academic distinction; it is critical to keep them separate to avoid double counting and therefore to inform policy making appropriately, as the second set of externalities may be policy irrelevant. For example, if GHGs worldwide were subject to a tax, then it would be appropriate to tax agricultural crops in the United States based on their GHG emissions and to tax agricultural crops grown elsewhere separately based on their emissions. In this case, it would be inefficient to tax U.S. agricultural crops for the sum of their own emissions plus those associated with land-use changes elsewhere—this would be double counting. Ideally, the U.S. policy would correct the market distortion for the production of externalities for crops grown in the United States, and other governments would do the same. On the other hand, if policy makers in the United States wish to set policy recognizing that GHGs are not optimally regulated elsewhere, then it may be appropriate to tax or regulate U.S. biofuel crops based on more than the direct externalities, taking into account some or all of the indirect externalities induced by market prices. In economics, this would be called policy design in a second-best setting. In this policy design, it generally would not be appropriate to add the damages from the indirect externalities to the direct externalities to form a basis for a tax. For purposes of this report, we do not attempt to explicitly inform decision making in a second-best setting, despite the presence of many distortionary tax elements in the U.S. economy (such as labor taxes and imperfect competitive sectors).

The committee's goal throughout this report is to define and estimate the externalities associated with the production of energy sources. By providing estimates of the direct effects of land use (as reported in GREET), we are providing an estimate of externalities that are consistent with those presented elsewhere. We recognize the important issue of indirect land uses, but we do not evaluate or incorporate them in our analysis.

## Land-Use Externalities from Biofuels: A Case Study of the Boone River Watershed

Given the relatively recent broad interest in biofuels, studies that assess the magnitude and value of externalities related to direct land-use changes and soil carbon provide incomplete coverage of the issues, particularly at the local landscape level where these effects may vary considerably across locations. A number of studies provide information on components of the externalities related to water quality. For example, Donner and Kucharik (2008), Simpson et al. (2008), and Secchi et al. (2009) examine the consequences of expanded corn production to produce ethanol and the amount of nitrogen and phosphorus entering the Gulf of Mexico, therefore potentially contributing to the recurrent hypoxic zone there. Other work addresses the consequences of higher corn prices on conservation reserve lands, and concerns have been expressed about the loss of habitat and local water quality. Much is still unknown about the set of externalities that a particular region or watershed might be expected to experience with expanded ethanol production.

To demonstrate one approach for estimating some of the externalities that are location-specific, the committee used an existing set of data and models for the Boone River watershed in central Iowa to perform a case study. The estimates for the externalities described here relate to water quality (nutrients and sediment). We stress that this exercise is meant to shed light on the process and approach needed to estimate these externalities associated with ethanol production rather than to provide firm estimates. Further, the estimates are unlikely to be transferable to other regions where biofuels may be produced and to other feedstocks grown for biofuel production.

To evaluate the water quality and carbon sequestration externalities associated with biofuels production in the Boone watershed, we analyzed three possible feedstocks: corn grain, corn stover, and switchgrass. To do so, we used a biophysical model, EPIC, to estimate the nitrogen, phosphorus, and erosion changes associated with different agricultural land uses and management at the field scale, then aggregated these to the watershed level. The EPIC model (Williams 1990, 1995; Williams et al. 1984, 1996, 2008) was designed with this purpose in mind, specifically to estimate the impacts of different cropping and management systems on a variety of environmental indicators, including soil erosion, nutrient losses, and soil carbon levels. EPIC is a field-scale model that functions on a daily time step and can simulate a wide range of crop rotations, tillage systems, and other management practices. More detailed discussion on modeling analysis is provided in Appendix E.

*EPIC Results*

Table 3-7 provides estimates of the average amounts of erosion, nitrogen, and phosphorus, and the amount of soil carbon sequestered for the baseline and for each of the scenarios. Recall that carbon sequestration is a positive externality where the nutrients (nitrogen and phosphorus) and sediment are negative externalities. It is also important to recall that EPIC is an "edge-of-field" model in that it predicts the amount of nutrients and sediment removed from each field under each scenario, but this does not necessarily mean that the pollutants will enter the waterways. (A fate-and-transport model that incorporates the hydrology of the region would be needed to estimate the waterway loadings.) For conciseness and ease of interpretation, several of the environmental indicators generated by EPIC have been combined. Specifically, the column entitled "Erosion" represents the sum of water and wind erosion predicted by EPIC. Likewise, "Nitro-

**TABLE 3-7** Water Quality and Externalities Estimated for Ethanol Scenarios[a]

|  | Erosion (tons/acre)[b] | Nitrogen (kg/acre)[c] | Phosphorus (kg/acre)[d] |
|---|---|---|---|
| Baseline | 0.31 | 20.11 | 0.29 |
| Corn stover: 50% | 0.44 | 19.62 | 0.35 |
| Corn stover: 80% | 0.69 | 21.09 | 0.48 |
| Corn stover: 100% | 1.23 | 24.53 | 0.72 |
| Corn, continuous planting | 0.45 | 30.68 | 0.29 |
| Corn stover, continuous planting: 50% | 0.78 | 29.12 | 0.43 |
| Corn stover, continuous planting: 80% | 1.16 | 30.46 | 0.61 |
| Corn stover, continuous planting: 100% | 1.55 | 32.19 | 0.79 |
| Switchgrass: 25% | 0.23 | 26.11 | 0.24 |
| Switchgrass: 50% | 0.16 | 31.93 | 0.18 |
| Switchgrass: 75% | 0.08 | 37.93 | 0.13 |
| Switchgrass: 100% | 0.01 | 43.79 | 0.08 |

[a]All values are annual averages.

[b]Erosion reports the sum of wind and water erosion.

[c]Nitrogen reports the sum of nitrogen loss with sediment, nitrate loss with runoff, and nitrate leached.

[d]Phosphorus reports the sum of the loss with sediment and runoff (labile phosphorus).

gen" represents the sum of soluble N loss, N leaching, and N loss via sediment. Although the pathways by which N leaves the field differ in each of those cases, most of the N losses that ultimately escape the crop fields and drainage ways will enter surface water because of the subsurface tile drains that capture the majority of leached N, so the aggregated N amounts are reasonable representations of the overall system losses. Finally, the numbers in the "Phosphorus" column represent the sum of sediment-bound and soluble phosphorus that is transported in surface runoff.

As expected, the water-quality externalities increase relative to the baseline when continuous corn planting becomes the predominant cropping system. This result reflects the fact that corn has high input requirements and is relatively "leaky."

In each of the first three stover scenarios, it is assumed that the baseline crop rotation is maintained but that some or all of the above-ground biomass is harvested for biomass to be used in ethanol production. Because the removal of stover (biomass) will generally increase erosion, three levels of removal are simulated for comparison: 50%, 80%, and 100%. As can be seen, model results predict that the average erosion per acre will increase from just under 1/3 ton/acre in the baseline to .45 tons/acre under a 50% removal, and well over 1 ton under 100% removal. The changes in nitrogen export are much less dramatic, which is expected, but larger for phosphorus. This result is expected, given that the majority of nitrogen is transported in the soluble phase while the phosphorus moves mainly with sediment.

Because stover removal could also occur under continuous corn planting, the committee evaluated the same three scenarios under continuous corn planting. The combination of continuous corn planting and stover removal at any of the three rates has fairly dramatic effects on the magnitude of both rates of erosion and phosphorus loss with the rate of nitrogen loss being lower.

The final four scenarios all relate to switchgrass produced as a feedstock. In this case, we evaluated four alternative levels of switchgrass planting in the watershed: 25%, 50%, 75%, and 100% of the acreage converted to the switchgrass production. The substitution of this perennial has notable effects on the erosion rates as well as on nutrient loss.

It is worth reiterating that the edge-of-field sediment loss indicators reported here cannot capture the complex watershed scale and in-stream sediment movement dynamics that have been reported in previous studies, such as Trimble (1999) and Simon and Rinaldi et al. (2006). Similar caution is stressed for the edge-of-field nutrient indicators.

*Ethanol Production and Monetization* Each of the scenarios presented are associated with different amounts of potential ethanol production. In Table 3-8, the committee presents estimates of the amount of ethanol that the feedstocks grown in the Boone watershed could produce so that the magnitude of the externalities reported can be compared with the fuel production with which they are associated. The first column of the table provides estimates of the total amount of ethanol that the identified scenario could produce, including the baseline. In each case, the predicted yield of corn grain is assumed to be convertible to ethanol at a rate of 105 gallons/metric ton. The predicted stover removed for biomass in the stover scenarios is assumed to be converted to ethanol at a rate of 100 gallons/metric ton. This is the same rate used for the switchgrass scenarios. These are the same values assumed in the GREET model transportation runs used in the rest of this report and are chosen for internal consistency.

The second column of the table shows the incremental amount of ethanol the scenario is predicted to produce above and beyond the production in the baseline. When the land use is changed to produce additional ethanol, it creates additional externalities. By computing the additional ethanol produced, those incremental externalities (a cost to society) can be compared with the incremental ethanol (a gain).

To demonstrate the monetization of land-use externalities, we focused on the erosion estimates reported in Table 3-7. We chose to monetize erosion only for several reasons. First, more information about the costs of

**TABLE 3-8** Estimated Ethanol Production from Feedstocks in the Boone River Watershed

| Scenarios | Potential,[a,b] Including Baseline Corn (gal/year) | Potential[c] Increment Over Baseline (gal/year) |
|---|---|---|
| Baseline | 112 | |
| Corn stover: 50% | 167 | 55 |
| Corn stover: 80% | 196 | 85 |
| Corn stover: 100% | 214 | 103 |
| Corn, continuous planting | 217 | 105 |
| Corn stover, continuous planting: 50% | 325 | 213 |
| Corn stover, continuous planting: 80% | 384 | 272 |
| Corn stover, continuous planting: 100% | 421 | 309 |
| Switchgrass: 25% | 150 | 39 |
| Switchgrass: 50% | 187 | 75 |
| Switchgrass: 75% | 226 | 115 |
| Switchgrass: 100% | 264 | 152 |

[a]These values assume that 105 gallons of ethanol can be produced per dry metric tonne of grain and 100 gallons/metric tonne of stover or switchgrass (GREET default values). Values in this column represent all the corn in the baseline that is used to produce ethanol as well as the addition of stover, corn, or switchgrass assumed in the scenario.

[b]Multiply the number of ethanol gallons by 0.6575 to convert to the gasoline gallon equivalent. That is the conversion factor used in the GREET model.

[c]Values in this column represent the additional ethanol produced by the scenario beyond the baseline: 0.6575.

erosion is available relative to the damages from nitrogen and phosphorus. Further, phosphorus and sediment tend to move together; therefore, the estimates of damages from erosion are already likely to include some of the costs associated with phosphorus. Likewise, the water-quality damages from all three (nitrogen, phosphorus, and erosion and sediment) are likely to be interrelated, and if separate values were added together for all three, we would risk double counting.

In a recent report, Hansen and Ribaudo (2008) provided a summary of studies that have valued erosion damages (or benefits from erosion reduction) from agricultural sources for numerous categories. They provided dollar per ton estimates of erosion reductions by 8-digit watershed code (the Boone River watershed represents HUC 07100005) for the following categories: sedimentation in reservoirs, navigation, water-based recreation, irrigation ditches, road drainage, municipal water treatment, flood damages, marine and freshwater fisheries, marine recreational fishing, municipal and industrial water use, and steam power plants. Estimates appropriate for the Boone River watershed indicate that the value of a 1-ton reduction in erosion is $4.43 (2007 USD). Hansen and Ribaudo noted that these values omit some potentially important categories of benefits, including effects on wetlands, endangered species, coastal recreation, and existence values, and they suggested that the numbers be viewed best as a lower bound.

Table 3-9 uses this value to monetize the erosion reductions on a per acre basis, and, in the final column, on a per gallon of ethanol basis. The scenarios that remove stover for ethanol production have fairly high costs when aggregated to the watershed level, particularly when stover removal is combined with continuous planting of corn. However, even in those cases, the costs on a per gallon of ethanol basis are quite small, averaging less than 1 cent per gallon in all cases except for 100% stover removal. It is worth bearing in mind that these results represent the externality costs associated with erosion only and are probably underestimated. We also note the need for enhanced capabilities for simulation of $N_2O$ and other GHG emissions in EPIC; such capabilities are now being tested and will be included in future releases of EPIC (Izaurralde et al. 2006). Nonetheless, the health-effect damages considered elsewhere in this report are significantly greater.

The scenarios that introduce switchgrass into the landscape yield gains in erosion—that is, total erosion is reduced relative to the baseline cropping pattern and, therefore, the costs are negative (a benefit). At a watershed level, the value of the benefits seems relatively large, while on a per gallon basis, these gains are again quite small.

**TABLE 3-9** Monetized Land-Use Damages of the Boone River Case Study[a]

| | Erosion Loss/Acre | $/Acre | $/Watershed | Damages $/gal Ethanol | Damages $/gge |
|---|---|---|---|---|---|
| Corn stover: 50% | 0.13 | $0.49 | $261,427 | $0.005 | $0.003 |
| Corn stover: 80% | 0.38 | $1.41 | $752,857 | $0.009 | $0.006 |
| Corn stover: 100% | 0.93 | $3.43 | $1,828,204 | $0.018 | $0.012 |
| Corn, continuous planting | 0.14 | $0.52 | $278,084 | $0.003 | $0.002 |
| Corn stover, continuous planting: 50% | 0.47 | $1.74 | $929,355 | $0.004 | $0.003 |
| Corn stover, continuous planting: 80% | 0.86 | $3.17 | $1,690,837 | $0.006 | $0.004 |
| Corn stover, continuous planting: 100% | 1.25 | $4.61 | $2,459,075 | $0.008 | $0.005 |
| Switchgrass: 25% | −0.08 | −$0.28 | −$149,076 | −$0.004 | −$0.003 |
| Switchgrass: 50% | −0.14 | −$0.53 | −$284,211 | −$0.004 | −$0.003 |
| Switchgrass: 75% | −0.23 | −$0.83 | −$444,829 | −$0.004 | −$0.003 |
| Switchgrass: 100% | −0.30 | −$1.09 | −$581,777 | −$0.004 | −$0.003 |

[a]Erosion monetized at $3.70 (2000 dollars). See Hansen and Ribaudo (2008), Appendix 1.

ABBREVIATION: gge = gasoline gallon equivalent.

### Modeled Estimates of Life-Cycle Emissions, and Damages from Biofuel Use in Light-Duty Vehcle Highway Transportation

Table 3-10 contains a brief summary of the modeling results from the GREET-APEEP modeling effort related to biofuels. The first row of the table contains the range and population-adjusted mean for conventional gasoline vehicles for 2005 and 2030, reported on a VMT basis. The remaining rows contain the same information for the three feedstocks (dry corn, herbaceous crops, and corn stover) used in production of E10 and E85, respectively.

The estimates do not differ significantly across the feedstock types, nor do the ethanol blends differ significantly from conventional gasoline. Given that only dry corn as a feedstock is truly a proven technology, the small differences in either the range across counties or the population-adjusted

**TABLE 3-10** Comparison of Health and Other Non-GHG Damages from Conventional Gasoline to Three Ethanol Feedstocks[a]

|  | 2005 | | | 2030 | |
|---|---|---|---|---|---|
|  | 5th and 95th Percentile Range[b] (Cents/VMT) | Population-adjusted Mean (Cents/VMT) | Population-adjusted Mean (Cents/gge)[c] | 5th and 95th Percentile Range[b] (Cents/VMT) | Population-adjusted Mean (Cents/VMT) |
| Conventional Gasoline | 0.34-5.07 | 1.34 | 29.20 | 0.45-4.87 | 1.35 |
| E10 (dry corn) | 0.35-5.26 | 1.35 | 29.18 | 0.44-4.87 | 1.32 |
| E10 (herbaceous) | 0.33-5.06 | 1.30 | 28.09 | 0.43-4.66 | 1.30 |
| E10 (corn stover) | 0.33-5.08 | 1.30 | 28.10 | 0.43-4.71 | 1.30 |
| E85 (dry corn) | 0.57-7.31 | 1.52 | 32.90 | 0.56-5.84 | 1.39 |
| E85 (herbaceous) | 0.40-5.45 | 1.20 | 25.89 | 0.47-4.06 | 1.22 |
| E85 (corn stover) | 0.39-5.78 | 1.21 | 26.13 | 0.47-4.63 | 1.22 |

[a]Costs are in 2007 USD.

[b]From the distribution of results for all counties in the 48 contiguous states in the United States.

[c]Cents/gge, calculated by multiplying average miles per gallon by per VMT damages. This calculation will show highest damages for the most fuel-efficient vehicles. Costs are in 2007 USD.

ABBREVIATIONS: GHG = greenhouse gas; VMT = vehicle miles traveled; gge = gasoline gallon equivalent.

mean should not be given much attention. Even the somewhat higher estimate for dry corn E85 of 1.52 cents is likely to contain enough error that
it should not be viewed as distinctly different from the other feedstocks of
conventional gasoline.

Several factors contribute to the aggregate damage estimates being
similar for ethanol blends and gasoline in Table 3-10. The GREET model
calculated similar estimates of vehicles emissions for all fuels shown in the
table; thus, the operational component of the aggregate damages are the
same. Because the E10 fuel is only 10% ethanol and 90% gasoline, similar
damage estimates were obtained across the entire life cycles for E10 and
gasoline. The damage costs for E85 (herbaceous and corn stover) are the
lowest for any of the vehicle-fuel life cycles when looking at the population-
adjusted means. A main reason is higher vehicle-fuel damages attributable
to the feedstock and fuel components of the other vehicle-fuel life cycles.

To aid in comparisons with other studies and policy uses, we converted
the costs/VMT into an equivalent costs per gallon. The mean-adjusted
population costs computed in cents/gge are reported for the 2005 results
in Table 3-10. These units are the same as those that Hill et al. (2009)
used to summarize their findings. A comparison of the results in the table
for 2005 with theirs is instructive. Hill et al. (2009) also use the GREET
model to estimate the health effects associated with conventional gasoline
and various forms of ethanol. They report estimates of health costs from
gasoline averaging $0.34/gallon. They contrast this estimate with estimates
of ethanol ranging from $0.16 for ethanol produced from prairie grasses
to $0.93 for ethanol produced from corn by using coal as the process heat.
As can be see via comparison with the results in Table 3-10, their estimates
are generally higher and somewhat more discouraging for corn ethanol
than our estimates.

One difference is that their results correspond to 2010 rather than our
2005 baseline. More important, the results that we report include emissions
from feedstock production, fuel production, vehicle operation, and vehicle
production. In contrast, Hill et al. (2009) focused only on fuel production
and use and did not consider vehicle production.

## ELECTRIC VEHICLES

### History and Current Status

The late 1990s saw the emergence—in large measure in response to so-
called zero-emission vehicle requirements of the California Air Resources
Board (CARB)—of both a small number of all-electric vehicles and the
first gasoline hybrid vehicles. Although the all-electric vehicles did not
continue in production, gasoline hybrid vehicles have continued to develop
and spread in the marketplace, more recently because of higher gasoline

prices and substantial tax incentives. Currently, such vehicles constitute approximately 1-2% of the U.S. light-duty vehicle fleet. Recently, there has been increased interest in developing different versions of "plug-in" hybrid electric vehicles (PHEVs) (which some are calling "extended-range electric vehicles"), although, other than after-market conversions, there are few such vehicles on the market.

There are two primary advantages that are usually cited for PHEVs. First, they will use electricity to power a portion of a vehicle's energy requirements and thus avoid some fraction of petroleum that would otherwise be consumed. This vehicle would presumably lead to reductions in petroleum imports. Second, although there would be some impact on emissions from electric power plants, vehicle emissions would be reduced especially in metropolitan and urban areas.

### Regulations and Technologies: Current and Anticipated in 2030

Although there are no formal national requirements for increased use of such vehicles (in the manner of the Renewable Fuel Standards that require increased use of biofuels), there are a number of regulatory and incentive programs that have the potential to affect the use of such vehicles. These programs have been put in place to address multiple objectives, including energy efficiency, reduced dependence on imported petroleum, and reduced GHG emissions. They include the following:

•   Continued regulation by CARB (and other states) requiring some number of so-called partial zero-emission vehicles (PZEV) as well as the pending CARB regulations for GHG emissions (which many other states have proposed to adopt as well).
•   Substantial tax credits for purchase of such vehicles, which, although they have been exhausted for some manufacturers (for example, Toyota), are still available for others (and could be revised and extended).
•   Substantial government-supported research and development of advanced battery technologies.

NAS/NAE/NRC 2009d estimates that gasoline and plug-in hybrids are likely to play an important role in the 2035 time frame that the committee is considering. (15-40% and 7-15%, respectively; see Table 3-11). Strictly speaking, the gasoline hybrids are more of a fuel-efficiency improvement than a new technology, placing new demands on the electricity grid. However, several important parts of the pathway described below concerning batteries are also relevant to this technology, especially if it expands dramatically.

On the basis of the AEF analysis, a significant market penetration of

**TABLE 3-11** Plausible Light-Duty-Vehicle Market Shares of Advanced Vehicles by 2020 and 2035

| Propulsion System | Plausible LDV Market Share by | |
|---|---|---|
| | 2020 | 2035 |
| Turbocharged gasoline SI | 10-15% | 25-35% |
| Diesels | 8-12% | 15-30% |
| Gasoline hybrid vehicles | 10-14% | 15-40% |
| Plug-in hybrid vehicles | 1-3% | 7-15% |
| Hydrogen fuel-cell vehicles | 0-1% | 3-6% |
| Battery electric vehicles | 0-2% | 3-10% |

ABBREVIATIONS: LDV = light-duty vehicle; SI = spark ignition.

SOURCE: NAS/NAE/NRC 2009d.

either fuel-cell or full-electric-battery vehicles is unlikely within the 2030 time frame.

### Technology and Fuel Pathways

Facilities involved with manufacture and assembly of motor vehicles are located throughout the United States, but many are clustered in the Great Lakes states, California, and Texas. Manufacturing and assembling the thousands of different parts that make up motor vehicles include the following processes: raw material recovery and extraction, material processing and fabrication, vehicle component production, finishing or electroplating metal surfaces, painting the vehicle body, vehicle assembly, and vehicle disposal and recycling. These processes are energy- and material-intensive, involving components made of metal (for example, steel, aluminum, or copper), glass, rubber, plastics, and fluids. Energy is required to transport the raw and processed materials along each process step. Some of the material production and transport takes place outside the United States.

Waste streams are generated by manufacturing and assembly facilities as a result of fuel combustion, materials used in processes that are not shipped out in product streams, and chemical reactions occurring within specific processes. Air pollutants include particulate matter, VOCs, $SO_2$, $NO_x$, and CO. GHG emissions are also produced. In addition, various manufacturing processes generate sludge or wastewater that contains toxic metals (for example, cadmium, lead, and chromium), oils, acids, and solvents.

The fuel cycle and potential effects pathways for electric vehicles are similar to other vehicles in a few respects (for example, manufacture of the vehicle) but substantially different in nearly all other respects. Major components of those pathways (for example, see Axsen et al. 2008; Samaras and Meisterling 2008) are the following:

• *Natural Resource Extraction.* The expanded use of electric batteries is likely to increase demand significantly for certain metals that come from relatively limited sites (some in unstable regions). The metals include lithium (major stocks in the Congo and Russia) and cobalt (major stocks in Bolivia). This use may pose national security costs (although they might not be an externality per se). It also would involve significant increases in worker exposure and emissions associated with transport.

• *Displacement of Imported Oil.* Increased use of hybrids could reduce dependence on imported petroleum. For example, a study by the Pacific Northwest National Laboratory (PNNL) (Kintner-Meyer et al. 2007) made aggressive projections for the introduction of PHEVs and estimated that "a shift from gasoline to PHEVs could reduce the gasoline consumption by 6.5 MMBpd, which is equivalent to 52% of the U.S. petroleum imports" (Kintner-Meyer et al. 2007).

• *Battery Manufacture.* This poses issues of worker exposure to metals as well as a potential for both conventional and GHG emissions from the manufacturing process. Table 3-12 provides estimates of the use of energy for the manufacture of batteries and other vehicle-related technologies.

• *Electric Power Grid Implications.* The PNNL study of the current capabilities of the electric power system in the United States analyzed 12

**TABLE 3-12** Energy Use During Vehicle Manufacturing and Disposal of Light-Duty Vehicles

| Propulsion System | Energy (gigaJoules/Vehicle)[a] |
|---|---|
| Current gasoline | 97-125 |
| Current diesel | 99-128 |
| Current gasoline hybrid | 114-144 |
| 2035 gasoline | 115-159 |
| 2035 diesel | 117-152 |
| 2035 plug-in hybrid vehicle (PHEV) | 138-175 |
| 2035 battery electric vehicle (BEV)[b] | |
| 2035 hydrogen fuel-cell vehicle (FCV) | 158-203 |

[a]Rounded estimates are presented. Lower values in each range are for cars; upper values are for light-duty trucks

[b]GREET 2.7 does not have the capability to estimate the BEV vehicle cycle impact accurately. The future versions of this model may include this capability.

SOURCE: Bandivadekar et al. 2008.

regions and estimated how many PHEV-33 vehicles could be supported and what impact they might have, for example, on emissions (summarized in Kintner-Meyer et al. 2007). This study was not a dynamic analysis, and there was no estimate of the market penetration of such vehicles. In some ways, it was a maximum estimate of what could be. Their conclusions were the following:

> The existing electricity infrastructure as a national resource has sufficient available capacity to fuel 84% of the nation's cars, pickup trucks, and SUVs (198 million) or 73% of the light-duty fleet (about 217 million vehicles) for a daily drive of 33 miles on average.

> Several other grid-related impacts are likely to emerge when adding significant new load for charging PHEVs. Higher system loading could impact the overall system reliability as the entire infrastructure is utilized near its maximum capability for long periods. "Smart" PHEV charging systems that recognize grid emergencies could mitigate the extent and severity of grid emergencies. Near maximum utilization of the nation's power plants is likely to affect wholesale electricity markets. The mix of future power-plant types and technologies may change as a result of the flatter load-duration curve, which favors more base-load power plants and intermittent renewable energy resources.

• *Vehicle Use.* The use of these vehicles is likely to involve three major externalities:

*Conventional pollutant and GHG emissions.* Potential reductions in urban emissions and exposures (a positive externality) from the use of HEVs and PHEVs and the potential increases in emissions from grid electricity are expected. NAS/NAE/NRC (2009d) (and other analyses reported below) estimated that the gradual expansion of the use of these technologies will result in emissions being representative of the average grid emissions (rather than the peak), although its assessment noted the probable unequal geographic distribution of these emissions.

*Safety.* Safety has been raised as a concern with a number of the battery formulations. This concern includes possible malfunction (with inappropriate chemical reactions, heat, and fire) and, probably most relevant for vehicles, potential exposures and impacts in vehicle accidents. Given the wide range of potential mixtures and significant uncertainty about which of these might become most prevalent, it is difficult to quantify these externalities at this time.

*Battery recycling and disposal.* With substantially increased use of batteries containing unusual metals, a key question will be where battery recycling and disposal will take place. In the United States and under U.S. regulatory requirements, improper emissions and worker exposures will probably be minimized (although at a minimum, there is a need for a review of current requirements to ensure their adequacy). If any significant portion of this activity takes place in the developing world, however, past experience suggests that there could be significant exposures of workers and even populations.

*Estimates of Effects and Monetized Damages for Electric Light-Duty Highway Vehicles*

The analysis of damages attributable to the operation of different electric technologies is highly dependent on the assumptions made about the energy mix and emissions from the electric utility system. The damage estimates for operation of hybrid and electric vehicles show significant lower damages than those for vehicles fueled by conventional gasoline (even when accounting for the uncertainty in the analysis). The difference is greatest when comparing damages resulting from the operation of electric vehicles to those resulting from the operation of vehicles fueled by conventional gasoline. Even damages resulting from the operation of grid-independent hybrid electric vehicles (which also consume gasoline) are approximately 20% lower compared with damages resulting from the operation of vehicles fueled solely by conventional gasoline.

However, emissions from electricity generation are included in the full life-cycle damages of the grid-dependent vehicles, specifically the emissions from the power plants as well as emissions from activities to produce the fossil fuels used in these plants. As shown in Table 3-13, when the damages attributable to other parts of the life cycle were included, especially the emissions from the feedstock and the fuel (emissions from electricity production), the aggregate damages for the grid-dependent and all-electric vehicles became comparable to, or somewhat higher than, those from gasoline.

Projections of the *Annual Energy Outlook* of the U.S. Energy Information Administration were used in this analysis and in Chapter 2 to estimate the electricity damages. Although very large decreases in emissions from fossil-fueled plants were projected for 2030 compared with current emissions (on a per kilowatt-hour basis), electricity from coal- and natural-gas-fired power plants would still account for 66% of total generation. This percentage is only a slight decrease from the 70% in 2005. Thus, although the committed estimates that the damages associated with electricity generated for use by the vehicle will decrease, the total life-cycle damages of the electric-vehicle technology are still estimated to be slightly greater than those of the conventional gasoline vehicle [by 1.49-1.35 = 0.14 cents/VMT (see Table 3-13)].

One or two important transformations would be needed for the (non-climate-change-related) life-cycle damages of electric vehicles to be equal

**TABLE 3-13** Comparison of Health and Other Non-GHG Damage Estimates for Hybrid- and Electric-Vehicle Types with Conventional Gasoline, 2005 and 2030[a]

| | 2005 | | | 2030 | |
|---|---|---|---|---|---|
| | 5th and 95th Percentile Range Aggregate Damages[b] (Cents/VMT) | Population-adjusted Mean Aggregate Damages (Cents/VMT) | Population-adjusted Mean Operations Only (Cents/VMT) | 5th and 95th Percentile Range Aggregate Damages[b] (Cents/VMT) | Population-adjusted Mean Aggregate Damages (Cents/VMT) |
| Conventional gasoline | 0.34-5.07 | 1.34 | 0.38 | 0.45-4.87 | 1.35 |
| Grid-independent HEV | 0.31-4.12 | 1.22 | 0.31 | 0.49-5.57 | 1.50 |
| Grid-dependent HEV | 0.27-8.90 | 1.46 | 0.22 | 0.45-9.20 | 1.62 |
| Electric | 0.20-15.0 | 1.72 | 0.05 | 0.35-12.2 | 1.49 |

[a]Costs are in 2007 USD.

[b]From the distribution of results for all counties in the 48 contiguous states in the United States.

ABBREVIATIONS: GHG = greenhouse gas; VMT = vehicle miles traveled; HEV = hybrid electric vehicle.

to or less than those of conventional vehicles. One of the transformations needed would be a dramatic shift to much greater nonfossil-fuel electricity generation—from renewable energy sources as well as nuclear power plants (for example, see Samaras and Meisterling, 2008). Instead of fossil fuels accounting for 66% of total generation in 2030, they would need to be lowered to about 37%. This estimated decrease is based on the assumption that no improvement in manufacturing efficiency will occur (see below) and that the fuel component of the damages would decrease by the 0.14 cents/VMT difference between gasoline and electric vehicles.

The other technological transformation would have to be a great improvement in energy efficiency in vehicle manufacture. As noted in Table 3-12, energy use in manufacturing a plug-in hybrid vehicle is about 13-23% greater than that for a gasoline vehicle in 2035, and both are greater than energy use to manufacture current gasoline hybrid and gasoline vehicles. Damages from the emissions associated with vehicle manufacture account for a large percentage of the overall life-cycle damages. Thus, even with the large decreases in emissions from generating electricity at fossil-fueled plants, the large damages from the vehicle-manufacture component mean that life-cycle damages for electric vehicles would probably be somewhat greater than those for conventional vehicles, unless there is significant reduction in energy use in manufacturing batteries and other electric vehicle components.

The aggregate damages also reflect approximately 20% higher energy use and emissions from the manufacture of the vehicles, based on higher estimated energy inputs in GREET for battery manufacture.

## NATURAL GAS

### Current Status

Natural gas vehicles (NGVs) are very similar to gasoline vehicles; the major difference is in fuel storage. Light-duty NGVs, and some heavy-duty vehicles like urban transit buses, use compressed natural gas (CNG). Heavy-duty vehicles can also use liquefied natural gas (LNG), which is denser but must be maintained below −260°F in very well insulated tanks (NGV America 2009; DOE 2009b).

In 2008, there were more than 150,000 NGVs in the United States. The main markets for NGVs are new transit buses and corporate fleet cars that are used mainly for short trips. That demand is due mainly to EPA's Clean-Fuel Fleet Program. NGVs are more expensive than hybrid vehicles or gasoline vehicles. For example, the Honda Civic GX NGV has an MSRP of $24,590 compared with $22,600 for the hybrid sedan, and $15,010 for the regular sedan (Rock 2008).

About 1,500 NGV fueling stations are in the United States as of 2008; a substantial portion is part of private company facilities and is not available to the general public. Natural gas is sold in units of gasoline gallon equivalent. One gasoline gallon equivalent represents the same energy content (124,800 British thermal units) as a gallon of gasoline. Natural gas for CNG is obtained directly from a distribution line. Stations require large, high-pressure compressors and storage tanks to fill a vehicle quickly. Alternatively, a small compressor can work overnight. Natural gas for LNG can also be taken from a gas pipeline and then liquefied on-site, but it also can be transported in liquid form to a refueling facility via tanker truck.

### Technology Development and Barriers

The main benefit of CNG has been its relatively low price (about 80% that of gasoline on a gasoline-gallon-equivalent basis). Also, transport and distribution can rely on an existing infrastructure for both industrial and household use (Yborra 2006). According to the AEF report (NAS/NAE/NRC 2009c), if natural gas were to be used for transportation instead of for electricity production, North American natural gas reserves could supply about 20-25% of transportation fuel needs by 2020 but only with investment in distribution infrastructure. To supply more would require importing natural gas and LNG to meet that increased demand. (Chapter 6 discusses hazards related to infrastructure for distribution of LNG in the United States.)

The AEF report indicates that the main challenges to increased use of NGVs include an insufficient number of refueling stations and inconvenient on-board CNG tanks that take up most of the trunk space. Another key disadvantage is a limited range. The average range of a gasoline or diesel vehicle is 400 miles, and the range of an NGV is only 100 to 150 miles, depending on the NG compression. The AEF report suggests that the most important barrier for NGVs could be a public perception that using CNG as a fuel would involve carrying a dangerous "explosive" on board a vehicle and that self-service refueling with a high-pressure gas would be too risky to offer to the general public.

### Fuel-Cycle Effects and Externalities

Natural gas has several significant advantages as a fuel for vehicles when compared with gasoline or diesel. Dedicated NGVs have the least exhaust emissions of CO, nonmethane VOCs, $NO_x$, and $CO_2$. NGVs emit unburned methane (which has a higher climate forcing potential than $CO_2$), but this might be compensated for by the substantial reduction in $CO_2$ emissions.

The choice of fuel pathway for CNG can have a large impact on GHG emissions over the fuel life cycle. If non-North American natural gas is imported as LNG via ocean tanker and then regasified and compressed to produce CNG, for example, CNG reduces life-cycle GHG emissions by only 5% compared with gasoline. If domestic gas is used, life-cycle GHG emissions are reduced by 15%. If gas that otherwise would be flared or landfill gas is used as the feedstock, net GHG emissions can be negative.

### Modeled Estimates of Damages from Light-Duty CNG Vehicles

Table 3-14 contains a summary of the modeling results from the GREET-APEEP modeling effort related to natural gas light-duty autos and trucks (with a row for reformulated gasoline autos for comparison purposes). Each row of Table 3-14 contains the range and population-adjusted mean for health damages on a VMT basis in 2005 and 2030. There is also a column showing the health costs per gasoline gallon equivalent. Because of population growth, other things being equal, damages would tend to increase from 2005 to 2030. So, decreases in damages mean that for a variety of reasons, emissions per VMT are diminishing over time faster than the population is growing.

**TABLE 3-14** Health and Other Non-GHG Damages from CNG Light-Duty Autos and Trucks (Values Reported in Cents/VMT)[a]

| | 2005 | | | 2030 | |
|---|---|---|---|---|---|
| | 5th and 95th Percentile Range[b] (Cents/VMT) | Population-adjusted Mean (Cents/VMT) | Population-adjusted Mean (Cents/gge)[c] | 5th and 95th Percentile Range[b] (Cents/VMT) | Population-adjusted Mean (Cents/VMT) |
| Conventional gasoline SI autos | 0.35-5.12 | 1.32 | 29.83 | 0.45-4.87 | 1.35 |
| CNG autos | 0.30-4.54 | 1.20 | 23.35 | 0.38-4.41 | 1.16 |

[a]Costs are in 2007 USD.

[b]From the distribution of results for all counties in the 48 contiguous states in the United States.

[c]Cents/gge, calculated by multiplying average miles per gallon by per VMT damages. Therefore the highest damages are shown for the most fuel-efficient vehicles.

ABBREVIATIONS: GHG = greenhouse gas; CNG = compressed natural gas; VMT = vehicle miles traveled; gge, gasoline gallon equivalent; SI = spark ignition.

In fact, damages for CNG autos are 1.2 cents per VMT or about 23 cents/gge. Emissions for trucks are much larger, reaching 28 cents a gallon for LDT2 in 2005. Emissions per VMT are increasing over time for all CNG vehicle types except for LDT2, where the population-adjusted means are 12% lower in 2030 than 2005. CNG autos outperform gasoline autos, with only 87% of the damages in both 2005 and 2030, implying that the emissions per VMT of CNG autos over the life cycle are that much lower than emissions of gasoline autos. On a per gasoline-gallon-equivalent basis, CNG autos do even better, with only 78% of the damages of gasoline vehicles.

By life-cycle stage, the difference in damages from CHG vehicles compared with gasoline vehicles is accounted for by lower operations emissions (particularly of $NO_x$ and VOCs) and lower emissions from the fuel stage for CNG, offset only somewhat by higher feedstock emissions (with identical emissions from the vehicle manufacturing stage).

Table 3-15 shows how the $CO_2$-equivalent emissions vary for CNG autos and reformulated-gasoline autos for the years 2005 and 2030 on a VMT basis. As can be seen, $CO_2$-equivalent emissions for CNG autos are about 89% of those for gasoline vehicles in 2005 but this advantage is greater in 2030, CNG emissions being only 79% of gasoline vehicle emissions in 2030. As expected, methane emissions for CNG vehicles are greater than those for gasoline, but $CO_2$ emissions are much lower, yielding a net decrease in $CO_2$-equivalent emissions for CNG vehicles.

**TABLE 3-15** Carbon Dioxide Equivalent ($CO_2$-eq) Emissions of GHGs from CNG Autos and Light-Duty Trucks Compared with Reformulated Gasoline Vehicles (Grams/VMT)

| Fuel-Vehicle Combination | $CO_2$-eq 2005 gal/VMT | $CO_2$-eq 2030 gal/VMT |
|---|---|---|
| RFG SI autos (conventional oil) | 552 | 365 |
| CNG autos | 492 | 280 |

NOTE: Costs are in 2007 USD.

ABBREVIATIONS: VMT = vehicle miles traveled; RFG = reformulated gasoline; SI = spark ignition; CNG = compressed natural gas.

One caveat with these estimates is that they take, as given, GREET default assumptions with respect to LNG imports. If LNG imports grow by more than assumed between 2005 and 2030, much, if not all, the gains from CNG vehicles relative to gasoline vehicles (at least from the perspective of GHG emissions) will be eroded.

## HYDROGEN FUEL-CELL VEHICLES

### Current Status

According to the AEF report (NAS/NAE/NRC 2009c), hydrogen fuel-cell vehicles (HFCVs) can yield large and sustained reductions in U.S. oil consumption and GHG emissions, but several decades will be needed to realize these potential long-term benefits. The NRC report *Transitions to Alternative Transportation Technologies—A Focus on Hydrogen* (NRC 2008c) estimates that the maximum practical number of HFCVs that could be operating in 2020 would be approximately 2 million in a fleet of 280 million light-duty vehicles. The number of HFCVs could grow rapidly to about 25 million by 2030 and account for more than 80% of new vehicles entering the fleet by 2050. These estimates assume that technical goals are met, consumers readily accept HFCVs, and policy instruments are in place to facilitate the introduction of hydrogen fuel and HFCVs through the market transition period.

### Modeled Estimates of Damages from Hydrogen Fuel-Cell Vehicles

Table 3-16 contains a summary of the modeling results from the GREET-APEEP modeling effort related to hydrogen fuel-cell autos relative to gasoline light-duty autos. GREET covers two technologies for fuel cells—one that assumes the vehicle uses hydrogen gas directly and another

**TABLE 3-16** Health and Other Non-GHG Damages from Hydrogen
Fuel-Cell Autos Compared with Reformulated Gasoline Autos[a]

|  | 2005 | | | 2030 | |
|---|---|---|---|---|---|
|  | 5th and 95th Percentile Range[b] (Cents/VMT) | Population-adjusted Mean (Cents/VMT) | Population-adjusted Mean (Cents/gge)[c] | 5th and 95th Percentile Range[b] (Cents/VMT) | Population-adjusted Mean (Cents/VMT) |
| Conventional gasoline SI autos | 0.35-5.12 | 1.32 | 29.83 | 0.45-4.87 | 1.35 |
| Hydrogen (gaseous) autos | 0.38-4.17 | 1.34 | 66.68 | 0.61-5.61 | 1.64 |

[a]Costs are in 2007 USD.

[b]From the distribution of results for all counties in the 48 contiguous states in the United States.

[c]Cents/gge, calculated by multiplying average miles per gallon per VMT damages. Therefore the highest damages are shown for the most fuel-efficient vehicles.

ABBREVIATIONS: GHG = greenhouse gas; VMT = vehicle miles traveled; gge = gasoline gallon equivalent; SI = spark ignition.

that assumes the vehicle carries a liquid fuel on the vehicle that is converted to hydrogen gas in a reformer. Because of the substantial uncertainties associated with the likely types and amounts of energy use for liquid hydrogen fuel, only results for hydrogen gas are included here. Each row of Table 3-16 contains the range and population-adjusted mean for health damages on a VMT basis in 2005 and 2030. There is also a column showing the health costs per gasoline gallon equivalent.

Table 3-16 shows that estimated damages for hydrogen (gaseous) and reformulated gasoline are similar in 2005. Yet, there are large differences in emissions over the life cycle. Hydrogen fuel cells have far larger emissions from the fuel stage and the vehicle-manufacturing stage than gasoline vehicles, which is about fully offset by lower emissions in the operation stage and to a lesser extent in the feedstock stage. By 2030, however, reformulated gasoline is less damaging than hydrogen (gaseous) owing to a bigger increase in emissions per VMT in the vehicle-manufacturing stage. Note that it is misleading to compare damages on a per gallon-gasoline-equivalent basis since hydrogen fuel cells use such a different means of propulsion and get such apparently "high" mileage per damage unit.

Table 3-17 shows how the $CO_2$-equivalent emissions vary among the different fuel vehicle types and between the years 2005 and 2030 on a VMT basis. As shown, the hydrogen (gaseous) vehicle fuel significantly

**TABLE 3-17** Carbon Dioxide Equivalent ($CO_2$-eq) Emissions of GHGs from Hydrogen Fuel-Cell Autos Compared with Reformulated Gasoline Autos

| Fuel-Vehicle Combination | $CO_2$-eq 2005 gal/VMT | $CO_2$-eq 2030 gal/VMT |
|---|---|---|
| RFG SI autos (conventional oil) | 552 | 365 |
| Hydrogen (gaseous) autos | 341 | 294 |

NOTE: Costs are in 2007 USD.

ABBREVIATIONS: GHGs = greenhouse gases; RFG = reformulated gasoline; SI = spark ignition.

outperforms gasoline vehicles for $CO_2$-equivalent, with only about 60% of the latter's emissions.

## SUMMARY AND CONCLUSIONS

The committee has presented here a detailed summary of the wide range of potential emissions and damages from the use of energy in transportation. Our discussion and analysis focus on the components of transportation energy use—for light- and heavy-duty on-road transportation—that account for the great majority of annual transportation energy use. Other transportation energy uses—for example, for nonroad vehicles, aircraft, locomotives, and ships—are not inconsequential, but they account for a smaller portion of transportation energy use and were beyond the scope of this analysis.

### Results of the Analysis: Health and Other Damages

Given these limitations, our analysis does provide some useful insight into the relative levels of damages from different fuel and technology mixes. Overall, we estimate that the aggregate national damages to health and other non-GWP effects would have been approximately $36.4 billion per year for the light-duty vehicle fleet in 2005; the addition of medium-duty and heavy-duty trucks and buses raises the aggregate estimate to approximately $56 billion. These estimates are probably conservative, as they do not fully account for the contribution of light-duty trucks to the aggregate damages and of course should be viewed with caution, given the significant uncertainties described above in any such analysis.

*Health and Other Non-GWP Damages on a per VMT Basis*

Although the uncertainties in the analysis preclude precise ranking of different technologies, Table 3-18 illustrates that on a cents per VMT basis

**TABLE 3-18** Relative Categories of Damages 2005 and 2030 for Major Categories of Light-Duty Fuels and Technologies[a]

| Category of Aggregate Damage Estimates (Cents/VMT) | 2005 | 2030 |
|---|---|---|
| 1.10-1.19 | | CNG<br>Diesel with low sulfur and biodiesel |
| 1.20-1.29 | E85 herbaceous<br>E85 corn stover<br>CNG<br>Grid-independent HEV | E85 corn stover<br>E85 herbaceous |
| 1.30-1.39 | Conventional gasoline and RFG<br>E10<br>Hydrogen gaseous | Conventional gasoline and RFG<br>E10<br>E85 corn |
| 1.40-1.49 | Diesel with low sulfur and biodiesel<br>Grid-dependent HEV | Electric vehicle |
| 1.50-1.59 | E85 corn | Grid-independent HEV<br>Grid-dependent HEV |
| >1.60 | Electric vehicle | Hydrogen gaseous |

[a]Costs are in 2007 USD.

ABBREVIATIONS: VMT = vehicle miles traveled; CNG = compressed natural gas; HEV = hybrid electric vehicle; RFG = reformulated gasoline.

there are some differences that provide useful insight into the levels of damages attributable to different fuel and technology combinations in 2005 and 2030. Overall, the damage levels illustrate several things:

•   Among the fuel and technology choices, there are some differences in damages, although overall, especially in 2030, the different fuel and technology combinations have remarkably similar damage estimates.

    o Some fuels—E85 from herbaceous and corn stover and CNG—have relatively lower damages than all other options in both 2005 and 2030
    o Diesel, which has relatively high damages in 2005, has one of the lowest levels of damage in 2030. This is due to the substantial reductions in both PM and $NO_x$ emissions that a 2030 diesel vehicle is required to attain.
    o Corn-based ethanol, especially E85, has relatively higher dam-

ages than most other fuels; this is in large measure due to the higher level of emissions from the energy required to produce the feedstock and the fuel.

○ Grid-dependent HEVs and electric vehicles have relatively higher damages in 2005. As noted above, these vehicles have significant advantages over all other fuel and technology combinations when considering only damages from operations. However, the damages associated with the current and projected mixes of electricity generation (the latter still being dominated by coal and natural gas in 2030, albeit at significantly lower rates of emissions) add substantial damages to these totals. In addition, the increased energy associated with battery manufacture adds approximately 20% to the damages from vehicle manufacture. However, further legislative and economic initiatives to reduce emissions from the electricity grid could be expected to improve the relative damages from electric vehicles substantially.

• Although the underlying level of aggregate damages in the United States could be expected to rise between 2005 and 2030 because of projected increases in population and to increases in the value of a statistical life, the results in our analysis for most fuel and technology examples in 2030 are very similar to those in 2005 in large measure because of the expected improvement in many fuel and technology combinations (including conventional gasoline) as a result of enhanced fuel efficiency (35.5 mpg) expected by 2030 from the recently announced new national standards for fuel efficiency. (It is possible, however, that these improvements are overstated somewhat, because there is evidence that improved fuel efficiency can also lead to increased travel, probably resulting in higher aggregate damages than would otherwise be seen.)

• As shown in Figure 3-7, these aggregate damages are not spread equally among the different life-cycle components. For example, in most cases, the actual operation of the vehicle is one-quarter to one-third of the aggregate damages, while the emissions incurred in creating the feedstock, refining the fuel, and making the vehicle are responsible for the larger part of aggregate damages.

## Health and Other Non-GHG Damages on a per Gallon Basis

As illustrated in Tables 3-3, 3-10, and 3-14, the committee also attempted to estimate the health and non-GHG damages on a per gallon basis. This estimate is made somewhat more complicated by the fact that simply multiplying expected miles per gallon for each fuel and vehicle type by the damages per mile will tend to make the most fuel-efficient vehicles,

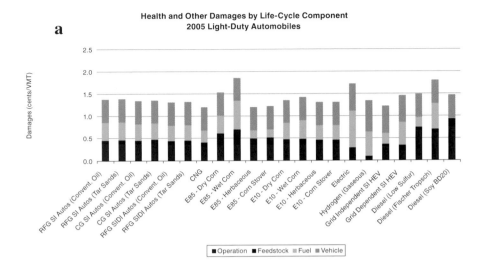

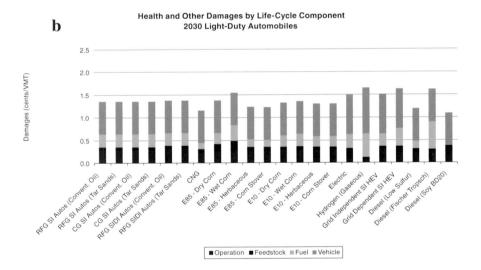

FIGURE 3-7 Health effects and other nonclimate damages are presented by life-cycle component for different combinations of fuels and light-duty automobiles in 2005 (*a*) and 2030 (*b*). Damages are expressed in cents per VMT (2007 U.S. dollars). Going from bottom to top of each bar, damages are shown for life-cycle stages as follows: vehicle operation, feedstock production, fuel refining or conversion, and vehicle manufacturing. Damages related to climate change are not included. ABBREVIATIONS: VMT, vehicle miles traveled; CG SI, conventional gasoline spark ignition; CNG, compressed natural gas; E85, 85% ethanol fuel; HEV, hybrid electric vehicle.

which travel the most miles on a gallon, appear to have higher damages per gallon than a less fuel-efficient vehicle. With that caveat in mind, the committee estimated that in 2005 the mean damages per gallon for most fuels ranged from 23 cents/gallon to 38 cents/gallon, the damages for conventional gasoline engines being in approximately the middle of that range at approximately 29 cents per gallon.

## Estimates of Aggregate National Health and Other Non-GHG Damages

Overall, and scaling up the per VMT damages reported here to reflect national VMT in 2005, we estimate that the aggregate national damages to health and other nonclimate-change-related effects would have been approximately $36 billion per year (2007 USD) for the light-duty vehicle fleet in 2005; the addition of medium-duty and heavy-duty trucks and buses raises the aggregate estimate to approximately $56 billion (2007 USD). These estimates are probably conservative, as they include but do not fully account for the contribution of light-duty trucks to the aggregate damages, and of course should be viewed with caution, given the significant uncertainties in any such analysis.

## Limitations in the Health and Other Non-GHG Damages Analysis

It is important in interpreting these results to consider two major limitations in the analysis:

• *Emissions and damages that were not quantifiable.* Although our analysis was able to consider and quantify a wide range of emissions and damages throughout the life cycle and included what arguably could be considered the most significant contributors to estimates of such damages (for example, premature mortality resulting from exposure to air pollution), there are many potential damages that could not be quantified at this time. Such damages include the following:

> ○ *Overall:* Impacts of hazardous air pollutants and damages to ecosystems (for example, from deposition), the full range of agricultural crops, and others.
> ○ *Biofuels:* Impacts on water use and water contamination, as well as any formal consideration of potential indirect land-use effects (see discussion of the latter in "Indirect Land Use and Externalities").
> ○ *Battery electric vehicles:* Potential effects from exposures to air toxics in battery manufacture, in battery disposal, and during accidents.

• *Uncertainty.* Any such analysis includes a wide set of assumptions and decisions about analytical techniques that can introduce uncertainty in the results. Although we did not attempt to conduct a formal uncertainty analysis, we have been cautious throughout our discussion of results—and urge the reader to be cautious—to not over-interpret small differences in results among the wide range of fuels and technologies assessed. Moreover, we engaged in limited sensitivity analyses to check the impacts of key assumptions.

### Results of the Analysis: GHG Emissions

• Similar to the damages estimates, the GHG emission estimates from each fuel and technology combination can provide relative estimates of GHG performance in 2005 and 2030. Although caution should be exercised in interpreting these results and in comparing the fuel and technology combinations, some instructive observations from Table 3-19 are possible:

> ○ Overall, the substantial improvements in fuel efficiency in 2030 (to a minimum of 35 mpg for light-duty vehicles) result in most technologies becoming much closer to each other in per VMT life-cycle GHG emissions. There are, however, some differences:
> ○ As with the damages reported above, the herbaceous and corn stover E85 have relatively low emissions; in terms of aggregate g/VMT of $CO_2$-equivalent emissions, E85 from corn also has relatively low emissions.
> ○ The tar-sand-based fuels have the highest GHG emissions of any of the fuels.
> ○ As shown in Figure 3-8 and in contrast to the damages analysis above, the operation of the vehicle is in most cases a substantial relative contributor to total life-cycle emissions. This is not the case, however, with either the grid-dependent technologies (for example, electric or grid-dependent hybrid) or the hydrogen fuel-cell vehicles, where the dominant contributor to life-cycle emissions is the processing of the fuel in the grid or in the production of hydrogen.

### Results of the Analysis: Heavy-Duty Vehicles

The committee also undertook a more limited analysis of the damages and GHG emissions associated with heavy-duty vehicles. Although this analysis included operations, feedstock, and fuel components of the life cycle, it could not include a vehicle-manufacturing component because of the

**TABLE 3-19** Relative Categories of GHG Emissions in 2005 and 2030 for Major Categories of Light-Duty Fuels and Technologies

| Category of Aggregate $CO_2$-Equivalent Emission Estimates (gal/VMT) | 2005 | 2030 |
|---|---|---|
| 150-250 | E85 herbaceous<br>E85 corn stover | E85 herbaceous<br>E85 corn stover |
| 250-350 | Hydrogen gaseous | E85 corn<br>Diesel with biodiesel<br>Hydrogen gaseous<br>CNG |
| 350-500 | E85 corn<br>Diesel with biodiesel<br>Grid-independent HEV<br>Grid-dependent HEV<br>Electric vehicle<br>CNG | Grid-independent HEV<br>SI conventional gasoline, RFG<br>Grid-dependent HEV<br>Electric vehicle<br>Diesel with low sulfur<br>E10 herbaceous, corn stover<br>SIDI conventional gasoline<br>E10 corn<br>SI tar sands |
| 500-599 | Conventional gasoline and RFG<br>E10<br>Low-sulfur diesel | |
| >600 | Tar sands | |

Costs are in 2007 USD.

ABBREVIATIONS: GHG = greenhouse gas; VMT = vehicle miles traveled; CNG = compressed natural gas; HEV = hybrid electric vehicle; RFG = reformulated gasoline.

wide range of vehicle types and configurations. In sum, and as illustrated in Figures 3-9 and 3-10, there are several conclusions that can be drawn:

- The damages per VMT in 2005 are significantly higher than those shown above for light-duty vehicles, although they accrue to a much higher weight of cargo and number of passengers being carried per mile as well.
- Damages drop significantly in 2030 because of the full implementation of the 2007-2010 Highway Diesel Rule, which requires substantial reductions in PM and $NO_x$ emissions.

GHG emissions are driven primarily in these analyses by the operations component of the life cycle and do not change substantially between 2005 and 2030 (except for a modest improvement in fuel economy). EPA

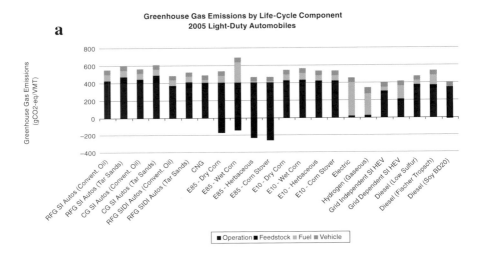

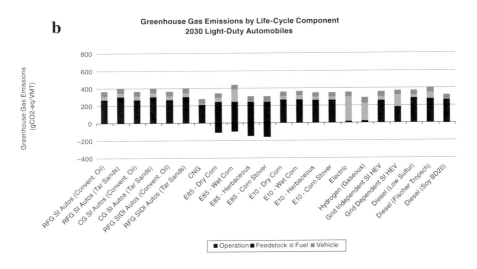

**FIGURE 3-8** Greenhouse gas emissions (grams $CO_2$-eq)/VMT by life-cycle component for different combinations of fuels and light-duty automobiles in 2005 (*a*) and 2030 (*b*). Going from bottom to top of each bar, damages are shown for life-cycle stages as follows: vehicle operation, feedstock production, fuel refining or conversion, and vehicle manufacturing. One exception is ethanol fuels for which feedstock production exhibits negative values because of $CO_2$ uptake. The amount of $CO_2$ consumed should be subtracted from the positive value to arrive at a net value. ABBREVIATIONS: g $CO_2$-eq, grams $CO_2$-equivalent; VMT, vehicle miles traveled; CG SI, conventional gasoline spark ignition; CNG, compressed natural gas; E85, 85% ethanol fuel; HEV, hybrid electric vehicle.

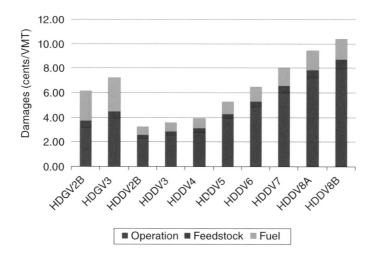

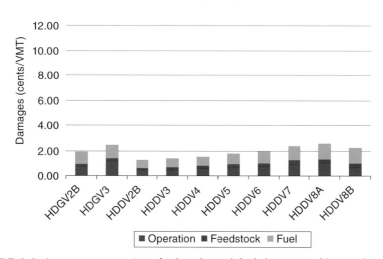

FIGURE 3-9 Aggregate operation, feedstock, and fuel damages of heavy-duty vehicles from air-pollutant emissions (excluding GHGs) (cents/VMT). (Top) Estimated damages in 2005; (Bottom) estimated damages in 2030.

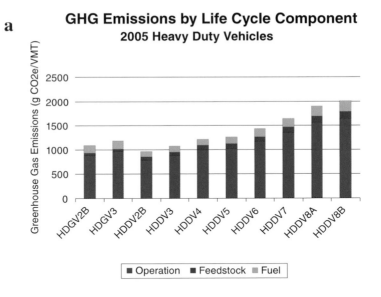

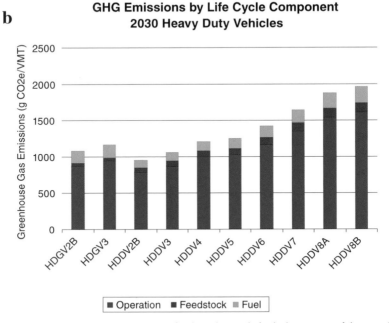

FIGURE 3-10 Aggregate operation, feedstock, and fuel damages of heavy-duty vehicles from GHG emissions (cents/VMT). (Top) Estimated damages in 2005; (Bottom) estimated damages in 2030.

and others are investigating possible future enhanced requirements for fuel economy among heavy-duty vehicles.

## Results of the Analysis: Damage and GHG Emission Comparisons

Although energy use and emissions generally track one another quite closely, the comparisons above indicate that they do not uniformly distinguish among the fuel and technology combinations. In general, there are few fuel and technology combinations that have significantly lower damages than gasoline in 2005 (Table 3-10), although several combinations have significant advantages in global warming potential (GWP). (The former is in part due to the GREET model, which assumes all fuel and vehicle combinations must at least meet similar emissions standards.) The electric and fuel-cell options have somewhat higher life-cycle damages than gasoline even though they have significantly lower GWP in most cases.

The conclusions to be drawn from the 2030 analysis are similar, although some diesel options begin to exhibit improvements in damages over gasoline damages because of the substantial mandated reduction in emissions, and the overall difference in damages is somewhat smaller as fuel efficiency among the fuel and technologies converge.

## Overall Implications of the Results

Perhaps the most important conclusion to be taken from these analyses is that, when viewed from a full life-cycle perspective, the results are remarkably similar across fuel and technology combinations. One key factor contributing to this result is the relatively high contribution of emissions to health and other non-GHG damages in life-cycle phases (such as those in the development of the feedstock, the processing of the fuel, and the manufacturing of the vehicle) other than in the phase of vehicle operation.

There some differences though, and from these, some conclusions can be drawn:

• The gasoline-driven technologies have somewhat higher damages and GHG emissions in 2005 than a number of other fuel and technology combinations. The grid-dependent electric options have somewhat higher damages and GWP than other technologies, even in our 2030 analysis, in large measure due to the continued conventional and GHG emissions from the existing and likely future grid at least as of 2030. (See below for mention of possible pathways for reducing those emissions.)

• In 2030, with the move to meet the enhanced 35 mpg requirements now being put in place, those differences among technologies tend to converge somewhat, although the fact that operation of the vehicle is generally less than a third of overall life-cycle emissions and damages tends to dampen the magnitude of that improvement. Further enhancements in fuel efficiency—the likely push for an extension beyond 2016 to further improvements—would improve the GHG emission estimates for all liquid-fuel-driven technologies.

- The choice of feedstock for biofuels can significantly affect the relative level of life-cycle damages, herbaceous and corn stover having some advantage in this analysis.
- Additional regulatory actions or changes in the mix of electricity generation can significantly affect levels of damages and GHG emissions. This result was illustrated in this analysis by the substantial reduction in diesel damages from 2005 to 2030. Similarly, major regulatory initiatives to reduce electricity-generation emissions or legislation to reduce carbon emissions would significantly improve the relative damages and emissions from the grid-dependent electric options. A shift to electricity generation with lower emissions (for example, natural gas, renewables, and nuclear) would also further reduce the life-cycle emissions and damages of the grid-dependent technologies.
- Overall, the differences are somewhat modest among different types of vehicle technologies and fuels, even under the likely 2030 scenarios, although some technologies (for example, grid-dependent electric) had somewhat higher life-cycle emissions. Therefore, some breakthrough technologies (such as cost-efficient conversion of advanced biofuels; cost-efficient carbon capture and storage, and much greater use of renewable resources for electricity generation) appear to be needed to dramatically reduce transportation-related externalities.

These results must be viewed in the context of a large number of potential damages noted above that cannot at this time be quantified and substantial continued uncertainties. There is a need for additional research to attain the following:

1. At the earliest possible stage in the research and development process, better understanding of the potential negative externalities for new fuels and technologies should be obtained to avoid these externalities as the fuels and technologies are being developed.

2. Understanding of the currently unquantifiable effects and potential damages should be improved, especially as they relate to biofuels (such as effects on water resources and ecosystems) and battery technology (such as effects throughout the battery life cycle of extraction through disposal).

3. More accurate emissions factors should be obtained for each stage of the fuel and vehicle life stages. In particular, there is a need, in the context of enhancing even further EPA's recent shift to the Motor Vehicle

Emission Simulator (MOVES) model for mobile-source emissions, to make measurements to confirm or refute the assumption that all vehicles will only meet but not exceed emission standards. In actual practice, there can be significant differences between on-road performance relative to emissions requirements and some alternative-fuel vehicles may do better or worse than expected.

4.   The issue of indirect land-use change is central to current debates about the merit of biofuels. Regardless of whether this impact is regarded as an externality associated with U.S. or foreign biofuels production, it is important to obtain more empirical evidence about its magnitude and causes, as well as to improve the current suite of land-use change models.

5.   Because a substantial fraction of life-cycle health impacts comes from both vehicle manufacture and fuel production, it is important to improve and expand the information and databases used to construct emissions factors for these life stages. In particular, there is a need to understand whether and how energy-efficiency improvements in these industrial components might change the overall estimates of life-cycle health damages.

6.   As better data become available, future studies should also focus on other transportation modes—both those that are alternatives to automobiles and light trucks (transit), as well as air, rail, and marine, which are alternatives for long-distance travel and for freight.

# 4

# Energy for Heat

## BACKGROUND

An evaluation of the externalities of energy used to produce heat as an end use is important because heat energy represents about 30% of U.S. primary energy usage.[1] Unlike the chapters on the electricity-production and transportation sectors, this chapter does not present a detailed assessment of externalities associated with all uses of all energy sources for heat. Rather, this chapter presents the committee's assessment of air-pollution damages from present uses (and uses expected in 2030) of natural gas for heat by residential and commercial sector buildings (see Box 4-1 for sector definitions) and presents some comparisons of natural gas and electricity use for heat. The industrial sector is discussed only qualitatively, because published statistics do not differentiate clearly between fuel used for heating and fuel used as feedstocks for manufacturing processes. Figures 4-1 and 4-2 show the overall energy use in the United States by sector, the details of building and industrial energy consumption by source, and the consumption by sector of electricity and natural gas (EIA 2008b). Natural gas is the

---

[1]"The energy that powers our civilization is obtained from a number of primary energy sources that exist in nature. These sources fall into two categories: flows of energy and stored energy. Examples of energy flows include sunlight, wind, and waves. Stored energy includes fossil energy (petroleum, natural gas, and coal), bioenergy (contained in biomass), and nuclear energy (stored in atomic nuclei in radioactive elements such as uranium) and the heat stored in Earth's upper crust. Primary energy sources can be converted into *useful* energy that, for example, powers a vehicle, lights a building, or supplies heat for an industrial process, although the conversion process inevitably involves energy losses (which can be quite considerable) and often entails substantial costs" (NAS/NAE/NRC, 2009a).

---

**BOX 4-1**
**Definition of Residential, Commercial, and Industrial Sectors**

*Residential sector:* An energy-consuming sector that consists of living quarters for private households. Common uses of energy associated with this sector include space heating, water heating, air conditioning, lighting, refrigeration, cooking, and running a variety of other appliances. The residential sector excludes institutional living quarters. NOTE: Various programs of the U.S. Energy Information Administration differ in sectoral coverage.

*Commercial sector:* An energy-consuming sector that consists of service-providing facilities and equipment of: businesses; federal, state, and local governments; and other private and public organizations, such as religious, social, or fraternal groups. The commercial sector includes institutional living quarters. It also includes sewage treatment facilities. Common uses of energy associated with this sector include space heating, water heating, air conditioning, lighting, refrigeration, cooking, and running a wide variety of other equipment. NOTE: This sector includes generators that produce electricity and useful thermal output primarily to support the activities of the above-mentioned commercial establishments.

*Industrial sector:* An energy-consuming sector that consists of all facilities and equipment used for producing, processing, or assembling goods. The industrial sector encompasses the following types of activity: manufacturing (NAICS codes 31-33); agriculture, forestry, fishing, and hunting (NAICS code 11); mining, including oil and gas extraction (NAICS code 21); and construction (NAICS code 23). Overall energy use in this sector is largely for process heat and cooling and powering machinery, with lesser amounts used for facility heating, air conditioning, and lighting. Fossil fuels are also used as raw material inputs to manufactured products. NOTE: This sector includes generators that produce electricity and useful thermal output primarily to support the above-mentioned industrial activities.

---

SOURCE: Glossary accessed at the Energy Information Administration (EIA) Web site http://www.eia.doe.gov/.

---

major fuel used for heat in buildings. However, buildings also consumed about 5% of the 39.7 quadrillion British thermal units (quads) of petroleum used in 2008; industry consumed about 25%. Industrial consumption of petroleum includes the petroleum refining industry, which in turn provides 70% of petroleum used as fuel in transportation.

Approximately 20% of total energy consumed in the United States is attributed to nonelectric use in the industrial sector (for both heating and feedstock); about 10% is attributed to nonelectric use in commercial and residential buildings. Building-sector energy is predominantly used for heating. The industrial and building sectors are also the consumers of almost all electricity generation—about 40% of the U.S. primary energy usage. Damages associated with electricity production were evaluated in Chapter 2.

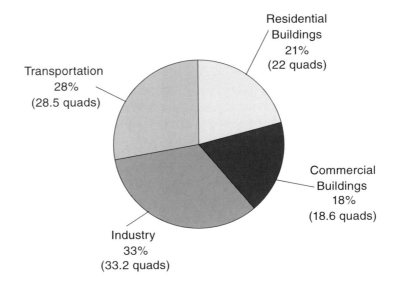

Notes:

- For each sector, "Total Energy Use" is direct (primary) fuel use plus apportioned purchased electricity and electricity system losses
- Economy-wide, total US primary energy use in 2008 was 102.3 quadrillion BTU (quads)
- Source: US Department of Energy, Energy Information Administration, *Annual Energy Outlook 2008*

**FIGURE 4-1** Total U.S. energy use by sector, 2008. SOURCE: EIA 2008b.

The types of damages considered by this committee and associated with end uses of electricity are relatively small compared with those associated with electricity generation.

This chapter provides approximate estimates of damages associated with the use of natural gas for heating applications in the industrial and building sectors. The technologies used in these sectors vary in type, size, and age and are widely distributed, but they mainly burn natural gas. The industrial sector uses some petroleum and small amounts of other primary fuels. The magnitude of associated externalities is strongly influenced by the amount of a particular fuel used and the locations of use.

Most industrial processes and buildings have operating lives of three or more decades, so, in addition to new installations to meet growth in demand, only a few percent of the existing stock is replaced each year. Much

225

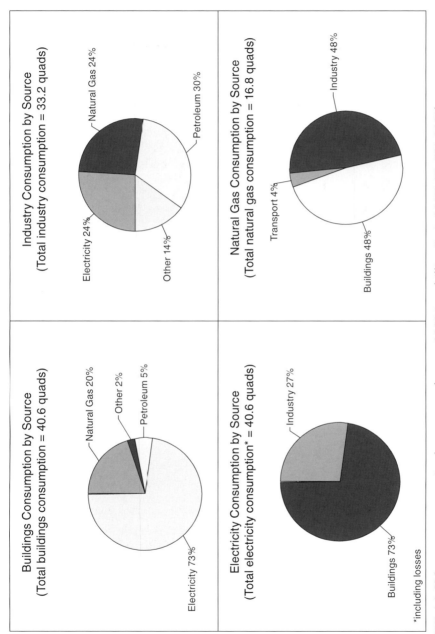

**FIGURE 4-2** U.S. energy consumption by source and sector, 2008 (quadrillion Btu). SOURCE: EIA 2008b.

of the existing building and industrial plant stock is thirty or more years old and employs older technologies. Therefore assessing externalities associated with future energy use needs to consider the upgrading of existing systems ("retrofits") as well as the introduction of new technologies. The America's Energy Future (AEF) report *Real Prospects for Energy Efficiency in the United States* (NAS/NAE/NRC 2009d) has been used as a major resource for the materials presented in this chapter.

Residential building sector emissions are generally distributed in the same manner that population is distributed. Commercial buildings are located in urban areas and suburban towns and villages. Industrial-fuel use is more concentrated in industrial areas, and varies by industry. Because of differences in scale and characteristics of the combustion processes, local health effects and other effects will be somewhat different, and these are identified and discussed in general. Greenhouse gas (GHG) emissions enter a common atmosphere and are not sensitive to location of the emission. Other externalities may exist but are not quantified in this chapter.

The committee used the methodology of the Air Pollution Emission Experiments and Policy (APEEP), with all the caveats described in detail in prior chapters and in Appendix C, to assess damages related to energy use for heat in the buildings and industrial sectors. The primary fuel, natural gas, is estimated to have relatively low nonclimate-change damages per kilowatt-hour compared with coal or wood, for example. We have not estimated damages associated with home heating by coal or biomass fuels because they are a relatively small part of the total energy mix for that use and because recent trends in increased use of natural gas fireplaces are expected to reduce damages related to coal or biomass use for space heating. Only about 12% of U.S. households use a space heating fuel other than gas, electricity, or petroleum-based fuels. At present, there is no other primary energy source that can be readily substituted for natural gas on a wide scale to provide further reduction of such damages. Therefore, opportunities for future reductions of nonclimate-change damages from energy use for heat in the building sectors, in particular, are likely to occur mainly through the incorporation of energy efficiency in the building structures and heat energy systems, as well as the inclusion of localized energy technologies, such as solar thermal water heating or geothermal heat pumps.

## HEAT IN RESIDENTIAL AND COMMERCIAL BUILDINGS

Buildings in the United States consume about 39% of U.S. primary energy, although 73% of this energy is delivered in the form of electricity. The remaining 27% of the energy is primarily used for heating purposes. NAS/NAE/NRC (2009d) provides a detailed description of buildings in the residential and commercial sectors in the United States and describes present status and a portfolio of future opportunities for reducing energy consumption in both sectors. In the past, energy for heating was a modest and affordable portion of annual building operating expenses. As energy prices have risen, more attention is now being given to energy conservation through investments in efficiency and in behavioral changes. Building codes are just starting to reflect this trend, but the construction industry, which uses many standardized building components and construction methods, is reluctant to change because a move toward building more innovative structures would require new investments and training and also increase costs. However, many "green" buildings are emerging from forward-thinking architects and design firms for wealthier clients, and the public is becoming more aware of these possibilities.

The residential building stock in the United States in 2005 consists of 111 million households, including 80 million single family homes, 24 million multifamily housing units, and nearly 7 million mobile homes (EIA 2009). Homes typically last 100 years or more; household electric appliances usually last for 10-20 years; and furnaces and water heaters last about 10 years.

There are about 5 million commercial buildings in the United States (2003) that have about 75 billion square feet of floor space (EIA 2008e, Table A1). Commercial buildings have life spans of 50 years or longer.

This section of the report looks separately at energy use for heat in residential and in commercial buildings for estimation of present externalities and implications for 2030 externalities. For each of these sectors, because of the large existing building stock, options for retrofit of old buildings and possibilities for improved new technologies and designs need to be considered.

The report *America's Energy Future: Technologies and Transformation* (NAS/NAE/NRC 2009a) finds the following:

- "Studies taking several different approaches are consistent in finding the potential for large, cost-effective energy savings in buildings . . . amounting to a 25-30% energy savings for the buildings sector as a whole over 20-25 years. If these savings were to be achieved, it would hold energy use in this sector about constant, in contrast to the current trend of continuing growth."

- "There are substantial barriers to widespread energy efficiency improvements in buildings. But a number of factors are counteracting these barriers. Drivers of increased energy efficiency include rising energy prices, growing concern about global climate change and the resulting willingness of consumers and businesses to take action to reduce emissions, the movement towards 'green buildings,' and growing recognition of the significant nonenergy benefits offered by energy efficiency measures."

## Residential Buildings

The major uses of energy for heating in residential buildings are water heating and space heating. The U.S. Energy Information Administration (EIA 2009i) data showed that in 2005, for 111 million households, energy use for space heating was 4.3 quads, and for water heating, it was 2.1 quads. Because 40% of water heaters are electric, roughly 1.2 quads can be attributed to the use of natural gas for water heating. This total energy-for-heat estimate for 2005 of about 5.5 quads is consistent with the 2007 sector-use numbers for liquid fuels and natural gas of 6.2 quads, allowing for some growth in the number and size of residential buildings. Renewable sources of energy provide a small part of the total and would be expected to produce smaller externalities than the fossil fuels in generating heat.

## Commercial Buildings

The major uses of energy for heat in commercial buildings are for space heating and water heating. EIA reported that 2.37 quads of energy were used for space heating in 2003 (EIA 2008e). About 0.5 quads were reported for water heating and 0.2 quads for cooking. Because some of these uses are provided by electricity, an estimate for nonelectric heating for water heating and cooking is about 0.4 quads. This results in an energy-for-heat estimate for commercial buildings of about 2.8 quads total in 2003 (for 58.5 billion square feet of floor space). A 2005 survey of industrial and commercial boilers in the United States (EEAI 2005) reported that there were almost 163,000 industrial and commercial boilers in the United States that consumed about 8.1 quads of fuel energy per year. The report stated that the total rated capacities for the 120,000 smaller commercial facility boilers was 1.1 million Btu (MMBtu)/h and estimated that these commercial boilers consumed about 3 quads of the 8.1 quads reported for total boiler usage. This estimate can be compared with the 2007 commercial buildings estimate in Table 4-1 of 3.9 quads (for 75 billion square feet of floor space), indicating that the 2007 increase mostly reflects the expansion in total floor space.

Much of the expansion in floor area is due to construction of strip malls. Looking forward, the DOE *Annual Energy Outlook* (EIA 2009e) estimates that commercial building energy use will increase 32% (1.1% per year) by 2030.

## HEAT IN THE INDUSTRIAL SECTOR

The U.S. industrial sector (see Box 4-2) consumes about one-third of the U.S. energy supply, but only about 21% of the total supply comes from nonelectric energy use. Of the 21 quads of nonelectric energy consumed in

**TABLE 4-1** U.S. Nonelectric Energy Consumption by Source and End-Use Sector: Years 2007 and 2030 (EIA Estimates) (in Quadrillion Btu)

| Energy Source | Industrial Sector | Residential Sector | Commercial Sector |
|---|---|---|---|
| Liquid fuels[a] | 9.96/8.35 | 1.35/1.10 | 0.63/0.59 |
| Natural gas | 8.02/8.47 | 4.86/5.06 | 3.10/3.53 |
| Coal | 1.83/2.23 | 0.01/0.01 | 0.07/0.06 |
| Renewables[b] | 2.07/3.89 | 0.43/0.50 | 0.12/0.12 |
| Total | 21.88/22.94 | 6.65/6.67 | 3.92/4.30 |

NOTE: Total U.S. primary energy consumption in 2007 was 101.92 quads; in 2030, total U.S. energy use is projected to be 112.35 quads.

[a]Liquefied petroleum gases, kerosene, distillate fuel oil, residual fuel oil, and gasoline.
[b]Hydropower, wood and wood waste, and municipal solid waste.

ABBREVIATION: EIA = Energy Information Administration.

SOURCE: EIA 2009e.

---

**BOX 4-2**
**Energy for Heat in Steel Manufacture**

Iron ores are mined as minerals in oxidized form. After cleaning and separation, the iron ore is reduced to pig iron in a coke-fueled blast furnace. Coke is the char material produced by heating bituminous coal in a sealed oven for 10 or more hours to drive off volatile "coal gases," resulting in a char material called coke. Without proper effluent treatment, coke ovens can emit substantial amounts of dust and a wide range of emissions that come from various criteria pollutants. In a blast furnace, iron ore is reduced to pig iron by reaction with the coke and the formation of $CO_2$. Energy was needed to produce the coke, but the coke reactions add some energy to the blast furnace. Further heat is required in additional refining steps in a basic oxygen furnace or an electric arc furnace.

When iron products are recycled, a much smaller amount of heat energy is needed to remelt them in an electric arc furnace than is needed in producing pig iron from mineral ores, partly because the reducing agents are not needed. Although it is difficult to compare "virgin" and "recycled" steel because nearly all steel is composed of some mix of recycled steel, the underlying processes are somewhat indicative of the difference between the two. Worrell et al. (2008, Table 1.1) gave "best practice" estimates of 14.8-17.8 GJ/tonne for a basic oxygen furnace and 2.6 GJ/tonne for a 100% scrap electric arc furnace.

---

the industrial sector in 2007, about 8 quads of it may be attributable to "nonfuel" purposes, such as the use of petroleum refining by-products in asphalt, feedstock for petrochemical products, and coal in the production of coke for steel making (EIA 2007, Table 1.5).[2] However, although asphalt,

---

[2]See EIA 2009j. These are the latest data available.

plastics, and similar products sequester some carbon from the feedstock, most of the industrial uses generate carbon dioxide ($CO_2$) emissions as a result of processing operations—in some cases more than would be generated from direct fuel combustion, if the process itself generates $CO_2$. In addition, as more industrial production has moved offshore, energy embodied in imported goods is not counted in the EIA statistics. Ultimately, some petrochemical products may end up in waste streams that are used as an energy feedstock.

### Estimation of Industrial Use of Energy for Heating

Table 4-1 presents EIA industrial energy use estimates by primary fuel type for 2007 and presents their use projections to 2030. DOE's *Annual Energy Review* suggests facility heating in the industrial sector consumes about 10% of electricity and natural gas (EIA 2008a). Figure 4-3 presents energy use, energy intensity, output, and structural effects in the industrial

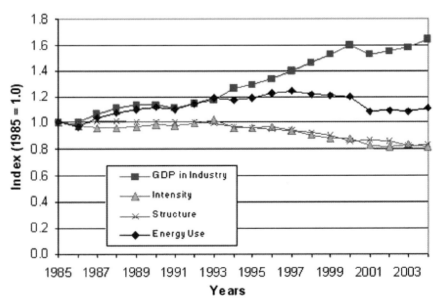

**FIGURE 4-3** Energy use, energy intensity, output, and structural effects in the industrial sector, 1985-2004. DOE uses input-output analyses to assess energy use across all U.S. industrial activities. Intensity is energy consumption per unit of demand for energy services (for example, per kilowatt hour, vehicle miles traveled, or, nationally, gross domestic product); structural effects attempt to account for variability across the spectrum of industry operations (see EIA 2003). SOURCE: DOE 2008b, Figure I1.

sector from 1985 to 2004. Table 4-1 indicates that the major industry-sector fuel sources generating externalities are about 10 quads of liquid fuels (liquefied petroleum gases, kerosene, distillate fuel oil, residual fuel oil, and gasoline), 8 quads of natural gas, and about 2 quads of coal. The mix includes about 2 quads of energy use from "renewables" (primarily hydropower, wood, and wood waste) and municipal solid waste. The two largest industrial sectors in terms of fossil-fuel consumption are the petroleum refining industry and the chemical industry. The petroleum industry in 2002 used about half of its 6 quads of total net energy use for feedstock (not associated with producing such energy products as gasoline and jet fuel); the chemical industry used more than half of its 6 quads of total net energy use for feedstock (NAS/NAE/NRC 2009d). Coal is mostly used to make coke and carbon black, but a portion of the coke ends up as heat energy in steel making. Therefore, it is only possible to make fairly crude estimates of energy use for heat in the industrial sector; estimating externalities associated with such uses is subject to even more uncertainty.

Industrial boilers are used to generate steam for a wide variety of industries and industrial processes. The 2005 survey of industrial and commercial boilers in the United States (EEAI 2005) cited earlier in the commercial sector discussion, reports that there are almost 163,000 industrial and commercial boilers in the United States that consume about 8.1 quads of fuel energy per year. Of that, about 5 quads is accounted for by industrial boilers; about 60% (3 quads) of this energy is supplied by natural gas. The industrial manufacturing sector accounts for 43,000 of the boilers, but these are of larger capacity than those used in commercial heating applications. The report presents total rated capacities for these industrial boilers at 1.6 MMBtu/h. Nonmanufacturing industrial boilers include those used in agriculture, mining, and construction. These 16,000 units have a nameplate capacity of 0.26 MMBtu/h. The remaining boilers use a diversity of fuels, often by-products of the industry involved. For example, the paper industry uses biomass waste streams to fuel 48% of its boilers (rated capacity 0.4 MMBtu/h); the primary metals industry utilizes process heat for 42% of their boilers (rated capacity 0.1 MMBtu/h); and the refining industry uses refinery by-products to fuel 49% of their boilers (rated at 0.2 MMBtu/h).

Because the U.S. industrial sector is so diverse and EIA energy statistics do not necessarily correspond to energy use for heat in this sector, the externalities attributable to industrial energy use for heat are difficult to separate from externalities associated with energy use for other industrial processes (Box 4-2).

Natural gas is the major fuel used for heating in the industrial sector. Figure 4-4 shows how its use was distributed among manufacturing sectors in 2002. Total industrial-sector natural gas consumption in 2002 was reported to be 6.47 quads with 5.8 quads used for heating and 0.67 quads used for feedstock purposes (for example, chemicals and fertilizer). Con-

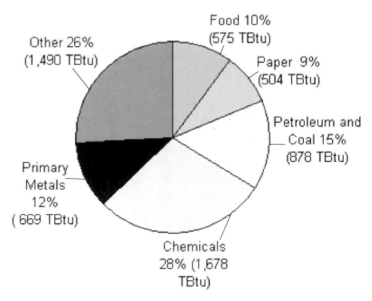

**FIGURE 4-4** Manufacturing sector consumption of natural gas as a fuel by industry, 2002. SOURCE: EIA 2006c, Figure 5.

sumption in 2007 was reported to be 8.02 quads, but the general distribution by industry sector is probably still representative.

The economic downturn at this time in the United States in progress as this report was being written is likely to reduce industrial activities to some extent. Nevertheless, the following adjusted 2007 EIA data (EIA 2009e) for the industrial sector are taken as the baseline for making rough estimates of externalities associated with heating uses in industry in our report.

- Petroleum: 10 quads – 3 quads used as feedstock = 7 quads net
- Natural Gas: 8 quads – 2 quads used as feedstock = 6 quads net
- Coal: 1.8 quads = 1.8 quads net
- Renewables (treat as biomass): 2.1 quads = 2.1 quads net

## ESTIMATES OF EXTERNALITIES ASSOCIATED WITH ENERGY USE FOR HEAT

It is much more difficult to make reliable estimates of the unpriced damages associated with energy use for heating in the buildings and industrial sectors than to evaluate such impacts for electricity generation or for transportation. However, because about 30% of U.S. primary energy is

used for heating purposes, it is important to attempt a quantification of associated damages even if detailed estimates are not possible. The residential
and commercial sector estimates are somewhat more tractable than those
for the industrial sectors, where some of the energy use reported by DOE
statistics does not sufficiently delineate fuels used as feedstocks.

Table 4-1 shows that 75% of the energy used in the residential and
commercial building sectors is natural gas. About 19% comes from liquid
fuels (with somewhat higher health impacts), about 5% comes from renewables; and less than 1% comes from coal. Detailed data were not available
on county-level consumption of wood for heating and so a comparable
damage estimation method for wood was not possible in this study. The
industrial-sector impacts involve more fuels and more diverse activities.

The focus of externalities considered in this chapter are health effects
associated with criteria-pollutant-forming emissions from fuel combustion.
GHG emission externalities that are linked to present and future changes in
climate and the associated impacts are discussed in Chapter 5.

### Residential Buildings: Damage Estimates for Criteria Air Pollutants

As shown in Table 4-1, iquid fuels and natural gas predominate in nonelectric energy consumption in residential buildings, and a small amount of
coal is used. Consumption associated with "renewables" are of a diverse
and smaller magnitude. Therefore, the focus here is on natural gas directly
burned for heating purposes; some comparisons are made with electricity
used for heating in buildings.

As noted previously, potential externalities from consuming natural gas
for heat could arise not only from the on-site combustion but also from the
upstream supply chain of extraction and distribution of the gas. The committee estimated damages attributable to criteria-pollutant-forming emissions from combustion on-site, but we were not able to estimate damages
from such emissions from upstream activities because of data and modeling
issues. Few studies have estimates these upstream emissions (for example,
Jaramillo et al. 2007), and these estimates were limited to only nitrogen
oxide ($NO_x$) and sulfur oxide ($SO_x$) emissions and showed large uncertainty
ranges. Because modeling the upstream extraction and distribution damages from criteria-pollutant-forming emissions would need to be allocated
to more than 300,000 wells and associated pipelines across the United
States (compared with the existing database of power plants for electricity
production in Chapter 2 and the roughly 100 plants for automobile production in Chapter 3), we elected not to estimate them. This is not to say the
externalities would be negligible; Jaramillo et al. (2007) estimated significant upstream emissions of $NO_x$ and $SO_x$ (ranges of 0.009-0.3 and 0.006-
0.03 lb/MMBtu, respectively) associated with North American natural gas

compared with the combustion emissions (0.094 and 0.0006 lb/MMBtu); however, the process of allocating natural gas use to the thousands of potential point and area sources, given such large uncertainty ranges in the literature, was deemed intractable because of the lack of time and resources available. Thus, we focused on criteria-pollutant-forming emissions from on-site combustion.

Our estimates of combustion externalities come from multiplying county-level consumption with county-level damages. The county-level consumption and criteria-pollutant-forming emissions of residential natural gas are taken from the EPA National Emissions Inventory (NEI) for 2002 (EPA 2008d). These emissions by county at ground level are multiplied by estimated county-level health and other damages per ton of criteria-pollutant-forming emissions from the APEEP model and are subject to the model's assumptions and limitations (see Chapters 1, 2, and 3).

The results are county-level health and ecosystem externality estimates, which are then normalized by the 2002 NEI's consumption of natural gas by county to estimate damages per thousand cubic feet (MCF) of gas.

Table 4-2 shows the national range across more than 3,000 U.S. counties of estimated damages from air pollutants (excluding $CO_2$, which is considered later), assuming the value of a statistical life (VSL) is $6 million ($7.2 million in 2007 U.S. dollars [USD]). Variability is a result of county differences. The median damage estimate is approximately 11 cents/MCF

TABLE 4-2 Residential Sector Natural Gas Use for Heat: National Damage Estimates from Air Pollutants (Excluding Greenhouse Gases) (Cents/MCF)[a] (Damage Estimated from 2002 NEI Data for 3,100 Counties)

|  | Mean | Standard Deviation | 5th Percentile | 25th Percentile | 50th Percentile | 75th Percentile | 95th Percentile |
|---|---|---|---|---|---|---|---|
| $SO_2$ | 0.37 | 2.4 | 0 | .06 | 0.16 | 0.27 | .9 |
| $NO_x$ | 26 | 180 | 1.7 | 4.9 | 8.3 | 13 | 48 |
| $PM_{2.5}$ | 0.8 | 5 | .05 | .12 | .23 | .50 | 2.1 |
| VOCs | 1.4 | 8.7 | 0.02 | 0.12 | 0.25 | 0.54 | 2.9 |
| $NH_3$ | 0.37 | 2.4 | 0 | 0 | 0 | 0 | 1.6 |
| Total (unweighted) | 35 | 230 | 3 | 7 | 11 | 18 | 72 |

NOTE: 200 counties, for which relatively few emissions data were available, were excluded to avoid skewing the distribution in an unrealistic way.

[a]Total damages (cents/MCF) are in 2007 U.S. dollars; other damages are in 2000 U.S. dollars.

ABBREVIATIONS: MCF = thousand cubic feet; NEI = National Emissions Inventory; $SO_2$ = sulfur dioxide; $NO_x$ = nitrogen oxides; $PM_{2.5}$ = particulate matter smaller than 2.5 microns; VOCs = volatile organic compounds; $NH_3$ = ammonia.

(the mean is approximately 35 cents/MCF). These estimates are unweighted; weighting the damages by county population of the source emissions would lead to an average of about 30 cents/MCF. As previously indicated, we do not include the upstream life cycle of sulfur dioxide ($SO_2$) emissions related to natural gas—such emissions are low. With the residential price of natural gas at about \$10/MCF,[3] the mean health-related externalities from criteria-pollutant-forming emissions are about 1% of the price. Aggregate damages (unrelated to climate change) were approximately \$500 million (2007 USD).

As done in the Chapter 2 on electricity, it is relevant to consider the regional variation within the United States for the externalities from heating. Table 4-3 illustrates the damage estimates on a census region basis). The median estimates of damages related to criteria-pollutant-forming emissions from different regions are similar to the national level, ranging from 6 cents/MCF to 14 cents/MCF (2007 USD). The regional breakdown highlights the large range of externalities in the South compared with other regions, but these outliers occur because of rounding errors in estimating externalities of counties with very low consumption of natural gas. Regardless, the 90th percentile values in the South still represent damages at only 5% of the price of natural gas.

These results can be used to compare the damages from natural gas combustion for heating with damages associated with using electricity for heat. From Chapter 2, production of coal-fired electricity has mean damages weighted by net generation from criteria-pollutant-forming emissions of \$0.032/kWh, and gas-fired electricity has a mean externality of at least \$.0016/kWh (excluding $CO_2$). Scaling these numbers using the weighted average national grid in Table 2-1 of Chapter 2 (48.5% coal and 21.3% gas) results in average damages near \$0.016/kWh (excluding $CO_2$; 2007 USD). Likewise the median damages from electricity would be estimated as $0.485 \times 1.8 + 0.213 \times .036 = 0.87 + \$0.0088/kWh$ (excluding GHGs; 2007 USD).

The obvious comparison to show is heating a house with natural gas versus heating a house with electricity at the national level (grid average). The average amount of electricity used to heat a house in the United States is 2,100 kWh (EIA 2009i, Table SH7). The average amount of natural gas used to heat a house is 49,000 cf. (49 MCF). Using the numbers above, we would say the estimated mean damages of electricity for heating, excluding GHGs, would be calculated as $\$0.016 \times 2,100 = \$34$/year (2007 USD), and the estimated mean damages for natural gas would be $\$0.35 \times 49 = \$17$/year (2007 USD). The estimated median damages of electricity

---

[3]A 2007 price for natural gas is used for consistency with the 2007 data on which the estimates of damages are based. Natural gas prices fluctuate and in 2009 are well below this price.

**TABLE 4-3** Residential Sector Natural Gas Use for Heat: Regional Damage Estimates (Excluding Greenhouse Gases)[a] (Cents/MCF) (Damage Estimated from 2002 NEI Data for 3,100 Counties)

|  | Mean | Standard Deviation | 5th Percentile | 25th Percentile | 50th Percentile | 75th Percentile | 95th Percentile |
|---|---|---|---|---|---|---|---|
| **MIDWEST** | | | | | | | |
| $SO_2$ | 0.4 | 2 | 0 | .05 | .15 | .22 | .49 |
| $NO_x$ | 35 | 290 | 4 | 7 | 11 | 15 | 36 |
| $PM_{2.5}$ | 0.8 | 7 | .05 | .1 | .18 | .34 | 1.1 |
| VOCs | 1 | 9 | .06 | .13 | .25 | .44 | 1.4 |
| $NH_3$ | 0.4 | 2 | 0 | 0 | 0 | .03 | 2.6 |
| Total (unweighted) | 46 | 370 | 5 | 9 | 14 | 19 | 47 |
| **NORTHEAST** | | | | | | | |
| $SO_2$ | 0.38 | 0.34 | 0 | .19 | .3 | .47 | 1.0 |
| $NO_x$ | 10 | 12 | 0 | 3.5 | 6.3 | 11 | 34 |
| $PM_{2.5}$ | 0.8 | 1.1 | 0 | 0.23 | 0.5 | 1 | 2.8 |
| VOCs | 1.1 | 1.3 | 0 | 0.35 | 0.71 | 1.5 | 3.4 |
| $NH_3$ | 0 | 0 | 0 | 0 | 0 | 0 | 0 |
| Total (unweighted) | 16 | 17 | 2 | 6 | 11 | 18 | 51 |
| **SOUTH** | | | | | | | |
| $SO_2$ | 0.4 | 1.1 | .03 | .09 | 0.2 | 0.34 | 1.1 |
| $NO_x$ | 24 | 79 | 2.1 | 4.8 | 7.8 | 13 | 64 |
| $PM_{2.5}$ | 0.92 | 3.1 | .09 | .17 | 0.31 | 0.64 | 2.5 |
| VOCs | 2 | 10 | 0 | 0.15 | 0.3 | 0.72 | 4.2 |
| $NH_3$ | 0.5 | 3 | 0 | 0 | 0 | 0 | 1.3 |
| Total (unweighted) | 33 | 107 | 3.4 | 6.8 | 11 | 18 | 93 |
| **WEST** | | | | | | | |
| $SO_2$ | .27 | 1.7 | 0 | .06 | .08 | .13 | .41 |
| $NO_x$ | 12 | 66 | .18 | 2.9 | 4.4 | 6.5 | 16 |
| $PM_{2.5}$ | 0.5 | 2.7 | 0.02 | 0.05 | .09 | .21 | 1.8 |
| VOCs | .32 | 1.6 | .02 | .06 | .09 | .18 | .83 |
| $NH_3$ | .03 | .3 | 0 | 0 | 0 | 0 | .05 |
| Total (unweighted) | 16 | 88 | 1.7 | 4 | 6 | 8.8 | 22 |

NOTE: This table reports the same data as in Table 4-2 aggregated by census region. Two hundred counties, for which relatively few emissions data were available, were excluded to avoid skewing the distribution in an unrealistic way.

[a]Total damages (cents/MCF) are in 2007 U.S. dollars; other damages are in 2000 U.S. dollars.

ABBREVIATIONS: MCF = thousand cubic feet; NEI = National Emissions Inventory; $SO_2$ = sulfur dioxide; $NO_x$ = nitrogen oxides; $PM_{2.5}$ = particulate matter smaller than 2.5 microns; VOCs = volatile organic compounds; $NH_3$ = ammonia.

and natural gas used for heat would be \$19/year and \$5/year respectively. Thus, the nonclimate damages from heating with gas instead of electricity are almost an order of magnitude less. Using the range of natural gas nonclimate-change damages would lead to results three times lower than electricity for heating at the 5th percentile, but at the 95th percentile, these natural gas damages would be about the same as nonclimate damages from using electricity for heating.

### Commercial Buildings: Damage Estimates for Criteria Air Pollutants

Following a similar method, the externalities for commercial sector heating from burning natural gas were estimated. Unlike data available for the residential sector, the 2002 NEI emissions inventory did not include corresponding estimates of natural gas consumption for the commercial sector in each county. Thus, the consumption of commercial sector natural gas was estimated by using AP-42 emissions factors for carbon monoxide (CO) of 84 lb/MCF, which do not vary drastically across combustion technologies (but adds some uncertainty to the committee's estimated consumption by county). When using this proxy, the total estimated consumption of natural gas in the commercial sector was 2.2 million MCF (or 2.2 thousand MMCF), somewhat lower than the EIA estimate for 2002 of 3.1 million MCF. Table 4-4 shows the national level range of externalities from

**TABLE 4-4** Commercial Sector Natural Gas Use for Heat: National Damage Estimates from Air Pollutants (Excluding Greenhouse Gases) (Cents/MCF)[a] (Damage Estimated from 2002 NEI Emission Data for 3,100 Counties)

| | Mean | Standard Deviation | 5th Percentile | 25th Percentile | 50th Percentile | 75th Percentile | 95th Percentile |
|---|---|---|---|---|---|---|---|
| $SO_2$ | 0.3 | 1.3 | .06 | 0.1 | 0.2 | 0.3 | 0.8 |
| $NO_x$ | 13 | 35 | 3.5 | 5.9 | 9.0 | 14 | 27 |
| $PM_{2.5}$ | 1.1 | 19 | .07 | .14 | .26 | .53 | 1.7 |
| VOCs | .65 | 2.7 | .08 | .16 | .28 | .53 | 1.7 |
| $NH_3$ | .68 | 2.6 | 0 | .03 | .13 | .44 | 2.7 |
| Total (unweighted) | 15 | 56 | 3.7 | 6.7 | 10 | 16 | 32 |

NOTE: Two hundred counties, for which relatively little emissions data were available, were excluded so as not to skew the distribution in an unrealistic way.

[a]Total damages (cents/MCF) are in 2007 U.S. dollars; other damages are in 2000 U.S. dollars.

ABBREVIATIONS: MCF = thousand cubic feet; NEI = National Emissions Inventory; $SO_2$ = sulfur dioxide; $NO_x$ = nitrogen oxides; $PM_{2.5}$ = particulate matter smaller than 2.5 microns; VOCs = volatile organic compounds; $NH_3$ = ammonia.

commercial (EIA 2009e) combustion of natural gas, the results being very similar to those for the residential sector. The median externality excluding GHG emissions, 11 cents/MCF, is plausible given its similarity to the residential damage estimate. Given the similarity, the externality estimates by census region for the commercial sector are not shown. Aggregate damages are about $300 million (excluding damages related to climate change) (2007 USD).

### Externalities Associated with Industrial Energy Use for Heating

Unfortunately a parallel analysis for externalities from heating in the industrial sector could not be undertaken because of several challenges. The level of detail for the residential and commercial sectors in the NEI, including fuel consumption by county, was not available for the industrial sector. While the NEI has estimates of emissions from industrial activities by county, disaggregating the emissions to include only estimates from the use of fuels for heating, and from industrial activities that would not be included elsewhere in this report, proved too problematic to overcome. Thus, the committee was able to make only qualitative assessments of these externalities.

Externalities associated with present energy uses for heat in the industrial sector might be approximated by using estimates of the externalities that are associated with the use of the particular energy source in electricity generation. These are the externalities caused by the production, processing, transportation, and combustion of the particular fuel (petroleum, natural gas, biomass, or coal) in electricity power plants, scaled by the annual use factors for the industrial sector and the electricity generating sector. This method gives approximate results for GHG emissions but is very crude for estimates of local health and environmental impacts because of large differences in emissions from power plants, and those associated with a wide range of industrial facilities that use energy for heat as a part of wider manufacturing activities. Specific local health and environmental effects are different for industrial locations, so the estimates are subject to considerable uncertainty. The EIA provides national and regional (Northeast, Midwest, South, and West) data for energy use by industry according to the North American Industry Classification System (NAICS) codes (EIA 2007, Table 3.2). Unfortunately many of the details are missing in regional summaries because errors are too high or because specific plants might be identified. Therefore, it was not possible to identify the locations of large emitters that might have more significant local health and environmental effects.

Table 4-1 shows that natural gas use in the industrial sector, less use for feedstock, was about 6 quads in 2007. This usage is actually smaller than the 8 quads of natural gas used in the residential and commercial building sectors. Therefore, in the absence of more detailed information, it may be assumed that the health and environmental externalities of this usage are probably of the same order or less than the impacts associated with natural gas use for heat in buildings. A very rough order of magnitude estimate of average externalities associated with the industrial sector use of natural gas is therefore 11 cents/MCF, excluding GHG damages. Thus, the 6 quads of natural gas used for industrial heat would generate about $600 million in damages.

The other large use of fuel for industrial energy is associated with liquid fuels (about 7 quads). Half of this use is associated with only three sectors—paper, chemicals, and petroleum refining—the last of which has already been included as part of the life-cycle upstream externality estimates made for the transportation sector use of petroleum fuels.

Table 4-1 infers that the nonelectric use of energy in the industrial sector is almost double that of the residential and commercial building sectors combined. However, when feedstock use is taken out of the fuel mix for the industrial sector, the remaining use of energy for heat is probably about equivalent to the residential and commercial building sectors. Figure 4-5 also shows that the GHG emissions from the industrial sector have been declining since 2000, and the building sectors show only small increases.

The EIA energy use projections for 2030 incorporate some consid-

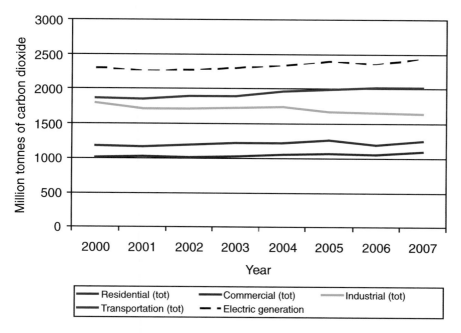

**FIGURE 4-5** Greenhouse gas emissions in the United States by sector. Totals include electric power use distributed across the end-use sectors. SOURCE: EIA 2008f.

eration for the incorporation of energy efficiency improvements to offset energy demand from growth in industrial capacity (EIA 2009e). These projections are based on present energy policies and could vary considerably if new policies are adopted. The estimates for nonelectric usage show a very small growth in energy use between now and 2030. Three efficiency studies, discussed later in this chapter, indicate that there is a good potential for achieving about 10-15% overall improvement in efficiency in the use of energy to produce heat in the industrial sector if more aggressive energy policies are adopted in the future (Interlaboratory Working Group 2000; IEA 2007; McKinsey 2007). Therefore, nonclimate externalities in 2030 might be about the same as those associated with the fuel uses that exist today—or they might be reduced by 10-15%. The most likely source of reducing externalities per unit of heat would be from changes in the electricity sector, as emissions from natural gas are relatively small and already well-controlled.

When considering externalities from heat in 2030, potential changes in energy sources should be considered. Over such a short time frame, substantial infrastructure replacements or changes (for example, by moving from electricity or gas to an alternative) are unlikely to occur. Instead, alternative sources of natural gas, such as those from shale deposits or from increased imports of liquified natural gas (LNG), are more likely to become prevalent. Our analysis has presumed domestically sourced natural gas for estimating externalities, both because LNG imports are currently small and because estimating the health damages from global operations was not possible. However the upstream emissions of LNG have been estimated to be somewhat higher. Thus, the externalities estimated here are low because they do not consider these upstream extraction emissions from any sources of natural gas.

## EMISSIONS OF GREENHOUSE GASES

Figure 4-5 shows the GHG emissions (in millions of metric tonnes[4] of $CO_2$ per year) for each of the end-use sectors. The sector estimates also include electricity use apportioned to use within the sector. To estimate the GHG emissions associated with heating uses, it is necessary to deduct the electricity component (also shown in Figure 4-5).

• 2007 residential and commercial sector emissions = $1,250 + 1,087 - (0.73)(2,433) = 561$ million tonnes of $CO_2$ (618 tons). This amount is roughly equivalent to about 10 quads of natural gas.
• 2007 industrial sector emissions = $1,640 - (0.27)(2,433) = 983$ million tonnes of $CO_2$ (1,084 tons).

---

[4]One tonne equals 2,200 pounds.

For natural gas, each MCF generates about 1,000 MMBtu on average and generates about 120 pounds of $CO_2$ (about 0.05 tonnes or 0.06 tons of $CO_2$). Methane, the major component of natural gas, is a GHG itself if it enters the atmosphere through leakage. It has a warming potential about 25 times that of $CO_2$.[5] However, EPA (2009i) estimated that such leakage amounts to about 3% of total U.S. $CO_2$-equivalent ($CO_2$-eq) emissions (excluding water vapor) attributable to energy-related activities in 2007. Nitrous oxide ($N_2O$) is also a GHG, but the emissions from its use are a very small share of total GHG emissions.

In the sections above, the committee estimated externalities from criteria-pollutant-forming emissions from the residential and commercial building sectors. Aggregate damages from combustion of natural gas for direct heat are estimated to be about $1.4 billion per year (2007 USD), assuming the magnitude of effects resulting from heat production in industrial activities is comparable to those of the residential and commercial sectors. These estimates did not include emissions of GHGs. Emissions of GHGs associated with burning natural gas can be estimated in a fairly straightforward manner, about 120 lb (0.06 tons) of $CO_2$-eq/MCF (EPA AP-42). Although the committee did not estimate damages related to criteria-pollutant-forming emissions from upstream activities because of spatial and geographical modeling concerns, we did estimate the emissions of upstream GHGs from natural gas from literature sources. Jaramillo et al.(2007) summarized estimates of upstream natural gas emissions of 15-20 lb $CO_2$-eq/MCF for North American sources or 30-70 lb $CO_2$-eq /MCF for LNG, adding about 15% and 40% to the emissions, respectively. Thus, in the near term, where domestic natural gas remains the dominant source, the emissions factor is likely to be approximately 140 lb $CO_2$/MCF (including upstream methane emissions), and in the long term, where LNG or shale gas is increased as part of the mix, the average emissions factor could be 150 lb $CO_2$-eq/MCF.

## POTENTIAL DAMAGES REDUCTIONS IN 2030

### Residential Buildings

The major options for reducing heating energy demand in the future are presented in more detail in NAS/NAE/NRC (2009d). The report focuses on the potential for reducing total energy consumption in the residential sector, where primary energy used to provide electricity is much greater than the consumption of nonelectric primary energy for heat. The ways to reduce energy use for heating mainly focus on better insulation of the building

---

[5]Based on 100-year GWP values.

envelope and use of higher efficiency methods for water heating. The main possibilities for existing and new buildings include the following:

- Existing buildings:
  o Addition of insulation to exterior walls and under roofs.
  o Replacement of old windows with high performance windows.
  o Replacement of old furnaces with higher efficiency devices.
  o Use of control systems to minimize heating of unoccupied spaces (except to prevent freezing of water pipes) and to lower temperatures at night.
  o Addition of solar thermal water heating.
  o Behavioral changes, such as reducing thermostat settings, sensible opening and closing of windows and shades, wearing warmer clothing to lower indoor set temperatures during the heating season (and cooler clothing to higher set temperatures during the cooling season).
  o Reducing air leakage from the building (with care to assuring that indoor air quality is not compromised—some heat exchangers are available to preheat inlet air with warm exhaust air).
- New construction:
  o Energy efficient design for site location—using passive solar heating, shading, geothermal heat pump systems, combined heat and power systems, high efficiency walls and windows, natural ventilation for warm climates, smart control systems, and many other techniques (see Box 4-3).
  o Behavioral changes to accommodate to smaller living spaces per capita or more desirable multiple dwelling units designed for energy efficiency.

Projecting how improvements in the heating demand for residences might evolve to a 2030 time period seems highly uncertain, although some improvements to at least offset sector growth appear feasible. As of 2008, the DOE's *Annual Energy Outlook* forecasted energy use in the residential sector to increase by 16% (0.4% per year) by 2030. The trends resulting from current financial issues are also likely to keep the 2030 energy use in this sector stable or reduced relative to 2007. This potential is discussed in the AEF.report (NAS/NAE/NRC 2009d).

### Commercial Buildings

Because space heating requirements are largely determined by the building envelope characteristics, there are limited opportunities for reducing the heating energy requirements for existing buildings. Replacement of

---

**BOX 4-3**
**Zero-Energy Concept Home**

Architects and engineers currently discuss the prospects of residences that have zero net demand for energy. This "zero-net" demand does not imply that the buildings have no energy demand; rather the buildings have technologies integrated into them (such as solar panels or geothermal wells); thus, they do not demand energy from beyond what the home is able to supply itself. Achieving such a goal depends on several innovative design changes in the residential housing industry. First, technologies need to be cost-effectively scaled to what can fit within the space and budget of a house. Second, to be able to generate enough of its own energy needs, a house needs to be designed to achieve much lower energy use regardless of energy source. Such designs include using advanced lighting (for example, solid-state lighting, also known as LED lighting), orienting the house to take advantage of sunlight, and improving insulation. Design for efficient utilization of space is also important to eliminate the need for heating and cooling of rarely used floor area. These measures reduce the demand for energy and allow the on-site energy-generating technologies to better supply this needed level of demand, making the "zero-energy" goal possible.

---

windows, upgrading of furnaces, boilers, and heat distribution systems may offer some improvement, but such investments may have limited cost-effectiveness except for much older or larger buildings. The AEF report (NAS/NAE/NRC 2009d) focuses primarily on larger efficiency gains that can be achieved through improvements in the use of electricity. Ironically, large amounts of waste energy from electric devices, such as those from inefficient electronic power supplies that give off heat, may reduce the heating load required for a commercial building (but increase the cooling load in warm weather).

New commercial buildings offer many more opportunities for investing in new designs to reduce heating energy requirements. Because commercial floor space has been expanding rapidly over the past decade, this space may be an area for innovative new buildings that greatly reduce their thermal energy (energy for heating) footprints. The U.S. Green Business Council has developed the Leadership in Energy and Environmental Design (LEED) Green Building Rating System, which uses third-party experts to evaluate new commercial buildings (or renovations of existing facilities) for their overall environmental and community performance and award ratings based on criteria that include energy efficiency of design and construction, as well as ease of maintenance, quality of working environment, and waste minimization (U.S. GBC 2008). The highest LEED rating is "platinum," followed by "gold," "silver," and "certified." The DOE has expanded its

Energy Star program to include green building design and makes available a variety of resources for the improved design, construction, and operation of commercial buildings (DOE 2009c). Investment in improved efficiency reduces energy use and eliminates any externalities that would be associated with the energy use avoided.

## Case Study of Passive Solar Design

One of the approaches listed in Box 4-3 for a hypothetical "zero-energy home" is the incorporation of passive solar design principles. Unlike most of the other technologies listed in this report (and the NAS/NAE/NRC reports), passive solar is not an energy-generating technology.

Passive solar design uses the light and heat of the sun to offset what would otherwise be energy or fuel use in a building. For example, using more skylights in the ceiling and arranging the layout so that the light is able to permeate wide areas of the living space reduces the amount of electricity needed for lighting. Similarly, creating south-facing windows allows the heat of the sun to enter the building and reduce needs for heating by other means.

There are multiple considerations for incorporating passive solar design and various cost trade-offs. It was not feasible for the committee to assess how the impacts of a passive solar house might compare with a traditional fossil-energy-fueled building, and no significant literature was found to have quantified the relevant trade-offs. For direct heat, south-facing windows with special glazing are used that then allow sunlight to enter and reflect off dark masonry floors, which absorb heat. This stored heat then is slowly released while the home cools later in the day (reducing the heating load). Heavier walls can be designed to store and release heat to even out variations between day and night outdoor temperatures. Solar heating can also drive natural circulations within buildings and provide more comfort without circulating devices requiring purchased energy. Case studies have shown massive energy reductions (up to 90%) for comparable new buildings; retrofits are more limited in their ability to reduce impacts and externalities because of various fixed design choices in the existing structures.

Although the operating phase of the building may require substantially less energy (and thus result in far lower externalities from heat production), the initial construction or renovation of the building along passive solar design principles may result in significant externalities from manufacturing new insulation, windows, or other intensive construction materials. These initial impacts can be apportioned over the total operating life of the building to provide life-cycle annual impacts. A full comparison of these externalities is outside the scope of this study but should be considered before viewing a passive solar house as being externality-free.

## Integrated Planning Opportunities

Construction and retrofit of residential and commercial buildings can use integrated design principles with a goal of improving the buildings' environmental and economic performance over their operating lifetimes. Integrated planning is more effective when initiated early in the design and construction process (for example, subsurface geothermal systems or siting for incorporation of photovoltaic or passive solar systems) but can also have benefits for retrofits. Combined heat and power (CHP) systems are now available that generate electricity and use the waste energy for heating purposes. When building designs reflect the extremes of local climate, operation of the building reduces both heating and cooling loads. Commercial buildings generate waste streams that can be used for energy on-site or off-site. Incorporation of recycled components into building materials is another way to reduce the life-cycle impacts of buildings. Construction debris represents a substantial component of the waste stream, and much of this debris can be recycled or converted to thermal energy and electricity with proper environmental control. Like energy, water conservation and reuse is another important component—as is indoor air quality and its associated health and productivity impacts.

## Industrial Facilities

NAS/NAE/NRC (2009d) devotes a chapter to the potential for energy efficiency improvements in the industrial sector and provides specific examples of how individual industry sectors can reduce energy demand through approaches including use of waste heat, and more efficient new technologies. The report looks at total energy use, including electricity and thermal energy use. It notes that the most energy-intensive industries (for example, petroleum refining, aluminum, iron and steel, and chemicals) have already placed high emphasis on efficiency of use, especially in domestic facilities.

Some improvements in these industries are possible; however, larger improvements seem available in the pulp and paper industry and through waste energy utilization and use of combined heat and power systems. When energy costs are a smaller fraction of total costs, companies may pay less attention to investing in efficiency. However, as energy prices increase, industry is geared to respond with innovations in efficiency, more utilization of waste heat, and new processes to reduce its energy consumption. Nevertheless, industry is more reluctant to initiate investment in energy efficiency when future energy prices are uncertain or volatile. Likewise, under poor market conditions, expenditures for process improvements are likely to be deferred.

Several studies cited in NAS/NAE/NRC (2009d) estimate the potential for further energy efficiency savings in the U.S. industrial sector as a whole:

- Savings potential of 18-26% (IEA 2007).
- Savings potential of 3.9 quads of energy reduction (about 12%) in 2020 (McKinsey 2007).
- Savings potential of 16.6% from 2000 to 2020 with advanced policies (Interlaboratory Working Group 2000).

With continuing emphasis on energy efficiency in industry and likely increases in the cost of energy, energy use for industrial heating in 2030 will probably be somewhat lower than levels in 2007. NAS/NAE/NRC (2009a) finds the following:

- "Independent studies using different approaches agree that the potential for improved energy efficiency in industry is large. Of the 34.3 quads of energy forecast to be consumed by U.S. industry in 2020 (EIA 2008b), 14 to 22% could be saved through cost-effective energy efficiency improvements (those with an internal rate of return of at least 10 percent or that exceed a company's cost of capital by a risk premium). These innovations would save 4.9 to 7.7 quads annually."
- "Additional efficiency investments could become attractive through accelerated energy research, development, and demonstration. Enabling and crosscutting technologies—such as advanced sensors and controls, . . . and high temperature membrane separation—could provide efficiency gains in many industries as well as throughout the energy system . . . ."

## SUMMARY

Externalities associated with heat production come from all sectors of the economy—residential and commercial buildings and industry. Most heat is generated from combustion of natural gas or from electricity. Combustion of natural gas results in relatively lower emissions compared with emissions from coal combustion, which is the main energy source for electricity generation. Therefore, damages related to providing heat directly from natural gas combustion are much less than damages related to use of electricity for heat. The better emissions performance of natural gas for direct heat also is reflected in the externality estimates of 11 cents/MCF (2007 USD) each for residential and commercial use, excluding GHGs. These results do not vary much regionally, although some counties have much higher externalities than others. Assuming industrial externality is 11 cents/MCF, aggregate damages from combustion of natural gas for direct heat is approximately $1.4 billion per year. The industrial sector contribution to this estimate reflects only natural gas use for heat generation. Including externalities from petroleum combustion, which is on the same scale of energy use as natural gas for industrial heat generation, would lead to a higher estimate of aggregate damages from energy use for heat. Available data are insufficient to conduct a parallel analysis of industrial activities that generate useful heat. This situation could be improved with greater attention by EIA to collecting fuel consumption data by county and to provide additional resolution to emissions from disaggregated industrial activities.

The results represented here are the result of an end-use assessment—that in terms of providing heat, natural gas has lower externalities than electricity. It is not an assessment of how or where to use natural gas, which can be used for direct combustion or indirectly as a fuel for generating electric power.

## Overall Implications of the Results

1. Aggregate damages associated with criteria-pollutant-forming emissions from the use of energy (primarily natural gas) for heating in the buildings and industrial sectors are low relative to damages from energy use in the electricity-generation and the transportation sectors.

2. GHG emissions associated with the use of energy (primarily natural gas) for heating in the buildings and industrial sectors are low relative to GHG emissions associated with transportation and electricity production because natural gas carbon intensity is lower than that of coal and gasoline.

3. The largest potential for reducing damages associated with the use of energy for heat lies in greater attention to improving the efficiency of energy use. NAS/NAE/NRC (2009a) suggests a potential for improving efficiency in the buildings and industrial sectors by 25% or more—with the likelihood that emission damages in these sectors could be held constant in spite of their growth between now and 2030.

## Future Research Needs

1. Assessment of energy use and its impacts in the industrial sector in particular (but in all sectors to some extent) could be improved by more extensive databases that contain details about specific forms of energy use and associated waste streams. Such databases should be designed so that life-cycle analysis of alternatives can be made without inadvertent double counting.

2. A more quantitative assessment of industrial sector externalities done collaboratively by the government and industry would be valuable in informing priority setting for future initiatives to reduce the externalities associated with industrial operations. Such an assessment was not possible in this study largely because of data limitations.

# 5

# Climate Change

## OVERVIEW OF QUANTIFYING AND VALUING CLIMATE-CHANGE IMPACTS

Burning fossil fuels creates externalities through its impact on the stock of greenhouse gases (GHGs) in the atmosphere and the subsequent effects of GHG concentrations on climate. This chapter provides a general overview of these effects and various attempts that have been made to quantify and monetize the damages associated with GHG emissions. The chapter begins by summarizing information on trends in Earth's temperature over the past century, the relationship between GHG concentrations and climate, and predictions of future changes in climate associated with various emissions trajectories. That summary is followed by an overview of the approach that economists have taken to quantifying the damages associated with GHG emissions, including a discussion of three integrated assessment models (IAMs), which provide estimates of the monetary impacts of GHG emissions. Given its resource constraints, it was not feasible for the committee to conduct a detailed critical review of the IAMs.

Estimates of the damages associated with GHG emissions in IAMs rest on estimates of the physical and monetary impacts of temperature changes in various market and nonmarket sectors. The next section of the chapter describes the physical impacts of climate change on weather, snow and ice formations, and water systems. That is followed by estimates of the physical and monetary impacts of climate change on individual market and nonmarket sectors, including water, agriculture, coastal infrastructure, health, and ecosystems. The next section discusses how monetary impacts

*248*

reported in the literature are aggregated across sectors and countries and presents estimates of the marginal damage of a ton of carbon dioxide equivalent[1] ($CO_2$-eq) from various IAMs. The committee did not conduct its own modeling analyses of damages related to climate change. We determined that attempting to estimate single values would be inconsistent with the rapidly changing nature of knowledge about climate change and the extremely large uncertainties associated with estimation of climate-change effects and damages.

### Climate-Change Observations, Drivers, and Future Projections

According to the Intergovernmental Panel on Climate Change (IPCC), scientists have documented that Earth's climate system is warming, the last decade was the warmest on record, global average temperatures have increased about 1.3°F since 1990, and sea levels at the end of the 20th century were rising almost twice as fast as over the century as a whole (IPCC 2007a,b).[2] Arctic sea ice and glaciers are rapidly shrinking. Economic losses from extreme weather events, such as tropical cyclones, heavy rain storms, flooding, severe heat waves, and droughts, are increasing rapidly (CCSP 2008).

The IPCC states that "most of the observed increase in global average temperatures since the mid-twentieth century is *very likely* due to the observed increase in anthropogenic GHG concentrations" (IPCC 2007a, p.5). With high and increasing confidence, a range of "fingerprinting" techniques attribute a substantial fraction of recent warming to anthropogenic causes (IPCC 2007a).

Although the greenhouse effect is a natural process necessary for life on Earth, humans have inadvertently intervened in this process so that the greenhouse effect is now trapping additional heat in Earth's atmosphere, which is driving climate change. Specifically, human activities have led to a significant increase in the amount of $CO_2$ and methane ($CH_4$) in the atmosphere. These additional GHGs absorb more energy and let less heat escape to space. Therefore, Earth's climate is warming.[3]

GHG emissions have steadily grown since the Industrial Revolution, with a 70% increase between 1970 and 2004. Burning fossil fuels, agri-

---

[1]$CO_2$-eq expresses the global warming potential of a GHG, such as methane, in terms of $CO_2$ quantities.

[2]The IPCC is an intergovernmental scientific body given to the assessment of climate change. It does not conduct research. IPCC estimates are derived from literature reviews and assessments, not from its independent predictions or projections.

[3]Airborne particles may have either a warming or cooling effect. Sulfate particles reflect incoming sunlight and cause a cooling effect at the surface. Other types of particles, referred to as carbon black, absorb incoming sunlight and trap heat in the atmosphere.

culture, and deforestation are the primary anthropogenic sources of these GHG emissions. In 2004, the burning of fossil fuels accounted for 56.6% of the GHGs emitted. Of the total anthropogenic emissions released in 2004, energy supply produced 25.9%, transportation produced 13.1%, and industry produced 19.4% (Figure 5-1) (IPCC 2007a).

### Future Projections

Using global climate models, scientists predict that, in the absence of concerted action to reduce GHG emissions, climate will warm substantially over the next century. The IPCC has developed scenarios that characterize a wide range of internally consistent, feasible alternative futures, characterized by trajectories in population, industrialization, governance, gross domestic product (GDP), and GHG emissions (IPCC 2000). By inputting these emission scenarios into global climate models, scientists have developed sophisticated estimates of what atmospheric temperatures could look like in 2100 (Figure 5-2).

If carbon concentrations were kept constant at the level produced in 2000, these models predict that Earth's climate would continue to warm (see Figure 5-2). Scenario A2 describes a heterogeneous world with a focus on self-reliance and regional identity and having relatively slow economic and technology growth. This scenario ends the 21st century with very high emissions and dramatic warming. Scenario A1B describes a future with rapid economic growth and human population that peaks around 2050 and then starts to decline. This scenario assumes significant interregional cooperation and a balanced portfolio of energy sources. A1B predicts continued warming that starts to slow by 2100. Scenario B1 describes the same population and economic trends as in scenario A1B. However, B1 incorporates a rapid shift toward a service and information economy, reduced material intensity, and the widespread adoption of efficient low-carbon energy technologies. B1 predicts a less dramatic increase in global average temperatures.

Since 2000, industrial carbon emissions have increased more rapidly than in any of the scenarios (Raupach et al. 2007). Moreover, natural feedback processes, such as melting permafrost and more extensive wild fires, are releasing carbon into the atmosphere more quickly than anticipated (IPCC 2007b). On the other hand, as of mid-2009, the carbon budget data have not yet been updated to reflect changes resulting from the global economic crisis of 2008-2009.

The U.S. Global Change Research Program (GCRP) concluded that climate-related changes are already under way in the United States and surrounding coastal waters, and the quantity and growth rate of these changes are dependent upon human choices in the present day (Karl et al. 2009).

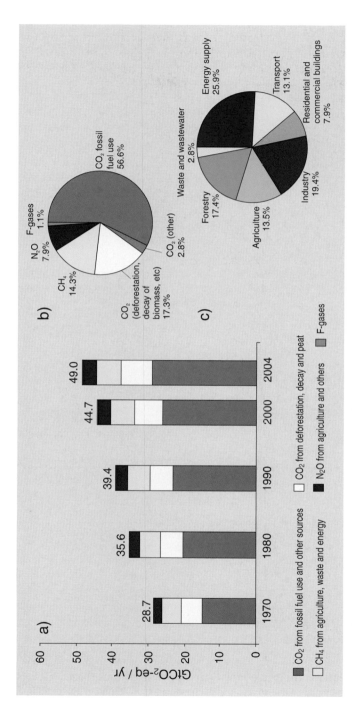

**FIGURE 5-1** Global anthropogenic greenhouse gas (GHG) emissions. (*a*) Global annual emissions of anthropogenic GHGs from 1970 to 2004; (*b*) share of different anthropogenic GHGs in total emissions in 2004 in terms of $CO_2$-equivalent; and (*c*) share of different sectors in total anthropogenic GHG emissions in terms of $CO_2$-equivalent (forestry includes deforestation.). SOURCE: IPCC 2007a, p. 5, Fig. SPM.3. Reprinted with permission; copyright 2007, Intergovernmental Panel on Climate Change.

252

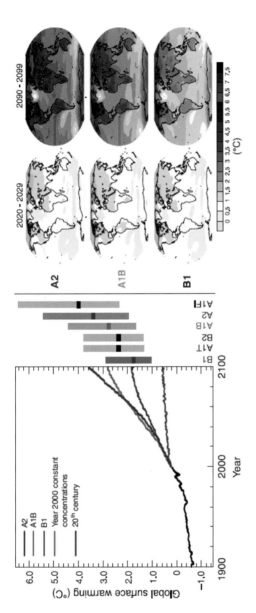

**FIGURE 5-2** Atmosphere-Ocean General Circulation Model (AOGCM) projections of surface warming. Left panel: Solid lines are multimodel global averages of surface warming (relative to 1980-1999) in the *IPCC Special Report on Emission Scenarios* (SRES) A2, A1B, and A1, shown as continuations of the 20th century simulations (IPCC 2000). The orange line is for the experiment where concentrations were held constant at year 2000 values. The bars in the middle of the figure indicate the best estimate (solid line within each bar) and the probable range assessed for the six SRES marker scenarios at 2090-2099 relative to 1980-1999. The assessment of the best estimate and probable ranges in the bars includes the AOGCMs in the left part of the figure, as well as results from hierarchy of independent models and observational constraints. Right panels: Projected temperature changes for the early and late 21st century relative to the period 1980-1999. The panels show the multi-AOGCM average projections for the A2 *(top)*, A1B *(middle)*, and B1 *(bottom)* SRES averaged over decades 2020-2029 *(left)* and 2090-2099 *(right)*. SOURCE: IPCC 2007b, p. 14, Figure SPM.5; Right Panel: IPCC 2007b, p. 15, SPM.6. Reprinted with permission; copyright 2007, Intergovernmental Panel on Climate Change.

## IPCC Mitigation Findings

The IPCC concluded that the impacts of climate change can be reduced, delayed, or avoided through mitigation strategies designed to stabilize atmospheric carbon concentrations. These concentrations can be stabilized primarily by reducing anthropogenic carbon emissions and, secondarily, by increasing carbon sinks (see Table 5-1). Figure 5-3 depicts the future carbon emission profiles needed to achieve the various stabilization concentrations and the global mean temperature associated with each stabilization concentration. The IPCC strongly suggests that the technology needed to achieve the needed stabilization levels is already or will very soon be available. They also claim that 60-80% of the needed emission reductions would have to come from the energy sector, via a shift to noncarbon-based energy sources and energy efficiency (IPCC 2007a,d)

In response to a request from Congress, the National Research Council (NRC) has undertaken America's Climate Choices (ACC), a suite of studies designed to inform and guide responses to climate change across the nation. A final ACC report, addressing strategies to reduce or adapt to the impacts of climate change, is expected to be complete in 2010.

## Overview of Quantification Methods, Key Uncertainties, and Sensitivities

### Defining the Marginal Damage of GHG Emissions

The combustion of fossil fuels is a major source of GHG emissions, which create externalities through their impact on the stock of GHGs in the atmosphere and the subsequent effects of GHG concentrations on climate. Evaluating the external costs of energy due to climate change is a daunting task. The principal difficulty is the complexity arising from the fundamental dimensionality of the climate problem. The relevant dimensions are time (indexed by $t$), location (indexed by $l$), the set of relevant climatic variables (indicated by the index $c$), the categories of physical impacts as a result of climate changes (indexed by $i$), and the categories of damage incurred by these impacts (indexed by $d$). The external cost of an additional ton of GHGs, $E_0$, emitted at time $t = 0$ depends on the following:

a.   The effect of emissions on Earth-system processes and, in turn, climatic variables in candidate locations over future time periods, $C_{c,l,t}$.

b.   The contemporaneous effect of climate changes in each location on various categories of physical impacts, $I_{i,l,t}$.

c.   The contemporaneous effect of impacts in each location on various categories of damage, $D_{d,l,t}$.

**TABLE 5-1** Characteristics of Post-Third Assessment Report (TAR) Stabilization Scenarios and Resulting Long-Term Equilibrium Global Average Temperature and the Sea-Level Rise Component from Thermal Expansion Only

| Category | $CO_2$ Concentration at Stabilization (2005 = 379 ppm)[b] | $CO_2$-Equivalent Concentration at Stabilization Including GHGs and Aerosols (2005 = 375 ppm)[b] | Peaking Year for $CO_2$ Emissions[a,c] | Change in Global $CO_2$ Emissions in 2050 (Percent of 2000 Emissions)[a,c] | Global Average Temperature Increase above Preindustrial Level at Equilibrium, Using "Best Estimate" Climate Sensitivity[d,e] | Global Average Sea-Level Rise above Preindustrial Level at Equilibrium from Thermal Expansion Only[f] | Number of Assessed Scenarios |
|---|---|---|---|---|---|---|---|
| | ppm | ppm | Year | Percent | °C | Meters | |
| I | 350-400 | 445-490 | 2000-2015 | −85 to −50 | 2.0-2.4 | 0.4-1.4 | 6 |
| II | 400-440 | 490-535 | 2000-2020 | −60 to −30 | 2.4-2.8 | 0.5-1.7 | 18 |
| III | 440-485 | 535-590 | 2010-2030 | −30 to +5 | 2.8-3.2 | 0.6-1.9 | 21 |
| IV | 485-570 | 590-710 | 2020-2060 | +10 to +60 | 3.2-4.0 | 0.6-2.4 | 118 |
| V | 570-660 | 710-855 | 2050-2080 | +25 to +85 | 4.0-4.9 | 0.8-2.9 | 9 |
| VI | 660-790 | 855-1130 | 2060-2090 | +90 to +140 | 4.9-6.1 | 1.0-3.7 | 5 |

[a]The emission reductions to meet a particular stabilization level reported in the mitigation studies assessed here might be underestimated because of missing carbon-cycle feedbacks.

[b]Atmosphere $CO_2$ concentrations were 379 parts per million (ppm) in 2005. The estimate of total $CO_2$-equivalent concentration in 2005 for all long-lived greenhouse gases (GHG) is about 455 ppm, and the corresponding value including the net effect of all anthropogenic forcing agents is 375 ppm $CO_2$-eq.

[c]Ranges correspond to the 15th to 85th percentile of the post-TAR scenarios distribution. $CO_2$ emissions are shown so multigas scenarios can be compared with $CO_2$-only scenarios.

[d]The best estimate of climate sensitivity is 3°C.

[e]Note that global average temperature at equilibrium is different from expected global average temperature at the time of stabilization of GHG concentrations due to the inertia of the climate system. For the majority of scenarios assessed, stabilization of GHG concentrations occurs between 2100 and 2150.

[f]Equilibrium sea-level rise is for the contribution from ocean thermal expansion only and does not reach equilibrium for at least many centuries. These values have been estimated using relatively simple climate models (one low-resolution AOGCM [Atmosphere-Ocean General Circulation Model] and several EMICs [Earth System Models of Intermediate Complexity] based on the best estimate of 3°C climate sensitivity) and do not include contributions from melting ice sheets, glaciers, and ice caps. Long-term thermal expansion is projected to result in 0.2 to 0.6 m per degree Celsius of global average warming above preindustrial levels.

SOURCE: IPCC 2007a, p. 20, Table SPM.6. Reprinted with permission; copyright 2007, Intergovernmental Panel on Climate Change.

255

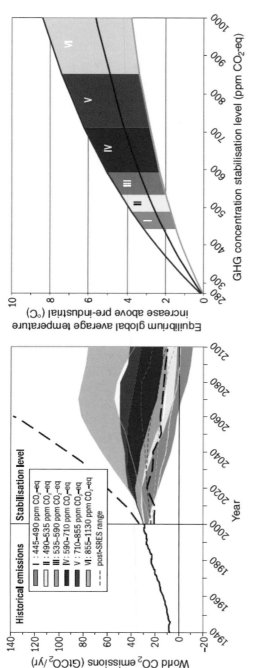

**FIGURE 5-3** Global $CO_2$ emissions for 1940 to 2000 and emission ranges for categories of stabilization scenarios from 2000 to 2100 (*left panel*); and the corresponding relationship between the stabilization target and the probable equilibrium global average temperature increase above preindustrial levels (*right panel*). Approaching equilibrium can take several centuries, especially for scenarios with higher levels of stabilization. Colored shadings show stabilization scenarios grouped according to different targets (stabilization categories I to VI). The right-hand panel shows ranges of global average temperature change above preindustrial levels, using (i) "best-estimate" climate sensitivity of 3°C (black line in middle of shaded area), (ii) upper bound of probable range of climate sensitivity of 4.5°C (red line at top of shaded area), and (iii) lower bound of probable range of climate sensitivity of 2°C (blue line at bottom of shaded area). Black dashed lines in the left panel give the emissions range of recent baseline scenarios published since IPCC (2000). Emission ranges of the stabilization scenarios comprise $CO_2$-only and multigas scenarios and correspond to the 10th to 90th percentile of the full scenario distribution. NOTE: $CO_2$ emissions in most models do not include emissions from decay of above-ground biomass that remains after logging and deforestation and from peat fires and drained peat soils. SOURCE: IPCC 2007a, p. 21, Figure SPM.11. Reprinted with permission; copyright 2007, Intergovernmental Panel on Climate Change.

The dependence can be summarized in Equation 5-1, which is the analogue of the impact-pathway approach used in the analysis in Chapters 2 and 3:

$$MD_0 = \sum_t \sum_l \sum_c \sum_i \sum_d \left[ \underbrace{\left(\frac{\partial C_{c,l,t}}{\partial E_0}\right)}_{\Delta EC} \underbrace{\left(\frac{\partial I_{i,l,t}}{\partial C_{c,l,t}}\right)}_{\Delta CI} \underbrace{\left(\frac{\partial D_{d,l,t}}{\partial I_{i,l,t}}\right)}_{\Delta ID} \right] \delta_t, \quad \text{Equation 5-1}$$

where $\delta_t = (1 + r)^{-t}$ is a factor that discounts damages in future year $t$ back to the present, and $r$ is the rate of discount. Effect b is captured by the term $\Delta CI$, which summarizes the results of physical impact models. These models suggest how changes in temperature and precipitation may affect agricultural yields or how changes in climate will affect biodiversity. Effect c is captured by the term $\Delta ID$, which captures the monetary damages associated with changes in agricultural yields or loss of species diversity. Attempts to measure these damages are reviewed briefly later in this section. In many cases, the relationships between climate impacts and damages are based on judgment, assumptions, or analogy because data are lacking.

This last point highlights the second difficulty facing any assessment of the costs of climate change: lack of information and uncertainty regarding effects a-c. The terms represented by $\Delta EC$, which include the extent of ice sheet melting and shifts in regional distribution of precipitation are still subject to considerable uncertainty. A vast amount of effort is actively being dedicated to elaborating the elements of $\Delta CI$. However, while natural scientists mostly agree on the climatic variables whose impacts should be studied, there is less consensus on what impacts are significant and should be examined. Moreover, even those impacts thought to be significant (for example, species loss) respond to changes in climate in ways that are poorly understood. The difficulties in estimating monetary damages are described in more detail below.

The climate equivalent of the Air Pollution Emission Experiments and Policy (APEEP) model would embody estimates of $\Delta EC$, $\Delta CI$, and $\Delta ID$ and combine them according to Equation 5-1 to produce a summary measure of marginal damage. The committee did not have access to an integrated assessment model (IAM) of this kind. Indeed, such a model does not exist—the IAMs used for climate studies are designed not to produce descriptively realistic, spatially disaggregate responses of climatic impact and damage variables but rather to bring together key stylized facts about these responses within the framework of Eq. 5-1 as a means of elucidating their joint implications. The benefits of such integration are often gained at the expense of introducing substantial theoretical and empirical weaknesses into IAMs. Although each element of $\Delta EC$, $\Delta CI$, and $\Delta ID$ can be thought of as a model in its own right, IAMs adopt reduced-form approaches, which reduce these complex relationships into simplified response surfaces, in the process oversimplifying the complexities of the underlying science. In addition, IAMs typically cope with the curse of dimensionality by considering only relatively narrow sets of impacts or types of damage. IAMs also tend to trade off coarse regional coverage in favor of broad global scope, so relationships validated for restricted geographic domains are implicitly scaled up to broader spatial scales. The remainder of this section describes in general terms how IAMs, such as the Regional Integrated Model of Climate and the Economy (RICE) and the Dynamic Integrated Model of Climate and the Economy (DICE), the Climate Framework for Uncertainty, Negotiation, and Distribution (FUND) model, and the Policy Analysis of the Greenhouse Effect (PAGE) model, evaluate marginal damages. In a typical IAM, the marginal damages from a ton of $CO_2$-eq emissions ($E$) emitted today (year 0) can be expressed as

$$MD_0 = \sum_{t=0}^{t_f} \frac{dT_t}{dE_0} \frac{dD_t}{dT_t} \delta_t , \qquad \text{Equation 5-2}$$

where

$t$ = year of impact,
$MD_0$ = marginal damage from GHG emissions in year 0 ($/ton $CO_2$-eq),
$T_t$ = mean global temperature in year $t$ relative to preindustrial levels (°C),
$E_0$ = GHG emissions in year 0 (ton $CO_2$-eq),
$D_t$ = total climate damages in year $t$ ($),
$\delta_t$ = discount factor from year $t$ to 0 = $(1 + r)^{-t}$, where $r$ is the discount rate, and
$t_f$ = final year for which climate damages are included.

The expression indicates that the marginal damages from GHG emissions ($MD$) depend on how much temperatures increase in response to a unit increase in emissions ($dT/dE$), how much additional climate damage results from this temperature increase ($dD/dT$), how one values future damages relative to the present ($\delta$), and how far into the future one aggregates impacts ($t_f$). In terms of the preceding discussion, climate effects have been reduced to temperature, and the link between climate and impacts and impacts and damages has been condensed into a single step.

The relationship between a ton of $CO_2$-eq emitted today and future temperature depends on the effects of GHG emissions on the concentration of GHGs in the atmosphere, and on the effects of GHG concentrations on temperature. Spatially detailed predictions of the impact of GHG concentrations on temperature and precipitation are provided by general circulation models. IAMs typically simplify these relationships and describe changes in mean global temperature corresponding to a ton of $CO_2$-eq emissions.

In the most disaggregated IAMs, the monetary damages associated with a change in temperature are calculated by estimating damages by sector (for example, energy, health, and agriculture) and geographic region. Damages are expressed as a percentage of GDP using methods described in the next subsection: Approach to Measuring Marginal Damages in Integrated Assessment Models. $dD_t/dT_t$ represents the aggregation of impacts across sectors and regions. How change in GDP are aggregated across regions—whether using equity weights or by summing the monetary changes in GDP—is discussed below.

Future monetary damages are discounted either at the market rate of interest (the revealed preference approach), or using the Ramsey formula (the prescriptive approach), which describes how the discount rate, $r$, varies along an optimal growth path (see Pearce et al. 2003). These two approaches are discussed in detail later in the chapter. Marginal damages are extremely sensitive to the choice of discount rate, given the fact that the climate impacts of a ton of $CO_2$-eq emissions will be felt for centuries ($t_f$ is typically 100 to 300 years).

The marginal-damage formula in Equation 5-2 assumes that the effect of a ton of $CO_2$-eq emissions on temperature and the effects of temperature on the economy are certain—which are clearly not the case. Indeed, a major difference between quantifying the local air pollution effects of fossil fuels and the impacts of GHG emissions is that the two differ significantly in their time dimension, their spatial scale, the variety of impacts, and, hence, in the certainty with which they can be estimated. In contrast to $SO_2$ or $NO_x$, $CO_2$ is a pollutant that resides in the atmosphere for centuries.[4] This factor implies that the effects of a ton emitted today must be estimated on a time scale (centuries) in which the state of the world is inherently more uncertain than the period during which effects of local air pollutants are estimated (months or years). Key sensitivities in Equation 5-2 include the impact of a change in atmospheric concentrations of $CO_2$ on temperature (termed climate sensitivity) and how $dD/dT$ varies with $T$. There is, in reality, a distribution of damages associated with any given temperature change.

IAMs typically handle this uncertainty in two ways. One is to calculate marginal damages using Monte Carlo analysis: Parameters used for dT/dE and dD/dT are drawn from probability distributions and used to calculate the corresponding distribution of marginal damages. A second approach is to acknowledge that, corresponding to each change in mean global temperature from pre-industrial levels, there is a probability of

---

[4]The atmospheric lifetime of $CO_2$ is complex. About half disappears in 40 years, but about 20% remains in the atmosphere for many centuries, essentially indefinitely.

abrupt, catastrophic events, such as the melting of the West Antarctic ice sheets or melting of permafrost—events that could result in huge declines in world GDP. Some models attempt to estimate what individuals would pay to avoid such events. The next section provides an overview of how the damages associated with various temperature changes are modeled in three prominent IAMs.

## Approach to Measuring Marginal Damages in Integrated Assessment Models

IAMs combine simplified global climate models with economic models in an effort to estimate the economic impacts of climate change and to identify emission paths that balance these economic impacts against the costs of reducing GHG emissions. Three of the most widely used IAMs are RICE and DICE (W. Nordhaus, Yale University), FUND (R.S.J. Tol, Economic and Social Research Institute, Dublin, Ireland), and PAGE (C. Hope, University of Cambridge). The goal of this section is to provide overviews of how each of these IAMs monetizes the impact of changes in mean global temperature.

### RICE and DICE Models

These models examine the links between economic growth, $CO_2$ emissions, the carbon cycle, the economic damages associated with climate change, and climate-change policies. These models incorporate the climate system's "natural capital" into a model based on traditional economic growth theory. They treat as exogenous global population, global stock of fossil fuels, and the pace of technological change, and they calculate world output and capital stock, $CO_2$ emissions and concentrations, global temperature change, and climate damages. RICE distinguishes various regions of the world (8 in some versions; 13 in others), and DICE has a single, aggregated global economy. The models are typically run from 1990 until 2100.

The approach to quantifying climate-change damages in each sector in RICE is as follows: (1) The percentage reduction in GDP associated with a mean global temperature increase of $T$ is calculated for the year 1995 for each sector ($i$) and region ($j$). (Call this $Q_{ij}(T)$.) (2) The impact of the $T$ °C temperature change is calculated for a future year, $t$, by multiplying $Q_{ij}(T)$ by the ratio of per capita GDP in year $t$ to per capita GDP in 1995 raised to a power $\eta$, where $\eta$ is the income elasticity of the impact index. The percentage change in GDP in year $t$ for temperature change $T$ is thus,

$$Q_{ij}(T) \, [y_j(t)/y_j(1995)]^{\eta} \qquad \text{Equation 5-3}$$

In practice $Q_{ij}(T)$ is calculated for benchmark warming—a 2.5°C increase in mean global temperature—based on a review of the literature. η is determined from a literature review or expert opinion. $Q_{ij}(T)$ changes as a function of $T$ according to a quadratic function. The sectors for which impacts are monetized in RICE and DICE are agriculture, sea-level rise, other market sectors, health, nonmarket amenities, human settlements and ecosystems, and catastrophic damages. The magnitudes of damages in these sectors are discussed in sections below.

### FUND Model

The Climate Framework for Uncertainty, Negotiation, and Distribution (FUND) model examines how a set of exogenous scenarios concerning economic growth, population growth, energy-efficiency improvements, decarbonization of energy use, and GHG emissions affect the concentration of atmospheric $CO_2$, global mean temperature, and the impacts of temperature change. FUND models these links for nine regions[5] over 250 years (1950 to 2200).

The sectors for which impacts are monetized in FUND are agriculture, forestry, water resources, energy consumption, sea-level rise, ecosystems, and human health. Monetization of impacts in FUND is slightly different for each sector and more detailed than the reduced-form approach used in RICE and DICE. For example, the impact of a temperature change on agricultural revenues consists of three components: One reflects the difference between future temperature and ideal growing temperature for the region; a second reflects the rate of increase in temperature, which captures opportunities for adaptation; and the third component reflects the carbon fertilization effect. To map the percentage change in agricultural revenues implied by these three effects into a change in GDP requires estimates of the share of agriculture in GDP. These estimates, in turn, are modeled as a function of per capita GDP and assumptions about the income elasticity of agriculture.

FUND differs from RICE and DICE in that the base year impacts in agriculture and ecosystems depend not only on the magnitude of temperature change but also on the rate of temperature change. It is also the case in FUND that the effect of a change in mean global temperature on marginal damages varies by sector

---

[5]OECD-America, OECD-Europe, OECD-Pacific, Central and Eastern Europe and the former Soviet Union, the Middle East, Latin America, South and Southeast Asia, Centrally Planned Asia, and Africa.

## PAGE Model

The Policy Analysis of the Greenhouse Effect (PAGE) model is a multi-regional model that models the impacts of climate change in three sectors—economic impacts, noneconomic impacts, and discontinuity impacts—that is, impacts associated with abrupt changes to the climate system.[6] Functions that describe the economic and noneconomic impacts of a given temperature change ($T$) in region $r$ are of the form

$$I(r) = A(r)T^{n(r)}, \qquad\qquad \text{Equation 5-4}$$

where $I(r)$ is the percentage change in GDP associated with the impact, $A(r)$ is a scaling factor and $n(r)$ lies between 1 and 3. The weights $A(r)$ represent the percent of GDP lost in region r relative to losses in the European Union. For a description of the modeling of discontinuity impacts see Hope (2006).

## IMPACTS ON PHYSICAL AND BIOLOGICAL SYSTEMS

Earth's climate system is integrally intertwined with many other global biological, chemical, and physical systems. Impacts on the weather, cryosphere, hydrosphere, coastal zones, and the biosphere are briefly discussed below. Impacts on human systems resulting from impacts on these physical systems are discussed in more detail in a subsequent section. This discussion is intended to provide a brief summary of recent knowledge. Another effort is under way within the NRC to study issues relating to global climate change.[7] It is important to keep in mind that none of the individual impacts described in this section have been monetized.

### Changes in the Weather

The most literal effect of climate warming is an increase in ambient air temperatures, particularly at night over land in the Northern Hemisphere (Figure 5-2). Global climate models predict an increase in the frequency of heat waves, heavy precipitation events, and the intensity of tropical cyclones (IPCC 2007b). The IPCC also predicts that precipitation will likely decrease in the subtropics and will very likely increase near the poles (Figure 5-4).

---

[6]Economic and noneconomic impacts are not disaggregated by sector, as in RICE and DICE and FUND.

[7]In response to a request from Congress, the NRC has launched America's Climate Choices, a suite of studies designed to inform and guide responses to climate change across the nation.

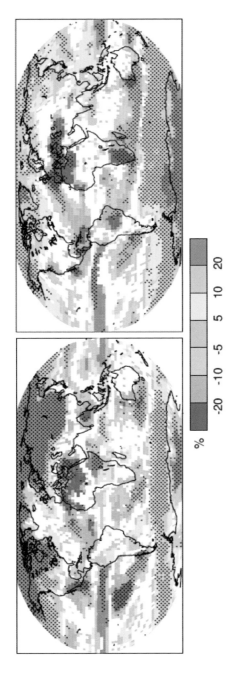

**FIGURE 5-4** Multimodel projected patterns of precipitation changes. Relative changes in precipitation (in percentage) for the period 2090-2099, relative to 1980-1999. Values are multimodel averages based on the *IPCC Special Report on Emission Scenarios* (SRES) A1B for December to February *(left)* and June to August *(right)*. White areas are where less than 66% of the models agree in the sign of the change, and stippled areas are where more than 90% of the models agree in the sign of the change. SOURCE: IPCC 2007b, P.16, Figure SPM.7. Reprinted with permission; copyright 2007, Intergovernmental Panel on Climate Change.

## Changes in the Cryosphere

The cryosphere, comprising all the permanent and seasonal snow and ice formations found on Earth, including the polar ice caps, sea ice, permafrost, glaciers, and seasonal snow and ice on land and water, is particularly sensitive to climate change, and a dramatic decrease is predicted in the amount of snow and ice on Earth as climate changes progress. In fact, scientists have already documented the following changes (Rosenzweig et al. 2007):

• The Arctic sea-ice extent has declined by about 10% to 15% since the 1950s, and the 2007 summer minimum is more than 35% smaller than the 1950-1980 average.
• Mountain glaciers have receded on all continents.
• Northern Hemisphere permafrost is thawing.
• Snowmelt and runoff have occurred increasingly earlier in Europe and western North America since the late 1940s.
• The annual duration of lake- and river-ice cover in Northern Hemisphere mid- and high latitudes has been reduced by about 2 weeks and become more variable.

IPCC (2007b) also predicts the complete disappearance of late-summer Arctic sea ice by the end of the 21st century.

## Changes in the Hydrosphere

The hydrosphere comprises all the liquid water systems on Earth, including the oceans, lakes, rivers, streams, and aquifers. The hydrosphere is tightly integrated with the climate system and the cryosphere. Climate change and consequent warming are linked to changes in the hydrologic cycle with increased evaporation from land and seas, changing precipitation patterns, and reduced snow cover. Specific observed changes in the hydrosphere include the following (Rosenzweig et al. 2007):

• The salinity of the North Atlantic is decreasing, most likely because of melting glaciers.
• Annual runoff is increasing in higher latitudes and decreasing in some parts of West Africa, southern Europe, and southern Latin America.
• Peak spring river flows are occurring earlier in areas with a seasonal snow pack. This causes less water to be available during the late summer and autumn when human and ecological demand tends to be the greatest.

•   The temperature, chemistry, and ultimately the structure of lakes and rivers are changing.

•   'Large' floods are occurring with more frequency around the globe.

•   Very dry areas have more than doubled since 1970, causing desertification and droughts.

Ultimately, between the changes in weather patterns and changes in the cryosphere and hydrosphere, scientists predict that Earth will become dryer in the subtropics, especially in the Northern Hemisphere, and much wetter and less frozen near the poles. In other words, climate change is likely to manifest in ways, with consequent impacts, that will not occur evenly across the globe. Global climate models (GCMs) predict increasing global precipitation, with important regional variation, including increases in high latitudes and parts of the tropics but decreases throughout the subtropics (Bates et al. 2008). The western United States, for example, is vulnerable to reduced water availability. Table 5-2 lists climate-related changes in the freshwater system presented in the fourth assessment report of the IPCC.

The physical impacts on water availability vary considerably geographically; some regions benefit from warming while other areas suffer. For example the Warren et al. (2006a) analysis shows water scarcity increasing on a global scale from 29% in 1995 to 39% in 2085 under the A1 and B1 scenarios, respectively, shown in Figure 5-2. Some areas see sharper increases in water scarcity under their analysis (South Asia more than doubles from 26% to 59%) while other areas see a decline in water scarcity (Europe falls from 38% to 26%). The United States and Canada see a modest increase in scarcity from 16% to 20% in their analysis.

## Changes in the Coastal Zones

Rising sea levels—among the best-documented impacts of climate change—are another consequence of the melting cryosphere. Sea level has been rising at the rate of 1.7 to 1.8 mm/yr over the past century. This rate increased to approximately 3 mm/yr over the past decade (IPCC 2007a). Rising sea levels and increased storm intensities are rapidly eroding coastlines around the globe. Seventy-five percent of the east coast of the United States and 67% of the east coast of the United Kingdom are thus affected (Rosenzweig et al. 2007). However, there is scientific consensus that over many centuries thermal expansion of the ocean due to global warming is very likely to cause much larger rises in sea levels than those observed over the 20th century. In the latest IPCC projections, thermal expansion contributes 70-75% of the best estimate of sea-level rise for each of the six *IPCC Special Report on Exposure Scenarios* (SRES) marker scenarios, in

**TABLE 5-2** Climate-Related Observed Trends of Various Components of the Global Freshwater Systems

|  | Observed Climate-Related Trends |
|---|---|
| Precipitation | Increasing over land north of 30°N over the period of 1901-2005 |
|  | Decreasing over land between 10°S and 30°N after the 1970s (WGI AR4, Chapter 3, Executive Summary) |
|  | Increasing intensity of precipitation (WGI AR4, Chapter 3, Executive Summary) |
| *Cryosphere* |  |
| Snow cover | Decreasing in most regions, especially in spring (WGI AR4, Chapter 4, Executive Summary) |
| Glaciers | Decreasing almost everywhere (WGI AR4, Chapter 4, Section 4.5) |
| Permafrost | Thawing between 0.02 m/yr (Alaska) and 0.4 m/yr (Tibetan Plateau) (WGI AR4, Charter 4, Executive Summary; WGII AR4, Chapter 15, Section 15.2) |
| *Surface Waters* |  |
| Streamflow | Increasing in Eurasian Arctic, significant increases or decreases in some river basins (WGII AR4 Chapter 1, Section 1.3.2) |
|  | Earlier spring peak flows and increased winter base flows in Northern America and Eurasia (WGII AR4, Chapter 1, Section 1.3.2). |
| Evapotranspiration | Increased actual evapotranspiration in some areas (WGI AR4, Chapter 3, Section 3.3.3). |
| Lakes | Warming, significant increases or decreases of some lake levels, and reduction in ice cover (WGII AR4, Chapter 1, Section 1.3.2). |
| Groundwater | No evidence for ubiquitous climate-related trend (WGII AR4, Chapter 1, Section 1.3.2) |
| *Floods and Droughts* |  |
| Floods | No evidence for climate-related trend (WGII AR4, Chapter 1, Section 1.3.2), but flood damages are increasing (WGII AR4, Chapter 3, Section 3.2) |
| Droughts | Intensified droughts in some drier regions since the 1970s (WGII AR4, Chapter 1, Section 1.3.2; WGI AR4, Chapter 3, Executive Summary) |
| Water quality | No evidence for climate-related trend (WGII AR4, Chapter 1, Section 1.3.2) |
| Erosion and sediment transport | No evidence for climate-related trend (WGII AR4, Chapter 3, Section 3.2) |
| Irrigation water demand | No evidence for climate-related trend (WGII AR4, Chapter 3, Section 3.2) |

NOTES: WGI AR4, Chapter 3, Trenberth et al. 2007; WGI AR4, Chapter 4, Lemke et al. 2007; WGII AR4, Chapter 1, Rosenzweig et al. 2007; WGII AR4, Chapter 3, Kundzewicz et al. 2007; and WGII AR4, Chapter 15, Anisimov et al. 2007.

SOURCE: Kundzewicz et al. 2007, p.177, Table 3.1. Reprinted with permission; copyright 2007, Intergovernmental Panel on Climate Change.

the most extreme case exhibiting a 5-95% confidence interval of 0.26-0.59 m by the year 2100 (IPCC 2007b, p. 820, Table 10.7; IPCC 2007b, p. 821, Figure 10.33). This projection is cause for concern, given that relative sea-level rises have exceeded 8 inches in some areas along the Atlantic and Gulf coasts (see Karl et al. 2009, p. 37, figure) Although the contributions to sea-level rise made by thermal expansion and melting glaciers are well understood, uncertainty remains about the magnitude of the ice sheets' effects, so much so that their impact was left unquantified in the most recent IPCC report. On the basis of several recent studies on sea-level rise, Karl et al. (2009) concluded that the IPCC predictions are likely to underestimate the impact and cite estimates by century's end of 0.9-1.2 m under higher emission scenarios, with an upper bound of 2 m.

### Changes in the Biosphere

Many plants and animals have relatively specific environmental conditions in which they can survive. Even small environmental changes, such as extremes in ambient temperature, or the availability of water, can make a region inhospitable to members of the existing flora and fauna. Ecologists are already documenting important shifts in ecosystem structures and functioning, such as the following (Rosenzweig et al. 2007):

•   Plant and animal ranges have shifted to cooler higher latitudes and altitudes. Therefore, as overall temperatures rise, plants and animals with very narrow temperature requirements will shift their ranges accordingly or become extirpated.
•   The timing of many life-cycle events, such as flowering, migration, and emergence, has shifted to earlier in the spring and often later in the autumn.
•   Different species change at different speeds and in different directions, causing a changing of species interactions (for example, predator-prey relationships).

## IMPACTS ON HUMAN SYSTEMS

Observed (and predicted) changes in Earth's global systems have significant ramifications for humans. The redistribution of water availability across the globe, for example, will amplify water conflicts, particularly in regions that are getting drier. Changes in the availability of water and in the length of growing seasons will affect which crops farmers can plant and how much those crops yield. Tropical diseases will start to affect more people as the ranges of disease vectors, such as mosquitoes, shift pole-ward.

Table 5-3 describes some of the many ways in which climate change may affect important human systems.

The impacts of climate change on humans will not be uniform throughout the world. Different regions will experience climate change somewhat differently. Southern Africa, for example, is predicted to become drier and will therefore need to cope with water scarcity. Northern Europe, on the other hand, is predicted to become wetter. Figure 5-5 summarizes some of the key regional impacts humans will experience. The rest of this section systematically explores the impacts of climate change on a variety of aspects of human life, including water resources; ecosystem services; food production and forest products; sea-level rise and coastal populations; and human health, industry, society, and security.

## Water Availability

A critical challenge facing the growing world population is access to water, which could be significantly affected by climate change. Warren et al. (2006a) noted that a country experiences water scarcity when available supply falls below 1,000 m3 per person per year and absolute scarcity when supply falls below 500 m3 per person per year. Globally, they estimate that roughly 30% of the world's population was "water stressed" (defined as experiencing water scarcity) in 1995 (Warren et al. 2006a, Table A2). By 2085, they project that 39-59% of the world's population could be water stressed, depending on economic and population growth. However, all analyses and predictions of physical impacts must be qualified by the great uncertainties about hydrologic cycles and their responses to warming. Additional caveats to estimates of impacts are the possibilities of adaptation and mitigation. For example, exposure to water scarcity will change as populations migrate for reasons related or unrelated to global warming.

Changes in water availability can lead to losses in crop production, premature deaths, and greater disease prevalence from water shortages in the short run and adjustment costs of population movements as people abandon areas that have become too dry and as people engineer new water transfers. However, measuring the impacts of increasing water scarcity is difficult. Among other issues, the value of losses is exacerbated by increasing demand for irrigation in agriculture (Mendelsohn and Williams 2007). Aldy et al. (2009) provide summaries of damages from climate change as measured in a number of studies. Table 5-4 reports estimates of damages arising from changes in water availability. These damages are reported as percentages of GDP at the end of this century. For the United States, damages range from a low of .01% to .03% for warming between 4.6 and 7.1°C to a high of .29% for 2.5°C.

**TABLE 5-3** Examples of Possible Impacts of Climate Change Due to Changes in Extreme Weather and Climate Events, Based on Projections to the Mid- to Late 21st Century

| Phenomenon[a] and Direction of Trend | Likelihood of Future Trends Based on Projections for 21st Century Using SRES Scenarios | Examples of Major Projected Impacts by Sectors |
|---|---|---|
| | | Agriculture, Forestry, and Ecosystems (WGII 4.4, 5.4) |
| Over most land areas, warmer and fewer cold days and nights, warmer and more frequent hot days and nights | *Virtually certain*[b] | Increased yields in colder environments; decreased yields in warmer environments; increased insect outbreaks |
| Warm spells and heat waves; frequency increases over most land areas | *Very likely* | Reduced yields in warmer regions due to heat stress; increased danger of wildfire |
| Heavy precipitation events; frequency increases over most areas | *Very likely* | Damage to crops; soil erosion, inability to cultivate land due to waterlogging of soils |
| Area affected by drought increases | *Likely* | Land degradation; lower yields and crop damage and failure; increased livestock deaths; increased risk of wildfire |
| Intense tropical cyclone activity increases | *Likely* | Damage to crops; windthrow (uprooting) of trees; damage to coral reefs |
| Increased incidence of extremely high sea level (excludes tsunamis[c] | *Likely*[d] | Salinization of irrigation water, estuaries, and freshwater systems |

[a]See WGI Table 3.7 for further details regarding definitions.

[b]Warming of the most extreme days and nights each year.

[c]Extreme high sea level depends on average sea level and on regional weather systems. It is defined as the highest 1% of hourly values of observed sea level at a station for a given reference period.

[d]In all scenarios, the projected global average sea level at 2100 is higher than it is in the reference period. The effect of changes in regional weather systems on sea level extremes has not been assessed (WGI 10.6).

| Water Resources (WGII 3.4) | Human Health (WGII 8.2, 8.4) | Industry, Settlement, and Society (WGII 7.4) |
|---|---|---|
| Effects on water resources relying on snowmelt; effects on some water supplies | Reduced human mortality from decreased cold exposure | Reduced energy demand for heating; increased demand for cooling; declining air quality in cities; reduced disruption to transport due to snow and ice; effects on winter tourism |
| Increased water demand; water quality problems, for example, algal blooms | Increased risk of heat-related mortality, especially for the elderly, chronically sick, very young and socially isolated | Reduction in quality of life for people in warm areas without appropriate housing; impacts on the elderly, very young. and poor |
| Adverse effects on quality of surface and groundwater; contamination of water supply; water scarcity may be relieved | Increased risk of deaths, injuries and infectious, respiratory and skin diseases | Disruption of settlements, commerce, transport, and societies due to flooding; pressures on urban and rural infrastructures; loss of property |
| More widespread water stress | Increased risk of food and water shortage; increased risk of malnutrition; increased risk of water- and food-borne diseases | Water shortage for settlements, industry, and societies; reduced hydropower-generation potentials; potential for population migration |
| Power outages causing disruption of public water supply | Increased risk of deaths, injuries, water- and food-borne diseases; post-traumatic stress disorders | Disruption by flood and high winds; withdrawal of risk coverage in vulnerable areas by private insurers; potential for population migrations; loss of property |
| Decreased freshwater availability due to saltwater intrusion | Increased risk of deaths and injuries by drowning in floods; migration-related health effects | Costs of coastal protection versus costs of land-use relocation; potential for movement of populations and infrastructure |

ABBREVIATION: SRES = Special Report on Emission Scenarios.

SOURCE: IPCC 2007d, p.18, Table SPM.1. Reprinted with permission; copyright 2007, Intergovernmental Panel on Climate Change.

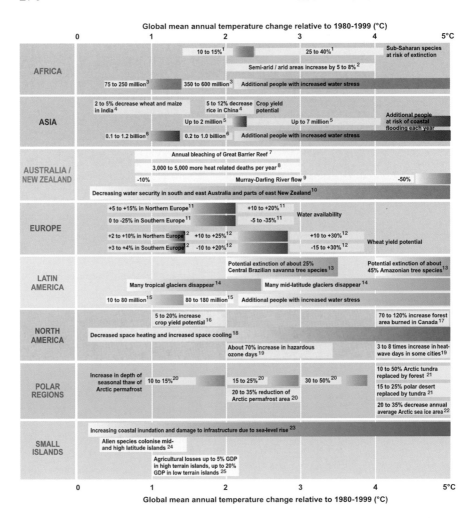

**FIGURE 5-5** Examples of regional impacts of climate change. SOURCE: Yohe et al. 2007, p. 829, Table 20.9.

World damages are modestly higher. Tol (2002a) reported the highest damages of .43% for 1°C warming. The range of estimates for the United States and for the world is great, especially when taking into account the different assumptions about global warming. This range speaks to the difficulties in making sharp impact predictions.

Regarding the three IAMs described earlier in the chapter, PAGE does not provide sector-specific estimates of damages, and RICE and DICE do not provide separate estimates for water resources. Indeed, Nordhaus and

**TABLE 5-4** Water Availability Effects from Climate Change for Selected Studies[a] (Percentage of Contemporaneous GDP Around 2100)

| | Cline 1992 | Fankhauser 1995a | Mendel-sohn and Neumann 1999 | Mendel-sohn and Williams 2004 | Mendel-sohn and Williams 2007 | Titus 1992 | Tol 1995 | Tol 2002a |
|---|---|---|---|---|---|---|---|---|
| Warming, C[b] | 2.5 | 2.5 | 2.5 | 4.6-7.1 | 2.5-5.2 | 4.0 | 2.5 | 1.0 |
| United States | .12 | .29 | .07 | .01-.03 | | .20 | n/a | .07 |
| World | n/a | .24 | n/a | .01-.03 | .00 -.02 | n/a | n/a | .43 |

NOTES: n/a = values not available or not estimated. In some cases, estimates for the United States also include Canada.

[a]Stern (2007) does not separate out individual categories within market and nonmarket impacts.

[b]Warming is relative to preindustrial (as opposed to current) temperatures.

SOURCE: Adapted from Aldy et al. 2009, with permission from the authors.

Boyer (1999, p. 4-13) argued that the damages from water availability can be set to zero, based on their survey of previous studies. The FUND model 3.0 measures water availability impacts for each of 16 regions using the following formula:

$$W_{t,r} = \min\left\{ \alpha_r Y_{1990,r} (1-\tau)^{t-2000} \left(\frac{Y_{t,r}}{Y_{1990,r}}\right)^{\beta} T_t^{\gamma}, \frac{Y_{t,r}}{10}\right\}, \quad \text{Equation 5-5}$$

where

$W$ = denotes the change in water resources in 1995 dollars in region $r$ in year $t$,

$Y$ = denotes income (in 1995 dollars),

$T$ = global mean temperature,

$\alpha$ = benchmarking parameter,

$\tau$ = parameter measuring technological progress in water supply and demand (ranges from 0 to .01 with a preferred estimate of .005,

$\beta$ = elasticity of impact with respect to income growth (ranging from .7 to 1 with a preferred estimate of .85),

$\gamma$ = elasticity of impact with respect to temperature change (ranging from .5 to 1.5 with a preferred estimate of 1).

The parameter choices are made by calibrating the FUND model to results from Downing et al. (1995, 1996a). The estimated impact of a 1°C increase in global temperature is –0.065% of GDP for the United States (FUND 2008, p. 33, Table EFW). A negative estimate indicates benefits to the United States from warming. This estimate is imprecisely estimated with a coefficient of variation equal to 1.0. The impact on other regions is small with a few exceptions. The former Soviet Union sees benefits as large as 2.75% of GDP, while China has losses of 0.57% of GDP. Overall, however, losses are small, and in all cases, the estimates have very large standard errors.

### Coastal Zone Impact of Climate Change

As previously mentioned, the coastal sector is one of best-documented areas of the impacts of climate change. However, it is difficult to assess with any confidence what the monetary damages of elevated seas might be for the United States, let alone globally. The only comprehensive assessment of the vulnerability of the U.S. coastline to sea-level rise (Thieler and Hammar-Klose 1999; 2000a,b) predates the latest IPCC estimates. Notwithstanding that, the fact that their methodology of assigning to segments of coastline an index of vulnerability calculated on the basis of rank-ordered attributes would suggest that updated data on sea-level rise will preserve the relative position of the coastline in the vulnerability hierarchy, at least over broad geographic scales.[8] Recent analyses at the regional scale indicate that sandy-shore environments, such as the Mid-Atlantic coastline, have a high likelihood of seeing more rapid erosion and segmentation of barrier islands, as well as wetland loss. For example, Figure 5-6 illustrates that for the Mid-Atlantic region an acceleration in sea-level rise of 2 mm/year over current rates will cause many wetlands to become stressed, while most wetlands probably will not survive a 7 mm/year acceleration (consistent with IPCC's upper-bound estimate (see section above "Changes in the Coastal Zones"). The value of these kinds of losses has not been rigorously quantified. Depending on the increase in sea level, the adaptation options confronting human populations in the coastal zone are to protect the shore, relocate inland, or do a combination of both, each of which is associated with forgone income and well-being—that is, damage. How much of each option to be chosen is essentially an economic decision, which is simulated within IAMs in the process of arriving at aggregate estimates of climate damages. The remainder of this section sheds light on the methodological details of

---

[8]The variables are geomorphology, shoreline erosion and accretion rates, coastal slope, rate of relative sea-level rise, mean tidal range, and mean wave height.

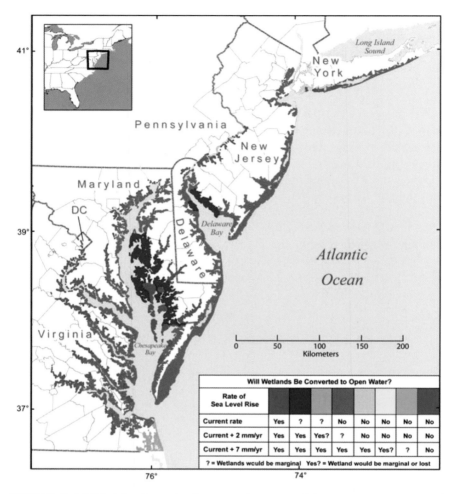

Scale bar: 0 — 50 — 100 — 150 — 200 Kilometers

| Will Wetlands Be Converted to Open Water? | | | | | | | | |
|---|---|---|---|---|---|---|---|---|
| Rate of Sea Level Rise | | | | | | | | |
| Current rate | Yes | ? | ? | No | No | No | No | No |
| Current + 2 mm/yr | Yes | Yes | Yes? | ? | No | No | No | No |
| Current + 7 mm/yr | Yes | Yes | Yes | Yes | Yes | Yes? | ? | No |
| ? = Wetlands would be marginal   Yes? = Wetland would be marginal or lost | | | | | | | | |

**FIGURE 5-6** Mid-Atlantic wetland marginalization and loss as a consequence of sea-level rise. SOURCE: CCSP 2009, Fig. ES.2.

this process, as a way of illustrating the large extent to which it is driven by assumptions on the part of IAM modelers.

There is a sizeable literature on the damages associated with sea-level rise. The differences in model results stem from different ways of representing the processes by which damages arise, including the level of detail in climate- and physical-impact modeling and the choice between a "process-based" and "reduced-form" approaches to representing impacts. The RICE and DICE models (Nordhaus and Boyer 2000) are typical of the reduced-form approach, while the detailed representations of damages in the FUND model (Tol 2002a,b) exemplify the process-based approach. Both authors develop damage estimates on a regional basis by extrapolating from studies of the United States and other countries, but to implement the process-based approach requires many more assumptions about the detailed impacts of sea-level rise and the character of affected individuals' adaptation responses.

Damages in the RICE model are constructed by developing a benchmark estimate of the cost of the sea level increase arising from 2°C warming in the United States (0.1% of GDP) and then applying this estimate to other regions using an index of coastal sensitivity. The benchmark estimate for the United States includes damages to developed and undeveloped land and damages from storms. The index of coastal sensitivity is constructed by dividing the ratio of coastal area to total area for a given region by the ratio for the United States (see Table 5-5). The income elasticity of coastal damages is assumed to be 0.2.

**TABLE 5-5** Values of the Benchmarking Parameter ($\alpha$)

|  | Coastal Impact[a] | Coastal Index (% of GDP, 1990) | $\alpha$ † (2.5°C Impact) |
|---|---|---|---|
| United States | 1.00 | 0.10 | 0.11 |
| China | 0.71 | 0.07 | 0.07 |
| Japan | 4.69 | 0.47 | 0.56 |
| Western Europe | 5.16 | 0.52 | 0.60 |
| Russia | 0.94 | 0.09 | 0.09 |
| India | 1.00 | 0.10 | 0.09 |
| Other high income | 1.41 | 0.14 | 0.16 |
| High-income OPEC | 0.52 | 0.05 | 0.06 |
| Eastern Europe | 0.14 | 0.01 | 0.01 |
| Middle income | 0.41 | 0.04 | 0.04 |
| Lower-middle income | 0.94 | 0.09 | 0.09 |
| Africa | 0.23 | 0.02 | 0.02 |
| Low income | 0.94 | 0.09 | 0.09 |
| Global |  |  |  |
|     Output weighted‡ |  |  | 0.32 |
|     Population weighted§ |  |  | 0.12 |

[a]Ratio of fraction of area in coastal zone in country to that fraction in the United States. "Coastal zone" is defined as that part of the region that lies within 10 kilometers of an ocean.

†Calibrated to impacts in the year 2100.

‡Output projections in 2100 from RICE model base case.

§ 1995 population.

SOURCE: Nordhaus and Boyer (2000: Tables 4-5 and 4-10). Reprinted with permission; copyright 2008, MIT Press.

For the FUND model, Tol (2002a,b) followed the method pioneered by Fankhauser (1995a,b) in estimating the costs of sea-level rise as the sum of the capital cost of structures for coastal protection and the cost of fore-gone services from "dry" and "wet" coastal land that is inundated. This method entails determining the optimal level of coastal protection, which determines the first component of cost and also the amount of coastal land that is inundated, for a given rise in sea level.

The cost of inundation of unprotected land depends on the extent of land loss and population displacement from the inundation. Tol estimates population displacement as the product of projected loss of dry lands and average population density and makes several assumptions about the destinations of the resulting migrants.[9] The next step is to monetize these impacts. The unit values of lost dry and wet land in countries of the Organization for Economic Co-operation and Development (OECD) are assumed to be $4 million/km$^2$ and $5 million/km$^2$, respectively, and are extrapolated to other regions by adjusting them according to the inundation probability-weighted population density in the coastal zone and per capita income. For population displacement, Tol assumes a cost of emigration from an affected zone equal to three times per capita income and an immigration cost equal to 40% of the per capita income in the host country.[10] The results are shown in Table 5-5.

The amount of land (percentage of the coast) that is protected is determined by comparing the costs and benefits of protection. Table 5-6 presents the optimal fraction of the coast protected by region, as well as the costs of that protection.

### Impacts on Ecosystems and Ecosystem Services

Without a solid, broadly accepted set of standards for the value of ecosystems, the external costs of climate change assigned to ecosystem effects tend to get categorized in one of two ways. Based on the IAMs that do incomplete and preliminary accounting, the damages are generally quite low, sometimes barely enough to register in the overall cost accounting for climate-change impacts. Other studies based on ecosystem services often start from the proposition that ecosystem services are critical for the maintenance of healthy people, communities, and people. As a consequence, they tend to assign large but rarely quantified amounts to the external impacts of

---

[9]Displaced persons in countries of the OCED, Central and Eastern Europe, and the former Soviet Union stay entirely within their own regions, and only 10% displaced persons in poorer regions emigrate from their own regions. A variety of assumptions about where the latter go are made.

[10]Compare Cline's (1992) rough estimate of $4,500 per migrant for the United States.

**TABLE 5-6** Benchmark Sea-Level Rise Estimates in FUND

|  | Cost Length ($10^3$ km) | Level of Protection (%) | Dry-land Loss ($10^3$ km$^2$) | Dry-land Value ($10^6$ km$^2$) | Wetland Loss ($10^3$ km$^2$) |
|---|---|---|---|---|---|
| OECD-A | 33 | 0.77 | 4.8 (2.4) | 1.3 (0.6) | 12.0 (8.6) |
| OECD-E | 59 | 0.86 | 0.7 (0.4) | 13.1 (6.6) | 4.0 (2.3) |
| OECD-P | 23 | 0.95 | 0.3 (0.4) | 13.7 (6.7) | 1.0 (1.1) |
| CEE&dSU | 25 | 0.93 | 1.2 (2.7) | 0.9 (0.5) | 0.0 (0.0) |
| ME | 6 | 0.30 | 0.6 (1.2) | 0.5 (0.3) | 0.0 (0.0) |
| LA | 39 | 0.86 | 7.8 (7.1) | 0.3 (0.2) | 50.2 (36.4) |
| S&SEA | 95 | 0.93 | 9.3 (9.6) | 0.5 (0.3) | 54.9 (48.0) |
| CPA | 33 | 0.93 | 8.4 (15.1) | 0.3 (0.2) | 15.6 (17.1) |
| AFR | 35 | 0.89 | 15.4 (18.4) | 0.4 (0.2) | 30.8 (14.8) |

NOTES: Definitions of the regions (which correspond to the regions of FUND) are as follows: Organization for Economic Co-operation and Development (OECD)-America (excluding Mexico) (OECD-A), OECD-Europe (OECD-E), OECD-Pacific (excluding South Korea) (OECD-P), Central and Eastern Europe and the former Soviet Union (CEE&fSU), Middle East (ME), Latin America (LA), South and Southeast Asia (S&SEA), Centrally Planned Asia (CPA), and Africa (AFR).

climate change. The general inclination of stakeholders who take this position to assign zero or even negative discount rates creates the foundation for extraordinarily large damages. The steps to quantitatively test or reconcile these perspectives will probably be numerous and challenging.

Four widely used IAMs (RICE and DICE, MERGE (model for evaluating regional and global effects), FUND, and PAGE) all estimate damage from climate change on the basis of willingness to pay for ecosystem services. An alternative approach, which calculates the economic value lost from the ecosystem services degraded by climate change, is addressed in the Millennium Ecosystem Assessment (2010), although these results are not quantitative in the sense that the output is damage per ton of $CO_2$. All the published representations of ecological damages from climate change are highly simplified. Willingness to pay is typically based on data from one or a few countries, often the United States, and then scaled to other countries on the basis of an assumed relationship with GDP.

In the RICE and DICE models, human settlements and ecosystems are treated together. They assume that the capital value of climate-sensitive human settlements and ecosystems ranges from 5% to 25% of regional output. For the United States, the number is 10%; for island countries, and for countries with sensitive ecosystems, the number is higher. Willingness to pay to avoid a 2.5°C temperature change is assumed to be equal to 1% of the capital value of the vulnerable system (Nordhaus and Boyer 1999).

| Wetland Value ($10^6$ km$^2$) | Protection Costs ($10^9$$) | Emigrants $10^6$ | Value $10^9$$ | Immigrants $10^6$ | Value $10^9$S | Total Costs $10^9$$/year |
|---|---|---|---|---|---|---|
| 5.4 (2.7) | 83 (74) | 0.13 (0.07) | 7.5 (5.3) | 0.0 (0.20) | 2.9 (2.1) | 1.6 (0.9) |
| 4.3 (2.2) | 136 (45) | 0.22 (0.10) | 8.2 (5.4) | 0.64 (0.32) | 3.1 (2.2) | 1.7 (0.5) |
| 5.9 (2.9) | 63 (38) | 0.04 (0.02) | 2.8 (2.0) | 0.18 (0.10) | 1.6 (1.2) | 0.8 (0.4) |
| 2.9 (1.5) | 53 (50) | 0.03 (0.03) | 0.7 (0.7) | 0.03 (0.03) | 0.0 (0.0) | 0.5 (0.5) |
| 1.3 (0.7) | 5 (3) | 0.05 (0.08) | 0.4 (0.6) | 0.04 (0.07) | 0.0 (0.0) | 0.0 (0.0) |
| 0.9 (0.5) | 147 (74) | 0.71 (1.27) | 3.9 (7.2) | 0.64 (1.14) | 0.5 (0.9) | 2.0 (0.9) |
| 0.3 (0.2) | 305 (158) | 2.30 (1.40) | 3.7 (2.9) | 2.07 (1.26) | 0.5 (0.4) | 3.3 (1.6) |
| 0.2 (0.1) | 171 (126) | 2.39 (3.06) | 2.5 (3.4) | 2.15 (2.75) | 0.3 (0.4) | 1.8 (1.30 |
| 0.4 (0.2) | 92 (35) | 2.74 (2.85) | 5.4 (6.3) | 2.47 (2.56) | 0.7 (0.8) | 1.1 (0.4) |

SOURCE: Tol 2002a. Reprinted with permission; copyright 2008, *Environmental and Resource Economics*.

The elasticity of willingness to pay with respect to income is assumed to be equal to 0.1.

FUND does a separate calculation for 16 regions. The impact of warming on ecosystems in FUND is calculated as a "warm-glow" effect in which people are assumed to assign value to biodiversity and other ecosystem services, independent of whether they receive any concrete benefits from those services (Tol 1999). The value of the damage function rises with the fraction of biodiversity lost, with the amount of warming, and with per capita income in each region (Warren et al. 2006b).

Approaches based on the valuation of ecosystem services typically calculate the cost of replacing natural services with human or industrial alternatives. Many studies of the value of ecosystem services, however, do not explicitly assess the vulnerability of the ecosystem services to climate change. Schröter et al. (2005) looked at the vulnerability of ecosystem services to climate change in Europe, but they did not calculate an explicit cost impact. Naidoo et al. (2008) concluded that, for a large set of ecosystem services, they could reliably estimate values for only four and that the values of these four ecosystem services do not align well with areas targeted for biodiversity conservation. Brauman et al. (2007) reviewed a number of approaches to assessing ecosystem services and concluded that, whether or not the services are monetized, trade-offs among them can provide a useful set of tools for evaluating policy options. This approach is used by the Millennium Ecosystem Assessment, which assessed the impacts of four future scenarios (including climate change) on the basis of the number of ecosystem services, in each of four categories, expected to increase or decrease.

Overall, estimates of the impacts of economic damage to ecosystems from climate change are more conceptual and heuristic than quantitatively meaningful. The approaches that generate explicit numbers are simple and nonmechanistic approaches, starting with the studies of a small number of ecosystem services for one region or country at one level of economic development and the willingness to pay. Even if these studies were accurate, they should not be assumed to cover the full suite of climate-sensitive ecosystem services or to capture effectively the extrapolation of willingness to pay to other services, regions, or levels of economic activity. Finally, the sensitivity of the ecosystem services to climate is not well-known. These factors combine to define an approach that can be very useful for understanding aspects of the way the system works but that are unlikely to provide values that can be robustly used for studies that address multiple sectors of the economy. Approaches based on valuing ecosystem services sometimes generate numerical values, but sometimes they do not. The approaches based on valuing ecosystem services are not yet integrated in any of the main IAMs. Realizing such integration would represent an important conceptual advance in the credibility of the modeling, but it might not yield dramatic improvements in model accuracy or utility.

### Impacts on Agriculture

The welfare effects of climate change on agriculture depend on the impacts of climate on crop yields and on how farmers adapt to the impacts. In many areas of sub-Saharan Africa, temperatures are predicted to exceed optimal temperatures for many crops currently grown, and even for crops that could be substituted for current crops. Yield losses will, however, be less when irrigation is possible. Farmers may also be able to reduce income losses from crops by raising cattle and thus diversifying their agricultural portfolios. In northern latitudes, yields are actually predicted to increase for many crops, and areas in which field crops, such as winter wheat, can be grown are likely to extend into higher latitudes. The magnitude of physical impacts, in addition to depending on adaptation to climate in the form of crop substitution and irrigation, will depend on the magnitude of the $CO_2$ fertilization effect: Increased carbon in the atmosphere will increase yields by promoting photosynthesis and reducing plant water loss.[11]

---

[11]This increase raises yields approximately 15% for such crops as rice, wheat, and soybeans (Cline 2007).

To estimate the GDP impacts of the effects of climate change on agriculture, economists predict the impact of temperature and precipitation on agricultural revenues. These estimates are based primarily on cross-sectional studies—often referred to as the Ricardian approach (Mendelsohn et al. 1994; Kurukulasuriya et al. 2006)—or on crop models (Parry et al. 2004). The Ricardian approach looks at variation in net revenues across different geographic areas that vary in climate. For example, in the Dinar et al. (1998) study of Indian agriculture, variation in the net revenue per hectare across districts in India is explained as a quadratic function of temperature and precipitation, measured during different seasons of the year. In principle, this captures adaptation to climate—farmers in North India, for example, are more likely to irrigate their crops than farmers in South India—a factor that is reflected both in revenues and in costs. Crop models examine the impact of changes in temperature and precipitation on yields in a controlled setting. The results can be used as inputs into models that simulate farmer adaptation changes in climate (for example, changing crop mix). With assumptions about food prices and input costs, crop models can also predict the impact of climate change on agricultural revenues (see Box 5-1).

To estimate the GDP impacts of a particular climate scenario—for example, an increase in mean global temperature of 2.5°C in the year 2100—researchers must predict the impact of a temperature change on agricultural revenues in the year 2100 as well as the share of agriculture in GDP in 2100. In practice, the percentage change in agricultural revenues associated with a climate scenario is multiplied by the share of agriculture in GDP to estimate the GDP impacts of the scenario. When percentage changes in agricultural revenues are predicted from Ricardian models, it is implicitly assumed that prices in the future will remain the same as they when the models were estimated. Yield changes predicted by crop models can, in principle, serve as inputs to world models of food trade that will predict future agricultural prices and, hence, revenue impacts in a future year. Models that produce country-level estimates of GDP impacts, such as FUND, RICE, and DICE, assume that the share of agriculture in GDP declines as per capita income rises.

What is the magnitude of estimates of the impact of climate on agriculture and how do they vary across countries? A recent study by Cline (2007) estimates the impact on agricultural yields of a 4.4°C increase in mean global temperature and a 2.9% mean increase in precipitation occurring during the period 2070-2099. As Figure 5-7 shows, the largest losses are predicted to occur in parts of Africa, in South Asia, and in parts of Latin America. In contrast, the United States and Canada, Europe, and China will, in general, benefit from an increase in mean global temperature. These are estimates of impacts on yields and do not represent impacts on GDP.

**BOX 5-1**
**Estimating the Impacts of Climate Change on Agriculture**

Estimates of the impacts of climate change on agriculture are based primarily on cross-sectional studies of land values or net revenues (the Ricardian approach; see Kurukulasuriya et al. 2006) or on crop models (Parry et al. 2004).

Crop models examine the impact of changes in temperature and precipitation on yields in a controlled setting, which can also control for the effects of $CO_2$ fertilization. The advantage of these models over statistical studies is that they allow for a much richer set of parameters that influence yields. Plant growth is modeled as a dynamic process of nutrient application, water balance, as well as many other factors. The potential pitfalls are that the sheer number of parameters makes it impossible to estimate them jointly in a regression model, and hence these models rely on calibration instead. Some authors are concerned about misspecification and omitted variable biases (Sinclair and Seligman 1996, 2000). The results can be used as inputs into models that simulate farmer adaptation changes in climate (for example, changing crop mix). Changes in yields predicted by these models are often used as inputs to world food-trade models to calculate the impacts of yield changes on prices and welfare. The effect of yield changes on world prices are not captured in the Ricardian framework and are ignored in Cline (2007).

The Ricardian approach looks at variation in land values or net revenues across different geographic areas that vary in climate. For example, in the Dinar et al. (1998) study of Indian agriculture, variation in the net revenue per hectare across districts in India is explained as a quadratic function of temperature and precipitation, measured during different seasons of the year. The Ricardian approach in principle captures adaptation to climate—farmers in North India, for example, are more likely to irrigate their crops than farmers in South India. This impact is reflected both in revenues and in costs: Farmers who irrigate have higher yields as well as higher costs. The Ricardian approach thus measures the impact of higher temperatures on net revenues, allowing for adaptation. The models also allow for crop substitution across different climate zones. If the results from such models are used to examine climate impacts, it is implicitly assumed that prices in the future will remain the same as they were when the model was estimated. Without additional adjustment, the predictions of Ricardian models will not capture $CO_2$ fertilization effects or the impact of international trade in food on welfare.

Other criticisms of the cross-sectional approach include the fact that climate variables may pick up other effects—for example, knowledge of farm practices—

Nordhaus and Boyer's (1999) estimates of the impact of agriculture on GDP corresponding to a doubling of $CO_2$ concentrations (estimated to occur in 2100) suggest increases in GDP of over 0.5% in China, Japan, and Russia but losses of over 1.5% of GDP in India. However, when weighted that also vary geographically. Any variable that is correlated with climate and that influences farmland values has to be accounted for in the analysis. For example, access to subsidized irrigation water in the United States is correlated with warmer temperatures and capitalizes into farmland values. Omitting irrigation from a hedonic analysis will wrongfully attribute these subsidies as a benefit of a warming climate (Schlenker et al. 2005). For example, an analysis that pools the entire United States in a regression analysis assumes that if Iowa were to become warmer, it would become like California, where farmers enjoy access to highly subsidized irrigation water. In reality, Iowa would probably become more like Arkansas, which is also warmer and more irrigated (72% of the corn acreage is irrigated), but does not have access to subsidized irrigation water. Although the decision to irrigate is endogenous, the access to water and its cost vary greatly in space. Irrigation is just one example of a potential variable that varies with climate and influences farmland values. Others might be soil quality and access to markets. It is difficult to account for all of them correctly.

Some authors have suggested using year-to-year weather fluctuations and examining how they affect yields or profits (Auffhammer et al. 2006, Deschenes and Greenstone 2007). The advantage is that a panel (a data set with repeated observations for each spatial unit, such as a county) allows for the use of fixed effects to capture all time-invariant factors, such as soil quality and access to irrigation. The potential problem is that year-to-year weather fluctuations are something fundamentally different from climate change. The former are inherently short term, examining how yields or profits change in response to weather fluctuations after the crop is planted. The latter are long-term responses to a permanent shift in climate, which include switching to other crops or production methods that are not available in the short term.

Both the Ricardian analysis and panel studies have distinct advantages and disadvantages. Research for the United States suggests that both approaches agree that primarily extremely warm temperatures have a negative influence on yields and farmland values. Yields of corn, soybeans, and cotton gradually increase with increasing temperature until a crop-specific threshold of 29°C to 32°C is reached (Schlenker and Roberts 2009). Further temperature increases quickly become very harmful. Hotter regions exhibit the same sensitivity to these high temperatures as cooler regions, suggesting that they were not able to adapt to the higher frequency of these warm-temperature events. Similarly, a Ricardian model of farmland that separates temperature into beneficial moderate temperatures and damaging extreme temperatures finds that the land values are most sensitive to extremely warm temperatures (Schlenker et al. 2006).

by GDP, the losses associated with a doubling of $CO_2$ concentrations, are less than 0.2% of world output (Warren et al. 2006b). Tol (2002a,b) found aggregate net benefits to agriculture from a doubling of $CO_2$ concentrations, although Warren et al. (2006b) criticized this finding as overly optimistic.

282

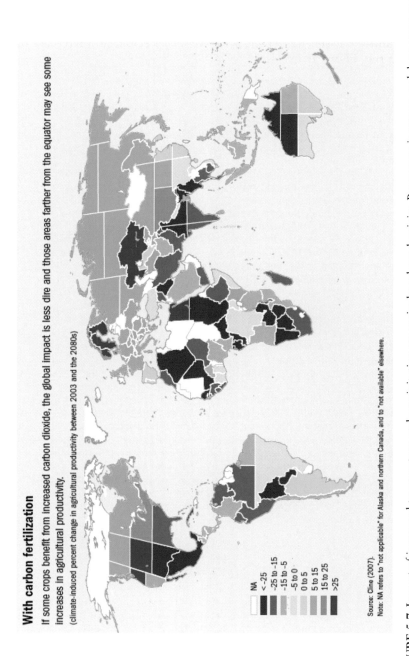

FIGURE 5-7 Impact of increased temperature and precipitation on agricultural productivity. Percentage increases and decreases were calculated assuming additional carbon fertilization. Negative values indicate percentage decreases in productivity. For example, the agricultural productivity in Mexico and the southwestern United States is predicted to decline by 25% or more. SOURCE: Cline 2007. Reprinted with permission; copyright 2008, Peterson Institute for International Economics.

## With carbon fertilization

If some crops benefit from increased carbon dioxide, the global impact is less dire and those areas farther from the equator may see some increases in agricultural productivity.

(climate-induced percent change in agricultural productivity between 2003 and the 2080s)

NA
< -25
-25 to -15
-15 to -5
-5 to 0
0 to 5
5 to 15
15 to 25
>25

Source: Cline (2007).
Note: NA refers to "not applicable" for Alaska and northern Canada, and to "not available" elsewhere.

## Impacts on Human Health

Theoretical analyses of the health consequences of rising average temperatures and the associated changes in average precipitation have led to research in the following five areas:

1. *Heat (and cold)-associated health conditions*, including the excess morbidity and mortality attributable to infectious, respiratory, and cardiovascular diseases and to over-exposure that occur after intense or prolonged cold weather and the heat-stress-related morbidity and mortality, especially excess cardiovascular disease mortality after intense or prolonged hot weather. This category could include the potential impacts on occupational health from working in hot and cold climates. These impacts are typically derived by looking at patterns of mortality either by day or season as a function of temperature for major cities and then using regression techniques to estimate temperature associated effects. Investigators differ in choice of daily changes—for example, heat waves, or average seasonal temperatures, the former providing higher estimates but with excess deaths typically limited to more vulnerable subpopulations.

2. *Vector-borne diseases*, especially malaria (mosquitoes), but also including dengue and yellow fever (mosquitoes), hanta and related viruses (rodents), Lyme and rickettsial diseases (ticks) and bird-borne viruses, such as West Nile and possibly influenza.

3. *Sanitation-related disorders*, including diarrheal diseases, such as cholera and others that occur with increased frequency in the setting of storms and prolonged droughts.

4. *Climate-associated changes in air-pollution health effects*, including atmospheric conversion of $NO_x$ and hydrocarbons to ozone and of $SO_2$ to its acid forms, which may be related to climate, although climate is not the source of air pollutants.

5. *Aeroallergen load* associated with altered ecosystems resulting from temperature and rainfall changes. As a consequence, potential increases in rates of upper and lower respiratory track allergies including asthma.

Substantial efforts have been made to model impacts in each category for the United States and other regions of the world based on the study of morbidity and mortality patterns in relation to climate patterns historically.

Accurate prediction of future impacts is substantially limited by the complexity of underlying assumptions about the populations at risk over time. The following factors complicate current efforts to estimate, based on various climate-change scenarios, what the impact on human health will be in the distant future:

1. *Demographics and development.* All the categories listed above affect different populations differentially, depending on such features as age, underlying health status, and stage of socioeconomic development. For example, the sanitation-related disorders are relevant only in the context of under-development; populations in advanced countries rarely suffer from these impacts under any climate conditions except in a hurricane Katrina-like disaster. Similarly, heat- and cold-associated disorders affect disproportionately the very young and very old, those with chronic health conditions, and those with resource limitations. Any predictions of the impact of a given climate-change scenario demands that explicit assumptions be made regarding the distribution of at-risk people in a given population, introducing more uncertainty.

2. *Adaptation.* The impact of many purported climate-associated health effects depends on the degree to which the affected population has become adapted to particular conditions. For example, heat-related morbidity and mortality are far more salient in populations living in temperate climates with large seasonal fluctuations in temperature than in those with year-round hot weather because of acclimatization; the *rate* of change may be a larger determinant than the extent of change in some of these estimates.

3. *Technology.* Separate from the impact of development on the underlying condition of populations is the potential impact of specific technological changes. For example, a successful malaria vaccine could neutralize the projected impact of increased malaria mortality even in the absence of underlying developmental change in regions of the world with endemic malaria. Likewise, advances in sanitation science and development of new antimicrobial techniques or agents or vector control technologies could substantially alter modeled impacts on sanitation-related effects.

4. *Mitigation.* Projected effects for each climate-change scenario could be substantially modified by efforts to anticipate the effects and mitigate them. Above and beyond the societal changes anticipated by societal development following its natural path, specific interventions could, in theory, reduce or eliminate effects due to any of the above categories. Interventions could include climate surveillance and institution of remedial steps under conditions of anticipated high risk or introduction of societal countermeasures, such as more stringent air-quality controls or provision of climate-controlled public shelters.

While the importance of each of those factors is widely acknowledged by investigators attempting multisector estimates of the (external) costs of damages related to human health under various scenarios (for example, RICE and DICE and FUND), the cost estimates for damages in each category have been developed generally under the default assumptions that development and demographic change over time would occur *unrelated* to intentional efforts to modify or mitigate the impacts of GCC. In other words, it is assumed that technologies and GDP will advance in parallel, and the underlying health status related to development will improve based on projections of regional GDP and global-warming potential, without taking into consideration specific efforts (or their costs) that might specifically offset or modify possible climate-related health effects.

The FUND and the RICE and DICE model estimates for health damages as percentage-lost GDP for the United States and globally reflect two somewhat divergent approaches. Tol (2002a,b) (FUND model), building on previous efforts, approaches the damages for the United States (and also other developed economies) by restricting attention to a single category of effects, namely, "heat-associated health conditions." FUND incorporates Martens (1998) meta-analysis of data from 17 European and American countries (20 cities)—a model based on seasonal averages—to calculate the impact of temperature under several climate scenarios. None of the other categories of potential health effects is added, possibly resulting in an underestimate. Nordhaus and Boyer (2002) (RICE and DICE), on the other hand, does not use this approach for a U.S. estimate and relies instead on deriving a temperature-associated estimate based on the WHO Global Burden of Disease (Murray and Lopez 1996) estimates for the region, resulting in a very small figure, more than 10-fold lower than the Tol figure.

It is noteworthy that both methods may have underestimated effects attributable to other categories. Most notably on the United States side is the possibility that pollution, interacting with climate, may have greater impact on ozone-related morbidity and mortality than estimates of temperature or pollution separately (for example, Knowlton et al. 2004). Globally, Kjellstrom et al. (2008) published a model that suggests substantial impact of climate in developing countries on the ability to work because of heat stress. The magnitude of this health effect has not been incorporated into either model, nor has either addressed the potential importance of allergy-related disease despite the global epidemic of asthma, still unexplained, already under way. Although Tol offers a range of estimates based on differing assumptions about the composition of the at-risk populations, neither model has tested divergent assumptions about the effects of mitigation, adaptation, or specific technological change.

### Other Impacts: Energy Production and Consumption, Socioeconomics, and National Security

Climate change is likely to result in many impacts that are poorly measured or difficult to quantify. In this section, the committee focuses on impacts to industry, population movements, energy supply and consumption, and national security. The greatest risks from climate change are likely to increase instability in vulnerable areas of the world with a consequent potential for increased risk of terrorism and political instability.

## Impacts on Energy Production and Consumption

In the United States, global warming will probably lead to modest decreases in heating demand in northern areas and modest increases in cooling demand in the southern area (CCSP 2007). Impacts on energy consumption primarily affect demand for electricity.

The CCSP study showed decreases in energy used in residential, commercial, and industrial space heating as possible effects of climate change. Estimates are quite imprecise. One study (Mansur et al. 2005) found a 2.8% decline in energy use for electricity-only customers, a 2% decline for gas customers, and a 5.7% decline for oil customers corresponding to a 1°C increase in January temperature in 2050. The variation in heating reduction is driven in part by regional variation in heating. Oil accounts for over one-third of heating in the Northeast, whereas electricity-only customers are likely to be located in the South or Southwest. Scott et al. (2005) found a stronger response relative to the results of the CCSP study.

Residential cooling impacts were stronger in the Mansur et al. (2005) study, which found a 4% increase in demand for electricity-only customers for a 1°C increase in July temperature in 2050. Increases for gas and fuel oil customers are 6% and 15%, respectively.

Annual energy consumption is affected by decreased heating and increased cooling costs. Mansur et al. found a 2% increase in residential expenditure and no impact on commercial expenditures at the national level. Others studies have found similar impacts, although regional variation is potentially significant.

Global impacts mirror regional impacts in the United States. Higher latitude regions benefit from reductions in heating, and lower latitude regions face higher costs of cooling (Stern 2007).

Beyond heating and cooling, the CCSP (2007) report found little impact on industrial energy demand (see studies by Amato et al. 2005; Ruth and Lin 2006). Industry may be affected in other ways. In particular, electric outages arising from extreme weather events would have significant impacts on energy-sensitive industries.

Similarly, impacts on energy production are likely to be modest in the aggregate. Regionally, certain areas may experience impacts. Reductions in water in the Northwest could reduce supply of hydroelectricity appreciably. Weather disruptions and extreme events could affect oil and gas supply and refining activities in the Gulf of Mexico. Similar impacts arise globally. Stern (2007, pp. 142-143) reported reductions in nuclear power production in France during the 2003 European heat wave due to overly warm river water that the plants rely on for cooling.

## Socioeconomic Impacts

It is difficult to predict the full range of potential socioeconomic impacts from global climate change. Stern (2007, pp. 128-129) reported that about 7 million people in sub-Saharan Africa have migrated to new regions to obtain food because of environmental stresses on agriculture. In addition to migration, climate change has the potential to create disruptions that affect education and gains in equality of women (Chew and Ramdas 2005). Coastal erosion, rising oceans, and extreme weather all disproportionately affect the most vulnerable members of society. (See a catalog of impacts in Leary et al. (2006).

## National Security Impacts

The socioeconomic instabilities described above have implications for national security of the United States, as well as global security. A recent report by the CNA Corporation (CNA 2007) found that climate change will add to instability in already volatile parts of the world (for example, Somalia and Darfur). In addition, the impacts will be felt globally and so create greater strains for the U.S. military as it stretches itself to cover conflicts in various parts of the world (acting either unilaterally or multilaterally).

Population migration will also affect currently stable countries and regions. The United States and Europe, for example, will face increased pressure from immigrant populations. Moreover while underdeveloped regions of the world are disproportionately affected by extreme weather (or are especially vulnerable), stable regions are not immune. The European heat wave of 2003 was estimated to have killed more than 50,000 people (Larsen 2006).

The CNA report highlights especially important regional impacts. Two-thirds of the Arab world currently relies on imported sources of water (CNA 2007, p. 30). Decreased precipitation exacerbates this problem and raises the specter of increased out-migration and land tension with neighbors. Nearly 40% of Asia's population lives no more than 45 miles from the coast. Sea-level rise could put millions of people at risk for inundation and increased risk of infectious disease (CNA 2007, p. 24).

The military implications of these impacts are twofold. First, U.S. military systems and bases will be stressed. Diego Garcia in the Indian Ocean serves as a major logistics hub for U.S. and British forces in that region of the world. The island at its highest point is only a few feet above sea level (CNA 2007, p. 37). In the event of significant sea-level rise, it may be possible to adapt by building dikes or other infrastructure but it would come at an economic cost, and the current capabilities of the base may be affected. In addition to impacts on military bases, climate change and severe weather make military missions much more challenging. Extreme weather also creates vulnerabilities for military energy supplies. Electricity systems are subject to outages in extreme weather, and the Department of Defense relies on electricity from the national grid to power critical infrastructure at installations (CNA 2007, p. 38).

A second concern identified by the CNA report is the Arctic. With global warming, retreat of the Arctic ice pack means that the U.S. Navy will have to expand its scope of operations to cover this area. In addition, increased access to the Arctic is likely to bring about increased competition for previously inaccessible resources, including potentially large reserves of oil.

### Estimates of Other Impacts

It is extremely difficult to monetize the external costs of political instability, population displacement, national security, and military costs arising from climate change. However, the committee can provide estimates of the external costs arising from increased demand for electricity due to climate change. We briefly discuss below the treatment of these costs in the FUND and DICE models.

The FUND model 3.0 does not provide estimates of most of the socio-economic costs discussed above, although resettlement costs are included in the costs of sea-level rise. FUND does provide estimates of increased heating and cooling costs based on an equation relating heating (or cooling) to increases in temperature, per capita income, population, and technological improvements. The elasticity of heating with respect to global mean temperature (relative to 1990 levels) is 0.5 for heating and 1.5 for cooling in the FUND base case. The income elasticity of space heating and cooling demand is 0.8 in the base case (Hodgson and Miller 1995, as cited in Downing et al. 1996b). Tol (2009) reported that globally the increased costs of electricity for cooling are the single largest component of the marginal damages from a ton of $CO_2$-eq emissions, while global reductions in heating costs reduce the marginal damages from a ton of $CO_2$-eq emissions.

Nordhaus and Boyer (1999) estimated that a 2.5°C increase in temperature would have negligible costs on energy and modest costs on human settlements ($6 billion in 1990 dollars; see Table 4-11, p. 4-45, of the study). These costs include costs of migration and adaptation as well as losses that occur because of difficulties in responding to sea-level rise or extreme weather. These costs are computed by estimating the capital value of vulnerable areas and assuming a willingness to pay to avoid damages equal to 1% of the value of at-risk capital (see section Coastal Zone Impact of Climate Change). The authors acknowledged that the method "is at this stage speculative and requires a detailed inventory and valuation of climatically sensitive regions for validation" (pp. 4-21 to 4-22). The authors noted, however, that this topic is likely to be of high impact and cost that will factor in to climate-change policy in important ways.

## ECONOMIC DAMAGE FROM IRREVERSIBLE AND ABRUPT CLIMATE CHANGE

The term "abrupt climate change" has several definitions (NRC 2002d; Clarke et al. 2003; Overpeck and Cole 2006). For the purposes of this assessment, a useful definition is articulated by CCSP (2008): "A large-scale change in the climate system that takes place over a few decades or less, that persists (or is anticipated to persist) for at least a few decades, and causes substantial disruptions in human and natural systems."

By contrast, irreversible climate changes represent fundamental regime shifts in major climatic variables that are likely to persist over hundreds to thousands of years. Irreversibilities are related to abrupt climate changes in that they embody the idea of thresholds or "tipping points" in physical or biogeochemical variables, which once crossed result in a large (implicitly rapid rate) and for all intents and purposes permanent change. In terms of climate responses, changes of this nature include such possibilities as the collapse of the Greenland or West Antarctic ice sheets, loss of coral reefs, or shutdown of the North Atlantic thermohaline circulation. However, incremental climate changes could eventually cross a threshold related to physical processes and thus result in irreversible climate change. One concern about such climate changes is the potential for them to trigger serious or catastrophic follow-on impacts, such as the release of methane and $CO_2$ trapped in ocean sediments and permafrost; loss of biodiversity through extinction, disruption of species' ecological interactions, and major changes in ecosystem structure; and disturbance regimes, such as wildfire and insects.

Another reflection of this concern is the tendency of these sorts of changes to be discussed in parallel with potential downstream consequences for human society, in particular the value of lost species, ecosystem services, arable land and attendant effects on food security, as well as adverse effects on human settlements, migration, and the potential human insecurity (i.e., refugees, violent conflict) arising therefrom.

An important feature of the preceding definitions is that they say little about the likelihood of occurrence of the events in question. Although it is tempting to view climate thresholds as a bright-line, there is little empirical basis for inferring how much of a change in probability they bring. On one hand, there is the probability of the threshold being reached, which depends on the trajectory of GHG emissions and consequent radiative forcing. Some indication of the relevant probabilities is given in Table 5-2. On the other hand, how the probability of occurrence of the impacts in question might change for increments in, say, temperatures in excess of the threshold is largely unknown. For example, IPCC (2007a) concluded that 20-30% of assessed species face about a 50% chance of increasingly high risk of extinction as global mean temperatures exceed 2-3°C above preindustrial levels but does not characterize the dependence of the probability of extinction on temperatures beyond the threshold.[12] This kind of impact is distinct from events that are low probability but very high consequence. It is unclear whether climate thresholds apply—such as a massive decadal-scale release of methane to the atmosphere from rapid clathrate destabilization.

Solomon et al. (2009) focused attention on atmospheric warming, precipitation changes, and sea-level rise driven by thermal expansion as adverse irreversibilities for which three criteria are met: (1) the relevant changes are already being observed, and there is evidence for their anthropogenic precursors; (2) the phenomena are based on physical principles that are thought to be well-understood; and (3) projections are available and are broadly robust across Earth-system models. These authors use results from a suite of models to construct ranges of long-run equilibrium changes in the climate. Their estimate of the irreversible temperature increase ranges from 1 to 4°C, with a corresponding 0.2-0.6 m sea-level rise per degree of global warming, for an irreversible global average sea-level rise of at least 0.4-1.0 m (and as much as 1.9 m for peak $CO_2$ concentrations in excess of 1,000 ppmv [parts per million by volume]) and complete losses of glaciers and small ice caps adding a further 0.2-0.7 m. The corresponding estimates of shifts in precipitation are subject to considerable uncertainty, but a robust change is an enhanced dry season in several regions (on the order of 20% in northern Africa, southern Europe, and western Australia) and 10% in southwestern North America, eastern South America, and southern Africa for 2°C of global mean warming. The equilibrium dependence of regional dry-season-precipitation impacts on $CO_2$ concentrations is illustrated in Figure 5-8.

At the other end of the spectrum, the implications of these climate changes for biodiversity loss, highly nonlinear impacts—such as ice sheet instability, thermohaline circulation (THC) collapse, and methane clathrate releases—and climate-induced violent conflict represent unknowns that derive from fundamental gaps in scientific understanding of the mechanisms of the relevant impact pathways in Figure 5-1 (NRC 2002d; Tol 2008). For

---

[12]IPCC (2007, Table 4.1) characterizes how the magnitude and geographic distribution of extinction impacts increase with temperature.

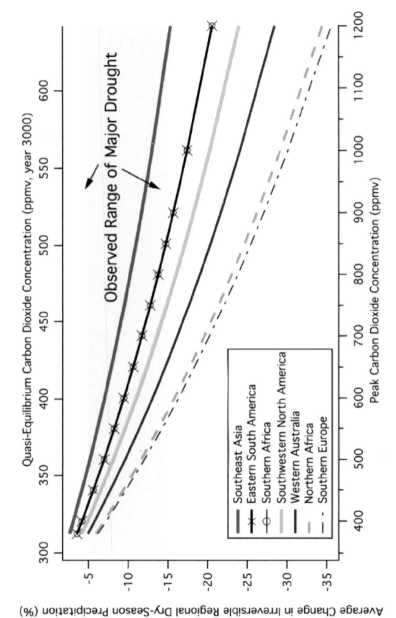

FIGURE 5-8 Irreversible precipitation changes by region. SOURCE: Solomon et al. 2009, Figure 4.

the sake of completeness, brief notes are provided about the current state of the literature in each of these areas.

There are comparatively few quantitative studies of the direct impacts of climate change on ecosystems and biodiversity at broad geographic scales,[13] and projecting changes in biodiversity at the regional and global levels is complicated by the need to account for large-scale, potentially non-linear interactions associated with such factors as shifting anthropogenic land use and invasive species (see Sala et al. 2000). The theoretical basis for valuation of ecosystem changes at these scales is very weak. Nicholls et al. (2008) explored a wide range of scenarios of resulting sea-level rise and assessed the impacts of a collapse of the West Antarctic Ice Sheet (WAIS) at 0.5-5 m per century. Their estimates of annual costs, which are not explicitly linked to temperature change, ranged from $0-28 billion in 2050 to $0.1-31 billion in 2100.

The full range of impacts arising from a slowdown or collapse of the Atlantic THC has yet to be systematically characterized. The only studies from which cost figures can be drawn are Keller et al. (2004), who arbitrarily assumed an uncertain cost in the range of 0-3% of gross world product, and Link and Tol (2004), who estimated that THC shutdown increases the marginal damage of GHG emissions by $0.1-2.2 per ton of carbon. The seemingly small magnitude of these figures may be appreciated when one considers that, in the Link and Tol study, THC collapse has a negligible influence on global average surface temperature but has a substantial impact on temperatures in the United States, Canada, and western Europe, inducing cooling of 0.5-1.5°C by 2150 and 1-3°C by 2300.

Considerable scientific uncertainty still besets the characterization of methane releases from clathrates in ocean sediments and permafrost. The most widely cited study by Harvey and Huang (1995) estimates a cumulative methane release of 53-887 gigatonnes of carbon after 2,000 years (see Table 9), resulting in an equilibrium atmospheric temperature rise of 1-9°C (see Figure 9). These authors find an amplification of global warming of 10-25%, but this range depends strongly on assumptions about the climate sensitivity and the warming due to projected anthropogenic $CO_2$ emis-

---

[13]Scientific studies have mostly focused on aggregate indicators of change. For example, Scholze et al. (2006) used the results of a suite of climate model runs as inputs to a dynamic global vegetation model and mapped the proportions of simulations that exhibit forest-nonforest shifts and exceedance of natural variability in wildfire frequency and freshwater supply. A landmark study by Thomas et al. (2004a) estimated that among the groups of organisms they assessed in regions covering 20% of Earth's land surface, warming by 2050 will cause extinction of 15-37% of species. However, the methods for combining climatic stressors with geographically localized data on individual species to characterize species extinctions are the subject of vigorous debate (Buckley and Roughgarden 2004, Harte et al. 2004, Lewis 2006, Thomas et al. 2004b, Thuiller et al. 2004).

sions, and is substantially outweighed by the latter uncertainties. Results by Renssen et al. (2004) highlight the importance of uncertainties regarding the "worst case" quantity of methane released. A massive emission (1,500 GtC) over the course of a millennium would entail large climate changes,[14] having peak additional surface warming of 2.6°C on average and up to 10°C at the poles, accompanied by regime shifts in the global overturning ocean circulation.

The discussion above describes the possibility of extreme climate changes that could result in large, irreversible economic damages to the planet. As noted at the beginning of the chapter, the possibility of extreme events is not handled well by IAMs in calculating the marginal damages of $CO_2$. The RICE and DICE models attempt to handle extreme events by calculating what a risk-averse individual would pay to avoid a catastrophic event (of given probability) that would reduce GDP from 22% to 44%, depending on the region of the world. The probability of such an event is calculated, for each T, on the basis of expert judgment. For a 2.5°C change in mean global temperature, willingness to pay to avoid catastrophic risk ranges from 1.9% of GDP in OECD-Europe and India to 0.45% of GDP in the United States. The corresponding figures are 10.79% and 2.53% of GDP to avoid catastrophic risk associated with a 6°C change in mean global temperature.

Weitzman (2009) demonstrated that, if one were to ask a risk-averse individual what he would pay to avoid the gamble described above, the amount would be infinite if the distribution over the catastrophic GDP loss were "fat-tailed" (formally, if the distribution has an infinite moment generating function). Clearly, the nature of the probability distribution of catastrophic outcomes matters and is handled only imperfectly by the willingness to pay calculations described in the preceding paragraph or by the Monte Carlo simulations performed to capture uncertainty in the key parameters of IAMs. The key problem here is that low-probability extreme-impact events located in the fat tails, which are extremely difficult to quantify, might drive the results of cost-benefit analysis. This possibility is disturbing because the answers to important questions about how much effort to put into climate-change mitigation can depend to an uncomfortable degree on subjective estimates about the likelihood of catastrophic outcomes.

---

[14]The authors design this scenario to be consistent with the Paleocene-Eocene thermal maximum, a period about 55.8 million years ago that experienced drastic changes in climate possibly as the result of releases of methane from hydrates.

## AGGREGATE IMPACTS OF CLIMATE CHANGE

To quantify the marginal impact of an additional ton of GHG emissions requires a number of steps beyond quantifying the individual impacts associated with a particular magnitude of climate change. Within each region, for a particular climate-change scenario (for example, doubling of $CO_2$ concentrations in the atmosphere), individual components of impact must be aggregated across sectors, raising difficulties for inclusion of impacts that have not been expressed in monetary metrics and interactive effects, such as that between water and agriculture. Next, because the impacts of GHG emissions emitted anywhere are felt globally, there is much interest in understanding the global impacts of climate change, not simply the effects of each country's emissions within its own borders. This factor requires aggregating impacts on people with widely differing incomes, raising questions about whether monetized impacts should be adjusted to account for differing marginal values of income across countries. Moreover, because the consequences of current GHG emissions are expected to persist for centuries, it is necessary to aggregate impacts on people living at different times in the future. Previous surveys of these issues include Pearce et al. (1996), Tol and Fankhauser (1998), Tol et al. (2000), Smith et al. (2001), Hitz and Smith (2004), Stern (2007), Yohe et al. (2007), and Tol (2008).

Tol (2008) identified 13 published studies that have estimated the monetized impacts of climate change at a global level; several of them also include total climate-change damage estimates individually for the United States and other regions (Fankhauser 1995a; Tol 1995, 2002b; Nordhaus 1994a,b, 2006; Nordhaus and Yang 1996; Plambeck and Hope 1996; Mendlesohn et al. 2000a,b; Nordhaus and Boyer 2000; Maddison 2003; Rehdanz and Maddison 2005; Hope 2006). In addition, the Nordhaus (2008) study contains an update based on Nordhaus and Boyer (2000), and at least another four published studies contain total-damage estimates for the United States alone (Nordhaus 1991; Cline 1992; Titus 1992; Mendelsohn and Neumann 1999). These estimates are not independent because many of them represent revised estimates by the same researchers over time (for example, Nordhaus, Tol, Hope, and Maddison), and other estimates are based on models that draw from several other prior studies (for example, impact valuation in Plambeck and Hope (1996) derives from Tol (1995) and Fankhauser (1995b); see discussion in Tol (2008). A review of impact estimates within IAMs (Warren et al. 2006b)—including the DICE and RICE models (Nordhaus), FUND model (Tol), and PAGE model (Hope)—found that the impacts in these models are based on literature from 2000 and earlier.

Table 5-7 summarizes results from many of these studies, for consistency expressed in terms of percentage loss in GDP. Most, but not all of the

**TABLE 5-7** Estimates of Total Damage Due to Climate Change from Benchmark Warming (Percentage Change in Annual GDP)

| Study | Temperature Change (°C) | Global[e] | United States | Range Across Regions |
|---|---|---|---|---|
| 2.5-3.0°C warming benchmark | | | | |
| Nordhaus (1991) | 3.0 | NA | −1.0 | NA |
| Cline (1992) | 2.5 | NA | −1.1 | NA |
| Nordhaus (1994a) | 3.0 | −1.3 | NA | NA |
| Nordhaus (1994b) | 3.0 | −1.9[c] | NA | NA |
| Fankhauser (1995b) | 2.5 | −1.4 | −1.3 | −4.7--0.7 |
| Tol (1995) | 2.5 | −1.9 | −1.5 | −8.7-0.3 |
| Nordhaus and Yang (1996) | 2.5 | −1.7[a] | −1.1 | −2.1-0.9 |
| Plambeck and Hope (1996) | 2.5 | −2.5[a] | −1.6 | −8.6-0.0 |
| Nordhaus and Boyer (2000) | 2.5 | −1.5 | −0.5 | −4.9-0.7 |
| Mendlesohn et al. (2000a,b)[b,d] | 2.5 | 0.00.1 | NA | −3.6-4.0[a], −0.5-1.7[a] |
| Hope (2006)[d] | 2.5 | −1.0 | −0.3 | −3.1-0.3 |
| Nordhaus (2006)[b] | 3.0 | −1.0 | NA | NA |
| Nordhaus (2008) | 2.5 | −1.8 | −0.7 | −20.0–16.4 |
| Other warming benchmarks | | | | |
| Titus (1992) | 4.0 | NA | −2.5 | NA |
| Tol (2002b) | 1.0 | 2.3 | 3.4 | −4.1–3.7 |

NOTES: Positive damage estimates indicate benefits from warming. NA indicates data are not available.

[a]As computed by Tol (2008).
[b]Estimate includes only market impacts; nonmarket impacts are not monetized.
[c]Median estimate from an expert opinion survey of 19 individuals.
[d]The study's mean estimates are given.
[e]Global GDP losses are simple (unweighted) sums of regional GDP losses.

scenarios are benchmarked to a 2.5-3°C temperature increase by 2100 associated with central estimates of the likely warming from a doubling of GHG concentrations in the atmosphere. Note that Mendlesohn et al. (2000a,b) and Nordhaus (2006) include only market impacts, while the other studies also include estimates of nonmarket impacts, at least to some degree.[15]

Table 5-7 shows that these studies typically estimate the aggregate global market plus nonmarket impact of doubling GHG concentrations at 1-2% of lost world GDP. The aggregate impacts mask significant differences in regional impacts and in the underlying impacts for individual damage

---

[15]Maddison (2003) and Rehdanz and Maddison (2005) estimates are not included in this table due to the incompleteness of the estimates relative to the others included. Maddison (2003) estimates the effect of temperature and precipitation on household market good impacts based on historical country-level demand data, and Rehdanz and Maddison (2005) estimate the effect of temperature and precipitation on historical country-level measures of "happiness."

categories estimated within each study. Estimated percentage damages tend to be lower in industrialized countries but significantly higher in many developing countries with relatively higher current temperatures, heavier dependence on agriculture, and lower adaptive capacity. No individual impact category consistently dominates other categories across studies. Previous surveys have also been careful to note the low quality of the numbers and the many shortcomings of the underlying studies.

In addition to differences across aggregate climate-impact studies in terms of methods and estimated regional and individual impact categories, there are a number of other key assumptions and sensitivities. One issue is whether GDP impacts in individual regions are weighted during aggregation to a global total. The global estimates noted above simply add up the estimated regional impacts in dollar terms (that is, they are output weighted) regardless of income levels in the different regions. However, it is widely accepted that individuals with low income tend to value a given dollar impact more heavily than a relatively high-income individual. This factor is known as the declining marginal utility of income, and approaches for incorporating it are often called "equity weighting" or "population weighting" (if global losses are based on regional percentage losses weighted by population shares as opposed to output shares). Estimates that allow for equity weighting typically find significantly more negative aggregate global impacts because regions with more substantial projected impacts are also relatively poor (Yohe et al. 2007).

Estimates of total climate damage also depend critically on the degree of temperature change that is being assessed. With the exception of the Titus (1992) and Tol (2002b) assessments of a 4°C and 1°C temperature increase, respectively, all the other studies in Table 5-8 focus on a benchmark warming scenario of 2.5-3.0°C, corresponding to best estimates of eventual temperature change from a doubling of GHG concentrations. Unsurprisingly, the pattern among available studies is that—beyond some amount of warming that is beneficial for certain regions and impact categories—greater degrees of temperature increase are associated with correspondingly higher damages.[16]

As an approximation, modeling assessments (for example, using DICE and RICE, FUND, and PAGE) that explore a range of emission, concentration, and temperature scenarios tend to assume that damages are proportional to the size of the world economy and that the fraction of world GDP lost (total or per capita) is a power function of temperature increase. The power function is calibrated to the damage estimate from benchmark

---

[16]Several studies have found positive impacts of climate change on agriculture in Canada, Europe, and parts of China (see Figure 5-7 from Cline [2007]). Heating requirements are also predicted to decline in Russia and parts of Europe.

**TABLE 5-8** Marginal Global Damages from GHG Emissions: Estimates from Widely Used Models

| Model | Study | Marginal GHG Damage ($/ton $CO_2$) | Discount Rate (%)[a] | Climate-Warming Scenario | Total Global Climate Damage (% GDP) |
|---|---|---|---|---|---|
| DICE | Nordhaus (2008) | 8[e] 8 | ~4.5 | No control: 3.1°C in 2100; 5.3°C in 2200 Optimal: 2.6°C in 2100; 3.5°C in 2200 | −1.8% at 2.5°C −4.5% at 4.0°C −7.1% at 5.0°C −10.2% at 6.0°C |
| FUND | Tol (2005a)[d] | 0 2 6 | 5 3 2 | No control: 3.7°C in 2100; 6.7°C in 2200 | ~0% at 2.5°C ~ −1% at 4.0°C ~ −1% at 5.0°C |
| PAGE[b] | Hope (2006)[c] Hope and Newbery (2008) | 6 (1-17) 22 (4-60) 108 (21-284) | ~4.5 ~3 ~1.5 | No control: 4.1°C in 2100; 7.9°C in 2200 | −1.0% at 2.5°C −2.6% at 3.9°C −11.3% at 7.4°C |
| | Stern (2007) | 102 36 | 1.4 | No control: 3.9°C in 2100; 7.4°C in 2200 Stabilize at 550 ppm $CO_2$-eq: eventual 3°C | −1.0% at 2.5°C −2.6% at 3.9°C −11.3% at 7.4°C |

NOTE: Negative numbers indicate a negative impact on GDP.

[a]Discount rate changes over time in Nordhaus (2008) and Hope (2006); the approximate effective discount rate is given.

[b]For PAGE model, mean global GDP impacts are given in Dietz et al. (2007), including market, nonmarket, and risk of catastrophic impacts.

[c]Mean estimate for 2001 emissions with 5th-95th percentile confidence interval from uncertainty analysis in parentheses.

[d]Estimate is for emissions in 2000 from FUND version 2.4.

[e]Estimate is for emissions in 2005.

warming (that is, it is one point on the function) and an assumed temperature level corresponding to zero damages. A linear relationship (that is, percentage of climate damages per degree temperature change is constant) corresponds to a power of 1, and a quadratic or cubic relationship corresponds to a damage exponent of 2 or 3. The DICE model (Nordhaus 2008), for example, assumes a quadratic damage function based on Nordhaus and Boyer (2000), yielding estimates of global climate damages increasing over fourfold from −1.8% to −8.2% of world GDP for a twofold increase in temperature from 3°C to 6°C. The PAGE model (Hope 2006), on the other hand, allows for a damage exponent ranging from 1 to 3, with an

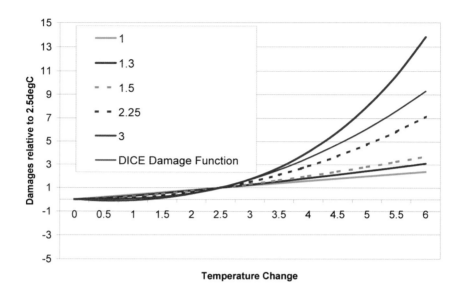

**FIGURE 5-9** Dependence of GHG damage on the amount of temperature change. The lines show the PAGE 2002 damages for damage exponents between 1 and 3. The damage function of the DICE model is also shown for comparison. In this figure, positive values indicate economic losses, and negative values indicate benefits from warming. SOURCE: Stern 2007, Technical Appendix.

associated range of global impact that varies by a factor of almost 6 for a 6°C temperature increase (see Figure 5-9, where the PAGE damage function for 6°C is about 2.4 times the level at 2.5°C for linear damages, while it is about 14 times as high assuming cubic damages).

Yet, in the absence of substantial mitigation action, projections of baseline GHG emissions tend to imply estimates of likely temperature increase that are significantly greater than that associated with a doubling of GHG concentrations. For example, the IPCC (2007a, p. 180, Figure 5.1) referenced plausible projections of GHG concentrations that go near to and beyond 1,000 ppm by 2100, with an associated best estimate global mean temperature increase above preindustrial levels of about 5-6°C and a likely range from just under 4°C to over 8°C. However, little is known about the precise shape of the temperature-damage relationship at such high temperatures.[17] Figure 5-10 illustrates the dependence of GHG damage, as a percentage of global GDP, on the amount of temperature change.

---

[17]See discussion in Stern (2007, pp. 659-662).

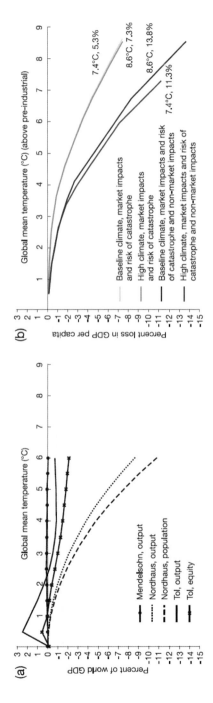

**FIGURE 5-10** Dependence of greenhouse gas (GHG) damage, as a percentage of global gross domestic product (GDP), on the amount of temperature change. (*a*) Damage estimates represented as a percentage of global GDP, as a function of increases in global mean temperature. (*b*) Damage estimates, as a percentage of global GDP per capita, are correlated with increases in global mean temperature. In this figure, positive values indicate benefits from warming. SOURCE: IPCC 2001, p. 1032 and Stern 2007 (as cited in Yohe et al. (2007, p. 822).

## MARGINAL IMPACTS OF GREENHOUSE GAS EMISSIONS

Given an estimate of the monetized global impact of a particular climate-change scenario at a particular future point in time, this total-damage estimate can then be translated into a marginal damage per ton estimate (often called the "social cost of carbon") by evaluating the linkage between current GHG emissions and future climate-change impacts (see Equation 5-2). It is usually estimated as the net present value of the impact over the next 100 years (or longer) of 1 additional ton of $CO_2$-eq emitted into the atmosphere. It is this marginal damage per ton of emissions that is normally used as a measure of the global climate externality. This measure requires assumptions about the emissions-temperature and temperature-damages linkages over time, as well as the rate at which future damages are discounted back to the present to account for differing valuation of monetary impacts felt at different points in time. Finally, uncertainties at each step of the analysis imply that different possible future conditions may yield widely differing impacts. The expected value of damages may be more sensitive to the possibility of low-probability catastrophic events than to the most likely or best-estimate values.

There have been many previous reviews of existing estimates of the marginal damages from GHGs, including Pearce et al. (1996), Tol (1999, 2005b, 2008), Clarkson and Deyes (2002), and Yohe et al. (2007). Tol (2008) identifies 211 marginal-damage estimates from 50 studies, although this number does not imply we know more about marginal than total damages (there are only 12 global total-damage estimates, as shown in Table 5-7). The explanation for how so many marginal costs can be generated from so few total-damage estimates lies in the variety of additional modeling assumptions that must be incorporated to translate total into marginal damages. As alluded to above, in addition to the benchmark estimate of total damages, important other assumptions include the change in damages with increased warming and with growth and changes in the composition of economic activity over time, the assumed emissions scenario, the climate sensitivity to GHG concentrations, the rate used to discount future impacts to the present, the timeframe over which impacts are considered, and the treatment of uncertainty and risk aversion. Box 5-2 discusses approaches used to determine a discount rate.

Pearce et al. (1996, p. 215, Table 6.11) summarized early estimates of marginal GHG damages, which ranged from $3 to $62 per ton of $CO_2$-eq for emissions occurring in the 2001-2010 decade.[18] As part of a United Kingdom effort to assess the social cost of carbon, Clarkson and Deyes (2002) suggested a pragmatic approach could be to use a central

---

[18]IAM results usually include $CO_2$ as the only GHG. Tol (2005b) refers to a cost per tonne of carbon.

estimate of $35 per ton of $CO_2$-eq, along with a sensitivity range of half and double this amount ($17-70 per ton of $CO_2$-eq). Tol (2005b) identified 103 marginal-climate-damage estimates from 28 published studies, finding a median estimate of $4 per ton $CO_2$, a mean of $25 per ton $CO_2$, and a 95th percentile of $96 per ton $CO_2$-eq across the estimates. Tol (2005b) also found that the subset of studies published in peer-reviewed journals reported lower estimates on average, with a mean of $12 per ton $CO_2$-eq. (The Tol [2005b, 2008] values are not adjusted for inflation.) Summarizing 211 estimates identified in Tol (2008) yields a median estimate of $8 per ton $CO_2$-eq, a mean of $29 per ton $CO_2$-eq, a 5th and 95th percentile of $0 and $105 per ton $CO_2$-eq, respectively, and a peer-reviewed mean of $14 per ton $CO_2$-eq (University of Hamburg 2009). In cases where a single study generated multiple estimates, Tol (2008) included a relative weight for each estimate that was provided by the author of each study. Using these weights, one can construct a single weighted estimate for each of the 50 studies. Summarizing these 50 estimates from individual studies yields a median estimate of $10 per ton $CO_2$eq, a mean of $30 per ton $CO_2$-eq, and a 5th and 95th percentile of $1 and $85 per ton $CO_2$-eq, respectively. Note, however, that due to the lack of necessary information, Tol did not adjust individual estimates for inflation, nor did he account for the timing of emissions (that is, the year they occur) or the GHG concentration and temperature scenario onto which those emissions are added. Adjusting for inflation from the study year to current dollars would make these figures higher. The underlying estimates also differ in terms of their assumed discount rates and how they aggregate regional impacts (using output, equity, or population weighting), among other factors.

To provide a more consistent comparison of marginal-damage estimates, it is helpful to focus on estimates using the most widely used impact assessment models, DICE, FUND, and PAGE, as shown in Table 5-9. The estimates represent the marginal damages from current emissions against an assumed reference case climate scenario without GHG mitigation. This subset of estimates spans approximately the same range as discussed above, from roughly 0 to $100 per ton of $CO_2$-eq. The table demonstrates that virtually all of the variation can be understood as a function of differences across the studies in what is assumed about the discount rate and the magnitude of GDP losses expected from uncontrolled warming. Nordhaus and Yang (1996) made the same point, noting that "the two crucial parameters are the discount rate (which indicates the relative importance of the future compared to the present) and the damages from climate change (which measure the willingness to pay to prevent or slow climate change). *It is interesting to note that both major uncertainties involve human preferences rather than pure questions of 'fact' about the natural sciences*" (emphasis in original).

## BOX 5-2
## Discounting and Equity Weighting

Quantifying the damages from GHG emissions requires aggregation of damages that occur at different times extending centuries into the future and to different populations across the globe at each point in time. The method chosen for aggregation has implications for how effects on different people are weighed.

Two methods for aggregating effects on different people are common: using monetary and utility measures. The monetary measure assumes that $1 of benefit to one person is equally as good as $1 of benefit to another. The utility measure assumes that the gain in utility (or well-being) from receiving $1 is larger for a poor person than a rich person because the poor person is likely to have more pressing needs.

Aggregating across people using the monetary measure is straightforward: One simply sums the monetary values of benefits and harms across the relevant population. To implement the utility-based approach, one needs to make some assumptions about how individual utility varies with income (or wealth). Often, it is assumed that utility is proportional to the logarithm of income or to a power function of income, where the power is less than 1. These functions have the property that utility increases with income but at a diminishing rate. After choosing a function, one can weight the monetary value of benefits and harms to each person by the incremental utility of income and sum these values. This "equity weighting" gives more weight to the same monetary value of damages when they are suffered by a poor person rather than a rich person.

For aggregating effects across time, it is conventional to discount the monetary value of future effects by a factor of $[1/(1 + r)]t$ that depends on the discount rate $r$ and number of years in the future $t$ at which the effect occurs. The present value of a stream of effects occurring at various times in the future is calculated by summing the discounted monetary values of the effects. In determining the appropriate discount rate to use for aggregating effects on the current and some future generation, one can distinguish between descriptive and prescriptive ap-

TABLE 5-9 Indicative Marginal Global Damages from Current GHG Emissions ($/Ton $CO_2$-eq)

|  | Damages from Benchmark Warming | |
| --- | --- | --- |
| Discount Rate | Relatively Low | Higher |
| 1.5% | 10 | 100 |
| 3.0% | 3 | 30 |
| 4.5% | 1 | 10 |

NOTE: Only order-of-magnitude estimates appear warranted.

proaches. The descriptive approach infers the rate at which society chooses between consumption at different times from market interest rates.

In contrast, the prescriptive approach derives the appropriate discount rate on monetary values (the consumption discount rate) as the sum of utility and growth discount rates. The utility discount rate is the rate at which the future generation's utility is discounted relative to the present generation's. Many scholars have suggested that it is inappropriate to value other people's well-being less simply because they come later in time and so argueed for a utility discount rate of zero. The growth discount rate accounts for differences in income between the current and future generation. If the future generation will have greater income than the current generation, it will lose less utility from $1 of damages than the current generation will. To aggregate the effects on utility, it is necessary to down-weight the monetary value of the damages to the future generation, just as one would down-weight the monetary value of effects on rich people at the same point in time in accordance with equity weighting. The extent of this growth discounting effect depends on the economic growth rate (that determines the difference in income between the two generations) and the utility function (that determines how much the incremental effect of income on utility falls). If the future generation is poorer than the present, the growth discounting effect will apply in the opposite direction and will give greater weight to the monetary value of damages suffered by the future generation.

Following the prescriptive approach, Stern (2007) adopt a near-zero utility discount rate of 0.1% per year, a relatively small value of the rate at which the incremental effect of income on utility falls to 1 (corresponding to a logarithmic utility function), and a low rate of economic growth, 1.3% per year. Together, these yield a consumption discount rate of 1.4%. In contrast, Weitzman (2007a,b) suggests that more plausible values are roughly 2%, 2, and 2%, yielding a much larger consumption discount rate of 6%. Nordhaus (2008) uses the descriptive approach; he calibrates his model parameters so that the consumption discount rate is consistent with market interest rates, yielding a discount rate of 4.5%.

Perhaps the clearest illustration of the influence of the discount rate is a comparison of the "no-control" relatively high (4.5%) discount rate scenario of Hope (2006) and the low discount rate (1.4%) of Stern (2007), which yield marginal damage estimates of $6 and $102 per ton, respectively: a 17-fold difference. Both studies used the same version of the PAGE model, so the only significant difference in assumptions is the discount rate. When Hope and Newbery (2008) applied approximately the same discount rate as Stern (2007) to the PAGE model, they found a similar marginal-damage estimate of $108 per ton $CO_2$. Similarly, the Nordhaus (2008) estimate of \$8 per ton $CO_2$, which also used a relatively high discount rate of about 4.5%, is quite close to the estimate of Hope (2006) using a 4.5% discount rate. Finally, when Nordhaus (2008) applied low discount rates similar to Stern's to the DICE model, he found a marginal-damage estimate similar in magnitude to Stern's (\$88 per ton $CO_2$-eq).

The rate at which future damages from current emissions are converted to present values in Stern (2007) is only slightly greater than the rate at which damages for an incremental increase in global temperature are projected to grow over time. In Stern (2007), damages are a fraction of world GDP per capita that depends on climate change. The rate at which GDP per capita is assumed to grow (1.3% per year) is nearly as large as the discount rate (1.4% per year). With impacts rising almost as fast as they are being discounted, it is primarily the limited time horizon (2200 in the PAGE model) that constrains the marginal-damage estimate from becoming virtually unbounded, given that the effects of current emissions on climate will persist for centuries. In contrast, the discount rates assumed in Nordhaus (2008) and Hope (2006) are high enough that even after accounting for these additional growth effects the present value of damages in the distant future is low.

The growth in incremental damages over time underpins the rationale for a marginal (per ton) GHG damage that rises over time. For example, in the PAGE model, marginal damages rise by about 2.4% per year (Hope and Newbery 2008), and in the DICE model, marginal damages rise by about 2.0% per year (Nordhaus 2008). Over a 20-year period (for example, from 2010 to 2030), marginal damages rising at a rate of 2-3% per year would increase in total by a factor of 50-80%. This estimate is due to a combination of a larger economy being affected and increasing proportionate impacts of increasing temperatures (that is, nonlinearity of the damage function).

The marginal damages from current emissions do not decrease appreciably for alternative scenarios with significantly lower GHG emissions and temperature increases in Nordhaus (2008) or in Hope and Newbery (2008). According to Hope and Newbery (2008), this finding is due to convexity of the damage function being roughly offset by concavity in the concentration-temperature relationship, which is logarithmic. Given that Stern also uses the PAGE model, it is surprising that Stern (2007) found that marginal damages fall dramatically from \$102 per ton under a no-control scenario to \$36 per ton $CO_2$-eq under a 550-ppm stabilization scenario. Although Stern (2007) does not provide an explanation for the derivation of these results, it appears to be a result of the much lower discount rate assumed in Stern, which gives higher weight to future damages and which are much lower in a stabilization than no-control scenario. In contrast, Nordhaus (2008) and Hope (2006) used significantly higher discount rates where these future damages (or lack thereof) matter less. One implication is that even low discount rate scenarios that give rise to high marginal damages with no climate mitigation may be consistent with substantially lower marginal-damage estimates (and corresponding Pigouvian emission prices) if, in fact, controls are undertaken. Put differently, even if one accepts marginal-damage estimates on the order of $100 per ton, the implication is not that emission prices at this level would be efficient.

At the other end of the range in Table 5-8 are the estimates from the FUND model. These estimates also demonstrate the importance of the discount rate for present value marginal GHG damages, implying that GHG emissions move from having negligible effects (and in some scenarios positive benefits) with relatively high discounting of 5%, to a larger impact of $6 per ton $CO_2$-eq with relatively low discounting of 2%. The generally lower estimates of FUND are clearly due to the assumed damage function, which specifies benefits to global GDP up until about a 2-2.5°C of warming. Even after this point, damages do not go much beyond about 1% of lost GDP even for large temperature increases, in contrast to the other models where damages increase nonlinearly.

The marginal damages of GHG emissions may be highly sensitive to the possibility of catastrophic events. Although a number of potentially catastrophic outcomes have been identified (for example, release of methane from permafrost that could rapidly accelerate warming, collapse of the West Antarctic or Greenland ice sheets raising sea level by several meters, and changes in North Atlantic currents that would dramatically alter European climate), the damages associated with these events and their probabilities are very poorly understood. Nordhaus and Boyer (2000) and Stern (2007) included some provision for catastrophic outcomes that could result in the loss of perhaps one-quarter of world GDP. Weitzman (2009) raised the even more sobering possibility that the probabilities of extreme outcomes are much larger than currently estimated. If taken into account, low-probability extreme outcomes, such as the possibility of a 10° or even 20°C increase in global mean temperature that could virtually destroy civilization as we know it, could dominate the expected value of damages, making it much greater than the values described above.[19]

Given the uncertainties and the still preliminary nature of the climate-damage literature, the committee finds that only rough order-of-magnitude estimates of marginal climate damages are possible at this time. Depending on the extent of future damages and the discount rate used for weighting future damages, the range of estimates of marginal global damages can vary by two orders of magnitude, from a negligible value of about $1 per ton to $100 per ton of $CO_2$-eq. Roughly an order of magnitude in difference

---

[19]For further discussion and alternative view, see Aldy et al. (2009).

can be attributed to discounting assumptions, and another to assumptions about future damages from current emissions. Table 5-9 summarizes these findings for discount rates of 1.5%, 3.0%, and 4.5%, respectively, and for relatively low and higher climate-damage assumptions (corresponding roughly to FUND-level damages versus DICE- or PAGE-level damages). For a discount rate of about 3%—a typical rate for use in long-term environmental analysis in the United States and elsewhere—the comparable marginal-damage estimates could be on the order of about $3 per ton to $30 per ton $CO_2$-eq for relatively low versus higher damage assumptions.[20] As discussed earlier, however, the damage estimates at the higher end of the range are associated only with emission paths without significant GHG controls. Therefore, care must be taken in translating these estimates for use in policies for decreasing GHG emissions. In Stern (2007), for example, the marginal-damage estimate is $36 per ton $CO_2$-eq for a stabilization trajectory associated with stabilization at about 550 ppm $CO_2$-eq, not the $102 per ton Stern found associated with uncontrolled emissions.

As described above, marginal-damage estimates for emissions in 2030 could be as much as 50-80% larger than those estimates. Estimates of the damages specifically to the United States would be a fraction of those levels because the United States is only about one-quarter of the world's economy, and the proportionate impacts on the United States are generally thought to be lower than for the world as a whole (see Table 5-7).

Table 5-10 presents three different estimates of external global damages from GHG emissions on a per unit basis. The damages were calculated by multiplying GHG emission rates from Chapters 2 (electricity), 3 (transportation), and 4 (heat) by each of the committee's assumed low, middle, and high marginal damages of $10, $30, and $100 per ton $CO_2$-eq.

In conclusion, the committee finds that the relative weight placed on potential impacts occurring decades to centuries in the future is absolutely central to the determination of a present value measure of the damages from current GHG emissions. Over these time horizons, the discount rate carries with it implications for intergenerational distribution. As with any social analysis involving significant distributional impacts, it is therefore

---

[20]To gain a rough sense for how marginal damages change as a function of growth and discounting, it is useful to consider the relative magnitude of the present value of a growing stream of damages discounted at different rates. As mentioned above, a typical climate economic model might imply marginal damages growing over time at about 2% per year due to economic growth and a convex damage function. Accumulated over several hundred years, the present value of a stream of damages growing at 2% per year increases by a factor of 2.5 using a discount rate of 3% rather than 4.5%. Using a discount rate of 1.5%, the cumulative value of a stream of damages growing at 2% per year is only bounded by the time horizon of the sum. As another point of reference, studies cited in reviews by Tol (2005b, 2008) using discounts rates of 3% also show a mean marginal damage in the range of $30 per ton $CO_2$-eq.

**TABLE 5-10** Illustration of Ranges of Climate-Related Damages for Selected Categories of Energy Use in the United States, 2005

| Sector | Fuel and Technology | Climate-Related Damages for $10-30-100/Ton $CO_2$-eq[a] |
|---|---|---|
| *Electricity* | | |
| | Coal plants (biomass)[b] | 1-3.0-10 ¢/kWh |
| | Natural gas plants | 0.5-1.5-5 ¢/kWh |
| | Nuclear, wind, solar | Much lower than natural gas |
| *Transportation* | | |
| | Cellulosic E85/car | 0.02-(0.15-0.25)-2 ¢/VMT |
| | CNG | 0.04-0.4-4.0 ¢/VMT |
| | Gasoline hybrid | 0.04-0.4-4.0 ¢/VMT |
| | Gasoline/car | 0.06-0.6-6.0 ¢/VMT |
| | E10 | 0.06-0.6-6.0 ¢/VMT |
| | $H_2(g)$ | 0.03-0.3-3.0 ¢/VMT |
| | Diesel/car | 0.05-0.5-5.0 ¢/VMT |
| | E85 corn/car | 0.05-0.5-5.0 ¢/VMT |
| | Grid-dependent HEV or EV[c] | 0.05-0.5-5.0 ¢/VMT |
| *Building and Industrial for Heating* | | |
| | Natural gas combustion[d] | 0.07-0.7-7.0 $/MCF |

[a]Rounded to one digit, 2007 USD.

[b]Biomass can be co-fired with coal in quantities up to about 20%.

[c]Ranges based on use of the fuel in a representative group of vehicles. Grid-electric cars are usually smaller than fleet average cars, so their better performance per vehicle mile traveled (VMT) is also dependent on use of smaller cars with lesser driving ranges.

[d]Future additions to supplies may include imported liquified natural gas, which will include nonclimate damages outside the United States at the source and will have increased climate damages in the range of 30% or more depending on the gas field and the liquefaction plant details.

ABBREVIATIONS: CNG = compressed natural gas; HEV = hybrid electric vehicle: EV = electric vehicle; MCF = thousand cubic feet.

crucial for decision makers not only to look at singular summary statistics (such as present value marginal damages) but also to understand the magnitude of impacts as individuals will bear them, both across time and at different points in time across regions. This concern is not particular to climate change, but the very long time frames associated with GHG residence in the atmosphere and with thermal inertia of the oceans raise the issue of discounting to a level that is present in few other problems. Nonetheless, the committee also finds that a consistent framework for discounting impacts occurring over similar time frames across all potential policy investments is essential for reasoned policy analysis.

## RESEARCH RECOMMENDATIONS

The committee makes the following recommendations to improve the understanding of physical, biological, and human impacts, as well as economic valuation aspects related to climate change.

- More research on climate damages is needed, as current valuation literature relies heavily on climate-change impact data from the year 2000 and earlier (see Tol [2008] for a number of fruitful areas).
- Marginal damages of GHG emissions may be highly sensitive to the possibility of catastrophic events. More research is needed on their impacts, the magnitude of the damage in economic terms, and the probabilities associated with various types of catastrophic events and impacts.
- Estimates of the marginal damage of a ton of $CO_2$ include aggregate damages across countries according to GDP, thus giving less weight to the damages borne by low-income countries. These aggregate estimates should be supplemented by distributional measures that describe how the burden of climate change varies among countries.

# 6

# Infrastructure and Security

## INTRODUCTION

The energy system depends on a massive infrastructure to produce and distribute energy to households and businesses. Table 6-1 gives a rough idea of the amount of fixed assets related to energy production. The infrastructure related to energy production amounted to nearly $2.9 trillion in 2007, or 12% of the value of the net stock of nonresidential fixed assets in that year.[1] The bulk of energy-related assets are structures—electricity-generation facilities and mining exploration, shafts, and wells.

In this chapter, the committee considers a variety of externalities that are associated with the energy infrastructure. In particular, we consider disruption externalities in the electricity-transmission grid, the vulnerability of energy facilities to accidents and possible attack, the external costs of oil consumption, supply security considerations, and national security externalities. Where possible, we quantify the externalities that we identify.

## DISRUPTION EXTERNALITIES IN THE ELECTRICITY-TRANSMISSION GRID

In the interconnected electric-power system, "reliability" is the degree to which the system delivers power to consumers within accepted standards

---

[1]This estimate does not include energy capital in the U.S. military, nor does it include the value of transportation assets or computers and other equipment used in the production and distribution of energy. Adding transportation-related equipment and structures alone would add $1.3 trillion to the value of energy-related fixed assets.

**TABLE 6-1** Net Stock of Energy-Related Fixed Assets in 2007 ($Billions)

| | | |
|---|---:|---:|
| *Private Fixed Assets* | | |
| Equipment and Software | | 523.9 |
| Engines and turbines | 83.5 | |
| Electrical transmission, distribution, and industrial apparatus | 358.4 | |
| Mining and oilfield machinery | 49.5 | |
| Electrical equipment, not elsewhere classified | 32.5 | |
| Structures | | 2,120.4 |
| Power | 1,230.6 | |
| Mining exploration, shafts, and wells | 889.8 | |
| *Government Fixed Assets* | | *241.5* |
| Power | 241.5 | |
| TOTAL | | 2,885.8 |

SOURCE: BEA 2009.

and in the amount desired (Abel 2006). Typically, reliability is good—the nation's electric-power grid delivers power when needed and within an acceptable quality range.

Occasionally, however, electric outages occur when the demand for electricity exceeds the supply. There are various causes of outages, including equipment failure, extreme weather events, such as ice storms and hurricanes; trees or animals physically damaging parts of the electric system; accidents that damage parts of the system; equipment failure; and operator error. Outages solely from overloads are rare. However, other things being equal, a greater load increases the likelihood of transmission congestion and of decreased reliability. Consequently, there are externalities associated with the consumption of electricity in the sense that when an electricity consumer draws from the grid, this increases the probability that demand will exceed supply and that an outage will occur.

In addition to outages or interruptions in electricity service, voltage sags, harmonic distortions, and other power-quality events occur. Although each event generally causes little damage, except for customers whose commercial activities depend on very-high-quality power, these events occur much more frequently and result in significant annual damage.

The possible externality that an individual consumer imposes when using additional electricity is the expected damage to all other users of the grid from an outage or power-quality event (the damages of an event weighted by the increased risk of an outage from the marginal consumer's use). The optimal price to internalize this externality would include this marginal damage (unless the costs of implementing such a pricing scheme exceed the benefits). This externality has long been recognized, and as we discuss later, various means of internalizing this externality are in place to varying degrees.

To the extent transmission externalities exist, they apply to all of the electricity-generation options, such as coal, natural gas, oil, wind, and hydropower. However, intermittency in generation has the potential to affect the frequency of outage events, as well as power quality.

### The Magnitude of the Electricity-Disruption Externality

To calculate the magnitude of this externality, one needs an estimate of the damages from outages and power-quality events and an estimate of the increased probability of these events occurring because of the consumption of additional electricity. The nature and severity of the impacts of an outage or power-quality disturbance vary and depend on the affected sector (manufacturing, commercial, or residential) and on the specific functions affected, as well as the availability of backup power, the duration of the outage, the time of the year, the time of day, the geographic region, and the extent to which customers are notified prior to the outage.

A number of empirical studies estimating the damages from outages and power-quality disturbances have been undertaken. Some are based on estimates of lost output and damage from actual outages; others value the prevention of an outage. Previous estimates are typically of the total annual cost, cost per kilowatt hour, value of lost load, or damage divided by total kilowatt hour (Primen 2001; Overdomain 2002; Lawton et al. 2003; LaCommare and Eto 2004; Layton and Moeltner 2005; van der Welle and van der Zwaan 2007; Mount et al. 2008). None of these estimates measures marginal damages per se. Table 6-2 provides estimates of average damages for different sectors.

In viewing the estimates in Table 6-2, it is important to note that SAIFI and SAIDI[2] estimates were unavailable for the different sectors, so the same overall averages were used for the residential, commercial, and industrial sectors. Anecdotal evidence strongly indicates, however, that commercial and industrial establishments are likely to have backup power and thus less frequent loss of power (lower SAIFI values). Thus, the estimates in Table 6-2 of the average damage per kWh for the commercial and industrial sectors have probably been significantly over-estimated.

The most striking observation from Table 6-2 is that smaller commercial businesses are most vulnerable to outages and power-quality disturbances, especially the latter. Assuming for the purpose of discussion that the estimates of average damage per kWh consumed are over-estimates of these damages in the commercial and industrial sectors by, say, an order of

---

[2]SAIFI is the System Average Interruption Frequency Index—the average number of interruptions a customer experiences in a year. SAIDI is the System Average Interruption Duration Index—the average cumulative duration of outages a customer experiences in a year.

**TABLE 6-2** Estimates of the Average Cost of Outages[a,b]

| Source | Estimate of Cost of Outages per kWh | Method Used for Estimate |
|---|---|---|
| Lawton et al. 2003 | *Averages* over all U.S. regions (costs vary by region) for studies done in 2001 and 2002<br><br>Large commercial and industrial—Outages<br>All durations: $0.0038<br>1-h duration: $0.0029<br><br>Small-medium commercial and industrial—Outages<br>All durations: $0.015<br>1-h duration: $0.0083 | Meta-analysis—Reviewed previous studies over the past 15 years ending in 2002. Studies were of outage costs conducted by seven electric utilities in 20 studies. Authors also provided ranges, regional breakdown, and duration. |

This committee

| Sector | Type of occurrence | Average damage per year per customer | Average damage in $ per kWh | |
|---|---|---|---|---|
| Residential | Power quality | $7.90-$40.00 | $0.00072-$0.0037 | Calculated based on estimates of costs per event and on ranges of the System Average Interruption Frequency Index, the System Average Interruption Duration Index, and the Momentary Average Interruption Frequency Index in LaCommare and Eto (2004), combined with the Energy Information Administration data on annual sales to each sector (in MWh). |
| | Outages | $2.80-$6.40 | $0.00025-$0.00072 | |
| Commercial | Power quality | $2,800-$9,100 | $0.038-$0.13 | Estimates do not account for anecdotal evidence that larger commercial and industrial establishments have backup power and thus experience fewer outages than smaller establishments and residential consumers. |
| | Outages | $1,200-$2,300 | $0.016-$0.03 | |
| Industrial | Power quality | $9,400-$25,000 | $0.0069-$0.018 | Thus, the estimates of average cost per kWh are significantly over-estimated for the larger establishments. |
| | Outages | $4,800-$8,500 | $0.0036-$0.0062 | |

NOTE: Estimates for the residential sector have probably been *under*-estimated, and for the commercial and industrial sectors, they have probably been *over*-estimated.

[a]2007 U.S. dollars per kilowatt hour (USD/kWh).

[b]The top part of this table provides a summary of a meta-analysis on outage costs based on studies done between 1987 and 2002. Outage costs are reported by utilities taking part in the studies. Costs are reported in USD/kWh consumed. The bottom part of the table reports estimates for power-quality events and outages for different customer classes. The right-hand column explains how the numbers in the table are obtained (top half) or constructed (bottom half).

magnitude, they are nevertheless sizeable compared with the average U.S. retail commercial electricity price of $0.094/kWh in January 2008 (EIA 2009k, Table 5.6a).

Although an outage, if it were to occur, would cause greater damage to industrial firms, there are far fewer of them compared with commercial establishments, and industrial firms use much more electric power than commercial establishments. The upshot of these differences is that average damage per kWh consumed in the industrial sector is almost an order of magnitude less than that in the commercial sector. Average damage per kWh consumed in the residential sector is, in turn, about an order of magnitude less than that in the industrial sector. As previously noted, these numbers measure average damages per kWh consumed rather than marginal damages. Not all damages are congestion-related network-disruption damages. It is not possible to disentangle the marginal effect of network congestion's contribution to outages, apart from other factors. Thus, the numbers in Table 6-1 are upper bounds on (and are probably significantly higher than) the marginal damages from outages and power-quality events due to congestion.

### Differences in the Effects of Alternative Electricity-Generation Technology and Fuel Options on Grid Reliability

The effects of renewable energy sources—especially wind and solar—on the reliability of the electric-power grid might be different from those of conventional sources. The timing and duration of wind and sunshine cannot be controlled. Because wind velocities and sun intensity determine power output, it is variable and not entirely predictable. Electricity generally cannot be stored, and transmission is costly. Although reserves and transmission must be provided with any type of electricity-generation technology, the issues are more prominent with some of the renewable energy resources.

There are concerns that, if interconnected to the grid, wind and solar facilities might reduce grid reliability or power quality. More backup sources and power-quality control devices, as well as additional wind or solar capacity to account for their lower capacity factors, might be needed. Transmission lines would be needed to carry power from more remote areas, where some of these facilities would be located, to where power is needed. These measures are costly but could be internalized in market transactions.[3]

---

[3]There has been an interesting debate about how great these costs would be (for example, Jacobson and Masters 2001; DeCarolis and Keith 2001).

## Extent to Which Grid Externalities Are Internalized

The degree to which externalities are internalized is difficult to determine, as there are several ways in which internalization could occur. First, distinguish between the local distribution network and the regional transmission grids. For the former, some of the externalities are probably internalized through pricing by the local distribution companies. Electricity pricing can take many forms (see Borenstein [2005] for a description of different forms of peak-load pricing). For example, it can be implemented through differing block prices for peak and off-peak periods. If these periods and tariffs are pre-established, then this is a form of time-of-use pricing and is common for commercial and industrial customers. Dynamic pricing, which itself has several forms, varies the price dynamically, depending on the load at different times of the day each day. As of 2004, over 70 utilities have experimented with dynamic pricing systems (see Barbose et al. 2004).

For the bulk-level transmission grid, the rule making, standards, and regulations set forth by the Federal Energy Regulatory Commission (FERC) and the North American Electric Reliability Corporation (NERC) help increase reliability and act to internalize some of the externalities. Utilities have long been required to have operating reserve margins. More recently, the Energy Policy Act of 2005 empowered FERC to enforce mandatory reliability standards, set by NERC, which apply to all participants in the bulk power system. In 2007, monetary penalties of up to $1 million per day were established for noncompliance with these standards. Performance-based incentives that allow increased rates of returns on transmission projects intended to reduce transmission congestion costs or increase reliability are another recent development. Regulators may also use performance-based rates that reward utilities for good reliability and penalize them for poor reliability.

Electricity markets are still in their infancy in providing incentives for socially efficient investment in grid infrastructure, especially transmission (Hogan 2008). Transmission-system operators responsible for providing transmission services have an incentive to reduce their private (as opposed to social) costs.[4] The transition toward more competitive power markets has resulted in declining reserve capacity because producers are striving to minimize their costs; this situation reflects the lack of adequate incentives to improve power reliability.[5]

---

[4]Social costs exceed private costs to the extent that disruption affects customers in other regions, and out-of-region impacts are not taken into account by system operators.

[5]Reserve capacity is the amount of generating capacity in excess of capacity required for electricity needs at any time. At periods of peak usage, reserve capacity is low. Lowering reserve capacity reduces costs for utilities but raises the risks of outages.

Reserve capacity also has public-good attributes. For technical and economic reasons, it is not possible to prevent customers from benefiting from it, even when they are not paying for the power delivered by that reserve capacity (van der Welle and van der Zwaan 2007).

Reserve requirements, the possibility of fines for noncompliance with NERC reliability standards, and performance-based incentives internalize some of the externalities. However, the extent to which these standards, associated fines, and incentives ensure a certain level of reliability—as measured by frequency and duration of outages—is uncertain. Compliance with many of the standards relies on self-certification, and penalties levied by FERC may be negotiated, possibly with economically inefficient agreements.

Many of the problems with the transmission grid are due to the age of the infrastructure.[6] Investing in a modern, "smart grid" will alleviate many of the problems that were noted above. New capital investment also will probably reduce risks associated with intermittent renewable electricity sources. This is an example where technological innovation can reduce or eliminate externalities.

## FACILITY VULNERABILITY TO ACCIDENTS AND ATTACKS

The United States has over 1.5 million miles of oil and gas pipelines, 104 operating nuclear plants, 9 liquefied natural gas (LNG) import facilities, 100 LNG peaking facilities, and over 17,000 non-nuclear electricity generators in the United States. The domestic infrastructure to deliver energy is complex and critical to the smooth functioning of the economy. The committee focuses in this section on the following questions:

1. To what extent is the U.S. energy infrastructure vulnerable to accidents and terrorist attacks?
2. To what extent should infrastructure vulnerability be regarded as an externality?

The following subsections explore these questions in the context of four infrastructure areas. Liquefied natural gas is a growing source of natural gas in the United States. LNG facilities are large and complex and have been the source of some community concern over safety. Oil transportation and storage are vulnerable to accidents and spills. The oil and gas pipeline network is extensive and critical to the flow of gas and oil around the United States. Finally nuclear power accidents continue to be of concern to the public.

---

[6]See the discussion in *America's Energy Future*, Chapter 9.

Analyzing these forms of energy infrastructure sheds light on the nature of externalities associated with our production and consumption of energy.

## LNG Infrastructure and Hazards

The infrastructure associated with LNG distribution in the United States includes tankers used to transport the gas from foreign ports, import terminals which are dedicated to LNG, and inland storage facilities. There are nine operating terminals within the United States (Parfomak 2008). The most significant hazards associated with LNG infrastructure include a pool fire, which occurs if there is a spill that is ignited, or flammable vapor clouds, which occur if there is a spill that is not immediately ignited. In this case, the evaporating natural gas can travel some distance and during that time is at risk of ignition. Other safety hazards include LNG spilled on water that may be able to regasify and cold LNG that can injure people and damage physical structures. Because LNG dissipates with no residue, the only environmental damages that might occur would be linked to fire or cold damage. Finally, terrorism hazards that are directed at ships carrying LNG and at land facilities are also a risk.

These damages can be associated with damage to a tanker, terminal, or inland storage facility. The most dangerous possibilities include a spill on water because such a spill could spread the farthest and, if ignited, seriously harm people and property at some distance. The safety record of tankers carrying LNG is quite good; since international shipping started in 1959, no spills have occurred, although a few groundings and collisions have occurred (Parfomak 2008). Tankers are double-hulled and engage GPS, radar, and other safety systems to reduce the risk of accidents and grounding.

In contrast to tanker safety, a number of incidents at terminals and onshore storage facilities have occurred. Worldwide, there are over 40 terminals and 150 onshore storage facilities, and there are 13 known accidents at those sites reported since 1944. These data are summarized in Table 6-3. No LNG tankers fly under the U.S. flag, so only a worldwide number is reported for number of tankers.

The number of serious accidents has been quite small, only 13 since 1959, with 29 fatalities and 74 injuries. With such a small number of incidents, it would be difficult to extrapolate from the historical data to estimate expected damages in the future. On average, over 260,000 million cubic feet of LNG has been imported or exported by the United States annually since 1985 (EIA 2009l). In this context, the apparent risk of accidental injury, death, or property damage is small.

A number of studies have quantitatively evaluated the hazards and risk of accidents or terrorist attacks, largely on a facility by facility basis. Parfomak and Vann (2008) summarized recent and most cited studies in

**TABLE 6-3** LNG Infrastructure and Safety Record[a]

| Infrastructure Type | Worldwide Numbers | U.S. Numbers | Expected Growth | Serious Incidents Since 1959 | Fatalities | Injuries |
|---|---|---|---|---|---|---|
| Tanker Ships | 200 | Not relevant | 200 more worldwide by 2013 | 0 | 0 | 0 |
| Terminals[d] | 40 | 10 | 21[b] | 13[c] | 29 | 74 |
| Storage Facilities | >150 | 103 | Not available | | | |

[a]The data in this table are taken from Parfomak and Vann (2008).

[b]Six terminals have been approved and are under construction in the United States; another 15 have been approved, but construction has not yet begun.

[c]Data on accidents, fatalities, and injuries are not separately reported for terminals and storage facilities. These numbers represent the worldwide sum since records are available.

[d]These numbers reflect import and export terminals.

their Table 6-2. A report prepared for the U.S. Department of Energy by Sandia National Laboratories summarizes and assesses four of those studies to develop estimates of the damages of a large pool spill from a tanker over water. The report does not monetize damages but provides estimates of numerous end points of damage (for example, asphyxiation, cryogenic burns, and structural damage).

Based on its review, the Sandia report provides estimates of the effects of various small- and large-scale accidents and intentional damage. For example, the report provides estimates of the size of a pool fire from different sizes of holes of a tanker breach, the size of the pool, the distance of thermal hazards, and the burn time. This information could be combined with data on population density and on monetary damages of death, injury, and property loss to quantify this externality at least for specific locations. However, the committee did not attempt such quantification. Further, the small risk of these incidents suggested by the historical record implies that the magnitude of this externality relative to other energy-based externalities is likely to be small.

The LNG industry faces unlimited liability for damages from accidents. In fact, fines have been large in the case of a pipeline fire in Bellingham, Washington, in 1999 and in training violations at the Everett, Massachusetts, LNG terminal. These facts suggest that the risk of accidents and spills is internalized in the LNG industry. In addition, private insurance held by facility owners contributes to internalization of the externality.

Finally, LNG tankers and facilities face considerable regulatory oversight. The Coast Guard has responsibility (and bears the cost) for shipping and terminal security, the Office of Pipeline Safety and the Transportation Security Administration have security authority for LNG storage plants and terminals, and the Federal Energy Regulatory Commission approves facility siting in conjunction with the other three agencies.

The Coast Guard estimated that it incurs about $62,000 to accompany a tanker through Boston harbor to the Everett facility. This figure, combined with estimates of policy, fire, and security costs incurred by the cities of Boston and Chelsea and the state of Massachusetts, suggests that these "shepherding" costs run to about $100,000 per tanker. Although public costs of providing safe passage to tankers or of assessing health and safety standards are not externalities (they represent the costs of mitigating the externalities), these costs will not be represented in the market price of energy unless there are requirements that industry pay some of the costs. This provides another example of a situation in which private costs do not reflect social costs in ways that are unrelated to externalities. Here, the divergence arises from the government provision of services that are not priced in the fuel.

### Oil Spills from Ships and Facilities

Oil is transported by tankers, barges, and other vessels where oil can be accidentally spilled during its transfer between vessels or where an accident can occur on a vessel itself. Oil is stored in a variety of facilities where spills are possible. A number of highly publicized oil spills have occurred in the previous two decades that have increased the public's awareness of the ecological harm and other damages that such spills can cause. Seabirds, marine mammals, a variety of reptiles and amphibians, fish, and invertebrates are all at risk of death or injury when they come into contact with oil. In addition, many coastal and wetland habitats can be significantly altered by the presence of large quantities of spilled oil, and oil spills deleteriously affect human use of the environment when recreational sites are damaged and when commercial activities are harmed (such as fishing or shrimping). Important cultural sites can also be damaged by these spills (Ramseur 2008).

Data on the prevalence and size of oil spills that occur on U.S. soil are collected and maintained by the U.S. Coast Guard. Table 6-4 reports the average number of spills in the United States between 1990 and 1998 as well as the average amount of oil spilled (in gallons).[7]

A number of studies of the damages caused by oil spills have been com-

---

[7]More detailed information on the source and type of spills can be obtained in Table 2-2 from *Oil in the Sea III: Inputs, Fates, and Effects* (NRC 2003c). The report also contains detailed estimates of spills worldwide.

**TABLE 6-4** Average Number and Volume of Oil Spills on U.S. Soil, 1990-1998[a]

|  |  | Annual Number per Year | Volume per Year (gallons) |
|---|---|---|---|
| Total |  | 8,831 | 2,645,247 |
| Spills by size | <100,000 gal | 8,828 | 1,163,484 |
|  | >100,000 gal | 3 | 1,481,763 |
| Spills by source | Tanks and barges | 506 | 1,273,950 |
|  | Other vessels | 4,214 | 344,621 |
|  | Facilities and unknown[b] | 4,055 | 776,263 |

[a]Data from the U.S. Census 2006.

[b]Data exclude spills from pipelines—information on pipeline spills is contained in the following section of the report.

pleted (for example, Cohen 1995; Garza-Gil et al. 2006). Talley (1999) estimated the property damage costs from tanker accidents but did not attempt to consider environmental damages. Overall, little research is available to estimate the expected damages from spills that could be appropriately attributed on a per gallon transported basis. One exception is the study by Cohen (1986) in which estimates of the per gallon benefits of avoiding oil spills are given. On the basis of the compensation payments from 11 spills and in-depth environmental damage studies from four large spills, Cohen estimated a $6.08/gallon benefit of avoided spillage for reduced environmental damages, $1.72/gallon benefit for avoided loss of oil, and $6.93/ gallon for avoided cleanup costs (values converted to 2007 dollars). The first of these categories clearly represents an externality of interest to this committee, and the second is a private cost and therefore not relevant for this study. The third component is also not an externality—it represents the costs of cleaning up an externality once it has occurred. However, if the optimal level of cleanup is chosen, then the marginal damages averted through cleanup would equal the marginal costs of cleanup and we could use this third component as a proxy for the damages averted through cleanup, giving aggregate marginal damages of $13.01 per gallon spilled.[8]

We need to convert from damages per gallon spilled to damages per gallon produced or consumed. According to data reported in Huijer (2005, Table 2),[9] tanker spills worldwide averaged 115,810 barrels between 2000

---

[8]To be clear, Cohen (1986) estimated average damages but assumed they were a proxy for marginal damages. A convex damage function using average damages underestimates marginal damages.

[9]Paper cited at http://www.itopf.com/information-services/data-and-statistics/statistics/ #quantities.

and 2004. Average global oil-trade movements over that period were 16.7 billion barrels per year (BP 2008, p. 20). Not all of these trade movements were over water. Casual inspection of the inter-area trade in oil suggests that nearly 80% of oil trade is by ship. To be conservative, assume that one-half of oil trade occurs by ship. Then the ratio of oil spilled to oil shipped is $115,810/(0.5 \times 16.7$ billion$) = 0.0000139$, or one barrel spilled for every 72,000 barrels shipped. Applying this percentage to the marginal damages per gallon spilled and converting gallons to barrels gives marginal damages of $0.0076 per barrel shipped.[10]

In 1990, the Oil Pollution Act was passed by Congress; it imposed comprehensive liability for spills. The U.S. Coast Guard promulgated a number of regulations as part of the act that included the requirement that all tanker ships have double hulls by 2015. There is also evidence that these requirements and negative publicity have resulted in the improved safety records of the industry. Etkin (2001) and Homan and Steiner (2008) reported data from the U.S. Coast Guard showing a general decline in ship and barge spills during the 1990s. Homan and Steiner's analysis of the data supports the interpretation that this decline is attributable to the requirements that came after the passage of the act. However, a GAO report (2007, p. 28) notes that limits to liability to the responsible parties exist, and although these limits have recently been increased, they may still not be high enough to cover all potential damages. For a complete description of the various federal and international laws and liability rules, see Ramseur (2008). On the basis of the above information, there has been at least partial internalization of the externalities from oil spills.

## Oil and Natural Gas Pipelines

The United States depends on a large network of pipelines to move natural gas and oil around the country. A National Research Council report (TRB 2004) notes that nearly all natural gas and roughly two-thirds of petroleum are moved through transmission pipelines in the United States. Transmission pipelines are only one part of the national pipeline network that includes gathering, transmission, and distribution pipelines.

Table 6-5 provides information on significant pipeline incidents in the United States on an annual basis for the period 2002 through 2006. A significant incident is an incident in which at least one of the following

---

[10]Cohen (1986) noted that the Coast Guard uses a "rule-of-thumb" cleanup cost for planning purposes of $20 per gallon for an oil spill of 500-1,000 gallons (in 2007 dollars). Using this value rather than the $6.93 cost figure used in the text doubles the marginal damage per barrel of oil shipped, raising the marginal damage to 1.5 cents per barrel. Even if marginal damages were four times the average damages due to convexity in the damage function, the marginal damages would only be raised to 6 cents per barrel of oil.

**TABLE 6-5** Annual Averages for Significant Pipeline Incidents, 2002-2006

| Pipeline Type | Significant Incidents | Fatalities | Injuries | Property Damage ($1,000) |
|---|---|---|---|---|
| Hazardous liquid | 124.4 | 1.6 | 5.0 | $8,729 |
| Natural gas transmission | 75.2 | 1.0 | 4.6 | $81,019 |
| Natural gas gathering | 9.8 | 0.0 | 1.0 | $40,875 |
| Natural gas distribution | 93.2 | 13.8 | 41.6 | $129,317 |
| Total | 302.6 | 16.4 | 52.2 | $349,940 |

NOTE: Significant incidents are those incidents reported by pipeline operators when any of the following conditions are met: (1) Fatality or injury requiring inpatient hospitalization. (2) $50,000 or more in total costs. (3) Highly volatile liquid releases of five barrels or more or other liquid releases of 50 barrels or more. (4) Liquid releases resulting in an unintentional fire or explosion. Property damage estimates are in 2007 dollars.

SOURCE: BTS 2009.

conditions occurs: (1) fatality or injury requiring inpatient hospitalization, (2) $50,000 or more in total costs, (3) highly volatile liquid releases of five barrels or more or other liquid releases of 50 barrels or more, or (4) liquid releases resulting in an unintentional fire or explosion (BTS 2009).

The table provides information on hazardous liquid pipelines. Hazardous liquids are defined by the Office of Pipeline Safety (OPS) as petroleum, petroleum products, and anhydrous ammonia. Most hazardous liquids moving through the pipeline are petroleum or petroleum products. Nearly half of the incidents reported to OPS are associated with hazardous liquids and are triggered by releases of 50 barrels of oil or more.

The table shows that the distribution of natural gas is responsible for the bulk of fatalities and injuries and over half of the property damage arising from significant incidents. Nearly 80% of natural gas pipeline stock in 2006 was distribution pipelines, transmission making up most of the rest (BTS 2009, Table1-10).

Table 6-6 provides information on incidents per mile of transit averaged over the 2002-2006 period. Scaling by the amount of gas traveling through pipelines indicates that the incidences of fatalities, injuries, and property damage are highest in the gathering pipelines.

An alternative way to scale the damages is to report fatalities, injuries, and property damage per unit of oil or natural gas consumed in the United States. Using the average fatalities, injuries, and damages from Table 6-5 over 2002 through 2006 and average consumption of oil and gas over that same period, we measure 0.29 fatalities and 0.90 injuries per billion barrels of oil that is delivered to refineries in the United States and $18 of

**TABLE 6-6** Annual Averages for Pipelines per Ton Miles, 2002-2006

| Pipeline Type | Ton Miles of Freight Per Year (millions) | Number of Incidents (Per Billion Ton Miles) | Fatalities (Per Billion Ton Miles) | Injuries (Per Billion Ton Miles) | Property Damage (Per Million Ton Miles) | Net Barrels Lost (Per Billion Ton Miles) |
|---|---|---|---|---|---|---|
| Hazardous liquid | 593,560 | 0.2 | 0.003 | 0.008 | $166 | 100 |
| natural gas | 336,493 | 0.5 | 0.044 | 0.140 | $747 | NA |
| Gathering | 3,365 | 22.3 | 0.297 | 1.367 | $24,077 | NA |
| Transmission | 67,299 | 0.1 | 0.000 | 0.015 | $607 | NA |
| Distribution | 265,829 | 0.4 | 0.052 | 0.156 | $486 | NA |
| Total | 930,053 | 0.7 | 0.047 | 0.149 | $913 | 100 |

NOTE: Estimates of ton miles for components of the natural gas pipeline system are based on distribution of pipeline miles from the Bureau of Transportation Statistics. Property damage estimates are in 2007 dollars.

SOURCE: BTS 2009.

property damage per thousand barrels of oil. For natural gas, the numbers are 0.72 fatalities and 2.30 injuries per trillion cubic feet of natural gas consumed and $12 of property damage per million cubic feet of natural gas consumed. The property damage and injuries are modest and probably internalized to a great extent. These damages are not considered in the subsequent analysis.

### Nuclear Power Accidents

In addition to potential damages associated with generation of electricity through nuclear technologies (see Chapter 2), there are several potential external costs associated with the potential for a nuclear accident. Unlike the situation with other potential damages associated with nuclear technologies, these possibilities are distinctive in that two well-studied accidents have already occurred (Three Mile Island and Chernobyl), providing the basis for widespread public concern about the issue. Specifically, the following considerations have been raised:

1. To what extent does the existing technology alter the probability and damage functions associated with an accident for the existing facilities or those under design.

2. To the extent that the above can be quantified, to what extent have they been internalized by existing regulations, insurance requirements (including liability costs required by regulations) or other market mechanisms.

Unlike the speculation surrounding nuclear-waste issues, the experience at Chernobyl highlighted both the extent of potential local damages and the spread of damages over a wide region, having health and other impacts demonstrated thousands of miles away. For example, Almond et al. (2007) found that Swedish children who were in utero during the Chernobyl accident had worse school outcomes than adjacent birth cohorts. Moreover, in the absence of technological change, it seems reasonable to assume that risks will increase in proportion to the expansion of nuclear power; that is, each new facility engenders an additional, theoretically calculable risk. These risks vary, depending on geography and population distribution but may affect large regions.

That recognized, there are abundant data to suggest that the risk going forward, at least in the United States, will be dramatically lower than it has been in the past based on advances in the technology and regulations, including most recently the Energy Policy Act of 2005. There appear to be no comprehensive direct estimates on which to base a numerical cost estimate.

The magnitude of the externality depends in large part on the extent to which insurance accounts for these costs. If the industry were fully insured against all risks from accidents, then there would be no external damages not reflected in market prices. The Price-Anderson Act regulates and establishes insurance pools and limits liability for the nuclear industry. The act, enacted in 1957 and revised in the Energy Policy Act of 2005, requires all commercial nuclear power plants to carry liability insurance in the amount of $300 million.[11] In the event of an accident that creates losses in excess of $300 million, each commercially active reactor is to be assessed an amount up to $95.8 million (payable over several years with annual payments capped at $15 million) for a total pool of approximately $10 billion. The industry is exempt from any liability in excess of this amount. Insurance covers "bodily injury, sickness, disease, or resulting death, property damage, and loss as well as reasonable living expenses for individuals evacuated." (U.S. NRC 2008d). Over its history, a total of $150 million has been paid in claims under the Price-Anderson Act, the accident at Three Mile Island in 1979 accounting for nearly half. As of 1997, over $70 million had been paid out in indemnity settlements and expenses from this accident (ANS 2005). Claims have been paid through the primary insurance held by each plant; the supplemental assessments have never been required.

The appropriate measure of any uninternalized externality arising from insufficient insurance depends critically on the distribution of damages from potential nuclear reactor accidents in excess of the supplemental assessment (approximately $10 billion). Estimation of these low-probability and high-

---

[11]These amounts are subject to inflation adjustments at 5-year intervals.

consequence events is difficult, in part because the events are sufficiently rare (fortunately) that empirical data are lacking.[12] Jones et al. (2001) suggested that uncertainty about key factors associated with population dose and human fatality risk in the event of a release of radioactivity is such that the true value could be 10 times larger or smaller than the central estimate. Because estimates of the consequences of low-probability events are calculated by multiplying several of those factors together, the resulting estimates could be wrong by several orders of magnitude.

The component of the externality that is internalized includes the primary insurance required of each operating unit plus the risk of the supplemental assessment that might be required in the event of an accident causing damages in excess of $300 million. The probability that this supplemental assessment will be levied is again difficult to estimate; to date, it has never been required. The component associated with primary insurance that covers the first $300 million of damages might be estimated from the premium paid—approximately $400,000 per reactor-year (U.S. NRC 2008d), roughly $0.50 per MWh of electricity production. (The 104 operating units produced an average of 80 GWh in 2007.) This quantity overestimates the expected value of the first $300 million of losses (because it includes insurers' administrative costs) but is likely to underestimate the total externality, as it excludes the expected value of damages exceeding $300 million per incident.

## EXTERNAL COSTS OF OIL CONSUMPTION

The United States is a large consumer of oil. In 2007, it consumed over 20 million barrels of oil a day, representing one-quarter of world oil supply (see Table 6-7). Imports as a share of domestic consumption have steadily risen over time to their current level of nearly 60% and are not projected to decline greatly over the next 20 years. Meanwhile, OPEC (Organization of Petroleum Exporting Countries) continues to be an important source of world oil supply with its share projected to rise to nearly 50% by 2030. While U.S. consumption continues to grow, the importance of oil in the

---

[12]Dubin and Rothwell (1990) constructed an estimate of the distribution of damages in the 1980s that was revised by Heyes and Liston-Heyes (1998). These estimates are based on insurance premiums and one estimate of a worst-case scenario (damages of $10 billion with annual probability 8/10 million). Note that the expected loss associated with this worst-case scenario is $8,000 per reactor-year, only 0.2% as large as the insurance premiums associated with the first $300 million of loss, suggesting that the contribution of losses in excess of $300 million to the total expected value of losses is negligible. Both papers contain logical errors (Dubin and Rothwell misinterpreted the insurance limit, and Heyes and Liston-Heyes estimated parameters that are inconsistent with the assumed distribution) and the committee does not rely on their estimates.

TABLE 6-7 U.S. Oil Dependence

|  | 1990 | 2000 | 2007 | 2030 |
|---|---|---|---|---|
| Net oil imports as percentage of total U.S. Supply | 42.2 | 52.9 | 58.2 | 55.5 |
| World oil price (2007 $/BBL) | 38 | 35 | 72 | 60 |
| World crude production (million BBD) | 65.5 | 74.9 | 81.5 | 102.9 |
| OPEC share (percentage) | 38.3 | 42.9 | 43.2 | 46.4 |
| U.S. petroleum consumption (million BBD) | 17 | 19.7 | 20.7 | 22.8 |
| U.S. share of world production (percentage) | 26.0 | 26.3 | 25.4 | 22.2 |
| Oil intensity (1,000 Btu/GDP) $2000 | 4.7 | 3.9 | 3.4 | 2.2 |
| Oil intensity (value of oil as a percentage of GDP) | 2.6 | 2.0 | 3.6 | 1.9 |

ABBREVIATIONS: BBL = billion barrels, BBD = barrels per day, Btu = British thermal units; GDP = gross domestic product.

SOURCES: BP 2008; EIA 2008a,b,g.

economy continues to decline. Oil intensity (measured as 1,000 British thermal units of oil consumption per dollar of gross domestic product [GDP]) has fallen by over one-quarter since 1990 and is projected to fall an additional one-third by 2030. Rising oil prices, however, offset the declining physical intensity; thus, the value of oil consumption in GDP is projected to remain at about 2% (although down from its anomalous level of 3.6% in 2007 and the sharp run-up in prices that year).

The importance of oil in the U.S. economy has given rise to a large amount of literature measuring the external costs of oil consumption.[13] Parry and Darmstadter (2003), for example, attempt to quantify the marginal external cost of petroleum consumption, defined as "the difference between the costs to the U.S. economy as a whole and that to individuals or firms from additional oil consumption. Marginal external costs, expressed in $/BBL [barrels], are referred to as the oil premium" (p. 11).

In this section, we consider three questions:

- What is the oil premium?
- Is the oil premium an externality?
- How does the oil premium relate to the optimal tax on oil consumption or imports?

### What Is the Oil Premium?

As the quotation from Parry and Darmstadter (2003) above indicates, the oil premium is a measure of the difference between the private and so-

---

[13]See, for example, Bohi and Toman (1993, 1995); Greene and Leiby (2006); Leiby (2007); Greene (2009).

cial costs of petroleum consumption measured in dollars per barrel. The literature identifies two major quantifiable sources of the discrepancy between private and social costs—U.S. monopsony power and economic disruptions arising from unanticipated price shocks. We discuss these in turn.

The literature on monopsony power takes as its point of departure the observation that the United States is a large consumer of oil. As such, any policy to reduce domestic oil demand reduces the world oil price and benefits the United States through lower prices on the remaining oil it imports. The oil premium arising from monopsony power reflects the lack of recognition by individual consumers to the buying power that the nation has if it acts in a coordinated fashion.

Figure 6-1 illustrates the idea of monopsony power. On the basis of individual demands for oil, aggregate demand is given by the downward sloping curve marked D. With an upward sloping supply curve for oil, the market equilibrium occurs at point $e$ where $Q_0$ barrels of oil are consumed at price per barrel $P_0$. If the U.S. government takes some action to reduce oil demand from D to D', the world oil price falls from $P_0$ to $P_1$. The gain to consumers from the fall in oil price is the rectangle $P_0P_1fg$. This is offset

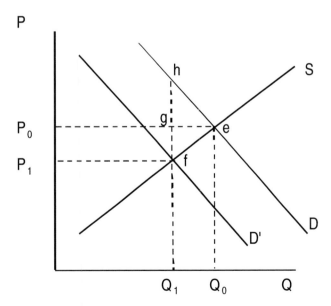

FIGURE 6-1 Illustration of monopsony. D = aggregate demand; e = market equilibrium $Q_0$ consumption and $P_0$ price; f = market equilibrium for $Q_1$ consumption and $P_1$ price; P= price per barrel of oil; Q = barrels of oil consumed.

by losses to producers in oil revenue (equal to the same rectangle).[14] If all supply comes from domestic production, there is no gain to the United States. The gain to producers is exactly offset by the loss to U.S. producers. If all supply comes from non-U.S. producers, the gain to U.S. consumers is financed by a transfer from other oil-producing countries. This gain is the monopsony benefit identified in the literature. The marginal oil premium is the incremental income transfer to U.S. consumers from foreign producers from a small reduction in demand for oil arising from a U.S. policy. One policy that would give rise to this income transfer is a tax on oil consumption. In Figure 6-1, an excise tax of $fh$ per barrel would shift demand from D to D' and lower demand from $Q_0$ to $Q_1$. Because the tax (ignoring the efficiency costs of taxation for the moment) is simply a transfer within the United States, the result follows.

The literature on oil consumption correctly notes that private markets in the United States do not account for the potential market power that U.S. consumers could wield in world oil markets. The literature also generally recognizes that any policy to take advantage of consumer purchasing power affects a transfer from foreign oil producing nations to the United States. Such a "beggar thy neighbor" policy has been justified on the grounds that OPEC is artificially inflating world oil prices at the expense of consuming nations and that the exercise of monopsony power is a countervailing policy.

Disruption costs have also been identified as a cost that is not incorporated into the price of oil. Leiby (2007) identifies two components of a disruption premium. First, a cost increase increases transfers of U.S. wealth from domestic consumers to foreign producers. The magnitude of this transfer depends on the price increase and the amount of oil imported into the United States. Second, cost increases induce shocks to the economy, resulting in losses in economic output, income, and jobs.

### Is the Oil Premium an Externality?

The committee considers the two components of the oil premium in turn. With respect to the issue of monopsony power, it is unquestionable that domestic policy can reduce aggregate demand and lead to a reduction in the world price of oil. Such a policy would generate a transfer in wealth from foreign oil producing nations to the United States.[15] However, the ability to exercise monopsony power is not the same as an externality.

---

[14]This ignores efficiency losses for the moment. We return to this issue in a moment.

[15]This transfer occurs even if foreign oil producing countries curtail production to stem the price reduction. Reductions in oil supply and demand from the countervailing policies will lead to a reduction in expenditures on imported oil in the United States.

Externalities create a market failure. Exercising monopsony power creates a market failure where one did not exist before. In Figure 6-1, the exercise of market power creates deadweight loss equal to the triangle *efh*.[16] In fact, this market failure is designed purely to transfer income from another country to the United States.

Bohi and Toman (1995) noted that the exercise of monopsony power is an example of a pecuniary externality designed to shift wealth from one nation to another.[17] They pointed out that the existence of market power on the part of energy producers complicates the analysis slightly. For one thing, market power leads to the creation of rents that transfer wealth from energy-consuming to energy-producing countries. The exercise of monopsony power can then perhaps be justified as a countervailing policy to prevent the excessive transfer of wealth from consuming to producing countries. Although this justification might provide a legitimate political reason to undertake such an action, we should stress that no externality in the sense considered in this report exists for this policy.[18] That the ability of the United States to exercise monopsony power is not an externality is recognized by Leiby (2007) and Greene (2009) among others. Green noted that the "costs of oil dependence are not external costs and neither a tax on oil nor a tax on imported oil is an adequate solution to the problem," although Green argued that either of these taxes can ameliorate the problem (see pp. 10-11 in Green 2009).

Turning to disruption costs, we consider the following questions. First, is macroeconomic disruption an externality? Second, if it is an externality, is the cost substantive and quantifiable? On the first question, most economists that have studied this issue would agree that abrupt increases in oil prices adversely impact the economy. Differences arise over the magnitude

---

[16]If OPEC is exercising cartel output restricting power, then the exercise of monopsony power adds to a pre-existing distortion, and the deadweight loss is slightly more complicated than suggested by the triangle in Figure 6-1.

[17]A pecuniary externality is not an externality in the sense defined in Chapter 1. Rather it is a transfer of income or wealth arising from some action or policy that is transmitted through the marketplace. Unlike standard externalities, pecuniary externalities do not involve any loss of efficiency.

[18]The use of monopsony power to extract rents from an energy cartel raises the question of the response by the cartel to the use of this power. An optimizing cartel will wish to raise the price to offset the exercise of monopsony power. However, it will be unable to recover all of the rents extracted by the use of monopsony power. A clumsy cartel (viz. Adelman [1980]) may be able to retaliate in a way that raises their profits. Such retaliation would simply reflect their previously nonoptimizing behavior. The analysis in Figure 6-1 assumes a competitive market supplying the product. How a cartel responds affects the welfare transfer to the United States. Alhajji and Huettner (2000) statistically rejected the hypothesis that OPEC acts as a cartel. They could not reject the hypothesis that Saudi Arabia acts as a dominant firm. In this view, Saudi Arabia can influence world oil prices but not OPEC member production decisions.

of the impact and the extent to which other events and actions play a role in magnifying the impacts.[19] The most recent run-up in oil prices in 2007 and 2008 was in large part a demand shock coupled with stagnation in supply, according to Hamilton (2009). Hamilton argued that the onset of the current recession would have been delayed from the fourth quarter of 2007 to the third quarter of 2008 in the absence of the price run-up (Hamilton 2009).

That there are links between oil shocks and economic performance is uncontroversial. Leiby (2007) estimated the macroeconomic disruption and adjustment costs for 2006 market conditions. Leiby reported a mean estimate of $5.14 per barrel (2007 dollars) and a range from $2.39 to $8.57 per barrel.[20] Does this imply an externality associated with oil consumption? The literature on the oil premium and the oil disruption component focuses on measuring the relationship between incremental oil consumption and its effect on disruptions to economic activity. We believe that oil disruption costs are not an externality. That said, it is certainly the case that policies that result in a reduction in oil consumption in the United States will most assuredly reduce vulnerability to future oil shocks.

In summary, quantifying this possible externality is a challenge. The cost depends importantly on the type of shock and policy response. Changes over time in economic institutions also pose a challenge to measuring the size of this externality. Given the conceptual difficulties in identifying the basis for and size of the externality, we do not think that it makes sense to include a disruption cost as a component in the list of externalities associated with the production or consumption of energy. We do recommend that further research be carried out to better understand this issue.

## SECURITY OF ENERGY SUPPLY

Concerns about the security of the energy supply (as distinct from national security as discussed in the next subsection) arise from the possibility that resources may become unavailable. Security concerns may pertain to energy sources (for example, oil, natural gas, and uranium) or materials that are critical for energy production, distribution, or consumption (for example, lithium for lithium-ion batteries). Risk of disruption exists when supply is dominated by one or a few countries (or facilities) that are unreliable (for example, unstable in ways that may disrupt operations, as from

---

[19]Bernanke et al. (1997), for example, argued that tightening of monetary policy exacerbated the output effects of the 1973 oil shock. Hamilton and Herrera (2004) presented results suggesting that monetary policy did not play a role. Blanchard and Gali (2008) argued that real wage rigidities in the 1970s exacerbated oil shocks.

[20]Leiby reported values in 2004 dollars.

civil strife within the country) or that may choose to restrict supply for political or other objectives (for example, the OPEC oil price shocks).

We argue that these sources of insecurity are not an externality. They are supply conditions that are presumably incorporated in market outcomes. For example, buyers of a resource that is subject to risk of supply interruption will seek ways to reduce the risk of disruption, or the harm if disruption occurs, through reducing reliance on the resource, seeking alternative suppliers, maintaining a stockpile, having financial insurance, and other measures. The expectation that demand for the resource would increase if its supply were more secure provides an incentive for suppliers to develop methods for enhancing security.

## NATIONAL SECURITY EXTERNALITIES

Energy is inextricably linked with national security. The U.S. demand for oil contributes to high oil prices that provide support for hostile foreign regimes with large reserves of oil. Second, dependence on foreign energy sources creates dependencies that may constrain foreign policy. Third, some have argued that the oil price paid by U.S. consumers does not reflect the true cost of oil, in particular the cost of U.S. military presence in the Middle East or of maintaining military readiness to protect oil supply lines. The committee discusses these issues in this section.

### Energy and Foreign Policy Considerations

High oil prices provide a source of revenue for countries with foreign policies at odds with the United States (for example, Iran and Venezuela). One could make the argument that U.S. consumers do not take into account that their oil consumption contributes to actions by foreign countries that negatively impact the United States.

A simple analogy illustrates the problem with viewing that situation as an externality. Let us assume that my neighbor burns trash in his backyard that causes pollution that adversely affects my household. This is a clear externality. Further assume that I purchase commodities in a store owned by my neighbor. My consumption thus provides income for my neighbor that leads him to purchase more commodities and produce more trash to be burned. My purchase of goods from my neighbor's store is not an externality. Rather, the neighbor's burning of trash is the externality. Restricting (or taxing) my purchases indirectly reduces the externality, but it does so in a highly inefficient manner. It would be more efficient to address the externality directly.

In a similar vein, U.S. oil consumption that enriches countries with which the United States has differences is not an externality. Rather, U.S.

consumption makes inimical actions possible. In the absence of any ability to address the foreign policy problem directly, it may be desirable to reduce oil consumption to lower world prices. However, such an effort would be an imperfect proxy for better targeted instruments and would hurt oil-producing friends and foe alike.[21]

In addition to funding activities that are inimical to U.S. interests, rising oil prices may weaken the instruments of economic statecraft. One could argue, for example, that high oil prices through the latter half of 2008 rendered economic sanctions on Iran for its nuclear activities ineffective.[22]

Dependence on foreign energy sources may constrain U.S. foreign policy. For example, the Bush administration's goal of furthering the spread of democracy in the world was constrained by U.S. ties to major oil-producing states with autocratic regimes in control. The 2006 report by the Council on Foreign Relations on U.S. oil dependency noted that oil dependence can cause "political realignments that constrain the ability of the United States to form partnerships to achieve common objectives. Perhaps the most pervasive effect arises as countries dependent on imports subtly modify their policies to be more congenial to suppliers. For example, China is aligning its relationships in the Middle East (e.g., Iran and Saudi Arabia) and Africa (e.g., Nigeria and Sudan) because of its desire to secure oil supplies" (Deutch and Schlesinger 2006, pp. 26-27).

Deutch and Schlesinger also noted that oil revenues can undermine efforts to support good governance. This is another example of the way in which oil revenues can undermine the tools of economic statecraft. Russia, for example, is less responsive to efforts to promote democracy when it has ample oil and gas revenues that reduce its reliance on Western economic assistance.

Having constraints placed on foreign policy goals because of oil dependency is arguably an externality that is not recognized in the price of oil, but consider two points. First, it is not clear what the incremental reduction in these costs would be were the United States to reduce its oil consumption by a modest amount (say, 10%). Second, it is not clear that this cost could be monetized even if the marginal cost were positive. Therefore, the committee notes the possibility of dependence on imported oil being an externality and recommends further research on this topic to better understand these important issues.

---

[21]See Fullerton et al. (2001) for a discussion of the efficiency of imperfectly targeted instruments.

[22]See Deutch and Schlesinger (2006) for further discussion of the role oil funds play in providing flexibility to countries to pursue policies at odds with those of the United States.

## Energy and Military Considerations

The argument has been made that the true cost of oil does not reflect the cost of maintaining a military presence in the Middle East or of maintaining a military preparedness. Parry and Darmstadter (2003) reported that analysts generally do not include this cost in any exercise to measure an oil premium for two reasons. First, it is difficult to disentangle military spending for such political goals as reducing terrorism or providing support for Israel from spending to protect oil supply routes. It is also unlikely that whatever spending is specific to securing the supply routes would change appreciably for a moderate reduction in oil flowing from that region to the United States. In other words, the marginal cost is essentially zero. This view is held by a number of other researchers in the area, including Bohi and Toman (1995). The committee adopts this position. We note, however, that military expenditures could be affected by a large drop in oil consumption—for example, a reduction in oil consumption to zero. Measuring the impacts on military spending (or for that matter on a host of economic and political responses) from a large change in oil consumption would require extrapolating existing statistical evidence well out of a sample. To do so would give rise to—at best—speculative estimates. We would go further and argue that military spending—to the extent it occurs to safeguard oil-production sites and transportation lanes—is a government subsidy to production. It replaces the need for private security expenditures that would otherwise have to be incurred to provide equivalent protection for oil production and transport.

## Nuclear Waste and Security

In addition to the potential health and environmental damages associated with generation of electricity through nuclear technologies (see Chapter 2), several potential external costs are associated with nuclear security. Specifically, the following considerations have been raised:

1. To what extent does the transportation and deposition of fissionable material post-use represent an increased opportunity for terrorists or other parties interested in unlawful use of the material?
2. To what extent does the long-term deposition of fissionable material create risk of catastrophic accidents above and beyond the theoretical risk of nuclear accidents at the sites?
3. To the extent that either of the above costs can be quantified, to what extent have they been internalized by existing regulations, insurance requirements (including liability costs required by regulations), or other market mechanisms?

Concerns about the environmental hazard of nuclear waste, which may continue to emit radioactive particles for thousands to millions of years, has been the subject of national debate as an environmental question for decades and came to a head after the incident at Three Mile Island. At that juncture, congressional legislation created an affirmative obligation of the federal government to provide long-term storage in the future, although this was not accomplished by 1998, and the U.S. government has been paying liability payments of approximately 0.5 billion dollars per year to the operators of the existing 104 nuclear power facilities for this failure. The facilities remain under private control but are regulated tightly by the U.S. Nuclear Regulatory Commission.

During the cold war, the Nuclear Regulatory Commission developed plans to protect sites from "enemies of the United States" based on "design-based threat" scenarios contained in a series of classified documents, but not until after 9/11 did the reference point change to consider nuclear waste as potential materials for harm—probably one of the reasons for legislative progress since that time, most notably the Energy Policy Act of 2005 (EPACT), which by all accounts has led to substantial upgrades in security procedures and oversight by the Nuclear Regulatory Commission (NRC 2005, GAO 2006). However, the NRC report in 2005 concluded that, despite progress, there remained at least some finite risk of security breaches at existing or planned sites and made a series of recommendations to further enhance security. However, neither the probability of such a breach nor the damages that might ensue, either locally or in the aggregate, have been estimated in quantitative terms.

Specific to the waste storage issue, DOE finally submitted to the Nuclear Regulatory Commission in 2008 the formal license application to operate a national waste-material repository at Yucca Mountain, Nevada, and provided extensive estimates of the cost in the range of $100 billion through 2030. Almost as quickly, the new administration announced its intent to abandon this project, but the future strategy for waste disposal remains open. The case for a deep central repository such as the one proposed, in addition to the potential economic efficiency and compliance with earlier legislation, is that security would be more readily achieved at a single site than at many, an argument that has not won favor among some near the site. As with the broader security issues at the nuclear facilities, probabilities of an adverse event involving waste storage, under the current (disseminated) or envisioned (single-site) schemes have not been quantified, nor have the potential damages under any scenario.

A further level of complexity relates to emerging technologies to modify the life cycle of nuclear fuels at the back end to reduce the long-term storage need. Simplifying the principle, reprocessing of spent fuel could, based on current knowledge, result in reuse of the material to extract almost all of its radioactivity; this approach is also being taken to achieve sustainability in all materials cycles. Although this may in the future displace in part or whole the security risks associated with storage, the technology itself, including the new facilities created and the likelihood that the reprocessed materials would need at various phases to be transported, creates new risks, for which neither the costs of appropriate controls nor any estimate of risk of breach has been calculated.[23]

Finally, it must be mentioned that there is also some potential in the security arena for external benefit from expanding the American nuclear energy capability, namely, the likelihood that the United States and its government could be proportionately more influential in global nuclear negotiations. Based on current developments, it is a certainty that many countries will turn to nuclear energy as the best solution to their energy needs, including many that are politically unstable or hostile to the United States; the potential that the United States could be a leader, both technologically but also politically, hinges, in the views of some (NEAC 2008), on the degree to which the United States also follows this energy pathway.

Taking the available information, the committee concludes as follows:

1.   The direct cost of nuclear storage under present and envisioned scenarios is high, but the potential for damages from security breaches not incorporated in these costs cannot be quantified. Even if the probability of such an event or its damages could be quantified, it would still be impossible to calculate the marginal cost—that is, the risk of an additional facility to a world still populated with nuclear warheads and with many foreign countries already committing to a nuclear energy future.

2.   As with other damage possibilities associated with the generation of electricity, the distribution of potential damages is certain to be unevenly shared. The move to Yucca Mountain or another centralized storage site, if approved, would probably reduce aggregate risk but obviously increase local and regional risk; conversely, a centralized site would reduce local and regional risks at the 100-plus sites where waste is currently disposed at U.S. government cost and at all future locations.

3.   It is also difficult to assess the extent to which the potential damage from security risk has already been internalized. Certainly, the net upgrade of security requirements brought about by EPACT and other post-9/11 Nuclear Regulatory Commission changes has internalized some of the costs. However, because taxpayers presumably bear some of the costs in the event of a high-cost security incident (through an implicit commitment to compensate victims of the event through government relief), the degree

---

[23]Note, however, that the MIT 2003 study concluded that once-through technology with permanent storage of waste material was preferable to closed fuel-cycle technologies.

to which the market has internalized these risks is difficult if not impossible to measure.

## CONCLUSION

In conclusion, the committee finds the following:

1.   The nation's electricity-transmission grid is vulnerable to failure at times because of transmission congestion and the lack of adequate reserve capacity. Electricity consumption generates an externality because individual consumers do not take into account the impact their consumption has on aggregate load. Damages from consumption could be significant, and it underscores the importance of investing in a modernized grid that takes advantage of new smart technology and that is better able to handle intermittent renewable power sources.

2.   Externalities from accidents at facilities are largely internalized and—in the case of the oil and gas transmission network—of negligible magnitude per barrel of oil or thousand cubic feet of gas trans-shipped. We find that the monopsony component of the oil consumption premium is not an externality.

3.   Although government policy may be desirable to serve as a countervailing force to monopoly or cartel producer power, it is a separate issue from the focus of this report.

4.   We find that macroeconomic disruptions from oil supply shocks are not an externality. We also find that sharp and unexpected increases in oil prices adversely affect the U.S. economy. Estimates in the literature of the macroeconomic costs of disruption and adjustment range from $2 to $8 per barrel in 2007 dollars.

5.   Dependence on imported oil has implications for foreign policy, and we find that some of the effects can be viewed as an externality. We find, however, that it is impossible to quantify these externalities. The role of the military in safeguarding foreign supplies of oil is often identified as a potential externality. We find it difficult if not impossible to disentangle nonenergy-related reasons for a military presence in certain regions of the world from energy-related reasons. Moreover much of the military cost is likely to be fixed in nature. A 20% reduction in oil consumption, for example, would probably have little impact on the strategic positioning of military forces in the world.

6.   Nuclear waste and proliferation raise important issues and pose difficult policy challenges. The extent to which uninternalized externalities exist is difficult to measure. Moreover, it is very difficult to quantify them. Thus, we do not report numerical values in this report but recognize the importance of studying this issue further.

# 7

# Overall Conclusions and Recommendations

In response to a charge from Congress, the committee defined and evaluated key external costs and benefits associated with the production, distribution, and consumption of energy from various selected sources. We were asked to focus on health, environmental, security, and infrastructure effects that are not—or may not be—fully incorporated into the market price of energy or into government policies related to energy production, distribution, or consumption. The external effects of energy are mostly negative, but the overall benefits of U.S. energy systems to society are enormous. However, the estimation of those benefits, which are mostly reflected in energy prices and markets, was not in the committee's charge.

The results of this study are intended to inform public policy choices, such as selecting among fuel types, or to help identify situations in which additional regulation may be warranted for reducing external costs produced by an energy-related activity. When sources with large aggregate damages are indentified, analysis of the costs and benefits of reducing the burdens resulting from those damages is warranted.

This chapter presents an overview of the results of the committee's analyses. It provides factors to keep in mind when interpreting the results of the evaluations, overall conclusions, and recommendations for research to inform future consideration of various issues.

## THE COMMITTEE'S ANALYSES

Our study examined external effects over the life cycle of electricity generation, transportation, and production of heat for the residential, commercial, and industrial sectors. We estimated damages that remained in 2005 after regulatory actions had taken place as well as damages expected to remain in 2030 in light of possible future regulations. Our boundaries for analysis were not identical in all sectors, but we sought to use existing data and methods for well-recognized externalities. We did not attempt to develop wholly new methods for estimating impacts and damages, but we did identify areas where additional research would be particularly valuable.

For electricity generation and production of heat, we focused on monetizing downstream effects related to air pollution from coal-fired and gas-fired processes. Upstream effects and other downstream effects have been quantified but not monetized or have been discussed in qualitative terms. We did not assess effects associated with power-plant construction, and we did not assess effects from methane emissions from transporting natural gas by pipeline for heat. For transportation, we monetized effects related to air pollution for essentially the full life cycle, including vehicle manufacture. We considered climate-change effects associated with energy production and use, and we reviewed various attempts that have been made in the literature to quantify and monetize the damages associated with the effects of climate change. We also considered the literature on a variety of damages that are associated with the nation's energy infrastructure: disruption in the electricity transmission grid, vulnerability of energy facilities to accidents and possible attack, external costs of oil consumption, supply security considerations, and national security externalities.

The committee focused its attention on externalities as generally defined by economists. As discussed in Chapter 1, there are many other distortions that occur in markets related to energy production and consumption that may create opportunities for improvement of social welfare but that are not externalities. There are also equity or "fairness" consequences of market activities. Although other distortions and equity concerns may be appropriate for policy formulation, they are beyond the scope of this study and were not considered.

## LIMITATIONS IN THE ANALYSES

Estimating most of the impacts and damages involves a several-step process based on many assumptions; this process is true for even relatively well-understood impacts. In summarizing our results, we attempt to convey the uncertainty surrounding our estimates. The results of the committee's study should be considered in light of important caveats. Although our analysis was able to consider and quantify a wide range of burdens and damages (for example, premature mortality resulting from exposure to air pollution), there are many potential damages that we did not quantify. Therefore our results should not be interpreted as a full accounting. As discussed in Chapter 1, studying selected sources was necessary because it would have been infeasible to evaluate the entire energy system with the time and resources available to the committee. Even within the sources selected by the committee, we were unable to monetize all externalities over a life cycle.

Our analysis required use of a wide set of assumptions and decisions about analytical techniques that can introduce uncertainty into the results. Although we did not attempt to conduct a formal uncertainty analysis, we have been cautious throughout our discussion of results—and urge the reader to be cautious—that is, not to over-interpret small differences in results among the wide range of energy sources and technologies assessed.

There is uncertainty in the analyses with respect to the quality of the data available, the completeness of the analyses (factors that may have been left out or have been unintentionally given inappropriate weight), and the degree to which computation models correctly include the most important variables. Uncertainty also involves unknowns. For example, some climate effects of greenhouse gas (GHG) emissions are poorly understood and might continue to be for some time. In some cases in which effects werre unknown, the committee was able to conclude that the effects were probably small compared with the known effects. In other cases, the committee was not able to provide even qualitative estimates of unknown effects; in such cases, we had to accept that we did not know. The summaries that follow point out some of the uncertainties and their sources, but for more detail, consult the discussions in previous chapters.

## ELECTRICITY GENERATION

Chapter 2 examines burdens, effects, and damages associated with electricity generation from coal, natural gas, nuclear power, wind, solar energy, and biomass. In the cases of fossil fuels (coal and natural gas) and nuclear power, the analysis includes externalities associated with upstream activities, exploration, fuel extraction and processing, and the transportation of fuel to generating facilities, as well as damages associated with downstream activities of electricity generation and distribution. Some effects are discussed in qualitative terms and others are quantified and, if possible, monetized. Although this section presents estimates of GHG emissions due to electricity generation, it does not present damages associated with effects related to climate change. Those damages are discussed in separate sections in this chapter.

### Electricity from Coal

For electricity generation from coal, the committee monetized effects on human health, visibility of outdoor vistas, agriculture, forestry, and damages to building materials associated with emissions of airborne particulate matter (PM), sulfur dioxide ($SO_2$) and oxides of nitrogen ($NO_x$) from 406 coal-fired power plants in the United States, excluding Alaska and Hawaii. More than 90% of monetized damages are associated with premature human mortality, and approximately 85% of damages come from $SO_2$ emissions, which are transformed into airborne PM. Aggregate damages (unrelated to climate change) in 2005 were approximately $62 billion (2007 U.S. dollars [USD]), or 3.2 cents per kilowatt hour (kWh) (weighting each plant by the electricity it produces); however, damages per plant varied widely. The distribution of damages across plants is highly skewed (see Figure 7-1). The 50% of plants with lowest damages per plant accounted for 25% of net electricity generation and produced 12% of damages. The 10% of plants with the highest damages per plant also accounted for 25% of net generation, but they produced 43% of the damages. Although damages are

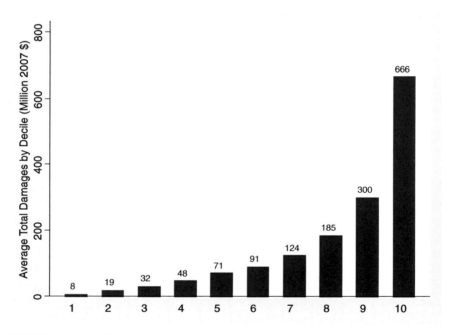

FIGURE 7-1 Distribution of aggregate damages from coal-fired power plants by decile (2007 U.S. dollars). In computing this chart, plants were sorted from smallest to largest based on aggregate damages. The lowest decile represents the 40 plants with the smallest aggregate damages per plant. The figure on the top of each bar is the average across all plants of damages associated with $SO_2$, $NO_x$, $PM_{2.5}$, and $PM_{10}$ (particles with diameters less than or equal to 2.5 and 10 microns, respectively). Damages related to climate-change effects are not included.

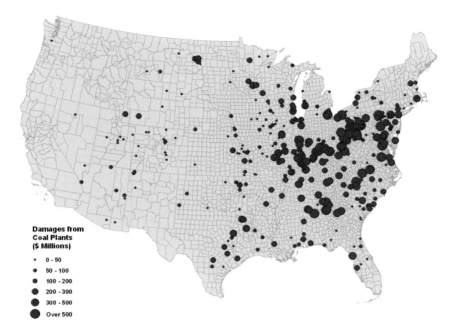

**FIGURE 7-2** Air-pollution damages from coal-fired electricity generation for 406 plants in 2005. Damage estimates are reported in 2007 U.S. dollars. Damages related to climate-change effects are not included.

larger for plants that produce more electricity, less than half of the variation in damages across plants is explained by differences in net generation. The map in Figure 7-2 shows the size of damages created by each of the 406 plants by plant location. Plants with large damages are concentrated to the east of the Mississippi River, along the Ohio River Valley, in the Middle Atlantic, and in the South.

Damages per kWh also varied widely across plants (Figure 7-3)—from over 12 cents per kWh (95th percentile) to less than a cent (5th percentile) (2007 USD).[1] Most of the variation in damages per kWh can be explained by variation in emissions intensity (emissions per kWh) across plants. In the case of $SO_2$ emissions, over 80% of the variation in $SO_2$ damages per kWh is explained by variation in pounds of $SO_2$ emitted per kWh. Damages per ton of $SO_2$, which vary by plant, are less important in explaining variation in $SO_2$ damages per kWh. (Damages per ton are capable of explaining only 24% of the variation in damages per kWh.)

---

[1]These estimates are not weighted by electricity generation.

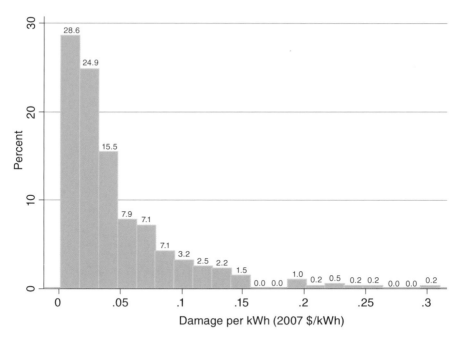

**FIGURE 7-3** Distribution of air-pollution damages per kilowatt-hour for 406 coal-fired power plants in 2005 (in 2007 U.S. dollars). All plants are weighted equally. Damages related to climate-change effects are not included.

For 2030, despite increases in damages per ton of pollutant due to population growth and income growth, average damages per kWh (weighted by electricity generation) at coal plants are estimated to be 1.7 cents per kWh, compared with 3.2 cents per kWh in 2005 (2007 USD). The fall in damages per kWh is explained by the assumption that pounds of $SO_2$ per megawatt hour (MWh) will fall by 64% and that $NO_x$ and PM emissions per MWh will fall by approximately 50% (see Chapter 2).

*Greenhouse Gas Emissions*

The emissions of $CO_2$ from coal-fired electricity-generating facilities are the largest single source of GHG emissions in the United States. Because the heat rate (energy from coal needed to generate 1 kWh of electricity) varies widely among coal-fired plants, the $CO_2$ emissions vary as well. The 5th-95th percentile range is 0.95-1.5 tons (the average being about 1 ton of $CO_2$ per MWh of power generated). The main factors behind the differences in the $CO_2$ emitted are the technology used to generate the power and the age of the plant.

## Electricity from Natural Gas

For estimating nonclimate-change-related damages for 498 facilities that generate electricity from natural gas in the United States, we used a similar approach as in the coal analysis. The gas facilities, which include electric utilities, independent power producers, and combined heat and power facilities, each generated at least 80% of their electricity from gas and had installed capacity of at least 5 MW. The aggregate damages associated with emissions of $SO_2$, $NO_x$, and PM from these facilities, which generated 71% of electricity from natural gas, were approximately \$0.74 billion (2007 USD), or 0.16 cents per kWh. Thus, on average, nonclimate-change damages associated with electricity generation from natural gas are an order of magnitude lower than damages from coal-fired electricity generation. The distribution of damages across plants is, however, highly skewed (see Figure 7-4). The 10% of plants with highest damages per plant

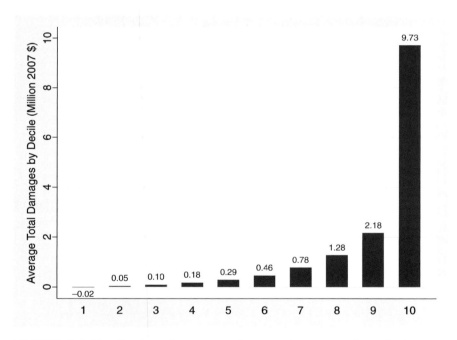

**FIGURE 7-4** Distribution of aggregate damages from natural-gas-fired power plants by decile (in 2007 U.S. dollars). Plants were sorted from smallest to largest based on aggregate damages to compute this chart. The lowest decile represents the 50 plants with the smallest aggregate damages per plant. The number on the top of each bar is the average across all plants of damages associated with $SO_2$, $NO_x$, $PM_{2.5}$, and $PM_{10}$ (particles with diameters less than or equal to 2.5 and 10 microns, respectively). Damages related to climate-change effects are not included.

accounted for 65% of the air-pollution damages produced by all 498 plants. The 50% of plants with lowest damages per plant accounted for only 4% of the aggregate damages. (Each group of plants, respectively, accounted for approximately one-quarter of the electricity generation.) Although damages were larger for plants that produced more electricity, less than 40% of the variation in damages across plants is explained by differences in net generation. The largest damages are produced by gas plants located in the Northeast (along the Eastern seaboard), and in Texas, California, and Florida (see Figure 7-5).

Damages per kWh also vary widely across plants: from more than 1.5 cents per kWh (95th percentile) to less than 0.05 cents (5th percentile) (2007 USD).[2] Most of the variation in $NO_x$ damages per kWh can be explained by variation in emission intensity across plants; however, for $PM_{2.5}$, which accounted for more than half of the monetized air-pollution damages, variation in damages per ton of $PM_{2.5}$ (that is, variation related to the location of the plant relative to population distribution and prevailing winds) are as important in explaining variation in $PM_{2.5}$ damages per kWh as differences in $PM_{2.5}$ emissions intensity.

Damages per kWh at the 498 facilities are predicted to be 30% lower in 2030 than in 2005; they are predicted to fall from 0.16 cents to 0.11 cents per kWh on average (2007 USD) (weighting each plant by electricity generation). The reduction is due to a predicted 19% fall in $NO_x$ emissions per kWh hour and a 32% fall in $PM_{2.5}$ emissions per kWh (see Chapter 2).

*Greenhouse Gas Emissions*

Natural gas plants on average emitted approximately half as much $CO_2$ at the generation stage as did coal-fired power plants in 2005—about half a ton of $CO_2$ per MWh. As the heat rate (energy from gas needed to generate 1 kWh of electricity) varied among gas-fired plants, so did $CO_2$ emissions, the 5th-95th percentile ranged from 0.3 to 1.1 tons per MWh. As discussed later in this chapter, nonclimate-change damages from natural-gas-fired electricity generation are likely to be much smaller than its damages related to climate change.

## Electricity from Nuclear Power

The committee did not quantify damages associated with nuclear power; however, we reviewed studies conducted by others and consider

---

[2]These estimates are not weighted by electricity generation.

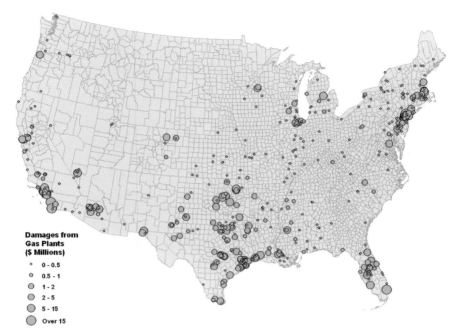

**Damages from Gas Plants ($ Millions)**

- 0 - 0.5
- 0.5 - 1
- 1 - 2
- 2 - 5
- 5 - 15
- Over 15

**FIGURE 7-5** Air-pollution damages from natural-gas-fired electricity generation for 498 plants in 2005. Damages are expressed in 2007 U.S. dollars. Damages related to climate-change effects are not included.

their conclusions relevant.[3] Overall, other studies have found that damages associated with the normal operation of nuclear power plants (excluding the possibility of damages in the remote future from the disposal of spent fuel) are low compared with those from fossil-fuel-based power plants.

For surface-mine workers, exposure to radon is generally less important than direct irradiation or dust inhalation, but radon exposure can be important for underground miners. However, if radiologic exposure is taken into account in miners' wages, it is not considered an externality. For members of the public, the most significant pathways from an operating uranium mine are radon transport and radionuclide ingestion following surface-water transport. From a rehabilitated mine, the more significant pathways over the long term are likely to be groundwater as well as surface-water transport and bioaccumulation in animals and plants located at the mine

---

[3]The committee did not quantify damages associated with nuclear power because the analysis would have involved power plant risk modeling and spent-fuel transportation modeling that would have taken far greater resources and time than were available for this study.

site or on associated water bodies. Little uranium is currently mined in the United States; most of the uranium supplied to U.S. nuclear power plants comes from Canada and Russia.

Downstream impacts are largely confined to the release of heated water used for cooling and the production of low-level radioactive wastes (LLRW) and high-level radioactive wastes (HLRW) from spent fuel. Release of highly radioactive materials has not occurred on a large scale in the United States (but obviously has occurred elsewhere). LLRW is stored for decay to background levels and then disposed of as nonradioactive waste (a practice possible with slightly contaminated materials), or it is disposed of in near-surface landfills designed for radioactive wastes. For spent nuclear fuel that is not reprocessed and recycled, HLRW is usually stored at the plant site. No agreement has been reached on a geologic repository for HLRW in the United States, and, therefore, little HLRW is transported for long distances. The issue of having a permanent repository is perhaps the most contentious nuclear-energy issue, and considerably more study on the externalities of such a repository is warranted.

### Electricity from Wind Energy

The committee relied on information in the scientific literature for its assessment of wind power for producing electricity; it focused on land-based wind turbines, because no offshore turbines have been permitted yet in the United States. Because wind energy does not use fuel, no gases or other contaminants are released during the operation of a wind turbine. Emissions of $SO_2$, $NO_x$, and PM and GHGs over the life cycle are much smaller per kWh than for coal or natural gas. Upstream effects are related to the mining, processing, fabrication, and transportation of raw materials and parts; those parts are normally transported to the wind-energy plant's site for final assembly. Effects related to downstream activities include visual and noise impacts, impacts on bird and bat species, and land-use effects that accompany the construction of any electricity-generating plant and transmission of electricity.

Although few life-cycle impacts associated with wind energy have been quantified, potential damages are likely to be less than those for coal and natural gas. For example, aggregate land-use damages over the entire life cycle are also likely to be smaller for electricity generation from wind than for coal and natural gas. However, better information is needed, especially in light of the probable increase in the number and density of wind turbines. Even if the expansion of wind energy is taken into account, the estimated number of birds killed by wind turbines is dwarfed by the number killed by transmission lines. On the other hand, bat deaths appear to be largely, if not uniquely, associated with wind generation, but good estimates of the numbers of bats killed are not available. In addition, the lack of under-standing of the demography and ecology of bats makes it difficult to assess the importance of bat deaths. Societal damages associated with the killing of bats by wind turbines are currently small by comparison with the ag-gregate damages associated with electricity generation by coal, natural gas, and the sum of all other sources.

### Electricity from Solar Energy

Much of the United States receives enough solar energy to produce about 1 kWh per square meter of panel area per day, with considerable lo-cal variability from north to south and regionally as a result of sun angles and weather patterns. At present, most solar panels are installed on build-ing roofs or immediately adjacent to buildings to provide electricity on site. When a site's electricity use exceeds solar energy availability, electricity is supplied from the grid (or from batteries, if electricity demand is low). In this case, solar panels reduce grid-based electricity demand at the end use, thus becoming similar to an energy efficiency improvement. Some solar panel installations also can feed excess electricity back into the grid during periods of peak solar or low local on-site demand periods.

Concentrating solar power (CSP)[4] and photovoltaic (PV) electricity generation by the electricity sector combined to supply 500 gigawatt hours (GWh) in 2006 and 600 GWh in 2007, which constitute about 0.01% of the total U.S. electricity generation. Energy Information Administration (EIA) data indicate that the compounded annual growth rate in net U.S. generation from solar was 1.5% from 1997 to 2007 (NAS/NAE/NRC 2009b). However, this estimate does not account for the growth in resi-dential and other small PV installations, which are applications that have displayed the largest growth rate for solar electricity. Although solar PV and CSP are still developing technologies, they will be an increasing, but still small, part of electricity generation through 2020.

Like wind power, solar power emits no gaseous pollutants during op-erations to produce electricity. Upstream life-cycle activities include mining of materials for solar panels and the balance-of-system components used to convert the electricity to alternating current. Downstream life-cycle ac-tivities include electricity generation, storage, and disposal or recycling of worn-out panels. Worn-out panels have the potential to produce a large

---

[4]CSP installations use arrays of mirrors to focus direct beam incident sunlight to heat a working fluid and generate electricity through a thermal power cycle. Desert locations with low humidity and high insolation could allow large-scale CSP electricity generation at lower costs than PV installations. Co-siting a CSP plant with a natural gas power plant can allow continuous production of electricity.

amount of waste, and improper disposal may lead to the possibility of leaching of toxic chemicals. If solar energy for electricity were to become an important part of the U.S. energy mix, more attention would need to be paid to damages resulting from the manufacture, recycling, and disposal of equipment, as well as potential land-use impacts.

### Electricity from Biomass

No attempt has been made to estimate damages associated with generating electricity using biomass feedstock derived from forestry practices, agricultural activities, and municipal solid waste because the amount of electricity generated from biomass feedstock is relatively small (total installed capacity is less than 1,600 MW) and is likely to remain so.[5] Many of the issues facing biomass combustors are similar to issues faced by large-scale fossil-fuel generation. Emissions from the combustion of biomass can include polychlorinated biphenyl compounds, although the focus of recent analysis has been primarily on enclosed systems. Nonclimate-change-related damages from biomass-generated electricity on a per-kWh basis might equal or even exceed those from coal in some cases. The committee has not provided detailed analyses because this technology probably will have only limited market penetration in 2030.

### Transmission and Distribution of Electricity

Transmission lines have raised concerns about health risks (for example, risks associated with exposure to extremely low-frequency [ELF] electromagnetic radiation), visual disamenities, and loss of property values. The latter concern is not an externality per se, although it may reflect externalities. Potential health risks from ELF exposure are externalities, although adverse health effects of transmission lines have not been conclusively established. Visual disamenities are also externalities and may become an increasing concern in association with renewable energy sources. Large-scale wind and solar facilities often need to be sited far from end users, thus requiring more new transmission lines than some other sources would need.

### TRANSPORTATION

We considered a wide range of potential emissions and damages related to air pollution from the use of energy in transportation. Our discussion

---

[5]Source: National Electric Energy Data System (NEEDS) for the Integrated Planning Model (EPA 2004b).

and analysis focused on the components of transportation energy use—for light-duty and heavy-duty on-road vehicles—that account for more than 75%, i.e. the great majority, of annual U.S. transportation energy use. Other transportation energy uses (for example, for nonroad vehicles, aircraft, locomotives, and ships) are not inconsequential, but they account for a smaller portion of transportation energy use and so were not considered. For each fuel and vehicle combination, the committee analyzed the life-cycle energy use and emissions, and then used those emissions data in a nationwide analysis of exposures and health effects as well as other nonclimate effects, and then developed estimates of monetized damages (described in Chapter 3). This section also presents estimates of GHG emissions due to transportation, but it does not present estimates of climate-change-related damages associated with those emissions. Those damages are discussed in a separate section of this chapter.

### Health Effects and Other Damages Not Related to Climate Change

Despite limitations, our analysis provides some useful insight into the relative levels of damages from different fuel and technology mixes. Overall, we estimate that the aggregate national damages in 2005 to health and other nonclimate-change-related effects were approximately $36 billion per year (2007 USD) for the light duty vehicle fleet; the addition of medium-duty and heavy-duty trucks and buses raises the aggregate estimate to approximately $56 billion (2007 USD). These estimates are probably conservative, as they include but do not fully account for the contribution of light-duty trucks to the aggregate damages, and should be viewed with caution, given the significant uncertainties in any such analysis.

### Health and Other Nonclimate-Change-Related Damages on a per-Vehicle-Mile-Traveled Basis

Although the uncertainties in the analysis preclude precise ranking of different technologies, Table 7-1 illustrates that, on a cents per-vehicle-mile-traveled (VMT) basis, there are some important differences in the levels of damages attributable to different fuel and technology combinations in 2005 and 2030.

Among the fuel and technology choices, there are some differences in damages, although overall, especially in 2030, the different fuel and technology combinations have remarkably similar damage estimates.

- Some fuels (E85 from herbaceous and corn stover feedstock) and compressed natural gas (CNG) have relatively lower damages than all other options in both 2005 and 2030.

**TABLE 7-1** Relative Categories of Health and Other Nonclimate-Change Damages 2005 and 2030 for Major Categories of Light-Duty Vehicle Fuels and Technologies (Damage Estimates Based on 2007 U.S. Dollars)

| Category of Aggregate Damage Estimates (Cents/VMT) | 2005 | 2030 |
|---|---|---|
| 1.10-1.19 | | CNG<br>Diesel with low sulfur and biodiesel |
| 1.20-1.29 | E85 herbaceous<br>E85 corn stover<br>CNG<br>Grid-independent HEV | E85 corn stover<br>E85 herbaceous |
| 1.30-1.39 | Conventional gasoline and RFG<br>E10<br>Hydrogen gaseous | Conventional gasoline and RFG<br>E10<br>E85 corn |
| 1.40-1.49 | Diesel with low sulfur and biodiesel<br>Grid-dependent HEV | Electric vehicle |
| 1.50-1.59 | E85 corn | Grid-independent HEV<br>Grid-dependent HEV |
| >1.60 | Electric vehicle | Hydrogen gaseous |

ABBREVIATIONS: VMT, vehicle miles traveled; E85, ethanol 85% blend; E10, ethanol 10% blend; HEV, hybrid electric vehicle; CNG, compressed natural gas; RFG, reformulated gasoline.

- Diesel, which has relatively high damages in 2005, has one of the lowest levels of damage in 2030. This result is due to the substantial reductions in both PM and $NO_x$ emissions that a diesel vehicle has been required to attain after the 2006 introduction of low-sulfur fuel.

- Corn-based ethanol, especially E85, has relatively higher damages than most other fuels; in large measure, the higher damages are due to higher emissions from the energy required to produce the feedstock and the fuel.

- Grid-dependent HEVs and electric vehicles have somewhat higher damages in both 2005 and 2030. As noted in Chapter 3, these vehicles have important advantages over all other fuel and technology combinations when only damages from operations are considered. However, the damages associated with the present and projected mixes of electricity generation (the latter still being dominated by coal and natural gas in 2030, albeit at significantly lower rates of emissions) add substantially to the life-cycle damages. In addition, the increased energy associated with battery manufacture adds approximately 20% to the damages from vehicle manufacture. However, further legislative and economic initiatives to reduce emissions from the electricity grid could be expected to improve the relative damages from electric vehicles substantially.

Although the underlying level of aggregate damages in the United States could be expected to rise between 2005 and 2030 because of projected increases in population and increases in the value of a statistical life, the results in our analysis of most fuel and technology examples in 2030 are very similar to those in 2005, in large measure because of the expected improvement in many technology and fuel combinations (including conventional gasoline) as a result of enhanced fuel efficiency (35.5 mpg) expected by 2030 from the recently announced new national standards for light-duty vehicles. (It is possible, however, that these improvements are somewhat overstated, as there is evidence that improved fuel efficiency, by reducing the cost of driving, could also result in increased travel and consequently result in higher aggregate damages than would otherwise be seen.)

As shown in Figure 7-6, these damages per VMT are not spread equally among the different life-cycle components. For example, in most cases the actual operation of the vehicle is one-quarter to one-third of the damages per VMT, and the emissions incurred in creating the feedstock, refining the fuel, and making the vehicle are responsible for the larger part of damages.

### Health and Other Nonclimate Damages on a Per-Gallon Basis

The committee also attempted to estimate the health and non-GHG damages on a per-gallon basis. This attempt was made somewhat more complicated by the fact that simply multiplying expected miles per gallon for each fuel and vehicle type by the damages per mile tend to make the most fuel-efficient vehicles, which travel the most miles on a gallon, appear to have higher damages per gallon than a less-fuel-efficient vehicle. With that caveat in mind, the committee analysis estimated that in 2005, the mean damages per gallon for most fuels ranged from 23 cents/gallon to 38 cents/gallon, the damages for conventional gasoline engines being in approximately the middle of that range at approximately 29 cents per gallon.

### Limitations in the Health and Other Nonclimate Damages Analysis

When interpreting these results, it is important to consider two major limitations in the analysis: the emissions and damages that were not quantifiable and the uncertainty in the analytical results that were obtained.

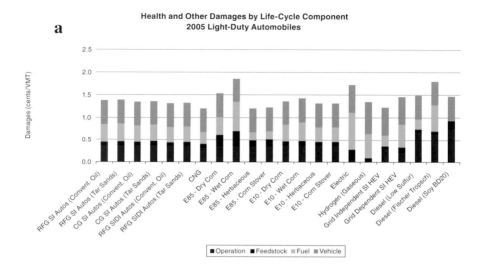

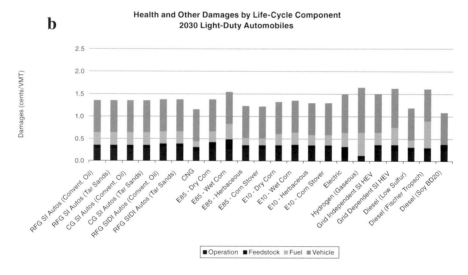

FIGURE 7-6 Health effects and other nonclimate damages are presented by life-cycle component for different combinations of fuels and light-duty automobiles in 2005 (*a*) and 2030 (*b*). Damages are expressed in cents per VMT (2007 U.S. dollars). Going from bottom to top of each bar, damages are shown for life-cycle stages as follows: vehicle operation, feedstock production, fuel refining or conversion, and vehicle manufacturing. Damages related to climate change are not included. ABBREVIATIONS: VMT, vehicle mile traveled; CG SI, conventional gasoline spark ignition; CNG, compressed natural gas; E85, 85% ethanol fuel; E10, 10% ethanol fuel; HEV, hybrid electric vehicle.

Although our analysis was able to consider and quantify a wide range of emissions and damages throughout the life cycle and included what arguably could be considered the most significant contributors to estimates of such damages (for example, premature mortality resulting from exposure to air pollution), many potential damages could not be quantified at this time. These damages include the following:

- *Overall:* Estimates of the impacts of hazardous air pollutants, estimates of damages to ecosystems (for example, from deposition), and estimates covering the full range of agricultural crops.
- *For biofuels:* Impacts on water use and water contamination, as well as any formal consideration of potential indirect land-use effects (see, however, the discussion of the latter in Chapter 3).
- *For battery electric vehicles:* Potential exposures to toxic contaminants as a result of battery manufacture, battery disposal, and accidents.

Any such analysis includes a wide set of assumptions and decisions about analytical techniques that can introduce uncertainty in the results. Although we did not attempt to conduct a formal uncertainty analysis, we engaged in limited sensitivity analyses to check the effects of key assumptions on the results. We urge the reader to be cautious when interpreting small differences in results among the wide range of fuels and technologies we assessed.

### Greenhouse Gas Emissions

Similar to the damage estimates presented above, the GHG emission estimates for each fuel and technology combination can provide relative estimates of GHG performance in 2005 and 2030. Although caution should be exercised in interpreting these results and comparing fuel and technology combinations, some instructive observations from Table 7-2 are possible. Overall, the substantial improvements in fuel efficiency in 2030 (to a minimum of 35.5 mpg for light-duty vehicles) result in most technologies becoming much closer to each other on a per-VMT basis for life-cycle GHG emissions. There are, however, some differences:

- As with damages above, the herbaceous and corn stover E85 have relatively low GHG emissions over the life cycle; in terms of aggregate grams per VMT of $CO_2$-equivalent ($CO_2$-eq)[6] emissions, E85 from corn also has relatively low GHG emissions.

---

[6]$CO_2$-eq expresses the global warming potential of a GHG, such as methane, in terms of $CO_2$ quantities.

**TABLE 7-2** Relative Categories of GHG Emissions in 2005 and 2030 for Major Categories of Light-Duty-Vehicle Fuels and Technologies

| Category of Aggregate CO$_2$-eq Emission Estimates (g/VMT) | 2005 | 2030 |
|---|---|---|
| 150–250 | E85 herbaceous<br>E85 corn stover | E85 herbaceous<br>E85 corn stover |
| 250–350 | Hydrogen gaseous | E85 corn<br>Diesel with biodiesel<br>Hydrogen gaseous<br>CNG |
| 350–500 | E85 corn<br>Diesel with biodiesel<br>Grid-independent HEV<br>Grid-dependent HEV<br>Electric vehicle<br>CNG | Grid-independent HEV<br>SI conventional gasoline, RFG<br>Grid-dependent HEV<br>Electric vehicle<br>Low-sulfur diesel<br>E10 herbaceous, corn stover<br>SIDI conventional gasoline<br>E10 corn<br>SI tar sands |
| 500–599 | Conventional gasoline and RFG<br>E10<br>Low-sulfur diesel | |
| >600 | Tar sands | |

ABBREVIATIONS: CO$_2$-eq, carbon dioxide equivalent; VMT, vehicle miles traveled; E85, ethanol 85% blend; E10, ethanol 10% blend; HEV, hybrid electric vehicle; CNG, compressed natural gas; RFG, reformulated gasoline; SI, spark ignition; SIDI, spark ignition direct injection.

- The tar-sands-based fuels have the highest GHG emissions of any of the fuels that the committee considered.

As shown in Figure 7-7, and in contrast to the damages analysis above, the operation of the vehicle is in most cases a substantial relative contributor to total life-cycle GHG emissions. That is not the case, however, with either the grid-dependent technologies (for example, electric or grid-dependent hybrid) or the hydrogen fuel-cell vehicles. In the latter vehicle technologies, the dominant contributor to life-cycle GHG emissions is electricity generation and the production of hydrogen rather than vehicle operation.

### Heavy-Duty Vehicles

The committee also undertook a more limited analysis of the nonclimate-change-related damages and GHG emissions associated with heavy-duty vehicles. Although this analysis included operation, feedstock, and fuel com-

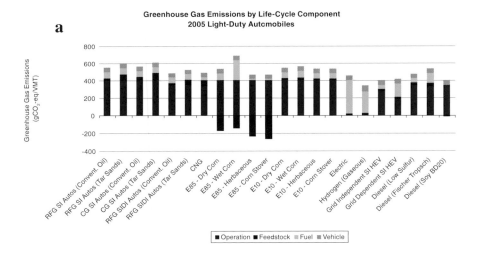

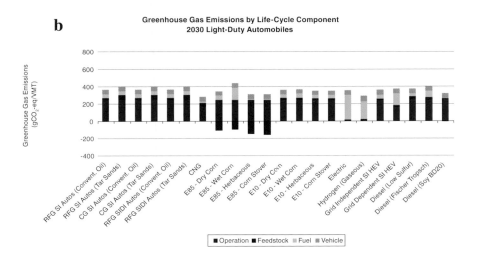

FIGURE 7-7 Greenhouse gas emissions (grams $CO_2$-eq)/VMT by life-cycle component for different combinations of fuels and light-duty automobiles in 2005 (*a*) and 2030 (*b*). Going from bottom to top of each bar, damages are shown for life-cycle stages as follows: vehicle operation, feedstock production, fuel refining or conversion, and vehicle manufacturing. One exception is ethanol fuels for which feedstock production exhibits negative values because of $CO_2$ uptake. The amount of $CO_2$ consumed should be subtracted from the positive value to arrive at a net value. ABBREVIATIONS: g $CO_2$-eq, grams $CO_2$-equivalent; VMT, vehicle mile traveled; CG SI, conventional gasoline spark ignition; CNG, compressed natural gas; E85, 85% ethanol fuel; E10. 10% ethanol fuel; HEV, hybrid electric vehicle.

ponents of the life cycle, it could not—because of the wide range of vehicle types and configurations—include a vehicle-manufacturing component. In sum, there are several conclusions that can be drawn:

• The nonclimate-change-related damages per VMT in 2005 are significantly higher than those for light-duty vehicles, as shown above, although they of course pertain to a much higher weight of cargo or number of passengers being carried per VMT.

• Damages not related to climate-change effects drop significantly in 2030 as a result of the full implementation of the 2007-2010 Highway Diesel Rule, which requires substantial reductions in PM and $NO_x$ emissions.

• Amounts of GHG emissions are driven primarily in these analyses by the operations component of the life cycle, and they do not change substantially between 2005 and 2030 (except for a modest improvement in fuel economy). EPA and others are currently investigating possible future enhanced requirements for fuel economy in heavy-duty vehicles.

## HEAT GENERATION

The committee conducted an assessment focused on air-pollution impacts associated with the present and future (2030) use of natural gas for heat in residential and commercial building sectors. The industrial sector was considered more qualitatively, as published statistics do not differentiate clearly between fuel used for heating and for process feedstocks. We focused our assessment on natural gas because it is the major energy source for heat in buildings, although buildings also consumed about 5% of the 39.7 quadrillion British thermal units (quads) of petroleum used in 2008. Only about 12% of U.S. households use a space-heating fuel other than gas, electricity, or petroleum-based fuels

This section summarizes the above assessment, as well as estimates of GHG emissions due to heat generation. Climate-change-related damages are discussed later in this chapter.

### Heat for Residential and Commercial Buildings

We estimated damages attributable to $SO_2$, $NO_x$, $PM_{2.5}$, VOC, and $NH_3$ emissions from on-site combustion across 3,100 U.S. counties. Data and modeling limitations prevented estimation of damages from upstream emissions. The median estimated damages (in 2007 USD) attributable to natural gas combustion for heat in residential buildings are approximately $0.11 per thousand cubic feet (MCF), or 1% of the 2007 residential price of natural gas. Aggregate damages (unrelated to climate change) were approximately $500 million (2007 USD). The median *regional* estimated damages from natural gas combustion for heat in residential buildings ranged from $0.06 to 0.14/MCF, the upper tail of the distribution was as much as 5% of the current residential price of natural gas when evaluating the 90th percentile value in the South region of the United States. We estimate that damages from combusting natural gas for direct heat are much lower than the damages related to heat produced from electricity, based on average values of the U.S. electricity grid.

Estimated damages from natural gas for heat in commercial buildings are very similar to the estimates for residential buildings. The median estimated externality of natural gas combustion for heat in the commercial building sector is approximately $0.11/MCF, and aggregate damages are about $300 million (excluding damages related to climate change) (2007 USD). The variation across U.S. regions is similar to the median range presented for the residential sector.

In 2007, the combined residential and commercial building sectors emitted an estimated 618 million tons of $CO_2$.

Damages associated with energy for heat in 2030 are likely to be approximately the same as those that exist today, contingent upon the development of additional sources to meet demand. Reduction would probably result from changes in the electricity sector, as emissions from natural gas are relatively small and already well-controlled. Increases are possible if new domestic development has higher emissions or if additional imports of liquefied natural gas are needed.

### Heat for Industry

Natural gas use for heating in the industrial sector (6 quads), excluding use for feedstock, is less than natural gas use in the residential and commercial building sectors (8 quads) for 2007; thus, health and environmental damages associated with industrial natural gas use are probably the same order of magnitude or less than the damages associated with natural gas use for heat in residential and commercial buildings. Therefore, a very rough order of magnitude estimate of average externalities associated with the industrial sector use of natural gas is $0.10/MCF, excluding GHG damages

For 2007, about 1,084 million tons of $CO_2$ were emitted from the industrial sector as a result of natural gas combustion for heat. That amount is greater than the combined amount of 617 million tons of $CO_2$ from the residential and commercial sectors. As discussed below, nonclimate-change damages from natural gas combustion for direct heat are likely to be much smaller than natural gas combustion damages related to climate change.

### In Sum

Aggregate damages from combustion of natural gas for direct heat are estimated to be about $1.4 billion per year (2007 USD), assuming the magnitude of effects resulting from heat production in industrial activities is comparable to those of residential and commercial sectors. Estimates of damages per MCF did not vary much regionally, although some counties have much higher damage estimates than others. The largest potential for reducing damages associated with the use of energy for heat lies in greater attention to improving the efficiency of energy use.

Damages associated with energy for heat in 2030 are likely to be about the same as those that exist today, assuming that the effects of additional sources to meet demand are offset by lower-emitting sources. *Reduction* in damages would only result from more significant changes—largely in the electricity-generating sector, as emissions from natural gas are relatively small and well-controlled. The greatest potential for reducing damages associated with the use of energy for heat lies in greater attention to improving efficiency. *Increased* damages would also be possible, however, if new domestic energy development resulted in higher emissions or if additional imports of liquefied natural gas, which would increase emissions from the production and international transport of the fuel, were needed.

Combustion of natural gas results in relatively lower GHG and criteria-pollutant-forming emissions, as compared with similar emissions from coal (the main energy source for electricity generation) and petroleum combustion.

## CLIMATE CHANGE

Energy production and use is a major source of GHG emissions, principally $CO_2$ and methane. Externalities are created as increased atmospheric GHG concentrations affect climate, and subsequently, weather, water quality and availability, sea-level rise, and biodiversity. Changes in these physical and biological systems in turn affect a variety of aspects of human life, including water resources, ecosystem services, food production, and health, among other impacts. Quantifying and valuing climate-change impacts to calculate the marginal damage of a ton of carbon, often referred to as the "the social cost of carbon," is an intricate process that involves detailed modeling and analysis. Integrated assessment models (IAMs), which produce such estimates, must make assumptions about the relationship between emissions and temperature change and temperature and economic impacts in multiple sectors. The magnitude of these impacts depends to a large extent on changes in climate and on human adaptation to climate change in the distant future. The discount rate used to determine present-day values of future impacts is thus of key importance, as is the extent to which various climatic changes are expected to be extreme and irreversible.

Given the complexity in evaluating the externalities of energy-induced climate change, the committee focused its efforts on a review of existing IAMs (specifically, the Dynamic Integrated Model of Climate and the Economy [DICE], the Climate Framework for Uncertainty, Negotiation, and Distribution [FUND] model, and the Policy Analysis of the Greenhouse Effect [PAGE] model) and the associated climate-change literature. The committee came to the following conclusions, as discussed in Chapter 5:

• The two features of IAMs that drive estimates of the marginal damage associated with emitting an additional ton of carbon are the choice of discount rate and the relationship between mean temperature change and the percentage change in world gross domestic product (GDP) (that is, the aggregate damage function).

• Holding the discount rate constant, the choice of damage function can alter estimates of marginal damages by an order of magnitude; for example, at a 3% discount rate, the marginal social cost of carbon is approximately $2 per ton of $CO_2$-eq using the FUND model and $22 per ton of $CO_2$-eq using the PAGE model. The differences between these two well-reviewed and respected IAMs illustrate the scientific uncertainties inherent in predicting the magnitude of climate-change damage functions.

• Holding the damage function constant, changing the discount rate from 4.5% to 1.5% will cause the marginal social cost of carbon to rise by an order of magnitude: in the PAGE model; for example, the marginal social cost of carbon is approximately $100 per ton of $CO_2$ at a 1.5% discount rate and $10 at 4.5% discount rate.

• In all IAMs, marginal damage estimates for 2030 GHG emissions are 50% to 80% larger than estimates of damages from emissions occurring within the past few years.

• The impacts of climate change are likely to vary greatly across countries. The estimates of the marginal damage of a ton of $CO_2$-eq, as cited in this report, sum damages across countries using relative GDP as weight, which gives less weight to the damages borne by low-income countries.

• There is great uncertainty about the impact of GHG emissions on future climate and about the impacts of changes in climate on the world economy. This uncertainty is usually handled in IAMs using Monte Carlo simulation. The model is run many times, drawing key parameters from their probability distributions that reflect the uncertainty about the values. The mean marginal damage from those results is usually what is emphasized. This approach does not adequately capture the small probability of catastrophic climate changes and impacts. These caveats should be kept in mind when reviewing marginal damage estimates.

## COMPARING CLIMATE AND NONCLIMATE DAMAGE ESTIMATES

Table 7-3 summarizes the results of the committee's quantitative analyses of damages related to the production and use of energy. The table presents the monetized health damages and other monetized damages not related to climate change that were presented in this report. In addition, for illustrative purposes, the table presents three different estimates of external global damages on a per-unit basis from effects related to climate change. The different estimates were obtained by selecting three alternative marginal GHG damage values ($10, $30, and $100 per ton $CO_2$-eq) and multiplying each of them by GHG emission rates for electricity generation (coal-fired and natural-gas-fired), for a range of transportation fuels and vehicle technologies, and for the production of heat by combusting natural gas. It is important to note that the damage estimates at the higher end of the range of marginal GHG damage values are associated only with emission paths without significant GHG controls.

The estimated damages related to climate change on a per-unit-of-fuel basis differ across various primary fuels and energy end uses. These estimates are summarized in Table 7-3. How the monetized value of damages related to climate change compares with the value of damages from $SO_2$, $NO_x$, and PM emissions depends on the value chosen for the social cost of carbon. If the social cost of carbon were $30 per ton of $CO_2$-eq, climate-change-related damages would be approximately 3 cents per kWh at coal-fired power plants and 1.5 cents per kWh at natural gas plants, equaling or exceeding in value the damages from $SO_2$, $NO_x$, and PM. For transportation, the value of climate-change damages begins to approach the value of nonclimate damages at $30 per ton of $CO_2$-eq. For direct heat, each estimate of climate-related damages substantially exceeds the damage estimate from nonclimate damages. Thus, in Table 7-3, damages related to climate change are dominant for electricity generated from natural gas and for heat production at all levels of the social cost of carbon. Climate damages for electricity generation from coal and for transportation can be larger than nonclimate damages if a high value is chosen for the social cost of carbon.

Estimates of damages presented in this report do not by themselves provide a guide to policy. Economic theory suggests that the damages associated with pollution emissions should be compared with the costs of reducing emissions: If distributional equity issues are put aside, the theory suggests that damages should not be reduced to zero, but only to the point where the marginal cost of reducing another ton of emissions or other type of burden equals the marginal damages avoided. Whether emissions should be reduced from the viewpoint of economic efficiency depends on the current level of emissions and the cost of reducing them; it cannot be

**TABLE 7-3** Monetized Damages Per Unit of Energy-Related Activity[a]

| Energy-Related Activity (Fuel Type) | Nonclimate Damage | Climate Damages (per ton CO$_2$-eq)[c] | | | |
|---|---|---|---|---|---|
| | | CO$_2$-eq Intensity | At $10 | At $30 | At $100 |
| Electricity generation (coal) | 3.2 cents/kWh | 2 lb/kWh | 1 cent/kWh | 3 cents/kWh | 10 cents/kWh |
| Electricity generation (natural gas) | 0.16 cents/kWh | 1 lb/kWh | 0.5 cent/kWh | 1.5 cents/kWh | 5 cents/kWh |
| Transportation[b] | 1.2 to >1.7 cents/VMT | 0.3 to >1.3 lb/VMT | 0.15 to >.65 cent/VMT | 0.45 to >2 cents/VMT | 1.5 to >6 cents/VMT |
| Heat production (natural gas) | 11 cents/MCF | 140 lb/MCF | 70 cents/MCF | 210 cents/MCF | 700 cents/MCF |

[a]Based on emission estimates for 2005. Damages are expressed in 2007 U.S. dollars. Damages that have not been quantified and monetized are not included.

[b]Transportation fuels include E85 herbaceous, E85 corn stover, hydrogen gaseous, E85 corn, diesel with biodiesel, grid-independent HEV, grid-dependent HEV, electric vehicle, CNG, conventional gasoline and RFG, E10, low-sulfur diesel, tar sands. (See Table 7-1 for relative categories of nonclimate damages and Table 7-2 for relative categories of GHG emissions.)

[c]Often called the "social cost of carbon."

ABBREVIATIONS: CO$_2$-eq, carbon dioxide equivalent; VMT, vehicle miles traveled; MCF, thousand cubic feet; E85, ethanol 85% blend; HEV, hybrid electric vehicle; CNG, compressed natural gas; RFG, reformulated gasoline.

determined from the size of damages alone. We emphasize, however, that economic efficiency is only one of several potentially valid policy goals that need to be considered in managing pollutant emissions and other damages.

## OVERALL CONCLUSIONS AND IMPLICATIONS

### Electricity Generation

Our analysis of the damages associated with energy for electricity focused on air-pollution damages—both local and global—associated with electricity generation. These estimates can be used to inform the choice of type of fuel used to generate electricity and to guide policies regarding the regulation of air emissions from electricity generation.

### Regarding Comparisons Among Fuels for Electricity Generation

•   In 2005 damages per kWh from $SO_2$, $NO_x$, and PM emissions were an order of magnitude higher for coal than for natural gas plants: on average, approximately 3.2 cents per kWh for coal and 0.16 cents per kWh for natural gas (2007 USD). $SO_2$, $NO_x$, and PM emissions per kWh were virtually nil for electricity generation from nuclear, wind, and solar plants and not calculated for plants using biomass for fuel.

•   Average figures mask large variations among plants in air-pollution damages per kWh, which primarily reflect differences in pollution control equipment. For coal plants, the 5th percentile of the distribution of damages was only 0.5 cents per kWh (2007 USD). Newer plants emit significantly less $SO_2$ and $NO_x$ per kWh than older plants.

### Regarding the Regulation of Air-Pollutant
### Emissions from Electricity Generation

•   Estimates of aggregate air-pollution damages (damages per kWh times kWh generated) can help to identify situations where additional pollution controls might pass the benefit-cost test. We note that the damages from $SO_2$, $NO_x$, and PM at all coal plants, conservatively calculated, were approximately $62 billion in 2005 (2007 USD). (This figure represents the damages from emissions in 2005 relative to zero emissions.) When considering regulations, these damages provide important information to be compared with the costs of controlling emissions related to criteria air pollutants—in particular, comparing the marginal damages per kWh or ton of pollutant with the marginal costs of reducing the emissions.

•   The distribution of damages associated with emissions of $SO_2$, $NO_x$, and PM is highly skewed for both coal-fired power plants and natural-gas-fired plants. The 10% of coal plants with the lowest damages produced 43% of air-pollution damages from all coal plants, while the 50% of the coal plants with the lowest damages produced less than 12% of the aggregate damages. (Each group of plants produced the same amount of electricity—about 25% of net generation from coal.) The 10% of natural gas plants with the highest damages per plant in our study produced 24% of the electricity but 65% of the damages.

•   For policy purposes, it is useful to know the damages associated with emitting an additional ton of a pollutant because the most economically efficient pollution-control policies are those that target emissions directly. These damages vary significantly depending on the pollutant ($NO_x$ vs. PM) and on where it is emitted. The damage associated with a ton of $SO_2$ varies from $1,800 to $10,700 (5th and 95th percentile) at coal plants and from $1,800 to $44,000 at natural gas plants (2007 USD). The differences reflect the fact that most coal-fired power plants are located farther away from population centers than natural gas plants are located from population centers. The highest damages per ton are associated with directly emitted PM. These damages vary from $2,600 to $160,000 (5th and 95th percentile) at natural gas plants and from $2,600 to $26,000 at coal-fired power plants (2007 USD).

## Transportation

Perhaps the most important conclusion to be taken from the transportation analyses is that, when viewed from a full life-cycle perspective, the results are remarkably similar across fuel and technology combinations. One key factor contributing to the similarity is the relatively high contribution to health and other non-GHG damages from emissions in life-cycle phases *other* than the operation of the vehicle. (These phases are the development of the feedstock, the processing of the fuel, and the manufacturing of the vehicle.) There are some differences, however, and some conclusions can be drawn from them:

•   The gasoline-driven technologies had somewhat higher damages related to air pollution (excluding climate change) and GHG emissions in 2005 than a number of other fuel and technology combinations. The grid-dependent electric vehicle options had somewhat higher damages than many other technologies, even in our 2030 analysis, in large measure because of continued conventional emissions and GHG emissions from the existing grid and the likely future grid.

•   The choice of feedstock for biofuels can significantly affect the relative level of life-cycle damages, and herbaceous and corn stover feedstock have some advantage in our analysis.

- Additional regulatory actions can significantly affect levels of damages and GHG emissions:

    o This is illustrated in the health and nonclimate damage analysis by the substantial reduction in diesel damages from 2005 to 2030. Major regulatory initiatives to reduce electricity-generation emissions or legislation to regulate carbon emissions would be expected to significantly reduce the relative damages and emissions from the grid-dependent electric-vehicle options. Similarly, a significant shift to lower-emitting grid technologies, such as natural gas, renewable sources, and nuclear, would also reduce these damages.

    o In 2030, with the implementation of enhanced 35.5 mpg requirements now being put in place for light-duty vehicles under CAFE and EPA GHG emission rules, the differences among technologies tend to converge somewhat, although the fact that operation of the vehicle is generally less than a third of overall life-cycle emissions and damages tends to dampen the magnitude of that improvement. Further enhancements in fuel efficiency, such as the likely push for an extension beyond 2016 to further improvements, would further improve the GHG emission estimates for all liquid-fuel-driven technologies.

Overall, there are somewhat modest differences among different types of vehicle technologies and fuels, even under the likely 2030 scenarios, although some technologies (for example, grid-dependent electric vehicles) had higher life-cycle emissions. It appears, therefore, that some breakthrough technologies, such as cost-efficient conversion of advanced biofuels, cost-efficient carbon capture and storage, and a shift to a mix of lower-emitting sources of electricity (such as natural gas, renewable sources, and nuclear) will be needed to dramatically reduce transportation-related externalities.

### Heat Generation

- The damages associated with criteria-pollutant-related emissions from the use of energy (primarily natural gas) for heating in the residential, commercial buildings, and industrial sectors are low relative to damages from energy use in the electricity-generation and the transportation sectors. This result is largely because natural gas has low rates of those emissions compared with emissions typically resulting from the electricity-generation and transportation sectors.
- The climate-change-related damages from the use of energy (primarily natural gas) for heating in the buildings and industrial sectors are low relative to climate-change-related damages associated with transpor-

tation and electricity production because natural gas carbon intensity is lower than that of coal and gasoline. Regarding energy use for heating, the climate-related damages are in general significantly higher than the nonclimate damages.

• The largest potential for reducing damages associated with the use of energy for heat lies in greater attention to improving the efficiency. The report *America's Energy Future: Technology and Transformation* suggests that the potential for improving efficiency in the buildings and industrial sectors is 25% or more—with the likelihood that emissions damages in these sectors could be held constant in spite of sectoral growth between now and 2030 (NAS/NAE/NRC 2009a).

## Climate Change

Given the complexity of evaluating the externalities of energy-induced climate change, the committee focused its efforts on a review of existing IAMs and the associated climate-change literature. The committee came to the following conclusions, as discussed in Chapter 5:

• The two features of IAMs that drive estimates of the marginal damage associated with emitting an additional ton of carbon (the marginal social cost of carbon) are the choice of discount rate and the relationship between mean temperature change and the percentage change in world GDP (that is, the aggregate damage function).

• Holding the discount rate constant, the damage function used in current IAMs can alter estimates of marginal damages by an order of magnitude.

• Holding the damage function constant, changing the discount rate from 4.5% to 1.5% in an IAM will cause the marginal social cost of carbon to change by an order of magnitude.

• In all IAMs, marginal damage estimates for 2030 GHG emissions are 50-80% larger than estimates of damages from emissions occurring within the past few years.

• There is great uncertainty about the impact of GHG emissions on future climate and about the impacts of changes in climate on the world economy. Mean values of marginal damage estimates are usually reported from integrated planning model simulations. This approach does not adequately capture the small probability of catastrophic climate changes.

## Infrastructure and Security

In Chapter 6, the committee considered damages related to disruptions in the electricity-transmission grid, the vulnerability of energy facilities to accidents and possible attack, the external costs of oil consumption, supply security considerations, and national security externalities. The committee strove to clarify approaches for considering security externalities and disentangle externalities from other motivations for energy policy. The committee concludes as follows:

- The nation's electricity-transmission grid is vulnerable to outages and to power quality degradation events because of transmission congestion and the lack of adequate reserve capacity. Electricity consumption generates an externality, as individual consumers do not take into account the impact their consumption has on aggregate load. Damages from this could be significant, and it underscores the importance of careful analysis concerning the costs and benefits of investing in a modernized grid that takes advantage of new smart technology and that is better able to handle intermittent renewable power sources.

- Externalities from accidents at facilities are largely internalized and—in the case of the U.S. oil and gas transmission network—of negligible magnitude per barrel of oil or thousand cubic feet of gas trans-shipped.

- The monopsony component of the oil consumption premium is not an externality. Government policy may be desirable as a countervailing force to monopoly or cartel producer power; however, this is a separate issue from the focus of this report.

- We find that macroeconomic disruptions from oil supply shocks are not an externality. We also find that sharp and unexpected increases in oil prices adversely affect the U.S. economy. Estimates in the literature of the macroeconomic costs of disruption and adjustment ranged from $2 to $8 per barrel in 2007 dollars

- Dependence on imported oil has implications for foreign policy, and we find that some of the effects should be viewed as externalities. We find, however, that it is impossible to quantify these externalities. The role of the military in safeguarding foreign supplies of oil is often identified as a potential externality. We find it difficult if not impossible to disentangle nonenergy-related reasons for a military presence in certain regions of the world from energy-related reasons. Moreover, much of the military cost is likely to be fixed in nature. A 20% reduction in oil consumption, for example, would probably have little impact on the strategic positioning of military forces in the world.

- Nuclear waste and security raises important issues and poses difficult policy challenges. The extent to which externalities exist is difficult to measure. Moreover, it is very difficult to quantify them. Thus, we do not report values in this report but recognize the importance of studying this issue further.

## RESEARCH RECOMMENDATIONS

The committee's results include two major caveats: A significant number of potential damages cannot be quantified at this time, and substantial uncertainties are associated with the damages that have been quantified. Developers of the committee's statement of task anticipated such circumstances, stating that when it is not feasible to assess specific externalities comprehensively, the committee should recommend assessment approaches and identify key information needs to inform future assessments. In response, the committee has developed a number of research recommendations specific to key topics in this report—electricity, transportation, heat generation, and climate change—as well as one overarching recommendation.

The overarching recommendation is as follows:

• Federal agencies should provide sufficient resources to support new research on the external costs and benefits of energy. In assembling its repository of literature, models, and data needed to carry out an assessment of externalities, the committee became aware that there is limited research funding available to address the topic of externality assessment. In particular, extramural funding from federal agencies provides little support or incentive to pursue this line of research. For example, the APEEP model used in our analysis was funded by a foundation. The GREET model, which we used to estimate transportation-related emissions, is federally supported, but does not explicitly address damages, so it must be coupled with a damage assessment model. EPA has had strong interest and ongoing programs in damage and benefit assessment of air pollution but offers limited resources for research to improve and evaluate its approaches or to develop and assess approaches for other environmental concerns. Because of the growing importance of impact assessment and impact valuation for policy decision making at all levels of government and to avoid a situation in which key uncertainties are addressed only as an adjunct to other research programs, the committee encourages federal agencies, such as the Department of Energy, the Department of Transportation, the National Institutes of Health, the National Science Foundation, and EPA, to support new research specific to externalities with financial resources that are sufficient to address the recommendations for the key topics below in a timely manner.

### Electricity

• Although life-cycle activities pre- and post-generation generally appear to be responsible for a smaller portion of the life-cycle externalities than electricity generation itself, it is desirable to have a systematic estimation and compilation of the externalities from these other activities that are comparable in completeness to the externality estimates for the generation part of the life cycle. In this compilation, it will be particularly important to take into account activities (for example, the storage and disposal of coal combustion by-products and the in situ leaching techniques for uranium mining) that may have locally or regionally significant impacts.

• The use of "reduced-form" modeling of pollutant dispersion and transformation is a key aspect in estimating externalities from airborne emissions on a source-by-source basis; these models should continue to be improved and evaluated.

• The health effects associated with toxic air pollutants, including specific components of PM, from electricity generation should be quantified and monetized. Given the importance of the "value of a statistical life" in determining the size of air-pollution damages, further exploration is needed to determine how willingness to pay varies with mortality-risk changes and with population characteristics, such as age and health status.

• Because current data on electricity-generation facilities are available mainly as national averages, improved data and methods are needed to characterize the mix of electricity-generation technologies (and their associated range of emissions per kWh) at city, state, and regional levels. The current disaggregation of national-level information to regional or state levels that are available from the Department of Energy and EPA are often not sufficiently detailed for impact or damage assessments within specific areas of the United States.

• Continued improvement is necessary in the development of methods to quantify and monetize ecological impacts of all stages of the life cycle of electricity generation, especially of fuel extraction, emission of pollutants, and land-use changes. Similar needs exist for other types of energy production and use.

• For fossil fuel options, more research is needed to quantify and monetize the ecological and socioeconomic impacts of fuel extraction, for example, of mountaintop mining and valley fill.

• For nuclear power, significant challenges in estimating potential damages include estimating and valuing risks when the probabilities of accidents and of radionuclide migration (for example, at a high-level waste repository) are very low but the consequences potentially extreme. It is important to assess how such risks would change based on advances in the technology and regulations and to determine whether the costs to utilities of meeting their regulatory requirements fully reflect these potential damages.

• The analysis of risks associated with nuclear power in the ORNL/ RFF (1992-1998) reports should be updated to reflect advances in technology and science.

• For wind technologies, the major issues lie in quantifying bird and bat deaths; in quantifying or otherwise systematically assessing disturbances to local landscapes, ecosystems, and human populations; and in valuing them in terms comparable to economic damages.

• For solar technologies, one of the greatest needs is an analysis of the upstream activities that quantifies the possible releases of toxic materials and their damages; other needs are a better understanding of the externalities that would accompany disposal or recycling of worn-out panels and dedicating tracts of land to solar power equipment.

• For the transmission lines needed in a transition to a national grid system, better estimates are needed of both the magnitude and the spatial distribution of negative and positive externalities that would accompany this transition.

## Transportation

• It is imperative to better understand potential negative externalities at the earliest possible stage in the research and development process for new fuels and technologies to avoid those externalities as the fuels and technologies are being developed.

• Improved understanding is needed of the currently unquantifiable effects and potential damages related to transportation, especially as they relate to biofuels (for example, effects on water resources and ecosystems) and battery technology (for example, effects throughout the battery life cycle of extraction through disposal).

• More accurate emissions factors are needed for each stage of the fuel and vehicle life cycle. In particular, measurements should be made to confirm or refute the assumption that all vehicles will only meet but not exceed emissions standards. In actual practice, there can be significant differences between on-road performance and emissions requirements, and some alternative-fuel vehicles may do better or worse than expected.

• Because a significant fraction of life-cycle health impacts comes from vehicle manufacture and fuel production, it is important to improve and expand the information and databases used to construct emissions factors for those life stages. In particular, there is a need to understand whether and how energy-efficiency improvements in these industrial components might change the overall estimates of life-cycle health damages.

• The issue of indirect land-use change is central to current debates about the merit of biofuels. Regardless of whether this impact is regarded as an externality associated with U.S. or foreign biofuels production, it is important to obtain more empirical evidence about its magnitude and causes, as well as to improve the current suite of land-use change models.

• As better data become available, future studies should take a range of transportation modes into account—not only those that are alternatives to automobiles and light trucks (for example, public transit), but also air, rail, and marine, which are alternatives for long-distance travel and freight.

## Heat Generation

• Assessment of energy use and its impacts in the industrial sector in particular (but in all sectors to some extent) could be improved by development of more extensive databases that contain details about specific forms of energy use and associated waste streams. Such databases should be designed so that life-cycle analysis of alternatives can be made without inadvertent double counting.

• A more quantitative assessment of industrial sector externalities, done collaboratively between the government and industry, would be valuable in informing priorities for future initiatives to reduce the externalities associated with industrial operations. Such an assessment was not possible in this study largely because of data limitations.

## Climate Change

• More research on climate damages is needed to estimate the impacts of climate change, especially impacts that can be expressed in economic terms, as current valuation literature relies heavily on climate-change impact data from the year 2000 and earlier.

• Marginal damages of GHG emissions may be highly sensitive to the possibility of catastrophic events. More research is needed on their impacts, the magnitude of the damages in economic terms, and the probabilities associated with various types of catastrophic events and impacts.

• Estimates of the marginal damage of a ton of $CO_2$-eq include aggregate damages across countries according to GDP, thereby giving less weight to the damages borne by low-income countries. This aggregate estimate should be supplemented by distributional measures that describe how the burden of climate change varies among countries.

## In Conclusion

In aggregate, the damage estimates presented in this report for various external effects are substantial. Just the damages from external effects that the committee was able to quantify add up to more than \$120 billion for the year 2005.[7] Although large uncertainties are associated with the committee's estimates, there is little doubt that the aggregate total substantially underestimates the damages because it does not include many other kinds of damages, such as those related to some pollutants, climate change, ecosystems, infrastructure, and security, that could not be quantified for reasons explained in the report. In many cases, we have identified the omissions in this report, with the hope that they will be evaluated in future studies.

Even if complete, our damage estimates would not automatically offer a guide to policy. From the perspective of economic efficiency, theory suggests that damages should not be reduced to zero but only to the point where the cost of reducing another ton of emissions (or other type of burden) equals the marginal damages avoided—that is, the degree to which a burden should be reduced depends on its current level and the cost of lowering it. The solution cannot be determined from the amount of damage alone. Economic efficiency, however, is only one of several potentially valid policy goals that need to be considered in managing pollutant emissions and other burdens. For example, even within the same location, there is compelling evidence that some members of the population are more vulnerable than others to a particular external effect.

Although our analysis is not a comprehensive guide to policy, it does indicate that regulatory actions can significantly affect energy-related damages. For example, the full implementation of the federal diesel-emissions rules would result in a sizeable decrease in nonclimate damages from diesel vehicles between 2005 and 2030. Similarly, major initiatives to further reduce other emissions, improve energy efficiency, or shift to a cleaner electricity-generating mix (for example, renewable sources, natural gas, and nuclear) could substantially reduce the damages of external effects, including those from grid-dependent hybrid and electric vehicles.

It is thus our hope that this information will be useful to government policy makers, even in the earliest stages of research and development on energy technologies, as an understanding of their external effects and damages could help to minimize the technologies' adverse consequences.

---

[7]These are damages related principally to emissions of $NO_x$, $SO_2$, and PM relative to a baseline of zero emissions from energy-related sources for the effects considered in this study.

# References

AAR (Association of American Railroads). 2009. Railroads and Coal. July 2008 [online]. Available: http://www.aar.org/InCongress/~/media/AAR/BackgroundPapers/Railroads%20and%20Coal%20%20July%202009.ashx [accessed July 24, 2009].

Abel, A. 2006. Electric Reliability: Options for Electric Transmission Infrastructure Improvements. Report No. RL32075. Congressional Research Service, Library of Congress, Washington, DC. September 20, 2006 [online]. Available: http://digital.library.unt.edu/govdocs/crs/permalink/meta-crs-10432:1 [accessed Aug. 5, 2009].

ACAA (American Coal Ash Association). 2008a. 2007 Coal Combustion Product (CCP) Production & Use Survey Results. ACCA, Aurora, CO. September 15, 2008 [online]. Available: http://www.acaa-usa.org/associations/8003/files/2007_ACAA_CCP_Survey_Report_Form%2809-15-08%29.pdf [accessed Apr. 20, 2009].

ACAA (American Coal Ash Association). 2008b. ACAA 2007 CCP Survey: 1966-2007 CCP Beneficial Use v. Production. ACCA, Aurora, CO [online]. Available: http://www.acaa-usa.org/associations/8003/files/Revised_1966_2007_CCP_Prod_v_Use_Chart.pdf [accessed Apr. 20, 2009].

Adelman, M.A. 1980. The clumsy cartel. Energy J. 1(1):43-53.

Alberta. 2008. Environmental Management of Alberta's Oil Sands. Government of Alberta, Canada. December 2008 [online]. Available: http://environment.gov.ab.ca/info/library/8042.pdf [accessed Apr. 16, 2009].

Aldy, J.E., A.J. Krupnick, RG. Newell, I.W.H. Parry, and W.A. Pizer. 2009. Designing Climate Mitigation Policy. NBER Working Paper No. 15022. National Bureau of Economic Research, Inc., Cambridge, MA.

Alhajji, A.F., and D. Huettner. 2000. OPEC and World Crude Oil Markets from 1973 to 1994: Cartel, oligopoly, or competitive? Energy J. 21(3):31-60.

Almond, D., L. Edlund, and M. Palme. 2007. Chernobyl's Subclinical Legacy: Prenatal Exposure to Radioactive Fallout and School Outcomes in Sweden. NBER Working Paper No. 13347. National Bureau of Economic Research, Inc., Cambridge, MA.

Amato, A.D., M. Ruth, P. Kirshen, and J. Horwitz. 2005. Regional energy demand responses to climate change: Methodology and application to the Commonwealth of Massachusetts. Climatic Change 71(1-2):175-201.

Anisimov, O.A., D.G. Vaughan, T.V. Callaghan, C. Furgal, H. Marchant, T.D. Prowse, H. Vilhjálmsson and J.E. Walsh. 2007. Polar regions (Arctic and Antarctic). Pp. 653-685 in Climate Change 2007: Impact, Adaptation and Vulnerability. Contribution of Working Group II to the Fourth Assessment Report of the Intergovernmental Panel on Climate Change, M. Parry, O. Canziani, J. Palutikof, P. van der Linden, C. Hanson, eds. Cambridge: Cambridge University Press [online]. Available: http://www.ipcc.ch/pdf/assessment-report/ar4/wg2/ar4-wg2-chapter15.pdf [accessed Apr. 19, 2010].

Ann, F., D. Gordon, H. He, D. Kodjak, and D. Rutherford. 2007. Passenger Vehicle Greenhouse Gas and Fuel Economy Standards: A Global Update. International Council on Clean Transportation, Washington, DC. July 2007 [online]. Available: http://www.theicct.org/documents/ICCT_GlobalStandards_2007_revised.pdf [accessed May 7, 2009].

ANS (American Nuclear Society). 2005. The Price-Anderson Act: Background Information. American Nuclear Society. November 2005 [online]. Available: http://www.ans.org/pi/ps/docs/ps54-bi.pdf [accessed Apr. 24, 2009].

API (American Petroleum Institutes). 2009. Offshore Access to America's Oil and Natural Gas Resources. America's Oil and Natural Gas Industry. April 24, 2009 [online]. Available: http://www.api.org/aboutoilgas/upload/access_primer.pdf [accessed May 7, 2009].

API /AOPL (American Petroleum Institutes and Association of Oil Pipe Lines). 2007. Pipeline 101: Overview: How Many Pipelines are There? American Petroleum Institutes and Association of Oil Pipe Lines, Washington, DC [online]. Available: http://www.pipeline101.com/Overview/energy-pl.html [accessed May 12, 2009].

Archer, C.L. 2000. A comparison of state legal approaches to CCB reuse. Pp. 87-92 in Proceedings of the Use and Disposal of Coal Combustion By-Products at Coal Mines: A Technical Interactive Forum, April 10-13, 2000, Morgantown, WV, K.C. Vories, and D. Throgmorton, eds. U.S. Department of Interior, Office of Surface Mining, Alton, IL, U.S. Department of Energy, National Energy Technology Center, and Coal Research Center, Southern Illinois University, Carbondale, IL [online]. Available: http://www.mcrcc.osmre.gov/PDF/Forums/CCB2/2f.pdf [accessed April 20, 2009].

Argonne National Laboratory. 2009. The Greenhouse Gases, Regulated Emissions, and Energy Use on Transportation (GREET) Model. U.S. Department of Energy, Argonne National Laboratory [online]. Available: http://www.transportation.anl.gov/modeling_simulation/GREET/ [accessed Oct. 12, 2009].

Arnett, E.B., W.K. Brown, W.P. Erickson, J.K. Fiedler, B.L. Hamilton, T.H. Henry, A. Jain, G.D. Johnson, J. Kerns, R.R. Koford, C.P. Nicholson, T.J. O'Connell, M.D. Piorkowski, and R.D. Tankersley. 2008. Patterns of bat fatalities at wind energy facilities in North America. J. Wildlife Manage. 72(1):61-78.

Arrow, K. 1975. Political and economic evaluation of social effects and externality. Pp. 47-64 in Frontiers of Quantitative Economics, M.D. Intriligator, ed. Amsterdam: North-Holland.

Auffhammer, M., V. Ramanathan, and J.R. Vincent. 2006. Integrated model shows that atmospheric brown clouds and greenhouse gases have reduced rice harvests in India. Proc. Natl. Acad. Sci. USA 103(52):19668-19672.

Australian Government. 2009. Protecting People from Impact of Uranium Mining. Australian Government, Department of the Environment, Water, Heritage, and the Arts, Canberra [online]. Available: http://www.environment.gov.au/ssd/research/protect/index.html [accessed May 21, 2009].

AWEA (American Wind Energy Association). 2009. U.S. and China in Race to the Top of Global Wind Industry. American Wind Energy Association Newsroom: February 2, 2009 [online]. Available: http://awea.org/newsroom/releases/us_and_china_race_to_top_of_wind_energy_02Fed09.html [accessed April 20, 2009].

Axsen, J., A.F. Burke, and K.S. Kurani. 2008. Batteries for Plug-in Hybrid Electric Vehicles (PHEVs): Goals and the State of Technology circa 2008. Research Report UCD-ITS-RR-

08-14. Institute of Transportation Studies, University of California, Davis. May 2008 [online]. Available: http://pubs.its.ucdavis.edu/publication_detail.php?id=1169 [accessed Oct. 12, 2009].

Bandivadekar, A., K. Bodek, L. Cheah, C. Evans, T. Groode, J. Heywood, E. Kasseris, M. Kromer, and M. Weiss. 2008. On The Road In 2035: Reducing Transportation's Petroleum Consumption and GHG Emissions. Report No. LFEE 2008-05 RP. Laboratory for Energy and the Environment, Massachusetts Institute of Technology. July 2008 [online]. Available: http://web.mit.edu/sloan-auto-lab/research/beforeh2/otr2035/ [accessed April 20, 2009].

Barbose, G., C. Goldman, and B. Neenan. 2004. A Survey of Utility Experience with Real Time Pricing. Paper LBNL-54238. Lawrence Berkeley National Laboratory, University of California. December 1, 2004 [online]. Available: http://repositories.cdlib.org/lbnl/LBNL-54238 [accessed May 14, 2008].

Bare, J.C., G.A. Norris, D.W. Pennington, and T. McKone. 2002. TRACI: The tool for the reduction and assessment of chemical and other environmental impacts. J. Ind. Ecol. 6(3-4):49-78.

Bartis, J.T., T. LaTourrette, L. Dixon, D.J. Peterson, and G. Cecchine. 2005. Oil Shale Development in the United States: Prospects and Policy Issues. Santa Monica, CA: RAND [online]. Available: http://www.rand.org/pubs/monographs/2005/RAND_MG414.pdf [accessed May 12, 2009].

Bates, B.C., Z.W. Kundzewicz, S. Wu, and J.P. Palutikof, eds. 2008. Climate Change and Water. Technical Paper of the Intergovernmental Panel on Climate Change. Geneva: IPCC Secretariat [online]. Available: http://www.ipcc.ch/pdf/technical-papers/climate-change-water-en.pdf [accessed Apr. 21, 2009].

Baumol, W.J., and W.E. Oates. 1988. On the theory of externalities. Pp. 7-154 in The Theory of Environmental Policy, 2nd Ed. New York: Cambridge University Press.

BEA (Bureau of Economic Analysis). 2009. National Economic Accounts: Fixed Assets Tables. U.S. Department of Commerce, Bureau of Economic Analysis [online]. Available: http://www.bea.gov/National/index.htm#fixed [accessed June 1, 2009].

Bennett, D.H., T.E. McKone, J.S. Evans, W.W. Nazaroff, M.D. Margni, O. Jolliet, and K.R. Smith. 2002. Defining intake fraction. Environ. Sci. Technol. 36(9):207A-211A.

Bernanke, B.S., M. Gertler, M. Watson, C.A. Sims, and B.M. Friedman. 1997. Systematic monetary policy and the effects of oil price shocks. Brookings Pap. Eco. Ac. 1:91-157.

Blanchard, O., and J. Gali. 2008. The Macroeconomic Effects of Oil Price Shocks: Why Are the 2000s Different from the 1970s? Economic Working Papers 1045 [online]. Available: http://www.econ.upf.edu/docs/papers/downloads/1045.pdf [accessed Oct. 12, 2009].

BLM (Bureau of Land Management). 2008. Oil Shale and Tar Sands Programmatic EIS Information Center: About Oil Shale. U.S. Department of the Interior, Bureau of Land Management [online]. Available: http://ostseis.anl.gov/guide/index.cfm [accessed Apr. 9, 2009].

BLS (Bureau of Labor Statistics). 2009a. Census of Fatal Occupational Injuries. U.S. Department of Labor, Bureau of Labor Statistics [online]. Available: http://www.bls.gov/iif/oshcfoi1.htm#19922002 [accessed Mar. 25, 2009].

BLS (Bureau of Labor Statistics). 2009b. Occupational Injuries/Illnesses and Fatal Injuries Profile. U.S. Department of Labor, Bureau of Labor Statistics [online]. Available: http://data.bls.gov:8080/GQT/servlet/InitialPage [accessed Mar. 25, 2009].

Bodek, K. 2006. U.S. Transportation Energy Fact Sheet. Massachusetts Institute of Technology. October 18, 2006 [online]. Available: http://www.mitenergyclub.org/assets/2008/11/15/TransportationUS.pdf [accessed July 30, 2009].

Bohi, D.R., and M.A. Toman. 1993. Energy security: Externalities and policies. Energ. Policy 21(11):1093-1109.

Bohi, D.R., and M.A. Toman. 1995. Energy Security as a Basis for Energy Policy. Prepared for the American Petroleum Institute, Washington, DC. April 1995.

Borenstein, S. 2005. Time-varying retail electricity prices: Theory and practice. Pp. 317-356 in Electricity Deregulation: Choices and Challenges, J.M. Griffin, and S.L. Puller, eds. Chicago: University of Chicago Press.

BP. 2008. BP Statistical Review of World Energy 2008. June 2008 [online]. Available: http://www.bp.com/liveassets/bp_internet/globalbp/globalbp_uk_english/reports_and_publications/statistical_energy_review_2008/STAGING/local_assets/downloads/pdf/statistical_review_of_world_energy_full_review_2008.pdf [accessed Aug.3, 2009].

Brauman, K.A., G.C. Daily, T.K. Duarte, and H.A. Mooney. 2007. The nature and value of ecosystem services: An overview highlighting hydrologic services. Annu. Rev. Environ. Resour. 32:67-98.

Brinkman, N., M. Wang, T. Weber, and T. Darlington. 2005. Well-to-Wheels Analysis of Advanced Fuel/Vehicle Systems—A North American Study of Energy Use, Greenhouse Gas Emissions, and Criteria Pollutant Emissions. May 2005 [online]. Available: http://www.transportation.anl.gov/pdfs/TA/339.pdf [accessed Apr. 20, 2009].

BTS (Bureau of Transportation Statistics). 2009. National Transportation Statistics 2009. Research and Innovative Technology Administration, Bureau of Transportation Statistics, U.S. Department of Transportation [online]. Available: http://www.bts.gov/publications/national_transportation_statistics/ [accessed Feb. 26, 2009].

Buckley, L.B., and J. Roughgarden. 2004. Biodiversity conservation: Effects of changes in climate and land use. Nature 430(6995):2, 33.

Buckley, T.D., and D.F. Pflughoeft-Hassett. 2007. National Synthesis Report on Regulation, Standards, and Practices Related to the Use of Coal Combustion Products. Final Report. 2007-EERC-12-11. Prepared for U.S. Department of Energy, National Energy Technology Laboratory, Pittsburgh, PA, and U.S. Environmental Protection Agency, Washington, DC, by Energy and Environmental Research Center, University of North Dakota, Grand Forks, ND. December 2007 [online]. Available: http://www.epa.gov/epawaste/partnerships/c2p2/pubs/natpraccc08.pdf [accessed Apr. 20, 2009].

Byun, D.W., and L.K. Schere. 2006. Review of the governing equations, computational algorithms, and other components of the models-3 Community Multiscale Air Quality (CMAQ) modeling system. Appl. Mech. Rev. 59(2):51-77.

CCSP (U.S. Climate Change Science Program). 2007. Effects of Climate Change on Energy Production and Use in the United States. A Report by the U.S. Climate Change Science Program, and Subcommittee on Global Change Research, T.J. Wilbanks, V. Bhatt, D.E. Bilello, S.R. Bull, J. Ekmann, W.C. Horak, Y.J. Huang, M.D. Levine, M.J. Sale, D.K. Schmalzer, and M.J. Scott, eds. U.S. Department of Energy, Office of Biological and Environmental Research, Washington, DC.

CCSP (U.S. Climate Change Science Program). 2008. Weather and Climate Extremes in Changing Climate, Region of Focus: North America, Hawaii, Caribbean, and U.S. Pacific Islands. Synthesis and Assessment Product 3.3. Report by the U.S. Climate Change Science Program and the Subcommittee on Global Change Research, T.R. Karl, G.A. Meehl, C.D. Miller, S.J. Hassol, A.M. Waple, and W.L. Murray, eds. U.S. Department of Commerce, NOAA's National Climatic Data Center, Washington, DC. June 2008 [online]. Available: http://www.climatescience.gov/Library/sap/sap3-3/final-report/sap3-3-final-all.pdf [accessed July 31, 2009].

CCSP (U.S. Climate Change Science Program). 2009. Coastal Sensitivity to Sea-Level Rise: A Focus on the Mid-Atlantic Region. Synthesis and Assessment Product 4.1. Report by the U.S. Climate Change Science Program and the Subcommittee on Global Change Research, [J.G. Titus (Coordinating Lead Author), K.E. Anderson, D.R. Cahoon, D.B. Gesch, S.K. Gill, B.T. Gutierrez, E.R. Thieler, and S.J. Williams (Lead Authors)]. U.S.

Environmental Protection Agency, Washington, DC. January 2009 [online]. Available: http://www.climatescience.gov/Library/sap/sap4-1/final-report/#finalreport [accessed Sept. 11, 2009].

Chew, L., and K.N. Ramdas. 2005. Caught in the Storm: Impact of Natural Disasters on Women. San Francisco: The Global Fund for Women [online]. Available: http://www.globalfundforwomen.org/cms/images/stories/downloads/disaster-report.pdf [accessed Apr. 21, 2009].

Clarke, G., D. Leverington, J. Teller, and A. Dyke. 2003. Superlakes, megafloods and abrupt climate change. Science 301(5635):922-923.

Clarkson, R., and K. Deyes. 2002. Estimating the Social Cost of Carbon Emissions. Government Economic Service Working Paper 140. The Public Enquiry Unit—HM Treasury, London [online]. Available: http://www.hm-treasury.gov.uk/d/SCC.pdf [accessed June 9, 2009].

Cline, W.R. 1992. The Economics of Global Warming. Washington DC: Institute for International Economics.

Cline, W.R. 2007. Global Warming and Agriculture: Impact Estimates by Country. Washington, DC: Peterson Institute for International Economics.

CNA Corporation. 2007. National Security and the Threat of Climate Change. CNA Corporation, Alexandria, VA [online]. Available: http://securityandclimate.cna.org/report/ [accessed Apr. 22, 2009].

Cohen, M.A. 1986. The costs and benefits of oil spill prevention and enforcement. J. Environ. Econ. Manage. 13(2):167-188.

Cohen, M.J. 1995. Technological disasters and natural resource damage assessment: An evaluation of the Exxon Valdez oil spill. Land Econ. 71(1):65-82.

Cohon, J.L. 2004. Multiobjective Programming and Planning. Mineola, NY: Dover Publications.

Consumer Energy Council of America Research Foundation. 1993. Incorporating Environmental Externalities into Utility Planning: Seeking a Cost-Effective Means of Assuring Environmental Quality. Washington, DC: Consumer Energy Council of America Research Foundation.

Cropper, M.L. 2000. Has economic research answered the needs of environmental policy? J. Environ. Econ. Manage. 39(3):328-350.

Darwin, R.F., and R.S.J. Tol. 2001. Estimates of the economic effects of sea level rise. Environ. Resour. Econ. 19(2):113-129.

Davis, S.C., S.W. Diegel, and R.G. Boundy. 2009. Transportation Energy Data Book, 28th Ed. ORNL-6984. Oak Ridge National Laboratory, U.S. Department of Energy [online]. Available: http://cta.ornl.gov/data/download28.shtml [accessed Sept. 17, 2009].

DeCarolis, J., and D.W. Keith. 2001. The real cost of wind energy. Science 294(5544): 1000-1003.

Delucchi, M.A. 1993. Emissions of Greenhouse Gases from the Use of Transportation Fuels and Electricity, Vol. 2: Appendixes A-S. ANL/ESD/TM-22, Vol. 2. Center for Transportation Research, Argonne National Laboratory, Argonne, IL. November 1993 [online]. Available: http://www.osti.gov/bridge/servlets/purl/10119540-bvi5FP/webviewable/10119540.pdf [accessed Apr. 22, 2009].

Delucchi, M.A. 2004. Social cost of motor vehicle use. Pp. 65-75 in Encyclopedia of Energy, Vol. 4. San Diego: Academic Press.

Delucchi, M.A. 2006. Lifecycle Analyses of Biofuels, Draft Report. UCD-ITS-RR-06-08. Institute of Transportation Studies, University of California, Davis, CA. May 2006 [online]. Available: http://www.its.ucdavis.edu/publications/2006/UCD-ITS-RR-06-08.pdf [accessed Apr. 22, 2009].

Dennis, S.M. 2005. Improved estimates of ton-miles. J. Transp. Stat. 8(1):23-44.

Deschenes, O., and M. Greenstone. 2007. The economic impacts of climate change: Evidence from agricultural output and random fluctuations in weather. Am. Econ. Rev. 97(1):354-385.

Deutch, J., and J.R. Schlesinger. 2006. National Security Consequences of U.S. Oil Dependency. Independent Task Force Report No. 58. New York: Council on Foreign Relations [online]. Available: http://www.cfr.org/content/publications/attachments/EnergyTFR.pdf [accessed May 18, 2009].

Dietz, S., C.W. Hope, N.H. Stern, and D. Zenghelis. 2007. Reflections on the Stern Review (1) A Robust Case for Strong Action to Reduce the Risks of Climate Change. World Econ. 8(1):121-168.

Dinar, A., R. Mendelsohn, R. Evenson, J. Parikh, A. Sanghi, K. Kumar, J. McKinsey, and S. Lonergan. 1998. Measuring the Impact of Climate Change on Indian Agriculture. World Bank Technical Paper No. 402. The World Bank, Washington, DC.

Dincer, I. 1999. Environmental impacts of energy. Energ. Policy 27(14):845-854.

Dockery, D.W., C.A. Pope, X. Xu, J.D. Spengler, J.H. Ware, M.E. Fay, B.G. Ferris, and F.E. Speizer. 1993. An association between air pollution and mortality in six U.S. cities. N. Engl. J. Med. 329(24):1753-1759.

DOE (U.S. Department of Energy). 2004. PV FAQs: What is the Energy Payback for PV. DOE/GO-102004-2040. Office of Energy Efficiency and Renewable Energy, U.S. Department of Energy, Washington, DC [online]. Available: http://www1.eere.energy.gov/solar/pdfs/37322.pdf [accessed July 28, 2009].

DOE (U.S. Department of Energy). 2006a. The History of Nuclear Energy. DOE/NE-0088. U.S. Department of Energy, Office of Nuclear Energy, Science, and Technology, Washington, DC [online]. Available: http://www.nuclear.gov/pdfFiles/History.pdf [accessed Mar. 12, 2009].

DOE (U.S. Department of Energy). 2006b. The U.S. Generation IV Fast Reactor Strategy. DOE/NE-0130. U.S. Department of Energy, Office of Nuclear Energy. December 2006 [online]. Available: http://nuclear.gov/pdfFiles/genIvFastReactorRptToCongressDec2006.pdf [accessed Apr. 22, 2009].

DOE (U.S. Department of Energy). 2008a. 20% Wind Energy by 2030: Increasing Wind Energy's Contribution to U.S. Electricity Supply. Energy Efficiency and Renewable Energy, U.S. Department of Energy. July 2008 [online]. Available: http://www1.eere.energy.gov/windandhydro/pdfs/41869.pdf [accessed July 27, 2009].

DOE (U.S. Department of Energy). 2008b. Industrial Total Energy Consumption. Energy Intensity Indicators in the U.S. Planning, Budget and Analysis. U.S. Department of Energy, Energy Efficiency and Renewable Energy [online]. Available: http://www1.eere.energy.gov/ba/pba/intensityindicators/printable_versions/total_industrial.html [accessed May 28, 2009].

DOE (U.S. Department of Energy). 2009a. GEN IV Nuclear Energy Systems [online]. Available: http://www.ne.doe.gov/genIV/neGenIV1.html [accessed May 6, 2009].

DOE (U.S. Department of Energy). 2009b. Alternative and Advanced Vehicles: Natural Gas Vehicles. Alternative Fuels and Advanced Vehicles Data Center, U.S. Department of Energy, Energy Efficiency and Renewable Energy [online]. Available: http://www.afdc.energy.gov/afdc/vehicles/natural_gas.html [accessed May 26, 2009].

DOE (U.S. Department of Energy). 2009c. Building Technologies Program. U.S. Department of Energy, Energy Efficiency and Renewable Energy [online]. Available: http://www.eere.energy.gov/buildings/ [accessed May 26, 2009].

Dones, R., T. Heck, M. Faist Emmenegger, and N. Jungbluth. 2005. Life cycle inventories for the nuclear and natural gas energy systems, and examples of uncertainty analysis. Int. J. LCA 10(1):10-23.

Donner, S.D., and C.J. Kucharik. 2008. Corn-based ethanol production compromises goal of reducing nitrogen export by the Mississippi River. Proc. Natl. Acad. Sci. USA 105(11): 4513-4518.

DOT/DOC (U.S. Department of Transportation and U.S. Department of Commerce). 2004. 2002 Economic Census: Transportation-2002 Commodity Flow Survey. EC02TCF-US. U.S. Department of Transportation, Bureau of Transportation Statistics, U.S. Department of Commerce, U.S. Census Bureau. December 2004 [online]. Available: http://www. bts.gov/publications/commodity_flow_survey/2002/united_states/pdf/entire.pdf [accessed Apr. 22, 2009].

Downing, T.E., R.A.Greener, and N. Eyre. 1995. The Economic Impacts of Climate Change: Assessment of Fossil Fuel Cycles for the ExternE Project. Environmental Change Unit, University of Oxford, and Eyre Energy Environment, Lonsdale.

Downing, T.E., N. Eyre, R. Greener, and D. Blackwell. 1996a. Full Fuel Cycle Study: Evaluation of the Global Warming Externality for Fossil Fuel Cycles with and Without $CO_2$ Abatement and for Two References Scenarios. Report to the International Energy Agency Greenhouse Gas R&D Programme, by Environmental Change Unit, University of Oxford. February 12, 1996 [online]. Available: http://www.mi.uni-hamburg.de/fileadmin/fnu-files/staff/tol/RM4722.pdf [accessed June 2, 2009].

Downing, T.E., N. Eyre, R. Greener, and D. Blackwell. 1996b. Projected Costs of Climate Change for Two References Scenarios and Fossil Fuels Cycles. Report to the European Commission ExternE Project, by Environmental Change Unit, University of Oxford. July 25, 1996 [online]. Available: http://www.mi.uni-hamburg.de/fileadmin/fnu-files/staff/tol/RM3157.pdf [accessed June 4, 2009].

Dubin, J.A., and G.S. Rothwell. 1990. Subsidy to nuclear power through the Price-Anderson Liability Limit. Contemp. Econ. Policy 8(3):73-79.

EC (European Commission). 1995a. ExternE: Externalities of Energy: Luxembourg: Office for Official Publications of the European Communities [online]. Available: http://www.externe.info/ [accessed Apr. 23, 2009].

EC (European Commission). 1995b. ExternE: Externalities of Energy, Vol. 5. Nuclear. Luxembourg: Office for Official Publications of the European Communities [online]. Available: http://www.externe.info/ [accessed May 12, 2009].

EC (European Commission). 2003. External Costs: Research Results on Socio-Environmental Damages Due to Electricity and Transport. EUR 20198. Luxembourg: Office for Official Publications of the European Communities [online]. Available: http://ec.europa.eu/research/energy/pdf/externe_en.pdf [accessed May 6, 2009].

EC (European Commission). 2005. ExterneE-Externalities of Energy: Methodology 2005 Update, P. Bickel and R. Friedrich, eds. EUR21951. Luxembourg: Office for Official Publications of the European Communities [online]. Available: http://www.externe.info/ [accessed Oct. 8, 2009].

EC (European Commission). 2008. ExternE-Externalities of Energy. A Research Project of the European Commission [online]. Available: http://www.externe.info/ [accessed May 26, 2009].

EC-JRC (European Commission Joint Research Center). 2008. Well-to-Wheels Analysis of Future Automotive Fuels and Powertrains in the European Context. WELL-TO-TANK Report Version 3.0. European Commission Joint Research Center. November 2008 [online]. Available: http://ies.jrc.ec.europa.eu/WTW [accessed Apr. 22, 2009].

EEAI (Energy and Environmental Analysis, Inc.). 2005. Characterization of the U.S. Industrial Commercial Boiler Population. Prepared for Oak Ridge National Laboratory, by Energy and Environmental Analysis, Inc., Arlington, VA. May 2005 [online]. Available: http://www.cibo.org/pubs/industrialboilerpopulationanalysis.pdf [accessed Apr. 22, 2009].

EIA (Energy Information Administration). 2003. Energy Efficiency Measurement Discussion. Energy Information Administration, U.S. Department of Energy [online]. Available: http://

www.eia.doe.gov/emeu/efficiency/measure_discussion.htm#Market-Basket%20Approach [accessed Oct. 9, 2009].

EIA (Energy Information Administration). 2004a. EIA-767 Data Files, Annual Stream-Electric Plant Operation and Design Data. Electricity Database Files. Energy Information Administration, U.S. Department of Energy [online]. Available: http://www.eia.doe.gov/cneaf/electricity/page/eia767.html [accessed Dec. 27, 2005] (as cited in NRC 2006a).

EIA (Energy Information Administration). 2004b. EIA-860 Database, Annual Electric Generator Report. Electricity Database Files. Energy Information Administration, U.S. Department of Energy [online]. Available: http://www.eia.doe.gov/cneaf/electricity/page/eia860.html [accessed Dec. 27, 2005] (as cited in NRC 2006a).

EIA (Energy Information Administration). 2006a. Annual Coal Report 2005. DOE/EIA-0584(2005). Energy Information Administration [online].Available: http://www.eia.doe.gov/cneaf/coal/page/acr/acr_sum.html [accessed March 2007} (as cited in NRC 2007a).

EIA (Energy Information Administration). 2006b. Renewable Energy Annual 2004 with Preliminary Data for 2004. Energy Information Administration, Office of Nuclear, Electric and Alternative Fuels, U.S. Department of Energy. June 2006 [online]. Available: http://tonto.eia.doe.gov/ftproot/renewables/060304.pdf [accessed May 22, 2009].

EIA (Energy Information Administration). 2006c. Energy Use in Manufacturing 1998-2002: Energy Consumption by the Manufacturing Sector-Cost of Energy. Energy Information Administration [online]. Available: http://www.eia.doe.gov/emeu/mecs/special_topics/energy_use_manufacturing/energyuse98_02/energy_cons.html#ce [accessed May 29, 2009].

EIA (Energy Information Administration). 2007. Manufacturing Energy Consumption Survey (MECS): Manufacturing and Industrial Energy Uses and Costs: 2002 Energy Consumption by Manufacturers-Data Tables [online]. Available: http://www.eia.doe.gov/emeu/mecs/mecs2002/data02/shelltables.html [accessed May 29, 2009].

EIA (Energy Information Administration). 2008a. Annual Energy Review 2007. Report No. DOE/EIA-0384(2007). Energy Information Administration, June 2008 [online]. Available: http://tonto.eia.doe.gov/FTPROOT/multifuel/038407.pdf [accessed May 1, 2009].

EIA (Energy Information Administration). 2008b. Annual Energy Outlook 2008. DOE/EIA-0383(2008). Energy Information Administration, U.S. Department of Energy, Washington, DC [onbline]. Available: http://www.eia.doe.gov/oiaf/aeo/pdf/0383(2008).pdf [accessed May 20, 2009].

EIA (Energy Information Administration). 2008c. Domestic Uranium Production Report (2003-2007). Energy Information Administration [online]. Available: http://www.eia.doe.gov/cneaf/nuclear/dupr/dupr.html [accessed May 20, 2009].

EIA (Energy Information Administration). 2008d. Petroleum Products: Supply. Energy Information Sheets. Energy Information Administration. September 2008 [online]. Available: http://www.eia.doe.gov/neic/infosheets/petroleumproducts.html [accessed: Mar. 18, 2009].

EIA (Energy Information Administration). 2008e. 2003 CBECS (Commercial Buildings Energy Consumption Survey) Detailed Tables. Energy Information Administration [online]. Available: http://www.eia.doe.gov/emeu/cbecs/cbecs2003/detailed_tables_2003/detailed_tables_2003.html#consumexpen03 {accessed May 28, 2009].

EIA (Energy Information Administration). 2008f. Emission of Greenhouse Gases in the United States 2007. DOE/EIA-573(2007). Energy Information Administration, U.S. Department of Energy, Washington, DC. December 2008 [online]. Available: http://www.eia.doe.gov/oiaf/1605/ggrpt/pdf/0573(2007).pdf [accessed July 31, 2009].

EIA (Energy Information Administration). 2008g. International Energy Outlook 2008. DOE/EIA-0484(2008), Energy Information Administration. September 2008 [online]. Available: http://www.eia.doe.gov/oiaf/ieo/pdf/0484(2008).pdf [accessed Aug. 3, 2009].

EIA (Energy Information Administration). 2009a. Electric Power Monthly, July 2009 Edition: with Data for April 2009. DOE/EIA-0226(2009/7. Energy Information Administration [online]. Available: http://www.eia.doe.gov/cneaf/electricity/epm/epm.pdf [accessed July 27, 2009].

EIA (Energy Information Administration). 2009b. Coal Basics 101. Energy Information Administration [online]. Available: http://www.eia.doe.gov/basics/coal_basics.html [accessed Apr. 22, 2009].

EIA (Energy Information Administration). 2009c. Annual Coal Report: 2007. DOE/EIA 0584 (2007). Energy Information Administration, Office of Coal, Nuclear, Electric, and Alternative Fuels, U.S. Department of Energy, Washington, DC. February 2009 [online]. Available: http://www.eia.doe.gov/cneaf/coal/page/acr/acr.pdf [accessed Apr. 22, 2009].

EIA (Energy Information Administration). 2009d. Electric Power Annual 2007. DOE/EIA-348(2007). Office of Coal, Nuclear, Electric and Alternative Fuels, Energy Information Administration, U.S. Department of Energy, Washington, DC. January 2009 [online]. Available: http://www.eia.doe.gov/cneaf/electricity/epa/epa.pdf [accessed July 27, 2009].

EIA (Energy Information Administration). 2009e. Annual Energy Outlook 2009, With Projections to 2030. DOE/EIA-0383(2009). Energy Information Administration, Office of Integrated Analysis and Forecasting, U.S. Department of Energy, Washington, DC. March 2009 [online]. Available: http://www.eia.doe.gov/oiaf/aeo/pdf/0383(2009).pdf [accessed Apr. 22, 2009].

EIA (Energy Information Administration). 2009f. Supplemental Tables to the Annual Energy Outlook 2009, With Projections to 2030. Energy Information Administration, Office of Integrated Analysis and Forecasting, U.S. Department of Energy, Washington, DC. March 2009 [online]. Available: http://www.eia.doe.gov/oiaf/aeo/supplement/ [accessed Oct. 7, 2009].

EIA (Energy Information Administration). 2009g. U.S. Overview, State Energy Profiles. Energy Information Administration [online]. Available: http://tonto.eia.doe.gov/state/ [accessed Apr. 17, 2009].

EIA (Energy Information Administration). 2009h. Petroleum Basics 101. Energy Information Administration [online]. Available: http://www.eia.doe.gov/basics/petroleum_basics.html [accessed Mar. 18, 2009].

EIA (Energy Information Administration). 2009i. 2005 Residential Energy Consumption Survey-Detailed Tables. Energy Information Administration [online]. Available: http://www.eia.doe.gov/emeu/recs/recs2005/c&e/detailed_tables2005c&e.html [accessed May 29, 2009].

EIA (Energy Information Administration). 2009j. Manufacturing Energy Consumption Survey (MECS): Manufacturing and Industrial Energy Uses and Costs [online]. Available: http://www.eia.doe.gov/emeu/mecs/ [accessed May 27, 2009].

EIA (Energy Information Administration). 2009k. Electric Power Monthly, April 2009 Edition: with Data for January 2009. DOE/EIA-0226(2009/04). Energy Information Administration [online]. Available: http://tonto.eia.doe.gov/ftproot/electricity/epm/02260904.pdf [accessed Aug. 3, 2009].

EIA (Energy Information Administration). 2009l. Independent Statistics and Analysis. Natural Gas. Imports/Exports & Pipelines. Energy Information Administration [online]. Available: http://www.eia.doe.gov/oil_gas/natural_gas/info_glance/natural_gas.html [accessed Apr. 14, 2010].

EPA (U.S. Environmental Protection Agency). 1995. Pp 1.1-5 in Compilation of Air Pollutant Emission Factors. Vol. 1. Stationary Point and Area Sources, 5th Ed. AP-42. Office of Air Quality Planning and Standards, Office of Air and Radiation, U.S. Environmental Protection Agency, Research Triangle Park, NC. January 1995 [online]. Available: http://www.epa.gov/ttn/chief/old/ap42/5th_edition/ap42_5thed_orig.pdf [accessed Apr. 23, 2009].

EPA (U.S. Environmental Protection Agency). 1999. Technical Background Document for the Report to Congress on Remaining Wastes from Fossil Fuel Combustion: Industry Statistics and Waste Management Practices. Supporting Materials, Fossil Fuel Combustion Waste, U.S. Environmental Protection Agency. March 15, 1999 [online]. Available: http://www.epa.gov/osw/nonhaz/industrial/special/fossil/ffc2_398.pdf [accessed Apr. 23, 2009].

EPA (U.S. Environmental Protection Agency). 2002. A Survey of the Water Quality of Streams in the Primary Region of Mountaintop/Valley Fill Coal Mining: October 1999-January 2001. Final Report. U.S. Environmental Protection Agency Region 3, Wheeling, WV. April 8, 2002 [online]. Available: http://www.epa.gov/region3/mtntop/pdf/appendices/d/stream-chemistry/MTMVFChemistryPart1.pdf [accessed Oct. 12, 2009].

EPA (U.S. Environmental Protection Agency). 2004a. Continuous Emission Monitoring System. Program and Regulation, U.S. Environmental Protection Agency. [online]. Available: http://www.epa.gov/airmarkets/monitoring/index.html [accessed Nov. 17, 2004] (as cited in NRC 2006a).

EPA (U.S. Environmental Protection Agency). 2004b. Documentation Summary for EPA Base Case 2004 (V.2.1.9) Using the Integrated Planning Model. EPA 430/R-04-008. Office of Air and Radiation, U.S. Environmental Protection Agency, Washington, DC. October 2004 [online]. Available: http://www.epa.gov/airmarkets/progsregs/epa-ipm/docs/basecase2004.pdf [accessed Apr. 20, 2010].

EPA (U.S. Environmental Protection Agency). 2005a. Mountaintop Mining/ Valley Fills in Appalachia Final Programmatic Environmental Impact Statement. EPA 9-03-R-05002. U.S. Environmental Protection Agency, Region 3, Philadelphia, PA [online]. Available: http://www.epa.gov/region03/mtntop/pdf/mtm-vf_fpeis_full-document.pdf [accessed May 3, 2009].

EPA (U.S. Environmental Protection Agency). 2005b. Regulatory Impact Analysis for the Final Clean Air Interstate Rule. EPA-452/R-05-002. Office of Air and Radiation, U.S. Environmental Protection Agency [online]. Available: http://www.epa.gov/cair/pdfs/finaltech08.pdf [accessed Sept. 10, 2009].

EPA (U.S. Environmental Protection Agency). 2005c. CMAQ Model Performance Evaluation for 2001: Updated March 2005. Office of Air Quality Planning and Standards, U.S. Environmental Protection Agency, Research Triangle Park, NC [online]. Available: http://www.epa.gov/scram001/reports/cair_final_cmaq_model_performance_evaluation_2149.pdf [accessed Sept. 18, 2009].

EPA (U.S. Environmental Protection Agency). 2008a. Landfills. Wastes-Non-Hazardous Waste-Municipal Solid Waste, U.S. Environmental Protection Agency [online]. Available: http://www.epa.gov/epawaste/nonhaz/municipal/landfill.htm [accessed Mar. 22, 2009].

EPA (U.S. Environmental Protection Agency). 2008b. FY 2008 End of Year Activity Report. Memorandum to UST/LUST Regional Division Directors, Regions 1-10, from Cliff Rothenstein, Director of Underground Storage Tanks, Office of Solid Waste and Emergency Response, U.S. Environmental Protection Agency, Washington, DC. November 20, 2008 [online]. Available: http://www.epa.gov/OUST/cat/ca_08_34.pdf [accessed May 12, 2009].

EPA (U.S. Environmental Protection Agency). 2008c. Lead Emissions from the Use of Leaded Aviation Gasoline in the United States. Technical Support Document. EPA420-R-08-020. Assessment and Standards Division, Office of Transportation and Air Quality, U.S. Environmental Protection Agency. October 2008 [online]. Available: http://www.epa.gov/OMS/regs/nonroad/aviation/420r08020.pdf [accessed May 12, 2009].

EPA (U.S. Environmental Protection Agency). 2008d. 2002 National Emission Inventory Data & Documentation. Technology Transfer Network Clearinghouse for Inventories and Emissions Factors, U.S. Environmental Protection Agency [online]. Available: http://www.epa.gov/ttnchie1/net/2002inventory.html [accessed May 29, 2009].

EPA (U.S. Environmental Protection Agency). 2009a. Mid-Atlantic Mountaintop Mining: Environmental Impacts. Region 3, U.S. Environmental Protection Agency [online]. Available: http://www.epa.gov/Region3/mtntop/#impacts [accessed Mar. 14, 2009].

EPA (U.S. Environmental Protection Agency). 2009b. Clean Air Mercury Rule. Office of Air and Radiation, U.S. Environmental Protection Agency [online]. Available: http://www.epa.gov/oar/mercuryrule/ [accessed July 27, 2009].

EPA (U.S. Environmental Protection Agency). 2009c. Environmental and Health Information. Wastes-Partnerships-Coal Combustion, Products Partnership, U.S. Environmental Protection Agency [online]. Available: http://www.epa.gov/epawaste/partnerships/c2p2/use/effects.htm [accessed Mar. 22, 2009].

EPA (U.S. Environmental Protection Agency). 2009d. Coal Combustion Residues (CCR)—Surface Impoundments with High Hazard Potential Ratings, Fact Sheet. EPA530-F-09-006. Office of Wastes, U.S. Environmental Protection Agency, Washington, DC. June 2009 (updated August 2009). [online]. Available: http://www.epa.gov/osw/nonhaz/industrial/special/fossil/ccrs-fs/index.htm [accessed Sept. 2, 2009].

EPA (U.S. Environmental Protection Agency). 2009e. Uranium Mining Waste. Radiation Protection Programs. U.S. Environmental Protection Agency [online]. Available: http://www.epa.gov/radiation/tenorm/uranium.html [accessed May 21, 2009].

EPA (U.S. Environmental Protection Agency). 2009f. The Green Book Nonattainment Areas for Criteria Pollutants. U.S. Environmental Protection Agency [online]. Available: http://www.epa.gov/oar/oaqps/greenbk/ [accessed Oct. 7, 2009].

EPA (U.S. Environmental Protection Agency). 2009g. FY 2008 Annual Report on the Underground Storage Tank Program. EPA-510-R-09-001. Office of Solid Waste and Emergency Response, office of Underground Storage Tanks, U.S. Environmental Protection Agency, Washington, DC. March 2009 [online]. Available: http://www.epa.gov/OUST/pubs/OUST_FY08_Annual_Report-_Final_3-19-09.pdf [accessed May 12, 2009].

EPA (U.S. Environmental Protection Agency). 2009h. EPA Lifecycle Analysis of Greenhouse Gas Emissions from Renewable Fuels. EPA-420-F-09-024. Office of Transportation and Air Quality, U.S. Environmental Protection Agency. May 2009 [online]. Available: http://www.epa.gov/oms/renewablefuels/420f09024.pdf [accessed July 29, 2009].

EPA (U.S. Environmental Protection Agency). 2009i. U.S. Greenhouse Gas Inventory Report, U.S. Environmental Protection Agency [online]. Available: http//epa.gov/climatechange/emissions/usinventoryreport09.html [Accessed Apr 12, 2010]

EPASAB (U.S. Environmental Protection Agency Science Advisory Board). 2007. SAB Advisory on EPA's Issues in Valuing Mortality Risk Reduction. EPA-SAB-08-001. Memorandum to Stephen L. Johnson, Administrator, U.S. Environmental Protection Agency, from M. Granger Morgan, Chair, EPA Science Advisory Board, and Maureen Cropper, Chair, Environmental Economics Advisory Committee, U.S. Environmental Protection Agency, Washington, DC. October 12, 2007 [online]. Available: http://yosemite.epa.gov/sab/sabproduct.nsf/4128007E7876B8F0852573760058A978/$File/sab-08-001.pdf [accessed Sept. 16, 2009].

EPASAB (U.S. Environmental Protection Agency Science Advisory Board). 2009. Valuing the Protection of Ecological Systems and Services. EPA-SAB-09-012. U.S. Environmental Protection Agency Science Advisory Board [online]. Available: http://yosemite.epa.gov/sab/sabproduct.nsf/WebBOARD/5A0F054E9847A796852575860062F1DB?OpenDocument [accessed May 6, 2009].

EPRI (Electric Power Research Institute). 2007. Program on Technology Innovation: Room at the Mountain. Report No. 1015046. EPRI, Palo Alto, CA. June 29, 2007.

Etkin, D.S. 2001. Analysis of oil spill trends in the United States and worldwide. Pp. 1291-1300 in Global Strategies for Prevention, Preparedness, Response, and Restoration:

Proceedings of the 2001 International Oil Spill Conference, March 26-29, 2001, Tampa, FL. API Publication 4710 B. Washington, DC: API [online]. Available: http://www.environmental-research.com/publications/pdf/spill_statistics/paper4.pdf, [accessed May 7, 2009]

Fankhauser, S. 1995a. Valuing Climate Change: The Economics of the Greenhouse. London: Earthscan.

Fankhauser, S. 1995b. Protection vs. retreat: The economic costs of sea level rise. Environ. Plann. A 27(2):299-319.

Fann, N., C.M. Fulcher, and B.J. Hubbell. 2009. The influence of location, source, and emission type in estimates of the human health benefits of reducing a ton of air pollution. Air Qual. Atmos. Health 2(3):169-176.

Finck, P.J. 2005. Statement of Dr. Phillip J. Finck, Deputy Associate Laboratory Director for Applied Science and Technology and National Security, Argonne National Laboratory, Before House Committee on Science, Energy Subcommittee, Hearing on Nuclear Fuel Reprocessing, June 16, 2005 [online]. Available: http://www.anl.gov/Media_Center/News/2005/testimony050616.html [accessed Apr. 23, 2009].

FRA (Federal Railroad Administration). 2008. Railroad Safety Statistics 2007 Preliminary Annual Report. U.S. Department of Transportation, Federal Railroad Administration [online]. Available: http://safetydata.fra.dot.gov/OfficeofSafety/publicsite/Prelim.aspx [accessed May 21, 2009].

Fthenakis, V., and E. Alsema. 2006. Photovoltaics energy payback times, greenhouse gas emissions and external costs: 2004-early 2005 status. Prog. Photovolt. Res. Appl. 14(3): 275-280.

Fthenakis, V.M., and H.C. Kim. 2007. Greenhouse-gas emissions from solar electric- and nuclear power: A life-cycle study. Energ. Policy 35(4):2549-2557.

Fthenakis, V.M., H.C. Kim, and E. Alsema. 2008. Emissions from photovoltaic life cycles. Environ. Sci. Technol. 42(6):2168-2174.

Fullerton, D., I. Hong, and G.E. Metcalf. 2001. A tax on output of the polluting industry is not a tax on pollution: The importance of hitting the target. Pp. 13-43 in Behavioral and Distributional Effects of Environmental Policy. National Bureau of Economic Research, Inc. [online]. Available: http://www.nber.org/chapters/c10604.pdf [accessed Aug. 10, 2009].

FUND (Climate Framework for Uncertainty, Negotiation and Distribution). 2008. FUND Model, Version 3.3 [online]. Available: http://www.mi.uni-hamburg.de/fileadmin/fnu-files/staff/tol/FundTechnicalTables.pdf [accessed June 2, 2009].

GAO (U.S. Government Accountability Office). 2006. Nuclear Power Plants: Efforts Made to Upgrade Security, but the Nuclear Regulatory Commission's Design Basis Threat Process Should be Improved. GAO-06-388. Washington, DC: U.S. Government Accountability Office. March 2006 [online]. Available: http://www.gao.gov/new.items/d06388.pdf [accessed Aug. 10, 2009].

GAO (Government Accounting Office). 2007. Maritime Transportation: Major Oil Spills Occur Infrequently, but Risks to the Oil Spill Fund Remain. GAO-07-1085. Washington, DC: GAO. September 2007 [online]. Available: http://www.gao.gov/new.items/d071085.pdf [accessed June 10, 2009].

Garza-Gil, M., A. Prada-Blanco, and M. Vazquez-Rodriguez. 2006. Estimating the short-term economic damages from the prestige oil spill in the Galician fisheries and tourism. Ecol. Econ. 58(4):842-849.

Gassman, P.W. 2008. A Simulation Assessment of the Boone River Watershed: Baseline Calibration/Validation Results and Issues, and Future Research Needs. Ph.D. Thesis, Iowa State University, Ames, IA.

Greco, S.L., A.M. Wilson, J.D. Spengler, and J.I. Levy. 2007. Spatial patterns of mobile source particulate matter emissions-to-exposure relationships across the United States. Atmos. Environ. 41(5):1011-1025.

Greene, D.L. 2009. Measuring energy security: Can the United States achieve oil independence? Energy Policy 38(4):1614-1621.

Greene, D.L., and P.N. Leiby. 2006. The Oil Security Metrics Model: A Tool for Evaluating the Perspective Oil Security Benefits of DOE's Energy Efficiency and Renewable Energy R&D Programs. ORNL/TM-2006/505. Oak Ridge National Laboratory, Oak Ridge, TN. May 2006 [online]. Available: https://apps3.eere.energy.gov/ba/pba/analysis_database/docs/pdf/Oil%20Metrics_FINAL.pdf [accessed May 18, 2009].

Greene, D.L., D.W. Jones, and M.A. Delucchi, eds. 1997. Measuring the Full Social Costs and Benefits of Transportation. Heidelberg, Germany: Springer.

GTCC LLRW EIS (Greater-Than-Class C Low-Level Radioactive Waste Environmental Impact Statement). 2009. Guide to Greater-Than-Class C Low-Level Radioactive Waste. Information Center [online]. Available: http://www.gtcceis.anl.gov/guide/index.cfm [accessed Apr. 27, 2009].

Guinée, J., and R. Heijungs. 1993. A proposal for the classification of toxic substances within the framework of life cycle assessment of products. Chemosphere 26(10):1925-1944.

Gullett, B.K., A. Touati, and M.D. Hays. 2003. PCDD/F, PCB, HxCBz, PAH, and PM emission factors for fireplace and woodstove combustion in the San Francisco Bay area. Environ. Sci. Technol. 37(9):1758-1765.

Hamilton, J. 2009. Causes and consequences of the oil shock of 2007-08. Brookings Pap. Eco. Ac. Spring 2009 [online]. Available: http://www.brookings.edu/economics/bpea/~/media/Files/Programs/ES/BPEA/2009_spring_bpea_papers/2009_spring_bpea_hamilton.pdf [accessed May 18, 2009].

Hamilton, J.D., and A.M. Herrera. 2004. Oil shocks and aggregate macroeconomic behavior: The role of monetary policy. J. Money Credit and Bank. 36(2):265-286.

Hansen, L., and M. Ribaudo. 2008. Economic Measures of Soil Conservation Benefits: Regional Values for Policy Assessment. Technical Bulletin No. 1922. Economic Research Service, U.S. Department of Agriculture. September 2008 [online]. Available: http://www.ers.usda.gov/Publications/TB1922/TB1922.pdf [accessed July 29, 2009].

Harte, J., A. Ostling, J.L. Green, and A. Kinzig. 2004. Biodiversity conservation: Climate change and extinction risk. Nature 430(6995):33.

Harvey, L.D.D., and Z. Huang. 1995. Evaluation of the potential impact of methane clathrate destabilization on future global warming. J. Geophys. Res. 100(D2):2905-2926.

Heggenstaller, A.H., K.J. Moore, M. Liebman, and R.P. Anex. 2009. Nitrogen influences productivity and resource partitioning by perennial, warm-season grasses. Agron. J. 101(6):1363-1371.

Heimbuch, D.G., E. Lorda, D. Vaughan, L.W. Barnthouse, J. Uphoff, W. Van Winkle, A. Kahnle, B. Young, J. Young, and L. Kline. 2007. Assessing coastwide effects of power plant entrainment and impingement on fish populations: Atlantic menhaden example. N. Am. J. Fish. Manage 27(2):569-577.

Hertwich, E.G., S.F. Mateles, W.S. Pease, and T.E. McKone. 2001. Human toxicity potentials for life cycle analysis and toxics release inventory risk screening. Environ. Toxicol. Chem. 20(4):928-939.

Heyes, A.G., and C. Liston-Heyes. 1998. Subsidy to nuclear power through Price-Anderson liability limit: Comment. Contemp. Econ. Policy 16(1):122-124.

Hill, J., S. Polasky, E. Nelson, D. Tilman, H. Huo, L. Ludwig, J. Neumann, H. Zheng, and D. Bonta. 2009. Climate change and health cost of air emissions from biofuels and gasoline. Proc. Natl. Acad. Sci. 106(6):2077-2082.

Hitz, S., and J. Smith. 2004. Estimating global impacts from climate change. Global Environ. Chang. A 14(3):201-218.

Hodgson, D., and K. Miller. 1995. Modelling UK energy demand. In Global Warming and Energy Demand, T. Barker, P. Ekins, and N. Johnstone, eds. London: Routledge (as cited in Downing et al. 1996b)

Hofstetter, P. 1998. Perspectives in Life Cycle Impact Assessment: A Structured Approach to Combine Models of the Technosphere, Ecosphere and Valuesphere. Boston: Kluwer.

Hogan, W.W. 2008. Electricity Market Design: Coordination, Pricing and Incentives. ERCOT Energized Conference, Austin, TX, May 2, 2008. [online]. Available: http://ksghome.harvard.edu/~whogan/Hogan_Ercot_050208.pdf [accessed Aug. 6, 2009].

Hohmeyer, O. 1988. Social Costs of Energy Consumption: External Effects of Electricity Generation in the Federal Republic of Germany. New York: Springer.

Homan, A.C. and T. Steiner. 2008. OPA 90's impact at reducing oil spills. Mar. Policy 32(4):711-728.

Hope, C. 2006. The marginal impact of CO2 from PAGE2002: An integrated assessment model incorporating the IPCC's five reasons for concern. Integr. Assess. J. 6(1):19-56.

Hope, C., and D. Newbery. 2008. Calculating the social cost of carbon. Pp. 31-63 in Delivering a Low Carbon Electricity System, M. Grubb, T. Jamasb, and M.G. Pollitt, eds. Cambridge: Cambridge University Press.

Horvath, A., C.T. Hendrickson, L.B. Lave, F.C. McMichael, and T.S. Wu. 1995. Toxic emissions indices for green design and inventory. Environ. Sci. Technol.29(2):86-90.

Huijer, K. 2005. Trends in Oil Spills from Tanker Ships 1995-2004. Presented at the 28th Arctic and Marine Oil Spill Program (AMOP) Technical Seminar, 7-9 June 2005, Calgary, Canada [online]. Available: http://www.itopf.com/_assets/documents/amop05.pdf [accessed Oct. 9, 2009].

Hung, M.F., and D. Shaw. 2005. A trading-ratio system for trading water pollution discharge permits. J. Environ. Econ. Manage. 49(1):83-102.

IAEA (International Atomic Energy Agency). 1999. Health and Environmental Impacts of Electricity Generation Systems: Procedures for Comparative Assessment. Technical Report No. 394. Vienna, Austria: IAEA [online]. Available: http://www-pub.iaea.org/MTCD/publications/PDF/TRS394_scr.pdf [accessed May 7, 2009].

IAEA (International Atomic Energy Agency). 2008. Uranium Report: Plenty More Where That Come From: Supply Sufficient for Next Century Amid Robust Demand Growth. IAEA News Centre, June 3, 2008 [online]. Available: http://www.iaea.org/NewsCenter/News/2008/uraniumreport.html [accessed May 20, 2008].

IARC (International Agency for Research on Cancer). 2008. World Cancer Report 2008, P. Boyle and B. Levin, eds. Lyon: IARC [online]. Available: http://www.iarc.fr/en/publications/pdfs-online/wcr/2008/index.php [accessed July 28, 2009].

IEA (International Energy Agency). 2007. Tracking Industrial Energy Efficiency and CO2 Emission. Paris, France: OECD/IEA [online]. Available: http://www.iea.org/textbase/nppdf/free/2007/tracking_emissions.pdf [accessed Apr. 24, 2009].

Interlaboratory Working Group. 2000. Scenarios for a Clean Energy Future. ORNL/CON-476 and LBNL-44029. Prepared for Office of Energy Efficiency and Renewable Energy, by Interlaboratory Working Group on Energy-Efficient and Clean Energy Technologies. Oak Ridge National Laboratory, Oak Ridge, TN, and Lawrence Berkeley National Laboratory, Berkeley, CA [online]. Available: http://www.nrel.gov/docs/fy01osti/29379.pdf [accessed Apr. 24, 2009].

IOM (Institute of Medicine). 2006. Valuing Health for Regulatory Cost-Effectiveness Analysis, W. Miller, L.A. Robinson, and R.S. Lawrence, eds. Washington, DC: The National Academies Press.

IPCC (Intergovernmental Panel on Climate Change). 2000. IPCC Special Report: Emission Scenarios: Summary for Policymakers. A Special Report of IPCC Working Group III [online]. Available: http://www.ipcc.ch/pdf/special-reports/spm/sres-en.pdf [accessed Apr. 22, 2009].

IPCC (Intergovernmental Panel on Climate Change). 2001. Climate Change 2001: Impacts, Adaptation and Vulnerability. Contribution of Working Group II to the Third Assessment

Report of the Intergovernmental Panel on Climate Change, J.J. McCarthy, O.F. Canziani, N.A. Leary, D.J. Dokken and K.S. White, eds. Cambridge: Cambridge University Press.

IPCC (Intergovernmental Panel on Climate Change). 2005. IPCC Special Report on Carbon Dioxide Capture and Storage. Prepared by Working Group III of the Intergovernmental Panel on Climate Change, B. Metz, O. Davidson, H. de Coninck, M. Loos, and L. Meyer, eds. New York: Cambridge University Press [online]. Available: http://arch.rivm.nl/env/int/ipcc/pages_media/SRCCS-final/SRCCS_WholeReport.pdf [accessed Apr. 24, 2009].

IPCC (Intergovernmental Panel on Climate Change). 2007a. Climate Change 2007: Synthesis Report—Summary for Policy Makers [online]. Available: http://www.ipcc.ch/pdf/assessment-report/ar4/syr/ar4_syr_spm.pdf [accessed Apr. 24, 2009].

IPCC (Intergovernmental Panel on Climate Change). 2007b. Climate Change 2007: The Physical Science Basis. Contribution of Working Group I to the Fourth Assessment Report of the Intergovernmental Panel on Climate Change, S. Solomon, D. Qin, M. Manning, M. Marquis, K. Averyt, M.M.B. Tingor, and H.L. Miller, and Z. Chen, eds. Cambridge: Cambridge University Press [online]. Available: http://www.ipcc.ch/ipccreports/ar4-wg1.htm [accessed Apr. 24, 2009].

IPCC (Intergovernmental Panel on Climate Change). 2007c. Climate Change 2007: Mitigation. Contribution of Working Group III to the Fourth Assessment Report of the Intergovernmental Panel on Climate Change, B. Metz, O. Davidson, P. Bosch, R. Dave, and L. Meyer, eds. Cambridge: Cambridge University Press [online]. Available: http://www.ipcc.ch/ipccreports/ar4-wg3.htm [accessed June 5, 2009].

IPCC (Intergovernmental Panel on Climate Change). 2007d. Climate Change 2007: Impact, Adaptation and Vulnerability. Contribution of Working Group II to the Fourth Assessment Report of the Intergovernmental Panel on Climate Change, M. Parry, O. Canziani, J. Palutikof, P. van der Linden, C. Hanson, eds. Cambridge: Cambridge University Press [online]. Available: http://www.ipcc.ch/ipccreports/ar4-wg2.htm [accessed June 1, 2009].

Izaurralde, R.C., J.R. Williams, W.B. McGill, N.J. Rosenberg, and M.C.Q. Jakas. 2006. Simulating soil C dynamics with EPIC: Model description and testing against long-term data. Ecol. Model. 192(3-4):362-384.

Jacobson, M.Z., and G.M. Masters. 2001. Exploiting wind versus coal. Science 293(5534): 1438.

Jaramillo, P., W.M. Griffin, and H.S. Matthews. 2007. Comparative life cycle air emissions of coal, domestic natural gas, LNG, and SNG for electricity generation. Environ. Sci. Technol. 41(17):6290-6296.

Jolliet, O., R. Müller-Wenk, J. Bare, A. Brent, M. Goedkoop, R. Heijungs, N. Itsubo, C. Pena, D. Pennington, J. Potting, G. Rebitzer, M. Stewart, H.U. de Haes, and B. Weidema. 2004. The LCIA midpoint-damage framework of the UNEP/SETAC life cycle initiative. Int. J. LCA 9(6):394-404.

Jones, J.A., J. Ehrhardt, L.H.J. Goossens, J. Brown, R.M. Cooke, F. Fischer, I. Hasemann, and B.C.P. Kraan. 2001. Probabilistic Accident Consequence Uncertainty Assessment Using COSYMA: Overall Uncertainty Analysis. EUR-18826. FZKA-6312. European Communities [online]. Available: ftp://ftp.cordis.europa.eu/pub/fp5-euratom/docs/eur18826_en.pdf [accessed June 10, 2009].

Karl, T.R., J.M. Melillo, and T.C. Peterson, eds. 2009. Global Climate Change Impacts in the United States: State Knowledge Report from the U.S. Global Change Research Program. Cambridge: Cambridge University Press [online]. Available: http://downloads.globalchange.gov/usimpacts/pdfs/climate-impacts-report.pdf [accessed Aug. 3, 2009].

Katzenstein, W., and J. Apt. 2009. Air emissions due to wind and solar power. Environ. Sci. Technol. 43(2):253-258.

Keller, K., B.M. Bolker, and D.F. Bradford. 2004. Uncertain climate thresholds and optimal economic growth. J. Environ. Econ. Manag. 48(1):723-741.

Keeney, R.L., and H. Raiffa. 1993. Decisions with Multiple Objectives: Preferences and Value Tradeoffs. Cambridge: Cambridge University Press.

Kintner-Meyer, M., K. Schneider, and R. Pratt. 2007. Impacts Assessment of Plug-in Hybrid Vehicles on Electric Utilities and Regional US Power Grids, Part 1. Technical Analysis Journal of EUEC Vol. 1, Paper No. 04 [online]. Available: http://www.euec.com/journal/documents/pdf/Paper_4.pdf [accessed Apr. 27, 2009].

Kjellstrom, T., R.S. Kovats, S.J. Lloyd, T. Holt, and R.S.J. Tol. 2008. The Direct Impact of Climate Change on Regional Labour Productivity. Working Paper No. 260. The Economic and Social Research Institute, Dublin. October 2008 [online]. Available: http://www.tara.tcd.ie/bitstream/2262/23779/1/WP260.pdf [accessed Oct. 12, 2009].

Knowlton, K., J.E. Rosenthal, C. Hogrefe, B. Lynn, S. Gaffin, R. Goldberg, C. Rosenzweig, K. Civerolo, J.Y. Ku, and P.L. Kinney. 2004. Assessing ozone-related health impacts under a changing climate. Environ. Health Perspect. 112(15):1557-1563.

Koch, F.H. 2000. Hydropower: Internalised Costs and Externalized Benefits. International Energy Agency (IEA), Ottawa, Canada.

Kochi, I., B. Hubbell, and R. Kramer. 2006. An empirical Bayes approach to combining and comparing estimates of the value of a statistical life for environmental policy analysis. Environ. Resour. Econ. 34(3):385-406.

Kruize, H., O. Hänninen, O. Breugelmans, E. Lebret, and M. Jantunen. 2003. Description and demonstration of the EXPOLIS simulation model: Two examples of modeling population exposure to particulate matter. J. Expo. Anal. Environ. Epidemiol. 13(2):87-99.

Kundzewicz, Z.W., L.J. Mata, N.W. Arnell, P. Döll, P. Kabat, B. Jiménez, K.A. Miller, T. Oki, Z. Sen, and I.A. Shiklomanov. 2007. Freshwater resources and their management. Pp. 173-210 in Climate Change 2007: Impacts, Adaptation and Vulnerability. Contribution of Working Group II to the Fourth Assessment Report of the Intergovernmental Panel on Climate Change, M.L. Parry, O.F. Canziani, J.P. Palutikof, P.J. van der Linden, and C.E. Hanson, eds. Cambridge, UK: Cambridge University Press [online]. Available: http://www.ipcc.ch/ipccreports/ar4-wg2.htm [accessed Apr. 27, 2009].

Kunz, T.H., E.B. Arnett, W.P. Erickson, A.R. Hoar, G.D. Johnson, R.P. Larkin, M.D. Strickland, R.W. Thresher, and M.D. Tuttle. 2007. Ecological impacts of wind energy development on bats: Questions, research needs, and hypotheses. Front. Ecol. Environ. 5(6):315-324.

Kurukulasuriya, P., R. Mendelsohn, R. Hassan, J. Benhin, T. Deressa, M. Diop, H.M. Eid, K.Y. Fosu, G. Gbetibouo, S. Jain, A. Mahamadou, R. Mano, J. Kabubo-Mariara, S. El-Marsafawy, E. Molua, S. Ouda, M. Ouedraogo, I. Séne, D. Maddison, S.N. Seo, and A. Dinar. 2006. Will African agriculture survive climate change? World Bank Econ. Rev. 20(3):367-388.

LaCommare, K.H., and J.H. Eto. 2004. Understanding the Cost of Power Interruptions to U.S. Electricity Consumers. Ernest Orlando Lawrence Berkeley National Laboratory, University of California Berkeley, Berkeley, CA. September 2004 [online]. Available: http://eetd.lbl.gov/ea/EMS/reports/55718.pdf [accessed Aug. 3, 2009].

Larsen, J. 2006. Setting the Record Straight: More than 52,000 People Died from Heat in Summer 2003. Earth Policy Institute, Eco–Economy Updates July 28, 2006 [online]. Available: http://www.earth-policy.org/Updates/2006/Update56.htm [accessed April 27, 2009].

Lawton, L., M. Sullivan, K. Van Liere, A. Katz, and J. Eto. 2003. A Framework and Review of Customer Outage Costs: Integration and Analysis of Electric Utility Outage Cost Surveys. LBNL 54365. Ernest Orlando Lawrence Berkeley National Laboratory, University of California, Berkeley, CA. November 2003 [online]. Available: http://certs.lbl.gov/pdf/54365.pdf [accessed Aug. 5, 2009].

Layton, D.F., and K. Moeltner. 2005. The cost of power outages to heterogeneous households—An application of the gamma-lognormal distribution. Pp. 35-54 in Applications of Simulation Methods in Environmental and Resource Economics, R. Scarpa, and A.A Alberini, eds. Dordrecht, The Netherlands: Springer.

Leary, N., J. Adejuwon, W. Bailey, V. Barros, M. Caffera, S. Chinvanno, C. Conde, A. De Comarmond, A. De Sherbinin, T. Downing, H. Eakin, A. Nyong, M. Opondo, B. Osman, R. Payet, F. Pulhin, J. Pulhin, J. Ratnasiri, E. Sanjak, G. von Maltitz, M. Wehbe, Y. Yin, and G. Ziervogel. 2006. For Whom the Bell Tolls: Vulnerabilities in a Changing Climate. A Synthesis from the AIACC Project. Working Paper No. 21. Washington, DC: Assessment of Impacts and Adaptations of Climate Change. January 2006 [online]. Available: http://www.aiaccproject.org/working_papers/Working%20Papers/AIACC_WP_21_Leary.pdf [accessed Apr. 27, 2009].

Leiby, P.N. 2007. Estimating the Energy Security Benefits of Reduced U.S. Oil Imports. ORNL/TM-2007/028. Oak Ridge National Laboratory, Oak Ridge, TN. July 23, 2007 [online]. Available: http://pzl1.ed.ornl.gov/Leiby2007%20 Estimating%20the%20Energy%20Security%20Benefits%20of%20Reduced%20 U.S.%20Oil%20Imports%20ornl-tm-2007-028%20rev2007Jul25.pdf [accessed May 18, 2009].

Lemke, P., J. Ren, R.B. Alley, I. Allison, J. Carrasco, G. Flato, Y. Fujii, G. Kaser, P. Mote, R.H. Thomas and T. Zhang. 2007. Observations: Changes in Snow, Ice and Frozen Ground. Pp. 337-383 in Climate Change 2007: The Physical Science Basis. Contribution of Working Group I to the Fourth Assessment Report of the Intergovernmental Panel on Climate Change, S. Solomon, D. Qin, M. Manning, M. Marquis, K. Averyt, M.M.B. Tingor, H.L. Miller, and Z. Chen, eds. New York: Cambridge University Press [online]. Available: http://www.ipcc.ch/pdf/assessment-report/ar4/wg1/ar4-wg1-chapter4.pdf [accessed Apr. 24, 2009].

Levy, J.I., L.K. Baxter, and J. Schwartz. 2009. Uncertainty and variability in health-related damages from coal-fired power plants in the United States. Risk Anal. 29(7):1000-1014.

Lewis, O.T. 2006. Climate change, species-area curves and the extinction crisis. Philos. T. Roy. Soc. B 361(1465):163-171.

Link, P.M., and R.S.J. Tol. 2004. Possible economic impacts of a shutdown of the thermohaline circulation: An application of FUND. Portuguese Econ. J. 3:99-114.

MacLean, H.L., and L.B. Lave. 2003a. Evaluating automobile fuel/propulsion system technologies. Prog. Energ. Combust. Sci. 29(1):1-69.

MacLean, H.L., and L.B. Lave. 2003b. Life cycle assessment of automobile/fuel options. Environ. Sci. Technol. 37(23):5445-5452.

Maddison, D. 2003. The amenity value of the climate: The household production function approach. Resour. Energy Econ. 25(2):155-175.

Mansur, E.T., R. Mendelsohn, and W. Morrison. 2005. A discrete-continuous choice model of climate change impacts on energy. Yale SOM Working Paper No. ES-43. March 14, 2005 [online]. Available: http://papers.ssrn.com/sol3/papers.cfm?abstract_id=738544# ( published as Mansur, E.T., R. Mendelsohn, and W. Morrison. 2008. Climate change adaptation: A study of fuel choice and consumption in the U.S. energy sector. J. Environ. Econ. Manage. 55(2):175-193).

Marshall, J.D., S.K. Teoh, and W.W. Nazaroff. 2005. Intake fraction of nonreactive vehicle emissions in U.S. urban areas. Atmos. Environ. 39(7):1363-1371.

Martens, W.J. 1998. Climate change, thermal stress and mortality changes. Soc. Sci. Med. 46(3):331-344.

McKinsey and Company. 2007. The Untapped Energy Efficiency Opportunity in the US Industrial Sector. Report prepared for the US Department of Energy, Office of Energy Efficiency and Renewable Energy.

Meade, J.E. 1952. External economies and diseconomies in a competitive situation. Econ. J. 62(245):54-67.

Meier, P.J., P.P.H. Wilson, G.L. Kulcinski, and P.L. Denholm. 2005. U.S. electric industry response to carbon constraint: A life-cycle assessment of supply side alternatives. Energ. Policy 33(9):1099-1108.

Mendelson, R., and J.E. Neumann. 1999. The Impact of Climate Change on the United States Economy. Cambridge: Cambridge University Press.

Mendelsohn, R., and L. Williams. 2007. Dynamic forecasts of the sectoral impacts of climate change. Pp. 107-118 in Human-Induced Climate Change: An Interdisciplinary Assessment, M.E. Schlesinger, H.S. Kheshgi, J. Smith, F.C. de la Chesnaye, J.M. Reilly, T. Wilson, and C. Kolstad, eds. Cambridge, MA: Cambridge University Press.

Mendelsohn, R., W.D. Nordhaus, and D. Shaw. 1994. The impact of global warming on agriculture: A Ricardian analysis. Am. Econ. Rev. 84(4):753-751.

Mendelsohn, R., W. Morrison, M.E. Schlesinger, and N.G. Andronova. 2000a. Country-specific market impacts of climate change. Climatic Change 45(3-4):553-569.

Mendelsohn, R., M.E. Schlesinger, and L.J. Williams. 2000b. Comparing impacts across climate models. Integr. Assess. 1(1):37-48.

Millennium Ecosystem Assessment. 2010. Millennium Ecosystem Assessment [online]. Available: http://www.millenniumassessment.org/en/index.aspx [accessed Apr. 12, 2010]

Mishan, E.J. 1965. Reflections on recent developments in the concept of external effects. Can. J. Econ. Polit. Sci. 31(1):3-34.

MIT (Massachusetts Institute of Technology). 2003. The Future of Nuclear Power: An Interdisciplinary MIT Study. Massachusetts Institute of Technology [online]. Available: http://web.mit.edu/nuclearpower/pdf/nuclearpower-full.pdf [accessed Aug. 3, 2009].

MMS (Minerals Management Service). 2009. Cape Wind Energy Project, Final Environmental Impact Statement, Vol. 2, Appendix A. Figures, Maps and Tables. MMS EIS-EA OCS 2008-040. Minerals Management Service, U.S. Department of the Interior. January 2009 [online]. Available: http://www.mms.gov/offshore/AlternativeEnergy/PDFs/FEIS/Appendix%20A%20-%20FiguresMapsTables/Appendix%20A_Table_of_Contents.pdf

Mount, T., A. Lamadrid, S. Maneevitjit, R. Thomas, and R.D. Zimmerman. 2008. The Economics of Reliability and the Importance of Events that didn't Happen. Presented at the Fourth Annual Carnegie Mellon Conference on the Electricity Industry: Future Energy Systems: Efficiency, Security, Control, Pittsburgh, PA, March 10, 2008.

Mrozek, J.R., and L.O. Taylor. 2002. What determines the value of life? A meta-analysis. J. Policy Anal. Manage. 21(2):253-270.

MSHA (Mining Safety and Health Administration). 2008. Mining Industry Accident, Injuries, Employment, and Production Statistics: All Coal Mining Data. U.S. Department of Labor, Mining Safety and Health Administration [online]. Available: http://www.msha.gov/ACCINJ/BOTHCL.HTM [accessed May 22, 2009].

MSHA (Mining Safety and Health Administration). 2009. Mine Injury and Worktime Quarterly Statistics: Coal Data-2008. U.S. Department of Labor, Mining Safety and Health Administration [online]. Available: http://www.msha.gov/ACCINJ/ALLCOAL.HTM [accessed May 22, 2009].

Mufson, S. 2009. USEC Denied Loan Guarantees: CEO Assails Obama After Energy Dept. Says Project is Unready. Washington Post, July 29, 2009 [online]. Available: http://www.washingtonpost.com/wp-dyn/content/article/2009/07/28/AR2009072802617.html [accessed Oct. 7, 2009].

Muller, N.Z., and R.O. Mendelsohn. 2006. The Air Pollution Emission and Policy Analysis Model (APEEP): Technical Appendix. Yale University, New Haven, CT. December 2006 [online]. Available: https://segueuserfiles.middlebury.edu/nmuller/APEEP_Tech_Appendix.pdf [accessed Oct. 7, 2009].

Muller, N.Z., and R.O. Mendelsohn. 2007. Measuring the damages from air pollution in the U.S. J. Environ. Econ. Manage. 54(1):1-14.

Muller, N., R. Mendelsohn, and W. Nordhaus. 2009. Environmental Accounting for Pollution: Methods with an Application to the United States Economy. May 26, 2009 [online]. Available: http://nordhaus.econ.yale.edu/documents/Env_Accounts_052609.pdf [accessed July 27, 2009].

Murray, C.J.L., and A.D. Lopez, eds. 1996. The Global Burden of Disease. Cambridge, MA: Harvard University Press.

Naidoo, R., A. Balmford, R. Costanza, B. Fisher, R.E. Green, B. Lehner, T.R. Malcolm, and T.H. Ricketts. 2008. Global mapping of ecosystem services and conservation priorities. Proc. Natl. Acad. Sci. USA 105(28):9495-9500.

NAS/NAE/NRC (National Academy of Sciences/National Academy of Engineering/National Research Council). 2009a. America's Energy Future: Technology and Transformation. Washington, DC: The National Academies Press.

NAS/NAE/NRC (National Academy of Sciences/National Academy of Engineering/National Research Council). 2009b. Electricity from Renewables: Status, Prospects and Impediments. Washington, DC: The National Academies Press.

NAS/NAE/NRC (National Academy of Sciences/National Academy of Engineering/National Research Council). 2009c. Liquid Transportation Fuels from Coal and Biomass: Technological Status, Costs, and Environmental Impacts. Washington, DC: The National Academies Press.

NAS/NAE/NRC (National Academy of Sciences/National Academy of Engineering/National Research Council). 2009d. Real Prospects for Energy Efficiency in the United States. Washington, DC: The National Academies Press.

NEAC (Nuclear Energy Advisory Committee). 2008. Nuclear Energy: Policies and Technology for the 21st Century. U.S. Department of Energy, Washington, DC. November 2008. [online]. Available: http://nuclear.energy.gov/neac/neacPDFs/NEAC_Final_Report_Web%20Version.pdf [accessed Apr. 15, 2010].

NEB (National Energy Board Canada). 2006. Canada's Oil Sand Opportunities and Challenges to 2015: An Update. An Energy Market Assessment. National Energy Board Canada. June 2006 [online]. Available: http://www.neb.gc.ca/clf-nsi/rnrgynfmtn/nrgyrprt/lsnd/pprtntsndchllngs20152006/pprtntsndchllngs20152006-eng.pdf [accessed May 12, 2009].

NEB (National Energy Board Canada). 2009. Total Crude Oil Exports. National Energy Board Canada [online]. Available: http://www.neb.gc.ca/clf-nsi/rnrgynfmtn/sttstc/crdlndptrlmprd ct/2008/ttlcrdlxprt2008.xls [accessed Apr. 8, 2009].

NEED (National Energy Education Development). 2008. Coal. Pp. 16-19 in Secondary Energy Infobook 2008-2009. National Energy Education Development Project, Manassas, VA [online]. Available: http://www.need.org/needpdf/Secondary%20Energy%20Infobook.pdf [accessed May 19, 2009].

NERA (National Economic Research Associates). 1993. External Costs of Electric Utility Resource Selection in Nevada. Prepared for Nevada Power Company. Cambridge, MA: NERA.

NETL (National Energy Technology Laboratory). 2007. Cost and Performance Baseline for Fossil Energy Plants, Vol. 1. Bituminous Coal and Natural Gas to Electricity. Final Report. DOE/NETL-2007/1281. U.S. Department of Energy, National Energy Technology Laboratory [online]. Available: http://www.netl.doe.gov/energy-analyses/pubs/Bituminous%20Baseline_Final%20Report.pdf [accessed Oct. 12, 2009].

NGVAmerica (Natural Gas Vehicles for America). 2009. Natural Gas for Transportation [online]. Available: http://www.ngvc.org/ [accessed May 26, 2009].

Nicholls, R.J., R.S.J. Tol, and A.T. Vafeidis. 2008. Global estimates of the impact of a collapse of the West Antarctic ice sheet: An application of FUND. Climatic Change 91(1-2):171-191.

Niemi, E., R. Mendelsohn, and E. Whitelaw. 1984. Economic Analysis of the Environmental Effects of a Combustion-Turbine Generating Station at Frederickson Industrial Park, Pierce County, Washington. Prepared for the Bonneville Power Administration, Portland, OR, by ECONorthwest, Portland, OR. March 1984.

Niemi, E., R. Mendelsohn and R. Gregory. 1987. Generic Coal Study: Quantification and Valuation of Environmental Impacts. Prepared for the Bonneville Power Administration, Portland, OR, by ECONorthwest, Portland, OR. January 1987.

Nishioka, Y., J.I. Levy, G.A. Norris, A. Wilson, P. Hofstetter, and J.D. Spengler. 2002. Integrating risk assessment and life cycle assessment: A case study of insulation. Risk Anal. 22(5):1003-1017.

Nordhaus, W.D. 1991. To slow or not to slow: The economics of the greenhouse effect. Econ. J. 101(407):920-937.

Nordhaus, W.D. 1994a. Managing the Global Commons: The Economics of Climate Change. Cambridge: The MIT Press.

Nordhaus, W.D. 1994b. Expert opinion on climate change. Am. Sci. 82(1):45-51.

Nordhaus, W.D. 2006. Geography and macroeconomics: New data and new findings. Proc. Natl. Acad. Sci. USA 103(10):3510-3517.

Nordhaus, W.D. 2008. A Question of Balance: Weighing the Options on Global Warming Policies. New Haven: Yale University Press.

Nordhaus, W.D., and J. Boyer. 1999. Impacts of climate change. Chapter 4 in Roll the DICE Again: Economic Models of Global Warming. Yale University [online]. Available: http://nordhaus.econ.yale.edu/rice98%20pap%20121898.PDF [accessed Apr. 29, 2009].

Nordhaus, W.D., and J. Boyer. 2000. Warming the World: Economic Models of Global Warming, 1st Ed. Cambridge: MIT Press.

Nordhaus, W.D., and Z. Yang. 1996. A regional dynamic general equilibrium model of alternative climate-change strategies. Am. Econ. Rev. 86(4):741-765.

NRC (National Research Council). 1998. Research Priorities for Airborne Particulate Matter: I. Immediate Priorities and a Long-Range Research Portfolio. Washington, DC: National Academy Press.

NRC (National Research Council). 1999. Research Priorities for Airborne Particulate Matter: II. Evaluating Research Progress and Updating the Portfolio. Washington, DC: National Academy Press.

NRC (National Research Council). 2000. Toxicological Effects of Methylmercury. Washington, DC: National Academy Press.

NRC (National Research Council). 2001. Research Priorities for Airborne Particulate Matter: III. Early Research Progress. Washington, DC: National Academy Press.

NRC (National Research Council). 2002a. Estimating the Public Health Benefits of Proposed Air Pollution Regulations. Washington, DC: The National Academies Press.

NRC (National Research Council). 2002b. Coal Waste Impoundments: Risk, Responses, and Alternatives. Washington, DC: National Academy Press.

NRC (National Research Council). 2002c. Effectiveness and Impact of Corporate Average Fuel Economy (CAFÉ) Standards. Washington DC: National Academy Press.

NRC (National Research Council). 2002d. Abrupt Climate Change: Inevitable Surprises. Washington DC: National Academy Press.

NRC (National Research Council). 2003a. Cumulative Environmental Effects of Oil and Gas Activities on Alaska's North Slope. Washington, DC: The National Academies Press.

NRC (National Research Council). 2003b. Managing Carbon Monoxide Pollution in Meteorological and Topographical Problem Areas. Washington, DC: The National Academies Press.

NRC (National Research Council). 2003c. Oil in the Sea III: Inputs, Fates, and Effects. Washington, DC: The National Academies Press.

NRC (National Research Council). 2004a. Valuing Ecosystem Services: Toward Better Environmental Decision-Making. Washington, DC: The National Academies Press.

NRC (National Research Council). 2004b. Research Priorities for Airborne Particulate Matter: IV. Continuing Research Progress. Washington DC: The National Academies Press.

NRC (National Research Council). 2004c. Air Quality Management in the United States. Washington, DC: The National Academies Press.

NRC (National Research Council). 2005. Safety and Security of Commercial Spent Nuclear Fuel Storage: Public Report. Washington, DC: The National Academies Press.

NRC (National Research Council). 2006a. New Source Review for Stationary Sources of Air Pollution. Washington, DC: The National Academies Press.

NRC (National Research Council). 2006b. Managing Coal Combustion Residues in Mines. Washington, DC: The National Academies Press.

NRC (National Research Council). 2007a. Models in Environmental Regulatory Decision Making. Washington DC: The National Academies Press.

NRC (National Research Council). 2007b. Environmental Impacts of Wind-Energy Projects. Washington, DC: The National Academies Press.

NRC (National Research Council). 2007c. Coal Research and Development To Support National Energy Policy. Washington, DC: The National Academies Press.

NRC (National Research Council). 2008a. Review of DOE's Nuclear Energy Research and Development Program. Washington, DC: The National Academies Press.

NRC (National Research Council). 2008b. Assessment of Technologies for Improving Light Duty Vehicle Fuel Economy: Letter Report. Washington, DC: The National Academies Press.

NRC (National Research Council). 2008c. Transitions to Alternative Transportation Technologies—A Focus on Hydrogen. Washington, DC: The National Academies Press.

NRC (National Research Council). 2008d. Water Implications of Biofuels Production in the United States. Washington, DC: The National Academies Press.

NYSERDA (New York State Energy Research and Development Authority). 2009. Comparison of Reported Effects and Risks to Vertebrate Wildlife from Six Electricity Generation Types in the New York/New England Region. Report 90-02. NYSERDA 9675. Prepared for the New York State Energy Research and Development Authority, Albany, NY, by Environmental Bioindicators Foundation, Inc., Fort Pierce, FL, and Pandion Systems, Inc, Gainesville, FL. March 2009 [online]. Available: http://www.nyserda.org/publications/Report%2009-02%20Wildlife%20report%20-%20web.pdf [accessed July 24, 2009].

Ogden, J.M., R.H. Williams, and E.D. Larson. 2004. Societal lifecycle costs of cars with alternative fuels/engines. Energ. Policy 32(1):7-27.

ORNL /RFF (Oak Ridge National Laboratory and Resources for the Future). 1995. Estimating Externalities of Nuclear Fuel Cycles. Report No.8 of External Costs and Benefits of Fuel Cycles: A Study by the U.S. Department of Energy and the Commission of the European Communities. Oak Ridge National Laboratory, Oak Ridge, TN, and Resources for the Future, Washington, DC. Washington, DC: McGraw-Hill/Utility Data Institute.

ORNL /RFF (Oak Ridge National Laboratory and Resources for the Future). 1992-1998. External Costs and Benefits of Fuel Cycles: Reports 1-8. Oak Ridge National Laboratory, Oak Ridge, TN, and Resources for the Future, Washington, DC. Washington, DC: McGraw-Hill/Utility Data Institute.

OSMRE (U.S. Office of Surface Mining Reclamation and Enforcement). 2009. CCB Information Network: Coal Combustion By-Products. U.S. Office of Surface Mining Reclamation and Enforcement Mid-Continent Region [online]. Available: http://www.mcrcc.osmre.gov/ccb/ [accessed May 20, 2009].

Ottinger, R.L., D.R. Wooley, N.A. Robinson, D.R. Hodas, and S.E. Babb. 1990. Environmental Costs of Electricity. New York, NY: Oceana Publications.

Overdomain. 2002. Electric Reliability for the End User: A Survey of the Literature. Prepared for Global Environment and Technology Foundation Contract: GETF/CECS 01-CA-002 by Overdomain, Santa Barbara, CA [online]. Available: http://www.overdomain.com/documents/ReliabilityFinal-v12v.pdf [accessed Aug. 7, 2009].

Overpeck, J.T., and J.E. Cole. 2006. Abrupt change in Earth's climate system. Annu. Rev. Environ. Resour. 31:1-31.

Ozment, S., and T. Tremwel. 2007. Transportation Management in the Wind Industry: Problems and Solutions Facing the Shipment of Oversized Products in the Supply Chain. Research Paper No. SCNR-WP025-1007. Supply Chain Management Research Center, University of Arkansas, Fayetteville, AR. October 15, 2007.

Pacca, S., and A. Horvath. 2002. Greenhouse gas emissions from building and operating electric power plants in the Upper Colorado River Basin. Environ. Sci. Technol. 36(14):3194-3200.

Parfomak, P.W. 2008. Liquefied Natural Gas (LNG) Infrastructure Security: Issues for Congress. CRC Report for Congress RL32073. May 13, 2008 [online]. Available: http://www.ncseonline.org/NLE/CRSreports/08Jun/RL32073.pdf [accessed May 15, 2009].

Parfomak, P.W., and A.S. Vann. 2008. Liquefied Natural gas (LNG) Import Terminals: Siting, Safety, and Regulation. CRS Report for Congress RL32205. May 15, 2008 [online]. Available: http://assets.opencrs com/rpts/RL32205_20080515.pdf [accessed May 15, 2009].

Parry, I.W.H., and J. Darmstadter. 2003. The Cost of U.S. Oil Dependency. Discussion paper 03-59. Resources for the Future, Washington, DC. December 2003 [online]. Available: http://www.rff.org/Documents/RFF-DP-03-59.pdf [accessed May 18, 2009].

Parry, I.W.H., M. Walls, and W. Harrington. 2007. Automobile Externalities and Policies. RFF Discussion Paper No. 06-26. Washington, DC: Resources for the Future. January 2007 [online]. Available: http://www.rff.org/RFF/Documents/RFF-DP-06-26-REV.pdf [accessed Sept. 10, 2009].

Parry, M.L., C. Rosenzweig, A. Iglesias, M. Livermore, and G. Fischer. 2004. Effects of climate change on global food production under SRES emissions and socio-economic scenarios. Global Environmental Change 14(1):53-67.

Paybins, K.S., T. Messenger, J.H. Eychaner, D.B. Chambers, and M.D. Kozar. 2000. Water Quality in the Kanawha-New River Basin: West Virginia, Virginia, and North Carolina, 1996-1998. U.S.G.S. Circular 1204. U.S. Department of the Interior, U.S. Geological Survey [online]. Available: http://pubs.usgs.gov/circ/circ1204/#pdf [accessed June 10, 2009].

Paydirt. 1999. Interesting Facts about Wyoming Uranium. Wyoming Mining Association [online]. Available: http://www.wma-minelife.com/uranium/articles/art392.htm [accessed May 20, 2009].

Pearce, D.W., C. Bann, and S. Georgiou. 1992. The Social Cost of Fuel Cycles. London: HMSO.

Pearce, D.W., W.R. Cline, A.N. Achanta, S. Fankhauser, R.K. Pachauri, R.S.J. Tol, and P. Vellinga 1996. The social costs of climate change: Greenhouse damage and the benefits of control. Pp. 179-224 in Climate Change 1995:Economic and Social Dimensions of Climate Change—Contribution of Working Group III to the Second Assessment Report of the Intergovernmental Panel on Climate Change, J.P. Bruce, H. Lee, and E.F. Haites, eds. Cambridge: Cambridge University Press.

Pearce, D.W., B. Groom, C. Hepburn, and P. Koundouri. 2003. Valuing the future: Recent advances in social discounting. World Econ. 4(2):121-141.

Pigou, A. 1920. The Economics of Welfare. London: Macmillan.

Plambeck, E.L., and C.W. Hope. 1996. PAGE95: An updated valuation of the impacts of global warming. Energy Policy 24(9):783-793.

Pope, C.A., R.T. Burnett, M.J. Thun, E.E. Calle, D. Krewski, K. Ito, and G.D. Thurston. 2002. Lung cancer, cardiopulmonary mortality, and long-term exposure to fine particulate air pollution. JAMA 287(9):1132-1141.

Primen. 2001. The Cost of Power Disturbances to Industrial and Digital Economy Companies. Primen, Madison, WI. June 2001 [online]. Available: http://www.epri-intelligrid.com/intelligrid/docs/Cost_of_Power_Disturbances_to_Industrial_and_Digital_Technology_Companies.pdf [accessed Aug. 5, 2009].

Ramseur, J.L. 2008. Oil Spills in U.S. Coastal Waters: Background, Governance, and Issues for Congress. CRC Report for Congress RL33705. September 2, 2008 [online]. Available: http://assets.opencrs.com/rpts/RL33705_20080902.pdf [accessed June 9, 2009].

Raupach, M.R., G. Marlan, P. Ciais, C. LeQuere, J.G. Canadell, G. Klepper, and C.B. Filed. 2007. Global and regional drivers of accelerating CO2 emissions. Proc. Natl. Acad. Sci. USA 104(24):10288-10293.

Regional Economic Research. 1991. Estimating the Air Quality Impacts of Alternative Energy Resources, Phase IV. Prepared for the California Energy Commission, Sacramento, CA.

Rehdanz, K., and D. Maddison. 2005. Climate and happiness. Ecol. Econ. 52(1):111-125.

Renssen, H., C.J. Beets, T. Fichefet, H. Goosse, and D. Kroon. 2004. Modeling the climate response to a massive methane release from gas hydrates. Paleoceanography 19:PA2010.

Rock, B. 2008. An Overview of 2007 American 2007 Natural Gas Vehicles. Helium [online]. Available: http://www.helium.com/items/451632-an-overviewof-2007-american-2007-natural-gas-vehicles [accessed Sept. 2, 2009].

Rosenzweig, C., G. Casassa, D.J. Karoly, A. Imeson, C. Liu, A. Menzel, S. Rawlins, T.L. Root, B. Seguin, and P. Tryjanowski. 2007. Assessment of observed changes and responses in natural and managed systems. Pp. 79-131 in Climate Change 2007: Impacts, Adaptation and Vulnerability. Contribution of Working Group II to the Fourth Assessment Report of the Intergovernmental Panel on Climate Change, M.L. Parry, O.F. Canziani, J.P. Palutikof, P.J. van der Linden, and C.E. Hanson, eds. Cambridge, UK: Cambridge University Press [online]. Available: http://www.ipcc.ch/pdf/assessment-report/ar4/wg2/ar4-wg2-chapter1.pdf [accessed Apr. 29, 2009].

Roumasset, J.A., and K.R. Smith. 1990. Exposure trading: An approach to more efficient air pollution control. J. Environ. Econ. Manage. 18(3):276-291.

RTI. 2007. Human and Ecological Risk Assessment of Coal Combustion Wastes. Draft. Prepared for Office of Solid Waste, U.S. Environmental Protection Agency, Research Triangle Park, NC, by RTI, Research Triangle Park, NC. August 6, 2007.

Ruether, J., M. Ramezan, and E. Grol. 2005. Life-Cycle Analysis of Greenhouse Gas Emissions for Hydrogen Fuel Production in the United States from LNG and Coal. DOE/NETL-2006/1227. U.S. Department of Energy, National Energy Technology Laboratory, Pittsburgh, PA. November [online]. Available: http://www.netl.doe.gov/energy-analyses/pubs/H2_from_Coal_LNG_Final.pdf [accessed Apr. 22, 2009].

Ruth, M., and A.C. Lin. 2006. Regional energy demand and adaptations to climate change: Methodology and application to the state of Maryland. Energ. Policy 34(17):2820-2833.

Sala, O.E., F.S. Chapin, J.J. Armesto, E. Berlow, J. Bloomfield, R. Dirzo, E. Huber-Sanwald, L.F. Huenneke, R.B. Jackson, A. Kinzig, R. Leemans, D.M. Lodge, H.A. Mooney, M. Oesterheld, N.L. Poff, M.T. Sykes, B.H. Walker, M. Walker, and D.H. Wall. 2000. Global biodiversity scenarios for the year 2100. Science 287(5459):1770-1774.

Salonen, J.T., K. Seppänen, K. Nyyssönen, H. Korpela, J. Kauhanen, M. Kantola, J. Tuomilehto, H. Esterbauer, F. Tatzber, and R. Salonen. 1995. Intake of mercury from fish, lipid peroxidation, and the risk of myocardial infarction and coronary, cardiovascular, and any death in Eastern Finnish men. Circulation 91(3):645-655.

Samaras, C., and K. Meisterling. 2008. Life cycle assessment of greenhouse gas emissions from plug-in hybrid vehicles: Implications for policy. Environ. Sci. Technol. 42(9): 3170-3176.

Schlenker, W., and M.J. Roberts. 2009. Nonlinear temperature effects indicate severe damages to U.S. crop yields under climate change. Proc. Natl. Acad. Sci. U.S.A. 106(37): 15594-15598.

Schlenker, W., W.M. Hanemann, and A.C. Fisher. 2005. Will U.S. agriculture really benefit from global warming? Accounting for irrigation in the hedonic approach. Am. Econ. Rev. 95(1):395-406.

Schlenker, W., W.M. Hanemann, and A.C. Fisher. 2006. The impact of global warming on U.S. agriculture: An econometric analysis of optimal growing conditions. Rev. Econ. Statist. 88(1):113-125.

Scholze, M., W. Knorr, N.W. Arnell, and C. Prentice. 2006. A climate-change risk analysis for world ecosystems. P. Natl. Acad. Sci. USA 103(35):13116-13120.

Schröter, D., W. Cramer, R. Leemans, I.C. Prentice, M.B. Araújo, N.W. Arnell, A. Bondeau, H. Bugmann, T.R. Carter, C.A. Gracia, A.C. de la Vega-Leinert, M. Erhard, F. Ewert, M. Glendining, J.I. House, S. Kankaanpää, R.J. Klein, S. Lavorel, M. Lindner, M.J. Metzger, J. Meyer, T.D. Mitchell, I. Reginster, M. Rounsevell, S. Sabate, S. Sitch, B. Smith, J. Smith, P. Smith, M.T. Sykes, K. Thonicke, W. Thuiller, G. Tuck, S. Zaehle, and B. Zierl. 2005. Ecosystem service supply and vulnerability to global change in Europe. Science 310(5752):1333-1337.

Schwartz, J., B. Coull, F. Laden, and L. Ryan. 2008. The effect of dose and timing of dose on the association between airborne particles and survival. Environ. Health Perspect. 116(1):64-69.

Scitovsky, T. 1954. Two concepts of external economies. J. Polit. Econ. 62(2):143-151.

Scott, M.J., J.A. Dirks, and K.A. Cort. 2005. The adaptive value of energy efficiency programs in a warmer world. Pp. 671-682 in Reducing Uncertainty through Evaluation, Proceedings of the 2005 International Energy Program Evaluation Conference, August 17-19, 2005, Brooklyn, New York. Madison, WI: OmniPress.

Searchinger, T., R. Heimlich, R.A. Houghton, F. Dong, A. Elobeid, J. Fabiosa, S. Tokgoz, D. Hayes, and T.H. Yu. 2008. Use of U.S. croplands for biofuels increases greenhouse gases through emissions from land-use change. Science 319(5867):1238-1240.

Secchi, S., P.W. Gassman, J.R. Williams, and B. Babcock. 2009. Corn-based ethanol production and environmental quality: A case of Iowa and the Conservation Reserve Program. Environ. Manage. 44(4):732-744.

Simon, A. and M. Rinaldi. 2006. Disturbance, stream incision, and channel evolution: The roles of excess transport capacity and boundary materials in controlling channel response. Geomorphology 79(3-4):361-383.

Simpson, T.W., A.N. Sharpley, R.W. Howarth, H.W. Paerl, and K.R. Mankin. 2008. The new gold rush: Fueling ethanol production while protecting water quality. J. Environ. Qual. 37:318-324.

Sinclair, T.R., and N.G. Seligman. 1996. Crop modeling: From infancy to maturity. Agron. J. 88(5):698-704.

Sinclair, T.R., and N.G. Seligman. 2000. Criteria for publishing papers on crop modeling. Field Crop. Res. 68(3):165-172.

Smith, J.B., H.J. Schellnhuber, M.M.Q. Mirza, S. Fankhauser, R. Leemans, L. Erda, L. Ogallo, B. Pittock, R. Richels, C. Rosenzweig, U. Safriel, R.S.J. Tol, J. Weyant, and G. Yohe. 2001. Vulnerability to climate change and reasons for concern: A synthesis. Pp. 913-967 in Climate Change 2001: Impacts, Adaptation, and Vulnerability, J.J. McCarthy, O.F. Canziani, N.A. Leary, D.J. Dokken, and K.S. White, eds. Cambridge: Cambridge University Press [online]. Available: http://www.ipcc.ch/ipccreports/tar/wg2/pdf/wg2TARchap19.pdf [accessed June 4, 2009].

Solomon, S., G.K. Plattner, R. Knutti, and P. Friedlingstein. 2009. Irreversible climate change due to carbon dioxide emissions. Proc. Natl. Acad. Sci. USA 106(6):1704-1709.

Sovacool, B.K. 2008. Valuing the greenhouse gas emissions from nuclear power: A critical survey. Energ. Policy 36(8):2950-2963.

Sovacool, B.K. 2009a. Running on empty: The electricity-water nexus and the U.S. electricity sector. Energy Law J. 30(1):11-51.

Sovacool, B.K. 2009b. Contextualizing avian mortality: A preliminary appraisal of bird and bat fatalities from wind, fossil-fuel, and nuclear electricity. Energ. Policy 37(6):2241-2248.

Spath, P.L., and M.K. Mann. 2000. Life cycle Assessment of a Natural Gas Combined-Cycle Power Generation System. NREL/TP-570-27715. National Renewable Energy Laboratory, Golden, CO. September 2000 [online]. Available: http://www.nrel.gov/docs/fy00osti/27715.pdf [accessed Apr. 30, 2009].

Spath, P.L., M.K. Mann, and D.R. Kerr. 1999. Life Cycle Assessment of Coal-Fired Power Production. NREL/TP-570-25119. National Renewable Energy Laboratory, Golden, CO. June 1999 [online]. Available: http://www.nrel.gov/docs/fy99osti/25119.pdf [accessed Apr. 30, 2009].

SRIC (Southwest Research and Information Center). 2009. Uranium Impact Assessment Program. Southwest Research and Information Center, Albuquerque, NM [online]. Available: http://www.sric.org/uranium/index.html [accessed Apr. 21, 2009].

Stern, N. 2007. The Economics of Climate Change: The Stern Review. Cambridge: Cambridge University Press.

Talley, W. K. 1999. Determinants of the property damage costs of tanker accidents. Transport. Res. D-Tr. E. 4:413-426.

Tester, J.W., E.M. Drake, M.J. Driscoll, M.W. Golay, and W.A. Peters. 2005. Sustainable Energy: Choosing among Options. Cambridge, MA: MIT Press.

Thieler, E.R., and E.S. Hammar-Klose. 1999. National Assessment of Coastal Vulnerability to Future Sea-Level Rise: Preliminary Results for the U.S. Atlantic Coast. Open-File Report 99-593. U.S. Geological Survey [online]. Available: http://pubs.usgs.gov/of/1999/of99-593/index.html [accessed Oct. 12, 2009].

Thieler, E.R., and E.S. Hammar-Klose. 2000a. National Assessment of Coastal Vulnerability to Future Sea-Level Rise: Preliminary Results for the U.S. Pacific Coast. Open-File Report 00-178. U.S. Geological Survey. [online]. Available: http://pubs.usgs.gov/dds/dds68/reports/westrep.pdf [accessed Oct. 12, 2009].

Thieler, E.R., and E.S. Hammar-Klose. 2000b. National Assessment of Coastal Vulnerability to Future Sea-Level Rise: Preliminary Results for the U.S. Gulf of Mexico Coast. Open-File Report 00-179. U.S. Geological Survey [online]. Available: http://pubs.usgs.gov/dds/dds68/reports/gulfrep.pdf [accessed Oct. 12, 2009].

Thomas, C.D., A. Cameron, R.E. Green, M. Bakkenes, L.J. Beaumont, Y.C. Collingham, B.F. Erasmus, M.F. de Siqueira, A. Grainger, L. Hannah, L. Hughes, B. Huntley, A.S. van Jaarsveld, G.F. Midgley, L. Miles, M.A. Ortega-Huerta, A.T. Peterson, O.L. Phillips, and S.E. Williams. 2004a. Extinction risk from climate change. Nature 427(6970):145-148.

Thomas, C.D., S.E. Williams, A. Cameron, R.E. Green, M. Bakkenes, L.J. Beaumont, Y.C. Collingham, B.F. Erasmus, M.F. de Siqueira, A. Grainger, L. Hannah, L. Hughes, B. Huntley, A.S. van Jaarsveld, G.F. Midgley, L. Miles, M.A. Ortega-Huerta, A.T. Peterson, and O.L. Phillips. 2004b. Biodiversity conservation: Uncertainty in prediction of extinction risk/Effects of changes in climate and land use/Climate change and extinction risk (reply). Nature 430(6995):33.

Thuiller, W., M.B. Araujo, R.G. Pearson, R.J. Whittaker, L. Brotons, and S. Lavorel. 2004. Biodiversity conservation: Uncertainty in predictions of extinction risk. Nature 430(6995):33.

Titus, J.G. 1992. The costs of climate change to the United States. Chapter 27 in Global Climate Change: Implications, Challenges, and Mitigation Measures, S.K. Majumdar, L.S. Kalkstein, B. Yarnal, E.W. Miller, and L.M. Rosenfeld, eds. Easton, PA: Pennsylvania Academy of Sciences.

Tol, R.S.J. 1995. The damage costs of climate change toward more comprehensive calculations. Environ. Resour. Econ. 5(4):353-374.

Tol, R.S.J. 1999. The marginal costs of greenhouse gas emissions. Energy J. 20(1):61-81.

Tol, R.S.J. 2002a. Estimates of the damage costs of climate change, Part I. Benchmark estimates. Environ. Resour. Econ. 21(1):47-73.

Tol, R.S.J. 2002b. Estimates of the damage costs of climate change, Part II. Dynamic estimates. Environ. Resour. Econ. 21(2):135-160.

Tol, R.S.J. 2005a. Emission abatement versus development as strategies to reduce vulnerability to climate change: An application of FUND. Environ. Dev. Econ. 10(5):615-629.

Tol, R.S.J. 2005b. The marginal damage costs of carbon dioxide emissions: An assessment of the uncertainties. Energy Policy 33(16):2064-2074.

Tol, R.S.J. 2008. Why worry about climate change? A research agenda. Environ. Value. 17(4):437-470.

Tol, R.S.J. 2009. Why worry about climate change? ESRI Research Bulletin 09/1, Spring 2009 [online]. Available: http://www.esri.ie/UserFiles/publications/20090428122502/RB20090101.pdf [accessed Oct. 12, 2009].

Tol, R.S.J., and S. Fankhauser. 1998. On the representation of impact in integrated assessment models of climate change. Environ. Model. Assess. 3(1):63-74.

Tol, R.S.J., S. Fankhauser, R.G. Richels, and J.B. Smith. 2000. How much damage will climate change do? World Economics 1(4):179-206.

TRB (Transportation Research Board). 2004. Transmission Pipelines and Land Use. Special Report 281. National Research Council. Washington, DC: The National Academies Press.

Trenberth, K.E., P.D. Jones, P. Ambenje, R. Bojariu, D. Easterling, A. Klein Tank, D. Parker, F. Rahimzadeh, J.A. Renwick, M. Rusticucci, B. Soden, and P. Zhai. 2007: Observations: Surface and Atmospheric Climate Change. Pp. 235-336 in Climate Change 2007: The Physical Science Basis. Contribution of Working Group I to the Fourth Assessment Report of the Intergovernmental Panel on Climate Change, S. Solomon, D. Qin, M. Manning, M. Marquis, K. Averyt, M.M.B. Tingor, and H.L. Miller, and Z. Chen, eds. New York: Cambridge University Press [online]. Available: http://www.ipcc.ch/pdf/assessment-report/ar4/wg1/ar4-wg1-chapter3.pdf [accessed Apr. 24, 2009].

Triangle Economic Research. 1995. Assessing Environmental Externality Costs for Electricity Generation. Prepared for Northern States Power Company, Minnesota, by Triangle Economic Research, Durham, NC.

Trimble, S.W. 1999. Decreased rates of alluvial sediment storage in the Coon Creek Basin, Wisconsin, 1975-93. Science 285(5431):1244-1246.

Union of Concerned Scientists. 1992. America's Energy Choices: Investing in a Strong Economy and Clean Environment. Cambridge, MA: Union of Concerned Scientists.

University of Hamburg. 2009. Social Cost of Carbon Meta-Analyses [online]. Available: http://www.fnu.zmaw.de/Social-cost-of-carbon-meta-analy.6308.0.html [accessed June 9, 20009].

U.S. GBC (U.S. Green Building Council). 2008. The Leadership in Energy and Environmental Design (LEED) Green Building Rating System™ [online]. Available: http://www.usgbc.org/ [accessed May 28, 2009].

USGS (U.S. Geological Survey). 1997. Radioactive Elements in Coal and Fly Ash: Abundance, Forms, and Environmental Significance. U.S. Geological Survey Fact Sheet FS 163-97. October 1997 [online]. Available: http://pubs.usgs.gov/fs/1997/fs163-97/FS-163-97.pdf [accessed Apr. 30, 2009].

USGS (U.S. Geological Survey). 2008. Mineral Commodity Summaries 2008. U.S. Geological Survey [online]. Available: http://minerals.usgs.gov/minerals/pubs/mcs/2008/mcs2008.pdf [accessed July 29, 2009].

USGS (U.S. Geological Survey). 2009a. Acid Mine Drainage: Introduction. U.S. Geological Survey [online]. Available: http://energy.er.usgs.gov/health_environment/acid_mine_
drainage/ [accessed Mar. 14, 2009].

USGS (U.S. Geological Survey). 2009b. National Water-Quality Assessment (NAWQA) Program. U.S. Geological Survey [online]. Available: http://water.usgs.gov/nawqa/ [accessed
July 27, 2009].

U.S. NRC (U.S. Nuclear Regulatory Commission). 2002. Radioactive Waste: Production,
Storage, Disposal. NUREG/BR-0216, Rev.2. U.S. Nuclear Regulatory Commission,
Washington, DC. May 2002 [online]. Available: http://www.nrc.gov/reading-rm/doc-
collections/nuregs/brochures/br0216/r2/br0216r2.pdf [accessed May 6, 2009].

U.S. NRC (U.S. Nuclear Regulatory Commission). 2008a. Operating Nuclear Power Reactors
[online]. Available: http://www.nrc.gov/info-finder/reactor/ [accessed May 22, 2009].

U.S. NRC (U.S. Nuclear Regulatory Commission). 2008b. Locations of Power Reactor
Sites Undergoing Decommissioning [online]. Available: http://www.nrc.gov/info-finder/
decommissioning/power-reactor/ [accessed May 22, 2009].

U.S. NRC (U.S. Nuclear Regulatory Commission). 2008c. Generic Environmental Impact
Statement for In-Situ Leach Uranium Milling Facilities, Vols 1-2. Draft Report for
Comment. NUREG-1910. Prepared by U.S. Nuclear Regulatory Commission, Office
of Federal and State Materials and Environmental Programs; Wyoming Department of
Environmental Quality. July 2008 [online]. Available: http://www.nrc.gov/reading-rm/
doc-collections/nuregs/staff/sr1910/ [accessed May 20, 2009].

U.S. NRC (U.S. Nuclear Regulatory Commission). 2008d. Nuclear Insurance and Disaster
Relief Funds. Fact Sheet. U.S. Nuclear Regulatory Commission. February 2008 [online].
Available: http://www.nrc.gov/reading-rm/doc-collections/fact-sheets/funds-fs.html [ac-
cessed Apr. 24, 2009].

U.S. NRC (U.S. Nuclear Regulatory Commission). 2009. Low-Level Waste Disposal Statistics.
U.S. Nuclear Regulatory Commission [online]. Available: http://www.nrc.gov/waste/llw-
disposal/licensing/statistics.html [accessed Apr. 27, 2009].

van der Welle, A., and B. van der Zwaan. 2007. An Overview of Selected Studies on the Value
of Lost Load (VOLL), Working Paper Final Version (November 15, 2007). Pp. 26-47
in WP5 Report (1) on National and EU Level Estimates of Energy Supply Externalities. Project No. 518294 SES6 Cases: Cost Assessment of Sustainable Energy Systems,
December 19, 2007 [online]. Available: http://www.feem-project.net/cases/documents/
deliverables/D_05_1%20energy%20supply%20externalities%20update_Dec_07.pdf [ac-
cessed May 15, 2009].

Viscusi, W.K. 1993. The value of risks to life and health. J. Econ. Lit. 31(4):1912-1946.

Viscusi, W.K., and J.E. Aldy. 2003. The value of statistical life: A critical review of market
estimates throughout the world. J. Risk Uncertainty 27(1):5-76.

Von Hippel, F.N. 2001. Plutonium and reprocessing of spent nuclear fuel. Science 293(5539):
2397-2398.

Waggitt, P. 2007. Uranium Mining Legacy Sites and Remediation—A Global Perspective.
InternationalAtomic Energy Agency, Namibia, October 2007 [online]. Available: http://
www.iaea.org/OurWork/ST/NE/NEFW/documents/RawMaterials/CD_TM_Swakop
mund%20200710/13%20Waggit4.PDF [accessed May 21, 2009].

Warren, R., N. Arnell, R. Nicholls, P. Levy, and J. Price. 2006a. Understanding the Regional
Impacts of Climate Change. Working Paper No. 90. Tyndall Centre for Climate Change
Research, University of East Anglia, UK. September 2006 [online]. Available: http://www.
tyndall.ac.uk/publications/working_papers/twp90.pdf [accessed Apr. 30, 2009].

Warren, R., C. Hope, M. Mastrandrea, R. Tol, N. Adger, and I. Lorenzoni. 2006b. Spotlight-
ing Impacts Functions in Integrated Assessment. Working Paper No. 91. Tyndall Centre
for Climate Change Research, University of East Anglia, Norwich, UK. September 2006

[online]. Available: http://www.tyndall.ac.uk/publications/working_papers/twp91.pdf [accessed May 1, 2009].

Weisser, D. 2007. A guide to life-cycle greenhouse gas (GHG) emissions from electric supply technologies. Energy 32(9):1543-1559.

Weitzman, M. 2007a. Structural Uncertainty and the Value of Statistical Life in the Economics of Catastrophic Climate Change. NBER Working Paper No. W13490. National Bureau of Economic Research, Inc., Cambridge, MA.

Weitzman, M.L. 2007b. A review of the "Stern Review on the Economics of Climate Change" J. Econ. Lit. 45(3):703-724.

Weitzman, M.L. 2009. On modeling and interpreting the economics of catastrophic climate change. Rev. Econ. Stat. 91(1):1-19.

WHO (World Health Organization). 2007. Extremely Low Frequency Fields. Environmental Health Criteria Monograph No. 238. Geneva: WHO Press [online]. Available: http://www.who.int/peh-emf/publications/Complet_DEC_2007.pdf [accessed July 28, 2009].

Wicke, L. 1986. Die ökologischen Milliarden: Das kostet die zerstörte Umwelt—so können wir sie retten. München: Kösel.

Williams, J.R. 1990. The erosion-productivity impact calculator (EPIC) model: A case history. Philos. T. Roy. Soc. Biol. Sci. 329(1255):421-428.

Williams, J.R. 1995. The EPIC model. Pp. 909-1000 in Computer Models of Watershed Hydrology, V.P. Singh, ed. Highlands Ranch, CO: Water Resources Publications.

Williams, J.R., C.A. Jones, and P.T. Dyke. 1984. Modeling approach to determining the relationship between erosion and soil productivity. Trans. ASAE 27(1):129-144.

Williams, J.R., M. Nearing, A. Nicks, E. Skidmore, C. Valentin, K. King, and R. Savabi. 1996. Using soil erosion models for global change studies. J. Soil Water Conserv. 51(5): 381-385.

Williams, J.R., J.G. Arnold, J.R. Kiniry, P.W. Gassman, and C.H. Green. 2008. History of model development at Temple, Texas. Hydrol. Sci. J. 53(5):948-960.

WISE-Uranium (World Information Service on Energy Uranium Project). 2009. Health and Environmental Impacts of Nuclear Fuel Production. WISE Uranium Project [online]. Available: http://www.wise-uranium.org/ [accessed Apr. 21, 2009].

Worrell, E., M. Neelis, L. Price, C. Galitsky, and N. Zhou. 2008. World Best Practice Energy Intensity Values for Selected Industrial Sectors. LBNL-62806. Ernest Orlando Lawrence Berkeley National Laboratory. February 2008 [online]. Available: http://ies.lbl.gov/iespubs/LBNL-62806.Rev.2.pdf [accessed May 1, 2009].

Yborra, S. 2006. Taking a second look at the natural gas vehicle. American Gas (Aug/Sept.): 32-36.

Yohe, G.W., R.D. Lasco, Q.K. Ahmad, N.W. Arnell, S.J. Cohen, C. Hope, A.C. Janetos, and R.T. Perez. 2007. Perspectives on climate change and sustainability. Pp. 811-841 in Climate Change 2007: Impact, Adaptation and Vulnerability. Contribution of Working Group II to the Fourth Assessment Report of the Intergovernmental Panel on Climate Change, M. Parry, O. Canziani, J. Palutikof, P. van der Linden, C. Hanson, eds. Cambridge: Cambridge University Press [online]. Available: http://www.ipcc.ch/pdf/assessment-report/ar4/wg2/ar4-wg2-chapter20.pdf [accessed June 1, 2009].

# Abbreviations

| | |
|---|---|
| AEF | America's Energy Future |
| Ag | silver |
| APEEP | Air Pollution Emission Experiments and Policy |
| As | arsenic |
| AWEA | American Wind Energy Association |
| | |
| BBL | barrel |
| BD2- | biodiesel 20% blend |
| BOS | balance of system |
| Btu | British thermal unit |
| | |
| CAFE | corporate average fuel economy |
| CAIR | Clean Air Interstate Rule |
| CAMR | Clear Air Mercury Rule |
| CARB | California Air Resources Board |
| CCB | coal combustion by-product |
| CCR | coal combustion residue |
| CCS | carbon capture and storage |
| Cd | cadmium |
| CdTe | cadmium-telluride |
| CFR | Code of Federal Regulations |
| CG | compressed gasoline |
| CH4 | methane |
| CHP | combined heat and power |
| CMAQ | Community Multiscale Air Quality model |

| CNG | compressed natural gas |
| CO | carbon monoxide |
| Co | cobalt |
| $CO_2$ | carbon dioxide |
| $CO_2$-eq | carbon dioxide equivalent |
| Cr | chromium |
| CRP | Conservation Reserve Program |
| CRS | Congressional Research Service |
| Cu | copper |
| | |
| DICE | Dynamic Integrated Model of Climate and the Economy |
| DOE | U.S. Department of Energy |
| DOT | U.S. Department of Transportation |
| | |
| E10 | ethanol 10% blend |
| E85 | ethanol 85% blend |
| EGR | enhanced gas recovery |
| EGU | electricity-generating unit |
| EIA | Energy Information Administration |
| EISA | Energy Independence and Security Act |
| EOR | enhance oil recovery |
| EPA | U.S. Environmental Protection Agency |
| ERR | estimated recoverable reserves |
| EtOH | ethanol |
| | |
| FBC | fluidized bed combustion |
| FERC | Federal Energy Regulatory Commission |
| FGD | flue gas desulfurization |
| FPEIS | Final Programmatic Environmental Impact Statement |
| FUND | Climate Framework for Uncertainty, Negotiation, and Distribution |
| | |
| GCM | global climate model |
| GDP | gross domestic product |
| gge | gasoline gallon equivalent |
| GHG | greenhouse gas |
| GWP | global-warming potential |
| GREET | Greenhouse Gases, Regulated Emissions, and Energy Use in Transportation |
| GTCC | greater than Class C |
| GW | gigawatt |
| GWh | gigawatt hour |

HAP          hazardous air pollutant
HDDV         heavy-duty diesel vehicle
HDGV         heavy-duty gasoline vehicle
HDV          heavy-duty vehicle
HDDV         heavy-duty diesel vehicle
HDGV         heavy-duty gasoline vehicle
HEV          hybrid electricity vehicle
HFCV         hydogren fuel-cell vehicle
Hg           mercury
HLRW         high-level radioactive wastes

IAM          integrated assessment model
IPCC         Intergovernmental Panel on Climate Change
IRIS         Integrated Risk Information System

kW           kilowatt
kWh          kilowatt hour

LCA          life-cycle assessment
LCIA         life-cycle impact assessment
LDV          light-duty vehicle
LEED         Leadership in Energy and Environmental Design
LLRW         low-level radioactive wastes
LNG          liquified natural gas
LWR          light water reactors

MCF          thousand cubic feet
MMBtu        million British thermal units
Mn           manganese
Mo           molybdenum
MOVES        Motor Vehicle Emission Simulator
MSRP         manufacturer's suggested retail price
MTM/VF       mountain top mining/valley fill
MW           megawatt
MWh          megawatt hour

$N_2O$       nitrous oxide
NEI          National Emissions Inventory
NEMS         National Energy Modeling System
NEPA         National Environmental Policy Act
NERC         North American Electric Reliability Corporation
NG           natural gas
NGV          natural gas vehicle

| | |
|---|---|
| $NH_3$ | ammonia |
| NHI | Nuclear Hydrogen Initiative |
| Ni | nickle |
| $NO_2$ | nitrogen dioxide |
| NOAA | National Oceanic and Atmospheric Administration |
| $NO_x$ | nitrogen oxides |
| NRC | National Research Council |
| NWPA | Nuclear Waste Policy Act |
| | |
| OECD | Organization for Economic Co-operation and Development |
| OMB | Office of Management and Budget |
| OPEC | Organization of Petroleum Exporting Countries |
| ORNL | Oakridge National Laboratory |
| | |
| PADD | Petroleum Administration for Defense Districts |
| PAGE | Policy Analysis of the Greenhouse Effect |
| Pb | lead |
| PC | pulverized coal |
| $PM_{10}$ | particulate matter less than or equal to10 micron in aerodynamic diameter (coarse particulate matter) |
| $PM_{2.5}$ | particulate matter equal to or small than 2.5 micron aerodynamic diameter (fine particulate matter) |
| PRB | Powder River Basin |
| PV | photovoltaic |
| | |
| QALY | quality-adjusted life year |
| quads | quadrillion British thermal units |
| | |
| R&D | research and development |
| RCRA | Resource Conservation and Recovery Act |
| RFF | Resources for the Future |
| RFG | reformulated gasoline |
| RICE | Regional Integrated Model of Climate and the Economy |
| | |
| SAB | Science Advisory Board |
| Se | selenium |
| $SF_6$ | sulfur hexafluoride |
| SI | spark-ignition |
| SIDI | spark-ignition, direct-injection |
| $SO_2$ | sulfur dioxide |
| $SO_x$ | sulfur oxides |
| SPR | strategic petroleum reserve |
| SRR | source-receptor relationships |

SUV          sports utility vehicle

THC          thermohaline circulation
Tl           thallium
TWh          terawatt hour

USD          U.S. dollars

VHTR         very-high-temperature reactor
VMT          vehicle miles traveled
VOC          volatile organic compound
VSL          value of a statistical life

WDL          workdays lost

Zn           zinc

# Common Units and Conversions

Energy use in the United States involves many diverse industries and sectors, each of which uses its own conventions and units to describe energy production and use. Although these units are in common usage throughout the energy industry, they are not always consistent and are not well understood by nonexperts. Similarly, different types of units are employed to describe emissions resulting from energy-related use activities. This appendix describes the units used for principal energy supply and consumption activities and provides some useful conversion factors. The U.S. Department of Energy's Energy Information Administration website provides additional information about energy (see www.eia.doe.gov/basics/conversion_basics.html) units and conversion factors, including easy-to-use energy conversion calculators. Total U.S. energy use in 2007 was 101.5 quadrillion ($10^{15}$) Btu or 96 Exa ($10^{18}$) Joules.

## Electricity

• **Electrical generating capacity** is *power* and expressed in units of *kilowatts* (kW), *megawatts* (MW = $10^3$ kW), and *gigawatts* (GW = $10^6$ kW). It is defined as the maximum electrical output that can be supplied by a generating facility operating at ambient conditions. Coal power plants typically have generation capacities of about 500 MW; nuclear plants about 1,000 MW (1 GW); intermittent sources (e.g., natural gas peaking plants and individual wind turbines) about one to a few megawatts; and residential roof-top installations of solar photovoltaics about a few kilowatts.

• **Electricity supply and consumption** is expressed in units of *kilowatt* *hours* (kWh), *megawatt hours* (MWh), *gigawatt hours* (GWh), or *terawatt hours* (TWh) ($10^9$ kWh). One kWh is equal to a power of 1,000 watts (the typical electricity that is consumed by a hand-held hair dryer) supplied or consumed over the period of an hour. Annual total delivered electricity in the United States is about 4,000 TWh and the average annual electricity consumption per U.S. household is about 11,000 kWh.

  o 1 kWh of electricity is equivalent to 3,410 Btu of thermal energy if the conversion has no inefficiencies.

  o In a 33% efficient power plant, 10,230 Btu of input primary energy are required to produce 1 kWh of electricity.

### Fossil Fuels and Other Liquid Fuels

• **Coal** supply and consumption in the United States is usually expressed in units of metric tons (sometimes written as tonnes and equal to 1,000 kg or 2,200 pounds [lb]) or short tons (2,000 lb); most of the rest of the world uses metric tons. This report uses short tons when discussing coal use in the United States.

  o A ton of typical coal contains about 22 MJ of energy.

  o A tonne of typical coal contains about 24 MJ of energy.

• **Petroleum and gasoline** supply and consumption quantities are expressed in the United States in gallons or barrels (1 barrel = 42 gallons) and internationally in liters (3.88 liters = 1 gallon). In the United States, the energy content of liquid fuel is expressed in British thermal units (Btu), million Btu (MMBtu or $10^6$ Btu), and quadrillion Btu (quad = $10^{15}$ Btu). The rest of world uses joules (J) to express the energy content of liquid fuels (1 Btu = 1,055 J). A Btu is defined as the amount of energy (in the form of heat) needed to raise the temperature of 1 lb of water by 1 degree Fahrenheit.[1] The energy content of different fuels can be converted to Btu using the following approximate factors:

  o 1 barrel crude oil = 5,800,000 Btu = 5.8 MMBtu

  o 1 barrel gasoline = 5.2 MMBtu

  o 1 barrel fuel ethanol = 3.5 MMBtu

• When different liquid fuels and blends are compared, this is often done on the basis of what volume would give the same energy as a gallon of gasoline. Therefore, about 1.5 gallons of ethanol would provide the energy equivalent of 1 gallon of gasoline.

• **Natural gas** supply and consumption usage is expressed in units of a thousand cubic feet (MCF or mcf). This is the equivalent volume of gas at atmospheric pressure and temperature. Here the prefix M stands for

---

[1]A joule is the amount of energy needed to heat a kilogram of water by 1 degree centigrade. 1,055 joules = 1 Btu.

a thousand, and MM is used to denote a million cubic feet. One MCF of natural gas contains about a million Btu of thermal energy.

### Basis for Quantifying Impacts

• **Activity-specific** impacts result from particular energy use. For example, impacts from the emissions from an electric power plant or impacts from tailpipe emissions from a passenger car.

• **Activity-aggregate** impacts are used to describe the impacts from energy use in a set of activities that include all impacts starting with the processing of primary energy, its conversions and its transportation to its end use point, its use to provide a set of energy services, and impacts associated with disposal of end use equipment. The aggregations are based on life-cycle assessment (LCA) methods and use a variety of data and models to estimate the impact. For example, electricity use to provide light in a building would include all the "upstream inputs" to produce feed energy for the power plant (mining, dams, etc.), the electricity production inputs to generate and distribute power to the site of the light bulb, and impacts associated with operation of the light bulb. Waste heat from the bulb and its disposal would be "downstream impacts." Larger downstream impacts would be associated with the health and other consequences from emissions at the power plant.

In this report, life-cycle impact assessment (LCIA) is a goal that can only be achieved incompletely due to limitations in data availability and complexity of the detailed systems, but where important impacts are present their magnitudes are estimated to the extent possible.

### Waste Streams and Hazardous Air Emissions

• **Solid and liquid wastes** are usually described using familiar units of volume or weight per unit time or quantity of energy produced. (cubic feet per minute [cfm]; tons per MWh; gallons per day; etc.). Where these waste streams contain contaminants, the concentration of the contaminant of concern is also important. (parts per million [ppm] by weight is the weight of contaminant in a million units of carrier weight; or pounds of contaminant per ton of carrier, or pounds of contaminant per gallon of liquid,)

• **Air emissions** are usually described by emissions per unit of energy produced or used—such as lb per MWh of electricity, lb per MCF of natural gas, or grams per vehicle miles traveled (VMT) and sometimes in terms of concentration of pollutants in emissions stream—such as parts per million (by volume) or pounds per cubic foot. The choice of a VMT basis is a compromise, since the more meaningful metric of passenger miles traveled would require information about the number of passengers per vehicle—and would only change the final result if more passengers on average travelled on vehicles powered by a particular fuel. Presentation of results per gallon of fuel makes for difficult comparisons since different fuels have different energy contents per gallon.

In this report, impacts are assessed nationally using detailed models for the overall activities. Using a VMT basis for the transportation emissions estimates includes not only the differences in the impacts for different fuels, but also includes differences in the size and weights of vehicles that constitute the national vehicle fleet.

## Greenhouse Gases

• **Carbon dioxide ($CO_2$)** emissions from energy production and use are expressed in tons (short tons) or metric tons (tonnes) of $CO_2$-equivalent ($CO_2$-eq). Although $CO_2$ is the principal greenhouse gas associated with energy use, other gases such as methane, nitrous oxide, black carbon, and $SF_6$, also make some contributions to warming potential. These other contributions are converted to an equivalent amount of $CO_2$ with a similar effect and the total is therefore expressed as tonnes of $CO_2$-equivalent. The United States emits about 7 billion tonnes of $CO_2$-equivalent per year, about 6 billion of which is $CO_2$ arising primarily from energy production and use. Average annual $CO_2$ emissions in the United States are about 20 tonnes per person. [Note: Sometimes greenhouse gas emissions are reported in terms of tonnes of "carbon." One tonne of carbon emissions equals 3.7 tonnes of $CO_2$ emissions, since the weight of $CO_2$ also includes the weight of the oxygen in the molecule.]

# Appendixes

# A

# Biographic Information on the Committee on Health, Environmental, and Other External Costs and Benefits of Energy Production and Consumption

**Jared L. Cohon** (*Chair*) is president of Carnegie Mellon University in Pittsburgh, Pennsylvania. His research interests focus on environmental system analysis, especially the development and application of techniques for planning and decision making in situations with multiple conflicting objectives. Dr. Cohon has been the author, co-author, and editor of more than 80 professional publications in this interdisciplinary field, which combines engineering, economics, and applied mathematics. He came to Carnegie Mellon in 1997 after serving as dean of the School of Forestry and Environmental Studies and professor of environmental systems analysis at Yale University from 1992 to 1997. Dr. Cohon began his teaching career in 1973 at Johns Hopkins University, where he served as assistant, associate, and full professor in the Department of Geography and Environmental Engineering and as assistant and associate dean of engineering and vice provost for research. Dr. Cohon is a member of the board of directors of Ingersoll-Rand Company (a producer of refrigeration and industrial equipment). The *Pittsburgh Post-Gazette* named him one of Pittsburgh's "Top 50" business leaders. In 1996, the National Audubon Society and the American Association of Engineering Societies jointly awarded him the Joan Hodges Queneau Medal for outstanding achievement in environmental conservation. He served on the U.S. Nuclear Waste Technical Review Board, appointed by President Clinton in 1995, and he served as the board's chairman from 1997 to 2002. Dr. Cohon also was appointed by President George W. Bush in 2002 to his Homeland Security Advisory Council. He served as chair of the council's Senior Advisory Committee on Academe, Policy, and Research, and he served as co-chair of the Council's Secure Borders/Open Doors Advisory

Committee. Dr. Cohon received his Ph.D. in civil engineering from the Massachusetts Institute of Technology (MIT).

**Maureen L. Cropper** (*Vice Chair*) is professor of economics at the University of Maryland, senior fellow at Resources for the Future, former lead economist at the World Bank, and research associate at the National Bureau of Economic Research. Dr. Cropper's research has focused on valuing environmental amenities (especially environmental health effects), on the discounting of future health benefits, and on the trade-offs implicit in environmental regulations. Her recent research analyzes the externalities associated with motorization and the interaction between residential location, land use, and travel demand. Dr. Cropper is a member of the National Academy of Sciences. She also is past president of the Association of Environmental and Resource Economists and a former chair of the Advisory Council for Clean Air Act Compliance Analysis, a subcommittee of the U.S. Environmental Protection Agency (EPA) Science Advisory Board. Dr. Cropper has served on the advisory boards of Resources for the Future, the Harvard Center for Risk Analysis, the Donald Bren School of the Environment, and the American Enterprise Institute Brookings Center on Regulation. She received her Ph.D. in economics from Cornell University.

**Mark R. Cullen** is professor of medicine and chief of the division of general internal medicine at Stanford University. His research interests are in occupational and environmental medicine, including isocyanate exposure in automobile shop workers, lung cancer in people exposed to asbestos, and lead toxicity in workers. He has published several textbooks, including *Clinical Occupational Medicine* and *Textbook of Clinical Occupational and Environmental Medicine*. He is a member of the DuPont Epidemiology Review Board, a member of the MacArthur Foundation Network on Socioeconomic Status and Health, and a corporate medical director for the Aluminum Company of America. Dr. Cullen is a member of the Institute of Medicine and served as a member of its Board on Health Sciences. He also served as a member of the National Research Council (NRC) Committee on Human Biomonitoring for Environmental Toxicants. Dr. Cullen received his M.D. from Yale University and did his residency in internal medicine.

**Elisabeth M. Drake** is emeritus staff affiliated with the MIT Energy Initiative. Prior to her retirement in 2001, she served as associate director for new energy technologies at the MIT Energy Laboratory. With other MIT colleagues, she developed an interdisciplinary graduate course on sustainable energy, and she was a co-author of the textbook. *Sustainable Energy: Choosing Among Options*, published in 2005 by the MIT Press. Her research interests focus on new and emerging energy supply-side and demand-side technologies that may facilitate a transition to a more sustainable future having lower greenhouse gas emissions at reasonable costs in both industrialized and developing countries. Early in her career she worked at Arthur D. Little, Inc. (ADL), initially as a cryogenic engineer working on liquid hydrogen refrigerators and the fueling system for the Saturn rocket. Later, she consulted on early liquefied natural gas (LNG) projects worldwide and on LNG safety research. She became a vice president for Technological Risk Management at ADL and later was vice president for their international Health, Safety, and Environmental Practice. She is a member of the National Academy of Engineering (NAE) and has served on a number of NRC committees, including topics as diverse as oversight of the U.S. Army's chemical weapons disposal operations and "Personal Cars and China" (a 2003 joint NRC report of the NAE and the Chinese Academy of Engineering). She currently serves as a member of the National Academies' Report Review Committee. She received a Sc.D. in chemical engineering from MIT.

**Mary R. English** is a research leader at the Institute for a Secure and Sustainable Environment and an associate of the Center for Applied and Professional Ethics at the University of Tennessee at Knoxville. Her current research interests include the socioeconomic impacts of biofuels production; the ethics of energy production, distribution, and consumption; state and local transportation and land-use planning; and participatory processes for environmental decision making. Over the past 20 years, her work has focused on the social and political aspects of various environmental management issues, such as siting controversial projects, cleanup of Superfund sites, the restoration and reuse of brownfield sites, and alternative mechanisms for involving stakeholders in environmental decisions. Dr. English is a member of the Tennessee Air Pollution Control Board. She previously served on the NRC Board on Radioactive Waste Management and several NRC study committees, such as the Committee on the Environmental Impacts of Wind-Energy Projects. She also served on the EPA National Environmental Justice Advisory Council and the National League of Women Voters' Advisory Committee to the Nuclear Waste Education Project. Dr. English received her Ph.D. in sociology from the University of Tennessee at Knoxville.

**Christopher B. Field** is the founding director of the Department of Global Ecology of the Carnegie Institution of Washington and professor of biology and of environmental earth system science at Stanford University. He is an ecosystem ecologist with research programs exploring the global carbon cycle and ecosystem responses to global change. Dr. Field is an author of more than 100 scientific papers, and he served on the editorial boards of

*Ecology, Ecological Applications, Ecosystems, Global Change Biology,* and *Proceedings of the National Academy of Sciences.* His recent NRC experience includes membership on the Board on International Scientific Organizations, Board on Environment Studies and Toxicology, Committee on Research at the Intersection of the Physical and Life Sciences, Committee on Earth-Atmosphere Interactions: Understanding and Responding to Multiple Environmental Stresses, Committee on Grand Challenges in the Environmental Sciences, Panel on Ecological Impacts of Climate Change, and chair of the U.S. National Committee for SCOPE. He is a member of the National Academy of Sciences. Dr. Field received his Ph.D. in biology from Stanford University.

**Daniel S. Greenbaum** is the president and chief executive officer of the Health Effects Institute (HEI), an independent not-for-profit research institute funded jointly by government and industry to provide trusted research on the health effects of air pollution. At HEI, Mr. Greenbaum has overseen the development and implementation of a strategic research plan that focuses the institute's efforts on providing critical research on and reanalysis of particulate matter, air toxics, and alternative fuels, increasingly on an international scale. He has served as chair of the EPA Blue Ribbon Panel on Oxygenates in Gasoline, as chair of EPA's Clean Diesel Independent Review Panel, and as a member of the national Clean Air Act Advisory Committee. Before joining HEI, he served as commissioner of the Massachusetts Department of Environmental Protection. In the NRC, he served on the Board on Environmental Studies and Toxicology, as vice chair of the Committee on Air Quality Management in the United States, and most recently as a member of the Committee on Estimating Mortality Risk Reduction and Economic Benefits from Controlling Ozone Air Pollution. Mr. Greenbaum earned a master's degree in city planning from MIT.

**James K. Hammitt** is professor of economics and decision sciences and director of the Harvard Center for Risk Analysis at the Harvard School of Public Health. His research and teaching concern the development of decision analysis, benefit-cost analysis, game theory, and other quantitative methods and their application to health and environmental policy in the United States and internationally. Dr. Hammitt is particularly interested in the management of long-term environmental issues, such as global climate change and stratospheric-ozone depletion, comprehensive evaluation of risk-control measures (including ancillary benefits and countervailing risks), and alternative methods for measuring the value of reducing health risks, including monetary and health-adjusted life-year metrics. He serves as a member of the EPA Science Advisory Board and its Environmental Economics Advisory Committee and chairs the EPA Advisory Council on Clear Air

Compliance Analysis. Dr. Hammitt previously served as a member of the American Statistical Association Committee on Energy Statistics (Advisory Committee to the U.S. Energy Information Administration) and the NRC panel studying the implications of dioxin in the food supply. He held the Pierre-de-Fermat Chaire d'Éxcellence at the Toulouse School of Economics and served as senior mathematician at the RAND Corporation and as a faculty member at the RAND Graduate School of Policy Studies. Dr. Hammitt received his Ph.D. in public policy from Harvard University.

**Rogene F. Henderson** is a senior biochemist and toxicologist in the Experimental Toxicology Program of the Lovelace Respiratory Research Institute. She is also a clinical professor in the College of Pharmacy at the University of New Mexico in Albuquerque. Her major research interests are in the use of bronchoalveolar lavage fluid analyses to detect and characterize biomarkers of developing lung disease, the toxicokinetics of inhaled vapors and gases, and the use of biological markers of exposure and of effects to link environmental exposure to disease. She has served on a number of scientific advisory boards, including those of DOE, EPA, NIEHS, and the U.S. Army. She was recently chair of EPA's Clean Air Scientific Advisory Committee. Dr. Henderson is a national associate of the National Academies and is a former member of the Board on Environmental Studies and Toxicology. She received her Ph.D. in chemistry from the University of Texas.

**Catherine L. Kling** is professor of economics and head of the Resource and Environmental Policy Division at the Center for Agricultural and Rural Development at Iowa State University. Her research has been in the areas of natural resource and environmental economics. Dr. Kling served on the EPA Science Advisory Board (SAB) and was a member of the SAB Environmental Economics Advisory Committee. She is currently a member of the NRC committee that is reviewing the U.S. Department of Agriculture Agricultural Resource Management Survey (ARMS). Dr. Kling received her Ph.D. in economics from the University of Maryland.

**Alan J. Krupnick** is a senior fellow and director of quality of the environment at Resources for the Future. His research focuses on analyzing environmental issues, in particular the benefits, costs, and design of air pollution policies in the United States and in developing countries. His research also addresses the valuation of health and ecological improvements and, more recently, the ancillary benefits of climate policy and urban transportation and development problems. Dr. Krupnick has served as a consultant to state governments, federal agencies, private corporations, the Canadian government, the European Union, the World Health Organization, and the World Bank. He co-chaired an advisory committee that counseled EPA on new ozone and particulate-matter standards. Dr. Krupnick participated in several NRC studies, including those of the Committee for the Evaluation of the Congestion Mitigation and Air Quality Improvement Program, the Committee on Research and Peer Review in EPA, the Surface Transportation Environmental Cooperative Research Program Advisory Board, and the Committee on Estimating Mortality Risk Reduction Benefits from Decreasing Tropospheric Ozone Exposure. He also served on a Royal Society of Canada committee analyzing ambient air-quality-standard setting in Canada. Dr. Krupnick received his Ph.D. in economics from the University of Maryland.

**Russell Lee** is a distinguished research and development (R&D) staff member in science and technology policy at Oak Ridge National Laboratory (ORNL). Previously, he was director of the Center for Energy and Environmental Analysis, and leader of the Resource Modeling Group at ORNL. Dr. Lee's research focus is on science and technology policy, energy supply and demand, environmental externalities associated with energy production and use, and transportation analysis. He was the U.S. lead person in the joint U.S.-European Commission study that resulted in the eight-volume report *Fuel Cycle Externalities* and senior author of the book *Health and Environmental Impacts of Electricity Generation Systems: Procedures for Comparative Assessment*. He is author of numerous other papers, book chapters and reports on the subject of energy externalities. Dr. Lee has been invited to present workshops, serve on expert panels, and consult on the subject of the impacts of electricity generation technologies, global climate change, environmental externalities, technology R&D policy, waste management, and related issues by the National Academies, the U.S. Department of Energy, the Organisation for Economic Cooperation and Development, the Commission of the European Communities, the International Energy Agency, the National Association of Regulatory Utility Commissioners, the International Atomic Energy Agency, the United Nations Framework Convention on Climate Change, foreign governments, universities, and industry. Prior to joining ORNL, he was an assistant professor at the University of Iowa and at Boston University. Dr. Lee is a member of the Committee on Social and Economic Factors of Transportation and of the Project Panel on Development of Research Performance Measurement Tool Box, both of the Transportation Research Board. Dr. Lee received his Ph.D. in geography from McMaster University in Canada.

**H. Scott Matthews** is the research director of the Green Design Institute and associate professor in the Department of Civil and Environmental Engineering and the Department of Engineering and Public Policy at Carnegie Mellon University. His work includes valuing the socioeconomic implications of environmental systems and infrastructure and industrial ecology. He focuses on using the internet to facilitate environmental life-cycle assessment of products and processes, estimating and tracking carbon emissions across the supply chain, and the sustainability of infrastructure. Dr. Matthews serves as chair of the Committee on Sustainable Systems and Technology with the Institute of Electrical and Electronic Engineers. He holds a Ph.D. in economics from Carnegie Mellon.

**Thomas E. McKone** is senior staff scientist and deputy department head at the Lawrence Berkeley National Laboratory and an adjunct professor and researcher at the University of California, Berkeley, School of Public Health. Dr. McKone's research interests include the development, use, and evaluation of models and data for human-health and ecological risk assessments; chemical transport and transformation in the environment; and the health and environmental impacts of energy, industrial, and agricultural systems. One of Dr. McKone's most recognized achievements was his development of the CalTOX multimedia risk- and impact-assessment framework for the California Environmental Protection Agency. He has been a member of several NRC committees, including the Committee on Improving Risk Analysis Approaches Used by the U.S. EPA, Committee on Environmental Decision Making: Principles and Criteria for Models, Committee on EPA's Exposure and Human Health Reassessment of TCDD and Related Compounds, Committee on Toxicants and Pathogens in Biosolids Applied to Land, and Committee on Toxicology. Dr. McKone was recently appointed by California Governor Arnold Schwarzenegger to the Scientific Guidance Panel for the California Environmental Contaminant Biomonitoring Program. He is a fellow of the Society for Risk Analysis, former president of the International Society of Exposure Analysis, and a member the Organizing Committee for the International Life-Cycle Initiative—a joint effort of the United Nations Environment Program and the Society for Environmental Toxicology and Chemistry. He earned his Ph.D. in engineering from the University of California, Los Angeles.

**Gilbert E. Metcalf** is professor of economics at Tufts University and research associate at the National Bureau of Economic Research. His primary research area is applied public finance with particular interests in taxation, energy, and environmental economics. His current research focuses on policy evaluation and design in the area of energy and climate change. Dr. Metcalf taught at Princeton University and the Kennedy School of Government at Harvard University and served as a visiting scholar at the Joint Program on the Science and Policy of Global Change at MIT. He served as a consultant to various organizations, including the Chinese Ministry of

Finance, the U.S. Department of the Treasury, and Argonne National Laboratory. He received his Ph.D. in economics from Harvard University.

**Richard G. Newell**[1] was appointed by President Obama as administrator of the Energy Information Administration and sworn into office on August 3, 2009. Previously, he was the Gendell Associate Professor of Energy and Environmental Economics at the Nicholas School of the Environment, Duke University. He is a research associate of the National Bureau of Economic Research, and a university fellow of Resources for the Future. He served as the senior economist for energy and environment on the President's Council of Economic Advisers, where he advised on policy issues ranging from automobile-fuel economy and renewable fuels to management of the Strategic Petroleum Reserve. He has been a member of expert committees, including the NRC Committee on Energy R&D, the Committee on Innovation Inducement Prizes, and the Committee on Energy Efficiency Measurement Approaches. Dr. Newell also served on the 2007 National Petroleum Council global oil and gas study. He currently serves on the boards of the *Journal of Environmental Economics and Management*, the journal *Energy Economics*, the Association of Environmental and Resource Economists, and the Automotive X-Prize. He served as an independent expert reviewer and advisor for many governmental, nongovernmental, international, and private institutions, including the OECD, Intergovernmental Panel on Climate Change, World Bank, National Commission on Energy Policy, EPA, U.S. Department of Energy, U.S. Energy Information Administration, U.S. National Science Foundation, and others. Dr. Newell received his Ph.D. from Harvard University.

**Richard L. Revesz** is dean at New York University Law School, where he also serves as the Lawrence King Professor of Law. His work focuses on five distinct areas: federalism and environmental regulation, design of liability regimes for environmental protection, positive political economy analysis of environmental regulation, analytical foundations of environmental law, and the use of cost-benefit analysis in administrative regulation. Dean Revesz also served as chair of the Committee on Judicial Review of the American Bar Association Section on Administrative Law and Regulatory Policy and as a member of the Environmental Economics Advisory Committee of the EPA Science Advisory Board. In addition, he has also been a visiting professor at the Woodrow Wilson School of Public and International Affairs at Princeton University, Yale Law School, Harvard University Law School, and University of Geneva School of Law. He holds a J.D. from Yale University and an M.S. in engineering from MIT.

---

[1] Dr. Newell resigned from the committee on August 2, 2009.

**Ian Sue Wing** is an associate professor in the Geography Department at Boston University (BU) and a research affiliate of the Center for Energy and Environmental Studies and the Center for Transportation Studies at BU and of the Joint Program on the Science & Policy of Global Change at MIT. In 2005-2006, he was a REPSOL-YPF energy fellow at Harvard's Kennedy School of Government. Dr. Sue Wing's research focuses on the economic analysis of energy and environmental policy, with an emphasis on climate change and computational general equilibrium analysis of economies' adjustment to policy shocks. He served on the Renewable Energy Modeling Analysis Partnership of the U.S. Department of Energy and on the Second Generation Model Advisory Panel of the EPA Science Advisory Board. He holds a Ph.D. in technology, management, and policy from MIT.

**Terrance G. Surles** is technology integration and policy analysis program manager for the Hawaii Natural Energy Institute. His research focuses on reducing dependence on petroleum from both a climate-change and energy-security perspective. Dr. Surles also serves as a senior advisor to the University of California's California Institute for Energy and Environment. In addition, he is director of the Pacific International Center for High Technology Research for which he works to develop and enhance collaborative programs between Japanese and American government- and private-sector clients. Previously, he was a vice president for the Electric Power Research Institute. He was a program director and assistant director for science and technology with the California Energy Commission. He was associate laboratory director for energy programs at Lawrence Livermore National Laboratory. Dr. Surles was deputy secretary for science and technology at the California Environmental Protection Agency. He was general manager of environmental programs at Argonne National Laboratory. He served as a member of the NRC Committee for Development of Methodology for Evaluating Prospective Benefits of U.S. Department of Energy Programs, Phases I and II. He holds a Ph.D. in analytical chemistry from Michigan State University.

# B

# A Simple Diagrammatic
# Example of an Externality

A simple stylized example helps illustrate the concept of an externality. Consider a firm generating electricity that releases air pollutants as a by-product. It can carry out various activities to abate pollution. Initial pollution reductions are relatively inexpensive to carry out, but costs rise as the firm reduces its pollution further. To illustrate that relationship, Figure B-1 diagrams pollution abatement along the horizontal axis and measures of cost per ton of abatement along the vertical axis. The figure provides an alternative approach to that shown in Figure 1-1 of Chapter 1 but leads to the same conclusion. Whereas Figure 1-1 focuses on optimal pollution levels, the discussion in this appendix focuses on optimal abatement activities. Both approaches are used in the literature.

The upward sloping line labeled "Marginal Abatement Cost" measures the cost to the firm for each additional ton of pollution reduction. In the absence of any policy intervention, the hypothetical firm will engage in no pollution abatement and incur no private abatement costs.

The horizontal line labeled "Marginal Benefit" is a measure of the reduction in aggregate damages across all people affected by pollution from this plant. The reduction could be a combination of reduced mortality risk and reduced morbidity summed over different populations. The marginal benefit of pollution abatement is simply a restatement of the marginal damages from pollution. Each ton of pollution avoided reduces incremental damages to society.

For purposes of this example, we assume that the marginal benefit of pollution abatement is constant and equal to $25 per ton of abated pollution. Equivalently, the dollar value of the marginal damages of pollution

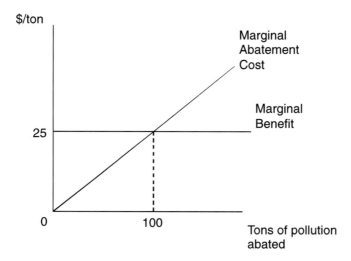

**FIGURE B-1** Pollution abatement (horizontal axis) and cost per ton of abatement (vertical axis).

for this plant is $25 per ton. In practice, the shape of this curve will be pollutant-specific (and might well be location- and time-specific).

Society is made better off if the firm increases its abatement from 0 to 100 tons. The benefit to society is the avoided damages of $25 per ton times the 100 tons abated, or $2,500. The cost to the firm of reducing its pollution is the sum of the incremental abatement costs. This is the area under the marginal abatement curve and it equals $1,250. The net gain to society following the firm's abatement action is $1,250.

Is 100 tons of pollution abatement the economically optimal amount? All other things being equal, the answer is yes. More generally, the economically optimal level of pollution abatement occurs at the point where marginal benefits equal marginal costs. To see why, consider an additional ton of abatement from 100 to 101 tons. The benefit to society is $25. The marginal cost, however, is an amount greater than $25 because the marginal abatement curve rises above $25 for abatement levels greater than 100. For abatement levels greater than 100 tons, the incremental abatement costs to the firm outweigh the incremental benefits to nearby residents. Similarly, any level of abatement below 100 tons are not economically optimal. At any level less than 100 tons, the cost to the firm of reducing pollution by 1 more ton is less than the benefit to nearby residents of that incremental pollution reduction.

Note, however, that the illustrative example does not include consideration of important distributional issues by simply summing the costs and benefits of pollution reduction. Distributional considerations can be taken into account. They will affect the economically optimal level of pollution abatement in the example but not the fundamental concepts that the example illustrates.

# C

# Description of the Air Pollution Emission Experiments and Policy (APEEP) Model and Its Application

The Air Pollution Emission Experiments and Policy (APEEP) analysis model (Muller and Mendelsohn 2006, 2009) is a traditional integrated assessment model (Mendelsohn 1980; Nordhaus 1992; Burtraw et al. 1998; EPA 1999). Like other integrated assessment models, APEEP connects emissions of air pollution through air-quality modeling to exposures, physical effects, and monetary damages. Making these links requires the use of findings reported in the peer-reviewed literature across several scientific disciplines.

APEEP is designed to calculate the marginal damage of emissions for nearly 10,000 distinct (individual and aggregated sources of air pollution in the contiguous United States. APEEP computes marginal damages of six pollutants: sulfur dioxide ($SO_2$), volatile organic compounds (VOCs), nitrogen oxides ($NO_x$), fine particulate matter ($PM_{2.5}$), coarse particulate matter ($PM_{10}$),[1] and ammonia ($NH_3$).

The individual and aggregate sources are defined by the U.S. Environmental Protection Agency (EPA 2009). Sources of emissions include both county-aggregated ground-level sources as well as point sources. Ground-level sources include vehicles, residences, and small industrial or commercial facilities without a smokestack. Emissions from individual ground-level sources are aggregated at the county level by EPA. Point sources are dif-

---

[1]The definition of $PM_{10}$ for the purpose of this modeling analysis is total particles less than 10 microns in size minus $PM_{2.5}$. This approach ensures that the consequences of $PM_{2.5}$ are not double counted. $PM_{10}$ is usually defined as particles that have an aerodynamic diameter less than or equal to 10 microns.

ferentiated by effective stack height and by location because the height of emissions affects the dispersion patterns from these sources. Emissions from point sources with an effective height of less than 250 m are aggregated to the county level, as are emissions from point sources with an effective height of 250 to 500 m. In contrast, point sources with an effective height of greater than 500 m, such as certain power plants and other large industrial facilities, are modeled individually—that is, APEEP does not aggregate emissions from these sources; they are modeled separately for each facility.

The air-quality models in APEEP use the emission data provided by EPA to estimate corresponding ambient concentrations in each county in the coterminous states. The accuracy of the estimated pollution levels produced by the APEEP model has been statistically tested against the Community Multiscale Air Quality (CMAQ) model (Byun and Schere 2006), which is considered the state-of-the-art air-quality model. The results of these statistical comparisons are shown in the accompanying materials to Muller and Mendelsohn (2006).

APEEP can be used to compute the marginal damage of emissions on a source-specific basis. This approach isolates the source-specific damage per ton for each of the six pollutants covered by the model. To calculate marginal damages, APEEP uses the following algorithm: First, APEEP estimates total damages due to all sources in the model, producing its baseline (observed) emissions (EPA 2009); next, APEEP adds 1 ton of one pollutant from one source and recomputes total damages. The marginal damage is the damage that occurs after adding 1 ton of pollutant minus the damages due to the baseline emissions. The algorithm isolates the contribution of a single ton of emissions from each source to total national damages. This approach captures the formation of secondary pollutants, such as sulfates and nitrates (constituents in $PM_{2.5}$) as well as tropospheric ozone ($O_3$) that are formed by the emissions of other substances. APEEP attributes the damage due to such secondary pollutants back to the source of emissions. As shown in Equation 1, the marginal damage is computed by adding the changes in damages across the complete set of receptor counties. (Receptor counties are those counties that receive emissions from a source.)

$$MD_{i,p} = \Sigma r \, Dr,ep - \Sigma r \, Dr,bp \qquad (1)$$

where
$MD_{i,p}$   = damage per ton of an emission of pollutant ($p$) from source ($i$).
$D_r$        = total dollar damage that occurs at receptor county ($r$).
$bp$         = 2002 baseline emissions of $p$.
$ep$         = 2002 baseline emissions plus 1 ton of $p$ from $i$.

After computing the marginal damage of emissions for a specific pollutant from source $i$, this experiment can be repeated for each of the six pollutants covered in APEEP and the approximately 10,000 distinct (individual and grouped) sources in the United States. The total 10,000 sources encompass all anthropogenic emissions of these six pollutants in the lower 48 states. It is important to note that the APEEP model, in its current form, does not test for interactions among emissions of multiple pollutants in terms of the damages that such emissions cause. The model is designed to simulate the emissions of 1 ton of one specific pollutant from a particular source and to estimate its impact rather than the emissions of multiple pollutants from a source and estimating their cumulative impact.

The following section briefly highlights the basic structure of the model and some of its most important assumptions. The model uses data on emissions (excluding carbon monoxide and lead and including ammonia) that contribute to the formation of criteria air pollutants. The data were provided by EPA's 2002 National Emission Inventory (EPA 2009). Concentrations due to the baseline levels of emissions are estimated by the air-quality models in APEEP. The air-quality modeling module makes use of a source-receptor matrix framework. That is, the marginal contribution of emissions in a source county ($s$) to the ambient concentration in a receptor county ($r$) is represented as the $s,r$ element in a matrix. Using a linear algebraic approach, APEEP multiplies the matrix times an emission vector to generate a vector of predicted ambient concentrations. When the emission vectors represent changes to existing emissions, the corresponding estimated concentrations reflect changes to the baseline levels, or existing concentrations. When the emission vectors represent the emission rates, then predicted concentrations reflect those rates, not changes to concentrations.

The model contains source-receptor matrices for the following pollutants in both summer and winter: $NO_x \rightarrow NO_x$, $SO_2 \rightarrow SO_2$. The matrix governing the relationship between $NO_x$ emissions, VOC emissions, and $O_3$ concentrations is calibrated to the summer season. The matrices representing formation and transport of particles ($PM_{2.5} \rightarrow PM_{2.5}$, $PM_{10} \rightarrow PM_{10}$, $NO_x \rightarrow PM$, $SO_2 \rightarrow PM$, $NH_3 \rightarrow NH_4$, $VOC \rightarrow PM$) produce annual means.[2] There is a specific matrix in APEEP for each of the emission-concentration relationships shown above.

The particulate matter source-receptor matrices compute the ammonium-sulfate-nitrate equilibrium, which determines the amount of ambient ammonium sulfate ($NH_4)_2$, $SO_4$, and ammonium nitrate ($NH_4NO_3$) at each receptor county. The equilibrium computations reflect several fundamental

---

[2]PM indicates that the contribution of $NO_x$, VOCs, and $SO_2$ is counted to both $PM_{2.5}$ and $PM_{10}$. Parameterization of the relationship between VOC emissions and the formation of PM is based on the work of Grosjean and Seinfeld (1989).

aspects of this system. First, ambient ammonium ($NH_4$) reacts preferentially with sulfate ($H_2SO_4$). Second, ammonium nitrate is only able to form if there is excess $NH_4$ after reacting with sulfate. Finally, particulate nitrate formation is a decreasing function of temperature, so the ambient temperature at each receptor location is incorporated into the equilibrium calculations. To translate VOCs emissions into secondary organic particulates, APEEP uses the fractional aerosol yield coefficients estimated by Grosjean and Seinfeld (1989). These coefficients represent the yield of secondary organic aerosols corresponding to emissions of gaseous VOCs.

APEEP simulates $O_3$ concentrations using an empirical model that translates ambient concentrations of VOC, CO, and $NO_x$ into ambient $O_3$ concentrations. The model captures many of the factors contributing to ambient concentrations of $O_3$, VOC, CO, and $NO_x$. These factors include forests and agricultural land uses, which produce biogenic hydrocarbons, as well as the ambient air temperature and several geographic variables. For a complete depiction of the $O_3$ modeling in APEEP, see Muller and Mendelsohn (2006). The inclusion of both linear and quadratic forms for $NO_x$, CO, and VOC concentrations in the $O_3$ models allows for the nonlinearity known to exist in $O_3$ production chemistry (Seinfeld and Pandis 1998). Specifically, the quadratic forms capture titration. This approach is critical to accurately predict $O_3$ levels in certain urban areas, where research has shown that additional emissions of $NO_x$ can result in reduced $O_3$ concentrations (Tong et al. 2006).

The source-receptor matrices in APEEP are derived from the Climatological Regional Dispersion Model (CRDM) (Latimer (1996). The original CRDM matrices have been calibrated to produce estimates of pollution levels that are in good agreement with the predictions produced by CMAQ. The correlations between APEEP's predicted surfaces and CMAQ's are especially strong for annual mean $PM_{2.5}$ levels and summer mean $O_3$ levels; the correlation coefficients are 0.82 and 0.77, respectively.[3] The matrices have been expanded in scope to encompass nearly 10,000 sources and source areas.

Following the estimation of ambient concentrations, exposures are computed by multiplying county-level populations times county-level pollution concentrations. In APEEP, populations include number of people (differentiated by age),[4] crops produced, timber harvested, an inventory of anthropogenic materials, visibility resources, and recreation usage (for each

---

[3]The correlations for $PM_{2.5}$ are expressed over n = 3,110, reflecting the 3,110 counties in the contiguous 48 states. The correlations for tropospheric $O_3$ are expressed over n = 24,880, reflecting eight hourly observations for the 3,110 counties in the coterminous U.S.

[4]Population data are provided by CDC Wonder, which is a database of the Centers for Disease Control and Prevention of the U.S. Department of Health and Human Services.

county in the contiguous United States). Each type of exposure is computed separately for each pollutant. The sources for each of these inventories are documented in Muller and Mendelsohn (2006).

In the next stage of the APEEP model, peer-reviewed concentration-response functions are used to translate exposures into the number of physical effects, including premature mortalities, cases of illness, reduced timber and crops yields, enhanced depreciation of anthropogenic materials, reduced visibility, and recreation usage. The studies that provide the concentration-response functions related to human health impacts are listed in Table C-1.

The final stage of the APEEP model attributes a dollar value to each of these physical effects. For effects on goods and services traded in markets (decreased crop yields, for example), APEEP multiplies the change in output due to exposures to air pollution times the market price. For nonmarket goods and services, APEEP uses valuation estimates from the nonmarket valuation literature in economics. APEEP values premature mortality risks using the value of a statistical life (VSL) approach (Viscusi and Aldy 2003). APEEP uses EPA's preferred VSL, which is equivalent to approximately $6 million (year 2000 real U.S. dollars). APEEP provides the option of using a VSL estimate of approximately $2 million from Mrozek and Taylor (2002) as an alternative to the EPA's VSL. The values attributed to chronic illnesses, such as bronchitis and asthma, are also derived from the nonmarket valuation literature. Acute illnesses are valued with cost of illness estimates. Each of the values applied to human health effects in APEEP are shown in Table C-3.

**TABLE C-1** Epidemiology Studies Used in APEEP

| Health Event | Pollutant | Study |
|---|---|---|
| All-cause adult chronic-exposure mortality[a] | $PM_{2.5}$ | Pope et al. 2002 |
| Infant chronic-exposure mortality | $PM_{2.5}$ | Woodruff et al. 2006 |
| Chronic bronchitis | $PM_{10}$ | Abbey et al. 1993 |
| Chronic asthma | $O_3$ | McDonnell et al. 1999 |
| Acute-exposure mortality | $O_3$ | Bell et al. 2004 |
| Respiratory admissions | $O_3$ | Schwartz 1995 |
| ER visits for asthma | $O_3$ | Steib et al. 1996 |
| COPD admissions | $NO_2$ | Moolgavkar 2000 |
| IHD admissions | $NO_2$ | Burnett et al. 1999 |
| Asthma admissions | $SO_2$ | Sheppard et al. 1999 |
| Cardiac admissions | $SO_2$ | Burnett et al. 1999 |

[a]Acute exposure mortality for $PM_{2.5}$ was not included in this analysis as a separate effect. See Muller and Mendelsohn (2007) for further discussion.

SOURCE: Muller and Mendelsohn 2006. Reprinted with permission; copyright 2007, *Journal of Environmental Economics and Management*.

The studies that provide the concentration-response functions for the remaining welfare effects are listed in Table C-2. Because $PM_{2.5}$ is a subset of $PM_{10}$, APEEP avoids double counting of damages due to $PM_{2.5}$ and $PM_{10}$. Specifically, APEEP estimates mortality impacts associated with emissions of $PM_{2.5}$, and the model measures chronic morbidity impacts of $PM_{10}$. In reporting the morbidity damages due to emissions of $PM_{10}$, APEEP nets out the mortality damages due to $PM_{2.5}$. In effect, the damages for $PM_{10}$ are expressed as $PM_{10}$-$PM_{2.5}$.

**TABLE C-2** Concentration-Response Studies Used in APEEP

| Welfare Effect | Pollutant | Study |
|---|---|---|
| Crop loss | $O_3$ | Lesser et al. 1990 |
| Timber loss | $O_3$ | Reich 1987; Pye 1988 |
| Materials depreciation | $SO_2$ | Atteras and Haagenrud 1982; ICP 2000 |
| Visibility | $PM_{10}$ | Muller and Mendelsohn 2006 |
| Forest recreation | $SO_2$, $NO_x$, $O_3$ | Muller and Mendelsohn 2006 |

**TABLE C-3** Value of Human Health Effects in APEEP[a]

| Health Event | Unit | U.S. Dollars |
|---|---|---|
| Chronic Exposure Mortality | Case | 5,910,000 |
| Chronic Bronchitis | Case | 320,000 |
| Chronic Asthma | Case | 30,800 |
| General Respiratory | Hospital Admission | 8,300 |
| General Cardiac | Hospital Admission | 17,526 |
| Asthma | Hospital Admission | 6,700 |
| COPD | Hospital Admission | 11,276 |
| Ischemic Heart Disease | Hospital Admission | 18,210 |
| Asthma | ER Visit | 240 |

[a]Values are in 2000 U.S. dollars; see Muller and Mendelsohn 2007.

SOURCE: Modified from Muller and Mendelsohn 2006.

**TABLE C-4** Value of Nonmarket Impacts of Air Pollution

| Welfare Effect | U.S. Dollars[a] | Location | Source |
|---|---|---|---|
| Recreation visibility (in-region) | 170 | Southwest | Chestnut and Rowe 1990 |
| Recreation visibility (out-region) | 135 | Southwest | Chestnut and Rowe 1990 |
| Recreation visibility (in-region) | 80 | Southeast | Chestnut and Rowe 1990 |
| Recreation visibility (out-region) | 50 | Southeast | Chestnut and Rowe 1990 |
| Residential visibility (in-region) | 174 | Eastern | McClelland et al. 1993 |
| Forest recreation visit | 63 | All | Kengen 1997 |

[a]Values are in 2000 U.S. dollars; see Muller and Mendelsohn 2007.

SOURCE: Modified from Muller and Mendelsohn 2006.

Each of the other nonmarket impacts of air pollution modeled by AP-EEP (impaired visibility and reduced recreation services) are also expressed in dollar terms. The values used in the APEEP model corresponding to these welfare effects are displayed in Table C-4.

## REFERENCES

Abbey, D.E., F. Peterson, P.K. Mills, and W.L. Beeson. 1993. Long-term ambient concentrations of total suspended particulates, ozone, and sulfur dioxide and respiratory symptoms in a nonsmoking population. Arch. Environ. Health 48(1):33-46.

Atteras, L., and S. Haagenrud. 1982. Atmospheric corrosion testing in Norway. Pp. 873-892 in Atmospheric Corrosion, W.H. Ailor, ed. New York: Wiley.

Bell, M.L., A. McDermott, S.L. Zeger, J.M. Samet, and F. Domenici. 2004. Ozone and short-term mortality in 95 U.S. urban communities, 1987-2000. JAMA 292(17):2372-2378.

Burnett, R.D., M. Smith-Doiron, D. Steib, S. Cakmak, and J. Brook. 1999. Effects of particulate and gaseous air pollution on cardiorespiratory hospitalizations. Arch. Environ. Health 54(2):130-139.

Burtraw, D., A. Krupnick, E. Mansur, D. Austin, and D. Farrell. 1998. Costs and benefits of reducing air pollutants related to acid rain. Contemp. Econ. Policy 16 (4):379-400.

Byun, D.W., and L.K. Schere. 2006. Review of the governing equations, computational algorithms, and other components of the models-3 Community Multiscale Air Quality (CMAQ) modeling system. Appl. Mech. Rev. 59(2):51-77.

Chestnut, L.G., and R.D. Rowe. 1990. Economic valuation of changes in visibility: A state of the science assessment for NAPAP. Pp. 27-153 to 27-175 in Report 27. Methods for Valuing Acidic Deposition and Air Pollution Effects. Acidic Deposition: State of Science and Technology, Vol. 4. Control Technologies, Future Emission, and Effects Valuation, P.M. Irving, ed. Washington, DC: U.S. National Acid Precipitation Assessment Program.

EPA (U.S. Environmental Protection Agency). 1999. The Benefits and Costs of the Clean Air Act 1990 to 2010, EPA Report to Congress. EPA-410-R-99-001. Office of Air and Radiation, Office of Policy, U. S. Environmental Protection Agency, Washington, DC. November 1999 [online]. Available: http://www.epa.gov/oar/sect812/1990-2010/chap1130.pdf [accessed Sept. 16, 2009].

EPA (U.S. Environmental Protection Agency). 2009. National Emissions Inventory (NEI) Data & Documentation. Office of Air Quality Planning and Standards, U. S. Environmental Protection Agency, Washington, DC [online]. Available: http://www.epa.gov/ttnchie1/net/2002inventory.html [accessed Sept 16, 2009].

Grosjean, D., and J. Seinfeld. 1989. Parameterization of the formation potential of secondary organic aerosols. Atmos. Environ. 23:1733-1747.

ICP (International Co-operative Programme). 2000. International Co-operative Programme on Effects of Air Pollution on Materials, including Historic and Cultural Monuments: Results [online]. Available: http://www.corr-institute.se/ICP-Materials/web/page.aspx?pageid=59263 [accessed Apr. 15, 2010].

Kengen, S. 1997. Forest Valuation for Decision-Making: Lessons of Experience and Proposals for Improvement. Food and Agriculture Organization of the United Nations, Rome, Italy. February 1997 [online]. Available: ftp://ftp.fao.org/docrep/fao/003/W3641E/W3641E00.pdf [accessed Sept. 15, 2009].

Latimer, D.A. 1996. Particulate Matter Source-Receptor Relationships Between all Point and Area Sources in the United States and PSD Class I Area Receptors. Prepared for Office of Air Quality Planning and Standards, U.S. Environmental Protection Agency, Research Triangle Park, NC. September 1996.

Lesser, V.M., J.O. Rawlings, S.E. Spruill, and M.C. Somerville. 1990. Ozone effects on agricultural crops: Statistical methodologies and estimated dose-response relationships. Crop Sci. 30(1):148-155.

McClelland, G.H., W.D. Schulze, D. Waldman, D. Schenk, J.R. Irwin, T. Stewart, L. Deck, and M.A. Thayer. 1993. Valuing Eastern Visibility: A Field Test of the Contingent Valuation Method. Prepared for Office of Policy Planning and Evaluation, U.S. Environmental Protection Agency, Washington, DC, by the University of Colorado. September 1993 [online]. Available: http://yosemite1.epa.gov/ee/epa/eerm.nsf/vwAN/EE-0008-1.pdf/$file/EE-0008-1.pdf [accessed Sept. 15, 2009].

McDonnell, W.F., D.E. Abbey, N. Nishino, and M.D. Lebowitz. 1999. Long-term ambient ozone concentration and the incidence of asthma in non-smoking adults: The AHSMOG study. Environ. Res. 80(1):110-121.

Mendelsohn, R. 1980. An economic analysis of air pollution from coal-fired power plants. J. Environ. Econ. Manage. 7:30-43.

Moolgavkar, S.H. 2000. Air pollution and hospital admissions for chronic obstructive pulmonary disease in three metropolitan areas in the United States. Inhal. Toxicol. 12(Suppl. 4):75-90.

Mrozek, J.R., and L.O. Taylor. 2002. What determines the value of life? A meta-analysis. J. Policy Anal. Manage. 21(2):253-270.

Muller, N.Z., and R.O. Mendelsohn. 2006. The Air Pollution Emission and Policy Analysis Model (APEEP): Technical Appendix. Yale University, New Haven, CT. December 2006 [online]. Available: https://segueuserfiles.middlebury.edu/nmuller/APEEP_Tech_Appendix.pdf [accessed Oct. 7, 2009].

Muller, N.Z., and R.O. Mendelsohn. 2007. Measuring the damages from air pollution in the U.S. J. Environ. Econ. Manage. 54(1):1-14.

Muller, N., and R. Mendelsohn. 2009. Efficient pollution regulation: Getting the prices right. Am. Econ. Rev. 99(5):1714-1739.

Nordhaus, W.D. 1992. An optimal transition path for controlling greenhouse gases. Science 258 (5086):1315-1319.

Pope, C.A., R.T. Burnett, M.J. Thun, E.E. Calle, D. Krewski, K. Ito, and G.D. Thurston. 2002. Lung cancer, cardiopulmonary mortality, and long-term exposure to fine particulate air pollution. JAMA 287(9):1132-1141.

Pye, J.M. 1988. Impact of ozone on the growth and yield of trees: A review. J. Environ. Qual. 17:347-360.

Reich, P.B. 1987. Quantifying plant response to ozone: A unifying theory. Tree Physiol. 3(1):63-91.

Schwartz J. 1995. Short term fluctuations in air pollution and hospital admissions of the elderly for respiratory disease. Thorax 50(5):531-538.

Schwartz, J., D. Slater, T.V. Larson, W.E. Pierson, and J.Q. Koenig. 1993. Particulate air pollution and hospital emergency room visits for asthma in Seattle. Am. Rev. Respir. Dis. 147(4):826-831.

Seinfeld, J.H., and S.N. Pandis. 1998. Atmospheric Chemistry and Physics. New York: John Wiley & Sons.

Sheppard, L., D. Levy, G. Norris, T.V. Larson, and J.Q. Koenig. 1999. Effects of ambient air pollution on nonelderly asthma hospital admissions in Seattle, Washington, 1987-1994. Epidemiology 10(1):23-30.

Steib, D.M., R.T. Burnett, R.C. Beveridge, and J.R. Brook. 1996. Association between ozone and asthma emergency department visits in St. Jon, New Brunswick, Canada. Environ. Health Perspect. 104(12):1354-1360.

Tong, D.Q., N.Z. Muller, D.L. Mauzerall, and R.O. Mendelsohn. 2006. Integrated assessment of the spatial variability of ozone impacts from emissions of nitrogen oxides. Environ. Sci. Technol. 40(5):1395-1400.

Viscusi, W.K., and J.E. Aldy. 2003. The value of a statistical life: A critical review of market estimates throughout the world. J. Risk Uncertain. 27(1):5-76.

Woodruff, T.J., J.D. Parker, and K.C. Schoendorf. 2006. Fine particulate matter (PM2.5) air pollution and selected causes of postneonatal infant mortality in California. Environ. Health Perspect. 114(5):786-790.

# D

# Description of GREET and Mobile6 Models and Their Applications

## BACKGROUND

### The Need for Emissions Data in the National Research Council Study

To evaluate the per vehicle miles traveled (VMT) total damages from transportation, the APEEP[1] county emission-unit-damage costs must be evaluated against vehicle emissions. Although the passenger and freight fleets are diverse, particular vehicle and fuel combinations dominated in 2005, and certain vehicles and fuels are of particular interest for 2030. It is important to acknowledge life-cycle considerations related to both the vehicles and the fuels. In particular, feedstock production, fuel production, and vehicle manufacturing could have significant emissions contributions in the life-cycle inventory. The vehicle-fuel inventory should include these life-cycle components in addition to vehicle operation.

### Available Options for Constructing Emissions Estimates

Although tools and data are available to evaluate the many vehicle and fuel operational emissions, GREET[2] stands as one of the few resources to evaluate life-cycle component emissions (Argonne National Laboratory 2009). The GREET life-cycle factors cover a range of light-duty vehicles and the fuels they consume. GREET evaluates the many processes involved from

---

[1] Air Pollution Emission Experiments and Policy.
[2] Greenhouse Gases, Regulated Emissions, and Energy Use in Transportation.

feedstock production through vehicle operation. Without using GREET, individual process assessments throughout the supply chain would need to be performed and combined for each vehicle and fuel of interest.

## EMISSIONS DATA AND MODELING

### The GREET Model

The Argonne GREET model is used to determine emissions from light-duty autos and trucks. The GREET model is a vehicle operation and fuel production life-cycle assessment tool, which captures fuel feedstock production, fuel refining, vehicle operation, and vehicle manufacturing. Feedstock production, fuel refining, and vehicle operation are estimated with the GREET 1.8b model; vehicle manufacturing is determined with GREET 2.7a. The version designations (1.8b and 2.7a) do not imply different generations of GREET but distinguish between a version developed for the fuel cycle (1.8b) versus a version developed for the vehicle cycle (2.7). The strength of the GREET model lies in its ability to estimate a variety of fuel inputs and vehicle combinations and their associated well-to-wheel life-cycle components. GREET allows for specification of critical inputs to these components (for example, emission factors, combustion technologies, energy efficiencies, and fuel types).

GREET evaluates several life-cycle components for the feedstock production, fuel production, and vehicle-manufacturing emissions inventory. For the feedstock and fuel production cycle, GREET captures extraction and creation of raw feedstock, transport to refineries, refinery processes, and transport to fueling stations. These constitute the well-to-pump components. On the vehicle-cycle side, GREET performs a materials-based life-cycle assessment capturing raw material extraction, processing, transport, and ultimately assembly into an automobile or light-duty truck. GREET does not estimate heavy-duty vehicle life-cycle factors, so additional data sources were needed to evaluate these vehicle classes.

GREET allows for the adjustment of many feedstock, fuel, and vehicle operation input parameters; however, particular inputs were targeted for the vehicle and fuel combinations evaluated. The evaluation year was toggled for the 2005 and 2030 scenarios to capture changes in both vehicle operational performance as well as efficiency changes in other devices, such as engines and turbines. The fraction of crude oil that comes from tar sands and the amount of reformulated gasoline were adjusted on the basis of the vehicle and fuel combination. For ethanol, GREET inputs for feedstocks (corn, herbaceous, and corn stover) and milling processes (dry or wet) were changed. Another critical input parameter for the assessment is the fraction of low-sulfur diesel. Last, the electricity mix for 2005 and 2030 were adjusted on the basis of the U.S. Energy Information Administration's *Annual Energy Outlook*, which reports historical mixes as well as future forecasts (EIA 2006, 2009a).

## Mobile6.2

To evaluate heavy-duty vehicle emissions, EPA's Mobile6.2 on-road emissions modeling tool was used (EPA 2009). Unlike GREET, Mobile6.2 is designed to evaluate the many different conditions under which vehicles may operate, and not feedstock production, fuel production, or vehicle-manufacturing life-cycle emissions. Mobile6.2 heavy-duty vehicle operational emission factors were used in combination with GREET feedstock and fuel production factors to create life-cycle inventories for several different vehicle classes.

GREET does not evaluate ammonia emissions, so Mobile6.2 is used to capture this pollutant for both light- and heavy-duty vehicles. Ammonia emissions, which result in secondary particle formation, were determined by Mobile6.2 for a set of vehicles that overlap with the vehicle and fuel combinations evaluated in GREET. Ammonia emissions were estimated for the vehicle operation component only; they were not estimated for the feedstock, fuel, and vehicle manufacturing components.

## THE EMISSIONS MODELING PROCESS

### Model Framework

The emissions model utilized GREET to generate feedstock, fuel production, operation, and vehicle-manufacturing factors for light-duty vehicles and Mobile6.2 to generate operational factors for heavy-duty vehicles. GREET feedstock and fuel production factors were applied to the heavy-duty vehicle Mobile6.2 operational factors, as described later in this appendix. For all vehicles, energy inputs, $CO_2$, $CH_4$, $N_2O$, VOC, CO, $NO_x$, $PM_{10}$, $PM_{2.5}$, and $SO_x$ emissions are determined for the life-cycle components. APEEP county unit damages are based on emissions of VOCs, $NO_x$, $PM_{2.5}$, and $SO_x$.

### GREET Temporal Boundaries

GREET can evaluate vehicles and life-cycle processes from 1990 through 2020. The tool has many time series for engines, turbines, and critical parameters that capture changes in efficiencies, emissions, and other parameters (for example, ethanol yields from corn and fuel sulfur levels) historically and up to 2020. GREET also makes the assumption that fleet age is 5 years. When evaluating life-cycle emissions in a year, GREET assumes that vehicles are 5 years older and assigns them the corresponding emissions. When using GREET to evaluate vehicles in 2005, emissions from vehicles correspond to year 2000. However, all other values in GREET's assessment (such as fuel sulfur levels or electricity mixes) correspond to 2005.

The 2030 assessment is outside the GREET temporal upper range, so 2020 is used as a baseline (although adjustments are made and described later in this section).

## GREET Vehicle-Manufacturing Emissions

The GREET 2.7a model was used to determine vehicle-manufacturing emissions. The model performs a life-cycle assessment from vehicle material inputs to determine emissions from manufacturing for cars and SUVs. The model distinguishes between internal-combustion-engine vehicles, hybrid electric vehicles, and fuel-cell electric vehicles from both conventional and light-weight materials. The material inputs are evaluated for the body, powertrain system, transmission system, chassis, battery, fluids, paint, traction motor, generator, electronic controller, and fuel-cell auxiliary system. These components are assessed from material extraction through assembly, and emissions are determined at each stage. Disposal is included.

There is no time dependency with GREET's vehicle-manufacturing assessment, so process changes from 2005 through 2030 are not captured. Energy and emission factors are determined for the vehicle size, power-delivery systems, and material-composition combinations, as shown in Table D-1 and Table D-2.

The car conventional-material factors are used for all light-duty autos, and the SUV conventional-material factors are used for light-duty trucks class 1 and 2.

## GREET Light-Duty Auto and Truck Energy and Emissions Factors

Light-duty automobile and truck life-cycle energy inputs and emissions are determined from GREET. GREET distinguishes between light-duty trucks class 1 and 2 to capture the increased energy requirements and resulting emissions of the larger vehicles. Class 1 trucks are between zero and 6,000 lb gross-vehicle-weight rating (GVWR) and less than 3,750 lb loaded vehicle weight (LVW), and class 2 trucks have the same GVWR and greater than 3,750 LVW. For each vehicle and fuel combination, GREET is used to determine feedstock, fuel, and operational factors for light-duty autos, trucks in class 1 (LDT1), and trucks in class 2 (LDT2).

GREET allows for the adjustment of many vehicle and fuel parameters; however, certain critical parameters are adjusted some of the vehicle and fuel combinations to estimate life-cycle emissions. For reformulated gaso-

**TABLE D-1** GREET 2.7a Vehicle-Manufacturing Results for Cars

|  | ICEV: Conventional Material | ICEV: Light-Weight Material | HEV: Conventional Material | HEV: Light-Weight Material | FCV: Conventional Material | FCV: Light-Weight Material |
|---|---|---|---|---|---|---|
| Lifetime VMT | 160,000 | 160,000 | 160,000 | 160,000 | 160,000 | 160,000 |
| Total energy | 633 | 619 | 645 | 669 | 792 | 797 |
| Fossil fuels | 592 | 573 | 600 | 618 | 732 | 735 |
| Coal | 223 | 164 | 235 | 190 | 264 | 218 |
| Natural gas | 243 | 226 | 243 | 241 | 308 | 298 |
| Petroleum | 126 | 183 | 122 | 187 | 160 | 219 |
| $CO_2$ | 47 | 46 | 50 | 50 | 62 | 60 |
| $CH_4$ | 0.082 | 0.077 | 0.083 | 0.083 | 0.102 | 0.099 |
| $N_2O$ | 0.001 | 0.001 | 0.001 | 0.001 | 0.001 | 0.001 |
| GHGs | 50 | 48 | 52 | 52 | 65 | 63 |
| VOC | 0.206 | 0.205 | 0.206 | 0.205 | 0.205 | 0.205 |
| CO | 0.250 | 0.093 | 0.226 | 0.098 | 0.217 | 0.089 |
| $NO_x$ | 0.075 | 0.080 | 0.077 | 0.085 | 0.094 | 0.099 |
| $PM_{10}$ | 0.082 | 0.066 | 0.080 | 0.071 | 0.092 | 0.081 |
| $PM_{2.5}$ | 0.033 | 0.028 | 0.031 | 0.029 | 0.036 | 0.033 |
| $SO_x$ | 0.137 | 0.147 | 0.228 | 0.213 | 0.286 | 0.259 |

**TABLE D-2** GREET 2.7a Vehicle-Manufacturing Results for SUVs

|  | ICEV: Conventional Material | ICEV: Light-Weight Material | HEV: Conventional Material | HEV: Light-Weight Material | FCV: Conventional Material | FCV: Light-Weight Material |
|---|---|---|---|---|---|---|
| Lifetime VMT | 180,000 | 180,000 | 180,000 | 180,000 | 180,000 | 180,000 |
| Total energy | 730 | 728 | 840 | 833 | 1030 | 970 |
| Fossil fuels | 683 | 672 | 780 | 766 | 951 | 890 |
| Coal | 263 | 194 | 316 | 244 | 350 | 271 |
| Natural gas | 280 | 266 | 318 | 299 | 399 | 361 |
| Petroleum | 140 | 212 | 146 | 223 | 203 | 259 |
| $CO_2$ | 54 | 54 | 65 | 62 | 80 | 73 |
| $CH_4$ | 0.095 | 0.090 | 0.108 | 0.103 | 0.132 | 0.120 |
| $N_2O$ | 0.001 | 0.001 | 0.001 | 0.001 | 0.001 | 0.001 |
| GHGs | 57 | 56 | 68 | 65 | 84 | 76 |
| VOC | 0.308 | 0.307 | 0.308 | 0.307 | 0.308 | 0.307 |
| CO | 0.298 | 0.105 | 0.316 | 0.116 | 0.297 | 0.103 |
| $NO_x$ | 0.085 | 0.092 | 0.096 | 0.102 | 0.118 | 0.117 |
| $PM_{10}$ | 0.095 | 0.078 | 0.107 | 0.089 | 0.121 | 0.100 |
| $PM_{2.5}$ | 0.038 | 0.033 | 0.041 | 0.036 | 0.047 | 0.041 |
| $SO_x$ | 0.151 | 0.166 | 0.297 | 0.267 | 0.373 | 0.317 |

line vehicles, GREET's share of reformulated gasoline in total gasoline factor was set to 100%. For conventional and reformulated gasoline vehicles using petroleum derived from tar sands oil, GREET's share of oil sands products in crude oil refineries was set to 100%. GREET assumes that in 2005 an 80% share of dry mill corn ethanol production (this increases to 90% by 2020). In evaluating E10 and E85 fueled vehicles from corn ethanol feedstock, this percentage was adjusted. For E10 and E85 from dry corn this was set to 100% while from wet corn, to 0% (or 100% wet milling plants). To evaluate the compression ignition direct injection low-sulfur diesel combination, the share of low-sulfur diesel in total diesel use was specified as 100% for 2005. For the other vehicle and fuel combinations, default GREET values were left unchanged. The ethanol yield factors were verified against existing literature and electricity mixes for the two time periods received slight adjustments based on the U.S. Energy Information Administration's Annual Energy Outlook. The energy and emission factors for the different vehicle types (LDA, LDT1, and LDT2) in 2005 are shown in Table D-3, Table D-4, and Table D-5 and for 2020 in Table D-6, Table D-7, and Table D-8.

### 2030 Fuel Economy and Emission-Factor Adjustments

The implementation of 35 miles per gallon fuel economy standards for 2030 requires an adjustment to GREET default 2020 emission factors. The GREET model assumes fuel economies between 20 and 30 miles per gallon for conventional gasoline and E85 light-duty automobiles in 2020. For light-duty trucks the fuel economy ranges are even lower (20-24 miles per gallon for LDT1 and 17-20 miles per gallon for LDT2). For 2030, all energy and emission factors are adjusted based on the GREET default fuel economies and the expected 35 miles per gallon standard. Fuel and feedstock factors from GREET are reduced by the percentage reduction of default and 35 mile per gallons economies (for example, if the 2020 fuel economy is specified as 24 miles per gallon then the fuel and feedstock emission factors for 2020 are multiplied by 24/35 to determine the adjusted 2030 factors). This is based on the assumption that with an increase in fuel economy, a proportional reduction is needed in fuel production, which results in lower feedstock requirements. Vehicle operation combustion factors are also reduced using the same methodology. VOC evaporative losses and PM tire and brake wear factors were left unchanged from GREET default values as well as vehicle manufacturing. Both automobiles and light-duty trucks were assessed the adjusted factors. Trucks show the largest changes from default to the 35 miles per gallon standard due to relatively low GREET estimated 2020 fuel economies. All vehicles that had fuel economies greater than 35 miles per gallon in GREET in 2020 were not adjusted.

438

TABLE D-3 GREET Energy and Emission Factors for Light-Duty Autos in 2005

| | | Total Energy Btu/VMT | Petroleum Btu/VMT | $CO_2$ g/VMT | $CO_{2e}$ g/VMT | VOC g/VMT | $NO_x$ g/VMT | $PM_{2.5}$ g/VMT | $SO_x$ g/VMT |
|---|---|---|---|---|---|---|---|---|---|
| RFG SI Autos (Conventional Oil) | Feedstock | 232 | 64 | 6 | 18 | 0.02 | 0.15 | 0.01 | 0.05 |
| | Fuel | 1241 | 457 | 72 | 77 | 0.13 | 0.17 | 0.02 | 0.11 |
| | Operation | 5259 | 5038 | 404 | 408 | 0.23 | 0.30 | 0.02 | 0.01 |
| | Vehicle | 633 | 126 | 47 | 50 | 0.21 | 0.08 | 0.03 | 0.14 |
| RFG SI Autos (Tar Sands) | Feedstock | 853 | 62 | 50 | 64 | 0.02 | 0.14 | 0.01 | 0.06 |
| | Fuel | 1295 | 457 | 72 | 78 | 0.13 | 0.17 | 0.02 | 0.11 |
| | Operation | 5259 | 5038 | 404 | 408 | 0.23 | 0.30 | 0.02 | 0.01 |
| | Vehicle | 633 | 126 | 47 | 50 | 0.21 | 0.08 | 0.03 | 0.14 |
| CG SI Autos (Conventional Oil) | Feedstock | 236 | 65 | 17 | 29 | 0.02 | 0.15 | 0.01 | 0.05 |
| | Fuel | 1007 | 447 | 68 | 71 | 0.13 | 0.14 | 0.02 | 0.10 |
| | Operation | 5334 | 5257 | 410 | 414 | 0.23 | 0.30 | 0.02 | 0.01 |
| | Vehicle | 633 | 126 | 47 | 50 | 0.21 | 0.08 | 0.03 | 0.14 |
| CG SI Autos (Tar Sands) | Feedstock | 865 | 63 | 62 | 76 | 0.02 | 0.14 | 0.01 | 0.06 |
| | Fuel | 1059 | 447 | 68 | 72 | 0.13 | 0.14 | 0.02 | 0.10 |
| | Operation | 5334 | 5257 | 410 | 414 | 0.23 | 0.30 | 0.02 | 0.01 |
| | Vehicle | 633 | 126 | 47 | 50 | 0.21 | 0.08 | 0.03 | 0.14 |
| RFG SIDI Autos (Conventional Oil) | Feedstock | 202 | 55 | 5 | 16 | 0.02 | 0.13 | 0.00 | 0.05 |
| | Fuel | 1080 | 397 | 62 | 67 | 0.12 | 0.15 | 0.02 | 0.10 |
| | Operation | 4573 | 4381 | 351 | 355 | 0.23 | 0.30 | 0.02 | 0.01 |
| | Vehicle | 633 | 126 | 47 | 50 | 0.21 | 0.08 | 0.03 | 0.14 |
| RFG SIDI Autos (Tar Sands) | Feedstock | 742 | 54 | 44 | 56 | 0.02 | 0.12 | 0.01 | 0.05 |
| | Fuel | 1126 | 397 | 63 | 68 | 0.12 | 0.15 | 0.02 | 0.10 |
| | Operation | 4573 | 4381 | 351 | 355 | 0.23 | 0.30 | 0.02 | 0.01 |
| | Vehicle | 633 | 126 | 47 | 50 | 0.21 | 0.08 | 0.03 | 0.14 |

| | | | | | | | | | |
|---|---|---|---|---|---|---|---|---|---|
| Diesel (Low Sulfur) | Feedstock | 193 | 53 | 19 | 29 | 0.02 | 0.12 | 0.00 | 0.04 |
| | Fuel | 613 | 309 | 46 | 48 | 0.02 | 0.09 | 0.01 | 0.07 |
| | Operation | 4383 | 4383 | 347 | 350 | 0.09 | 0.30 | 0.07 | 0.00 |
| | Vehicle | 633 | 126 | 47 | 50 | 0.21 | 0.08 | 0.03 | 0.14 |
| Diesel (Fischer Tropsch) | Feedstock | 280 | 18 | 22 | 35 | 0.02 | 0.11 | 0.00 | 0.05 |
| | Fuel | 2882 | 68 | 106 | 114 | 0.03 | 0.25 | 0.06 | 0.09 |
| | Operation | 4383 | 0 | 334 | 338 | 0.09 | 0.30 | 0.07 | 0.00 |
| | Vehicle | 633 | 126 | 47 | 50 | 0.21 | 0.08 | 0.03 | 0.14 |
| Diesel (Soy BD20) | Feedstock | 427 | 213 | -30 | -13 | 0.03 | 0.23 | 0.01 | 0.13 |
| | Fuel | 9433 | 14 | 12 | 5 | 0.28 | -0.03 | 0.00 | -0.06 |
| | Operation | 4383 | 3555 | 344 | 348 | 0.09 | 0.30 | 0.07 | 0.04 |
| | Vehicle | 633 | 126 | 47 | 50 | 0.21 | 0.08 | 0.03 | 0.14 |
| CNG | Feedstock | 436 | 24 | 31 | 64 | 0.03 | 0.16 | 0.00 | 0.07 |
| | Fuel | 415 | 20 | 35 | 36 | 0.00 | 0.05 | 0.01 | 0.12 |
| | Operation | 5615 | 0 | 333 | 342 | 0.15 | 0.30 | 0.02 | 0.00 |
| | Vehicle | 633 | 126 | 47 | 50 | 0.21 | 0.08 | 0.03 | 0.14 |
| E85 (Dry Corn) | Feedstock | 822 | 229 | -225 | -173 | 0.06 | 0.42 | 0.02 | 0.19 |
| | Fuel | 6277 | 345 | 72 | 82 | 0.26 | 0.50 | 0.05 | 0.18 |
| | Operation | 5334 | 1412 | 402 | 406 | 0.22 | 0.30 | 0.02 | 0.00 |
| | Vehicle | 633 | 126 | 47 | 50 | 0.21 | 0.08 | 0.03 | 0.14 |
| E85 (Wet Corn) | Feedstock | 898 | 498 | -212 | -144 | -0.14 | 0.58 | 0.03 | 0.29 |
| | Fuel | 6386 | 201 | 221 | 233 | 0.25 | 0.41 | 0.12 | 0.40 |
| | Operation | 5334 | 1412 | 402 | 406 | 0.22 | 0.30 | 0.02 | 0.00 |
| | Vehicle | 633 | 126 | 47 | 50 | 0.21 | 0.08 | 0.03 | 0.14 |
| E85 (Herbaceous) | Feedstock | 570 | 216 | -282 | -237 | 0.06 | 0.29 | 0.02 | 0.06 |
| | Fuel | 4618 | 179 | 3 | 15 | 0.19 | 0.45 | 0.03 | -0.02 |
| | Operation | 5334 | 1412 | 402 | 406 | 0.22 | 0.30 | 0.02 | 0.00 |
| | Vehicle | 633 | 126 | 47 | 50 | 0.21 | 0.08 | 0.03 | 0.14 |

continued

TABLE D-3 Continued

| | | Total Energy Btu/VMT | Petroleum Btu/VMT | CO$_2$ g/VMT | CO$_{2e}$ g/VMT | VOC g/VMT | NO$_x$ g/VMT | PM$_{2.5}$ g/VMT | SO$_x$ g/VMT |
|---|---|---|---|---|---|---|---|---|---|
| E85 (Corn Stover) | Feedstock | 407 | 229 | -263 | -261 | 0.03 | 0.24 | 0.02 | 0.11 |
| | Fuel | 4206 | 179 | 3 | 14 | 0.19 | 0.43 | 0.03 | -0.02 |
| | Operation | 5334 | 1412 | 402 | 406 | 0.22 | 0.30 | 0.02 | 0.00 |
| | Vehicle | 633 | 126 | 47 | 50 | 0.21 | 0.08 | 0.03 | 0.14 |
| E10 (Dry Corn) | Feedstock | 287 | 79 | 1 | 17 | 0.02 | 0.17 | 0.01 | 0.06 |
| | Fuel | 1343 | 430 | 66 | 69 | 0.13 | 0.16 | 0.02 | 0.11 |
| | Operation | 5334 | 4990 | 409 | 413 | 0.23 | 0.30 | 0.02 | 0.01 |
| | Vehicle | 633 | 126 | 47 | 50 | 0.21 | 0.08 | 0.03 | 0.14 |
| E10 (Wet Corn) | Feedstock | 294 | 103 | 2 | 19 | 0.01 | 0.19 | 0.01 | 0.07 |
| | Fuel | 1352 | 417 | 79 | 82 | 0.13 | 0.15 | 0.03 | 0.12 |
| | Operation | 5334 | 4990 | 409 | 413 | 0.23 | 0.30 | 0.02 | 0.01 |
| | Vehicle | 633 | 126 | 47 | 50 | 0.21 | 0.08 | 0.03 | 0.14 |
| E10 (Herbaceous) | Feedstock | 265 | 78 | -4 | 11 | 0.02 | 0.16 | 0.01 | 0.05 |
| | Fuel | 1197 | 416 | 60 | 63 | 0.13 | 0.15 | 0.02 | 0.09 |
| | Operation | 5334 | 4990 | 409 | 413 | 0.23 | 0.30 | 0.02 | 0.01 |
| | Vehicle | 633 | 126 | 47 | 50 | 0.21 | 0.08 | 0.03 | 0.14 |
| E10 (Corn Stover) | Feedstock | 251 | 79 | -2 | 9 | 0.02 | 0.16 | 0.01 | 0.06 |
| | Fuel | 1161 | 416 | 60 | 63 | 0.13 | 0.15 | 0.02 | 0.09 |
| | Operation | 5334 | 4990 | 409 | 413 | 0.23 | 0.30 | 0.02 | 0.01 |
| | Vehicle | 633 | 126 | 47 | 50 | 0.21 | 0.08 | 0.03 | 0.14 |
| Electric | Feedstock | 159 | 56 | 12 | 25 | 0.03 | 0.08 | 0.13 | 0.04 |
| | Fuel | 2734 | 81 | 380 | 381 | 0.01 | 0.53 | 0.02 | 1.31 |
| | Operation | 1778 | 84 | 0 | 0 | 0.00 | 0.00 | 0.01 | 0.00 |
| | Vehicle | 645 | 122 | 50 | 52 | 0.21 | 0.08 | 0.03 | 0.23 |

| | | | | | | | | | |
|---|---|---|---|---|---|---|---|---|---|
| Hydrogen (Gaseous) | Feedstock | 180 | 10 | 13 | 27 | 0.01 | 0.07 | 0.00 | 0.03 |
| | Fuel | 1539 | 29 | 243 | 250 | 0.01 | 0.14 | 0.04 | 0.16 |
| | Operation | 2319 | 0 | 0 | 0 | 0.00 | 0.00 | 0.01 | 0.00 |
| | Vehicle | 792 | 160 | 62 | 65 | 0.21 | 0.09 | 0.04 | 0.29 |
| Grid-Independent SI HEV | Feedstock | 159 | 44 | 11 | 20 | 0.01 | 0.10 | 0.00 | 0.04 |
| | Fuel | 680 | 302 | 46 | 48 | 0.09 | 0.10 | 0.01 | 0.07 |
| | Operation | 3604 | 3552 | 277 | 281 | 0.16 | 0.25 | 0.02 | 0.00 |
| | Vehicle | 645 | 122 | 50 | 52 | 0.21 | 0.08 | 0.03 | 0.23 |
| Grid-Dependent SI HEV | Feedstock | 159 | 48 | 12 | 21 | 0.02 | 0.09 | 0.04 | 0.04 |
| | Fuel | 1358 | 229 | 156 | 158 | 0.06 | 0.24 | 0.01 | 0.48 |
| | Operation | 3002 | 2408 | 185 | 188 | 0.10 | 0.17 | 0.01 | 0.00 |
| | Vehicle | 645 | 122 | 50 | 52 | 0.21 | 0.08 | 0.03 | 0.23 |

**TABLE D-4** GREET Energy and Emission Factors for Light-Duty Trucks 1 in 2005

| | | Total Energy | Petroleum | $CO_2$ | $CO_{2e}$ | VOC | $NO_x$ | $PM_{2.5}$ | $SO_x$ |
|---|---|---|---|---|---|---|---|---|---|
| | | Btu/VMT | Btu/VMT | g/VMT | g/VMT | g/VMT | g/VMT | g/VMT | g/VMT |
| RFG SI Autos (Conventional Oil) | Feedstock | 297 | 82 | 7 | 23 | 0.02 | 0.19 | 0.01 | 0.07 |
| | Fuel | 1587 | 584 | 92 | 98 | 0.17 | 0.22 | 0.03 | 0.15 |
| | Operation | 6722 | 6439 | 516 | 520 | 0.32 | 0.52 | 0.02 | 0.01 |
| | Vehicle | 730 | 140 | 54 | 57 | 0.31 | 0.08 | 0.04 | 0.15 |
| RFG SI Autos (Tar Sands) | Feedstock | 1090 | 79 | 65 | 82 | 0.03 | 0.18 | 0.01 | 0.08 |
| | Fuel | 1655 | 584 | 93 | 99 | 0.17 | 0.22 | 0.03 | 0.15 |
| | Operation | 6722 | 6439 | 516 | 520 | 0.32 | 0.52 | 0.02 | 0.01 |
| | Vehicle | 730 | 140 | 54 | 57 | 0.31 | 0.08 | 0.04 | 0.15 |
| CG SI Autos (Conventional Oil) | Feedstock | 301 | 83 | 22 | 37 | 0.02 | 0.19 | 0.01 | 0.07 |
| | Fuel | 1287 | 572 | 87 | 91 | 0.16 | 0.18 | 0.02 | 0.13 |
| | Operation | 6818 | 6719 | 524 | 528 | 0.32 | 0.52 | 0.02 | 0.01 |
| | Vehicle | 730 | 140 | 54 | 57 | 0.31 | 0.08 | 0.04 | 0.15 |
| CG SI Autos (Tar Sands) | Feedstock | 1106 | 80 | 80 | 97 | 0.03 | 0.18 | 0.01 | 0.08 |
| | Fuel | 1353 | 572 | 87 | 92 | 0.16 | 0.18 | 0.02 | 0.13 |
| | Operation | 6818 | 6719 | 524 | 528 | 0.32 | 0.52 | 0.02 | 0.01 |
| | Vehicle | 730 | 140 | 54 | 57 | 0.31 | 0.08 | 0.04 | 0.15 |
| RFG SIDI Autos (Conventional Oil) | Feedstock | 258 | 71 | 6 | 20 | 0.02 | 0.16 | 0.01 | 0.06 |
| | Fuel | 1380 | 508 | 80 | 86 | 0.15 | 0.19 | 0.02 | 0.13 |
| | Operation | 5845 | 5599 | 449 | 453 | 0.32 | 0.52 | 0.02 | 0.01 |
| | Vehicle | 730 | 140 | 54 | 57 | 0.31 | 0.08 | 0.04 | 0.15 |
| RFG SIDI Autos (Tar Sands) | Feedstock | 948 | 69 | 56 | 71 | 0.02 | 0.16 | 0.01 | 0.07 |
| | Fuel | 1439 | 508 | 80 | 86 | 0.15 | 0.19 | 0.02 | 0.13 |
| | Operation | 5845 | 5599 | 449 | 453 | 0.32 | 0.52 | 0.02 | 0.01 |
| | Vehicle | 730 | 140 | 54 | 57 | 0.31 | 0.08 | 0.04 | 0.15 |

| | | | | | | | | | | |
|---|---|---|---|---|---|---|---|---|---|---|
| Diesel (Low Sulfur) | Feedstock | 247 | 68 | 24 | 37 | 0.02 | 0.15 | 0.01 | 0.06 |
| | Fuel | 783 | 394 | 59 | 61 | 0.02 | 0.12 | 0.02 | 0.09 |
| | Operation | 5602 | 5602 | 443 | 447 | 0.19 | 0.67 | 0.08 | 0.00 |
| | Vehicle | 730 | 140 | 54 | 57 | 0.31 | 0.08 | 0.04 | 0.15 |
| Diesel (Fischer Tropsch) | Feedstock | 358 | 23 | 28 | 45 | 0.03 | 0.14 | 0.00 | 0.07 |
| | Fuel | 3684 | 87 | 136 | 146 | 0.04 | 0.32 | 0.08 | 0.11 |
| | Operation | 5602 | 0 | 427 | 431 | 0.19 | 0.67 | 0.07 | 0.00 |
| | Vehicle | 730 | 140 | 54 | 57 | 0.31 | 0.08 | 0.04 | 0.15 |
| Diesel (Soy BD20) | Feedstock | 546 | 273 | −39 | −17 | 0.04 | 0.30 | 0.02 | 0.16 |
| | Fuel | 12057 | 17 | 16 | 7 | 0.35 | −0.04 | 0.00 | −0.07 |
| | Operation | 5602 | 4544 | 440 | 444 | 0.19 | 0.67 | 0.08 | 0.04 |
| | Vehicle | 730 | 140 | 54 | 57 | 0.31 | 0.08 | 0.04 | 0.15 |
| CNG | Feedstock | 557 | 31 | 39 | 82 | 0.04 | 0.21 | 0.00 | 0.09 |
| | Fuel | 530 | 25 | 44 | 46 | 0.00 | 0.07 | 0.02 | 0.15 |
| | Operation | 7177 | 0 | 426 | 436 | 0.21 | 0.52 | 0.02 | 0.00 |
| | Vehicle | 730 | 140 | 54 | 57 | 0.31 | 0.08 | 0.04 | 0.15 |
| E85 (Dry Corn) | Feedstock | 1051 | 292 | −288 | −222 | 0.08 | 0.53 | 0.03 | 0.24 |
| | Fuel | 8022 | 440 | 92 | 105 | 0.33 | 0.64 | 0.06 | 0.24 |
| | Operation | 6818 | 1804 | 514 | 518 | 0.30 | 0.52 | 0.02 | 0.00 |
| | Vehicle | 730 | 140 | 54 | 57 | 0.31 | 0.08 | 0.04 | 0.15 |
| E85 (Wet Corn) | Feedstock | 1148 | 636 | −271 | −184 | −0.18 | 0.74 | 0.04 | 0.38 |
| | Fuel | 8162 | 256 | 282 | 298 | 0.32 | 0.53 | 0.15 | 0.51 |
| | Operation | 6818 | 1804 | 514 | 518 | 0.30 | 0.52 | 0.02 | 0.00 |
| | Vehicle | 730 | 140 | 54 | 57 | 0.31 | 0.08 | 0.04 | 0.15 |
| E85 (Herbaceous) | Feedstock | 728 | 276 | −360 | −302 | 0.07 | 0.37 | 0.02 | 0.07 |
| | Fuel | 5903 | 228 | 4 | 19 | 0.24 | 0.57 | 0.04 | −0.03 |
| | Operation | 6818 | 1804 | 514 | 518 | 0.30 | 0.52 | 0.02 | 0.00 |
| | Vehicle | 730 | 140 | 54 | 57 | 0.31 | 0.08 | 0.04 | 0.15 |

continued

**TABLE D-4** Continued

| | | Total Energy Btu/VMT | Petroleum Btu/VMT | $CO_2$ g/VMT | $CO_{2e}$ g/VMT | VOC g/VMT | $NO_x$ g/VMT | $PM_{2.5}$ g/VMT | $SO_x$ g/VMT |
|---|---|---|---|---|---|---|---|---|---|
| E85 (Corn Stover) | Feedstock | 520 | 293 | -337 | -333 | 0.04 | 0.31 | 0.02 | 0.14 |
| | Fuel | 5376 | 228 | 4 | 18 | 0.24 | 0.55 | 0.03 | -0.03 |
| | Operation | 6818 | 1804 | 514 | 518 | 0.30 | 0.52 | 0.02 | 0.00 |
| | Vehicle | 730 | 140 | 54 | 57 | 0.31 | 0.08 | 0.04 | 0.15 |
| E10 (Dry Corn) | Feedstock | 367 | 101 | 1 | 21 | 0.03 | 0.22 | 0.01 | 0.08 |
| | Fuel | 1716 | 550 | 84 | 88 | 0.17 | 0.20 | 0.02 | 0.13 |
| | Operation | 6818 | 6378 | 523 | 527 | 0.32 | 0.52 | 0.02 | 0.01 |
| | Vehicle | 730 | 140 | 54 | 57 | 0.31 | 0.08 | 0.04 | 0.15 |
| E10 (Wet Corn) | Feedstock | 375 | 131 | 3 | 25 | 0.01 | 0.24 | 0.01 | 0.09 |
| | Fuel | 1728 | 534 | 101 | 105 | 0.17 | 0.19 | 0.03 | 0.16 |
| | Operation | 6818 | 6378 | 523 | 527 | 0.32 | 0.52 | 0.02 | 0.01 |
| | Vehicle | 633 | 126 | 47 | 50 | 0.21 | 0.08 | 0.03 | 0.14 |
| E10 (Herbaceous) | Feedstock | 339 | 100 | -5 | 14 | 0.03 | 0.20 | 0.01 | 0.07 |
| | Fuel | 1530 | 531 | 77 | 80 | 0.16 | 0.20 | 0.02 | 0.11 |
| | Operation | 6818 | 6378 | 523 | 527 | 0.32 | 0.52 | 0.02 | 0.01 |
| | Vehicle | 730 | 140 | 54 | 57 | 0.31 | 0.08 | 0.04 | 0.15 |
| E10 (Corn Stover) | Feedstock | 320 | 101 | -3 | 12 | 0.03 | 0.20 | 0.01 | 0.07 |
| | Fuel | 1484 | 531 | 77 | 80 | 0.16 | 0.19 | 0.02 | 0.11 |
| | Operation | 6818 | 6378 | 523 | 527 | 0.32 | 0.52 | 0.02 | 0.01 |
| | Vehicle | 730 | 140 | 54 | 57 | 0.31 | 0.08 | 0.04 | 0.15 |
| Electric | Feedstock | 203 | 72 | 16 | 32 | 0.04 | 0.11 | 0.16 | 0.05 |
| | Fuel | 3494 | 103 | 485 | 487 | 0.01 | 0.68 | 0.02 | 1.68 |
| | Operation | 2273 | 108 | 0 | 0 | 0.00 | 0.00 | 0.01 | 0.00 |
| | Vehicle | 840 | 146 | 65 | 68 | 0.31 | 0.10 | 0.04 | 0.30 |

| | | | | | | | | | |
|---|---|---|---|---|---|---|---|---|---|
| Hydrogen (Gaseous) | Feedstock | 230 | 13 | 16 | 34 | 0.02 | 0.09 | 0.00 | 0.04 |
| | Fuel | 1967 | 37 | 310 | 320 | 0.02 | 0.18 | 0.06 | 0.21 |
| | Operation | 2964 | 0 | 0 | 0 | 0.00 | 0.00 | 0.01 | 0.00 |
| | Vehicle | 1030 | 203 | 80 | 84 | 0.31 | 0.12 | 0.05 | 0.37 |
| Grid-Independent SI HEV | Feedstock | 203 | 56 | 15 | 25 | 0.02 | 0.13 | 0.00 | 0.05 |
| | Fuel | 870 | 386 | 59 | 62 | 0.11 | 0.12 | 0.02 | 0.09 |
| | Operation | 4607 | 4540 | 354 | 358 | 0.22 | 0.44 | 0.02 | 0.01 |
| | Vehicle | 840 | 146 | 65 | 68 | 0.31 | 0.10 | 0.04 | 0.30 |
| Grid-Dependent SI HEV | Feedstock | 203 | 61 | 15 | 27 | 0.02 | 0.12 | 0.06 | 0.05 |
| | Fuel | 1736 | 293 | 199 | 202 | 0.08 | 0.31 | 0.02 | 0.61 |
| | Operation | 3836 | 3077 | 237 | 240 | 0.15 | 0.29 | 0.02 | 0.00 |
| | Vehicle | 840 | 146 | 65 | 68 | 0.31 | 0.10 | 0.04 | 0.30 |

TABLE D-5 GREET Energy and Emission Factors for Light-Duty Trucks 2 in 2005

| | | Total Energy | Petroleum | $CO_2$ | $CO_{2e}$ | VOC | $NO_x$ | $PM_{2.5}$ | $SO_x$ |
|---|---|---|---|---|---|---|---|---|---|
| | | BTU/VMT | BTU/VMT | g/VMT | g/VMT | g/VMT | g/VMT | g/VMT | g/VMT |
| RFG SI Autos (Conventional Oil) | Feedstock | 324 | 89 | 8 | 25 | 0.03 | 0.20 | 0.01 | 0.07 |
| | Fuel | 1730 | 637 | 100 | 107 | 0.18 | 0.24 | 0.03 | 0.16 |
| | Operation | 7329 | 7021 | 562 | 567 | 0.82 | 1.03 | 0.02 | 0.01 |
| | Vehicle | 730 | 140 | 54 | 57 | 0.31 | 0.08 | 0.04 | 0.15 |
| RFG SI Autos (Tar Sands) | Feedstock | 1189 | 86 | 70 | 89 | 0.03 | 0.20 | 0.01 | 0.08 |
| | Fuel | 1805 | 637 | 101 | 108 | 0.18 | 0.24 | 0.03 | 0.16 |
| | Operation | 7329 | 7021 | 562 | 567 | 0.82 | 1.03 | 0.02 | 0.01 |
| | Vehicle | 730 | 140 | 54 | 57 | 0.31 | 0.08 | 0.04 | 0.15 |
| CG SI Autos (Conventional Oil) | Feedstock | 328 | 90 | 24 | 41 | 0.03 | 0.20 | 0.01 | 0.07 |
| | Fuel | 1403 | 623 | 95 | 99 | 0.18 | 0.20 | 0.03 | 0.14 |
| | Operation | 7433 | 7326 | 571 | 576 | 0.82 | 1.03 | 0.02 | 0.01 |
| | Vehicle | 730 | 140 | 54 | 57 | 0.31 | 0.08 | 0.04 | 0.15 |
| CG SI Autos (Tar Sands) | Feedstock | 1206 | 88 | 87 | 106 | 0.03 | 0.20 | 0.01 | 0.08 |
| | Fuel | 1476 | 623 | 95 | 100 | 0.18 | 0.20 | 0.03 | 0.15 |
| | Operation | 7433 | 7326 | 571 | 576 | 0.82 | 1.03 | 0.02 | 0.01 |
| | Vehicle | 730 | 140 | 54 | 57 | 0.31 | 0.08 | 0.04 | 0.15 |
| RFG SIDI Autos (Conventional Oil) | Feedstock | 324 | 89 | 8 | 25 | 0.03 | 0.20 | 0.01 | 0.07 |
| | Fuel | 1730 | 637 | 100 | 107 | 0.18 | 0.24 | 0.03 | 0.16 |
| | Operation | 7329 | 7021 | 562 | 567 | 0.82 | 1.03 | 0.02 | 0.01 |
| | Vehicle | 730 | 140 | 54 | 57 | 0.31 | 0.08 | 0.04 | 0.15 |
| RFG SIDI Autos (Tar Sands) | Feedstock | 1189 | 86 | 70 | 89 | 0.03 | 0.20 | 0.01 | 0.08 |
| | Fuel | 1805 | 637 | 101 | 108 | 0.18 | 0.24 | 0.03 | 0.16 |
| | Operation | 7329 | 7021 | 562 | 567 | 0.82 | 1.03 | 0.02 | 0.01 |
| | Vehicle | 730 | 140 | 54 | 57 | 0.31 | 0.08 | 0.04 | 0.15 |

| | | | | | | | | | |
|---|---|---|---|---|---|---|---|---|---|
| Diesel (Low Sulfur) | Feedstock | 269 | 74 | 26 | 40 | 0.02 | 0.17 | 0.01 | 0.06 |
| | Fuel | 854 | 430 | 64 | 66 | 0.03 | 0.13 | 0.02 | 0.10 |
| | Operation | 6108 | 6108 | 483 | 487 | 0.30 | 1.03 | 0.08 | 0.00 |
| | Vehicle | 730 | 140 | 54 | 57 | 0.31 | 0.08 | 0.04 | 0.15 |
| Diesel (Fischer Tropsch) | Feedstock | 391 | 26 | 30 | 49 | 0.03 | 0.15 | 0.00 | 0.07 |
| | Fuel | 4016 | 95 | 148 | 159 | 0.05 | 0.35 | 0.09 | 0.13 |
| | Operation | 6108 | 0 | 466 | 470 | 0.30 | 1.03 | 0.07 | 0.00 |
| | Vehicle | 730 | 140 | 54 | 57 | 0.31 | 0.08 | 0.04 | 0.15 |
| Diesel (Soy BD20) | Feedstock | 595 | 297 | −42 | −18 | 0.04 | 0.32 | 0.02 | 0.18 |
| | Fuel | 13146 | 19 | 17 | 7 | 0.38 | −0.05 | 0.00 | −0.08 |
| | Operation | 6108 | 4955 | 480 | 483 | 0.30 | 1.03 | 0.08 | 0.05 |
| | Vehicle | 730 | 140 | 54 | 57 | 0.31 | 0.08 | 0.04 | 0.15 |
| NG | Feedstock | 607 | 34 | 43 | 90 | 0.05 | 0.23 | 0.00 | 0.10 |
| | Fuel | 578 | 27 | 48 | 50 | 0.00 | 0.08 | 0.02 | 0.17 |
| | Operation | 7825 | 0 | 463 | 480 | 0.54 | 1.03 | 0.02 | 0.00 |
| | Vehicle | 730 | 140 | 54 | 57 | 0.31 | 0.08 | 0.04 | 0.15 |
| E85 (Dry Corn) | Feedstock | 1146 | 319 | −314 | −242 | 0.09 | 0.58 | 0.03 | 0.26 |
| | Fuel | 8747 | 480 | 101 | 114 | 0.36 | 0.69 | 0.07 | 0.26 |
| | Operation | 7433 | 1967 | 561 | 565 | 0.78 | 1.03 | 0.02 | 0.00 |
| | Vehicle | 730 | 140 | 54 | 57 | 0.31 | 0.08 | 0.04 | 0.15 |
| E85 (Wet Corn) | Feedstock | 1252 | 693 | −295 | −200 | −0.19 | 0.81 | 0.04 | 0.41 |
| | Fuel | 8899 | 279 | 308 | 324 | 0.35 | 0.58 | 0.16 | 0.55 |
| | Operation | 7433 | 1967 | 561 | 565 | 0.78 | 1.03 | 0.02 | 0.00 |
| | Vehicle | 730 | 140 | 54 | 57 | 0.31 | 0.08 | 0.04 | 0.15 |
| E85 (Herbaceous) | Feedstock | 794 | 301 | −392 | −330 | 0.08 | 0.40 | 0.02 | 0.08 |
| | Fuel | 6436 | 249 | 5 | 21 | 0.26 | 0.63 | 0.04 | −0.03 |
| | Operation | 7433 | 1967 | 561 | 565 | 0.78 | 1.03 | 0.02 | 0.00 |
| | Vehicle | 730 | 140 | 54 | 57 | 0.31 | 0.08 | 0.04 | 0.15 |

continued

TABLE D-5 Continued

| | | Total Energy BTU/VMT | Petroleum BTU/VMT | $CO_2$ g/VMT | $CO_{2e}$ g/VMT | VOC g/VMT | $NO_x$ g/VMT | $PM_{2.5}$ g/VMT | $SO_x$ g/VMT |
|---|---|---|---|---|---|---|---|---|---|
| E85 (Corn Stover) | Feedstock | 567 | 320 | -367 | -364 | 0.05 | 0.33 | 0.02 | 0.15 |
| | Fuel | 5861 | 249 | 5 | 20 | 0.26 | 0.60 | 0.04 | -0.03 |
| | Operation | 7433 | 1967 | 561 | 565 | 0.78 | 1.03 | 0.02 | 0.00 |
| | Vehicle | 730 | 140 | 54 | 57 | 0.31 | 0.08 | 0.04 | 0.15 |
| E10 (Dry Corn) | Feedstock | 400 | 110 | 1 | 23 | 0.03 | 0.24 | 0.01 | 0.09 |
| | Fuel | 1871 | 599 | 92 | 96 | 0.19 | 0.22 | 0.03 | 0.15 |
| | Operation | 7433 | 6954 | 570 | 575 | 0.82 | 1.03 | 0.02 | 0.01 |
| | Vehicle | 730 | 140 | 54 | 57 | 0.31 | 0.08 | 0.04 | 0.15 |
| E10 (Wet Corn) | Feedstock | 409 | 143 | 3 | 27 | 0.01 | 0.26 | 0.01 | 0.10 |
| | Fuel | 1884 | 582 | 110 | 114 | 0.19 | 0.21 | 0.04 | 0.17 |
| | Operation | 7433 | 6954 | 570 | 575 | 0.82 | 1.03 | 0.02 | 0.01 |
| | Vehicle | 633 | 126 | 47 | 50 | 0.21 | 0.08 | 0.03 | 0.14 |
| E10 (Herbaceous) | Feedstock | 369 | 109 | -5 | 16 | 0.03 | 0.22 | 0.01 | 0.07 |
| | Fuel | 1668 | 579 | 83 | 88 | 0.18 | 0.21 | 0.02 | 0.12 |
| | Operation | 7433 | 6954 | 570 | 575 | 0.82 | 1.03 | 0.02 | 0.01 |
| | Vehicle | 730 | 140 | 54 | 57 | 0.31 | 0.08 | 0.04 | 0.15 |
| E10 (Corn Stover) | Feedstock | 349 | 110 | -3 | 13 | 0.03 | 0.22 | 0.01 | 0.08 |
| | Fuel | 1618 | 579 | 83 | 87 | 0.18 | 0.21 | 0.02 | 0.12 |
| | Operation | 7433 | 6954 | 570 | 575 | 0.82 | 1.03 | 0.02 | 0.01 |
| | Vehicle | 730 | 140 | 54 | 57 | 0.31 | 0.08 | 0.04 | 0.15 |
| Electric | Feedstock | 221 | 78 | 17 | 35 | 0.04 | 0.12 | 0.18 | 0.06 |
| | Fuel | 3810 | 112 | 529 | 531 | 0.01 | 0.74 | 0.02 | 1.83 |
| | Operation | 2478 | 117 | 0 | 0 | 0.00 | 0.00 | 0.01 | 0.00 |
| | Vehicle | 840 | 146 | 65 | 68 | 0.31 | 0.10 | 0.04 | 0.30 |

| | | | | | | | | | |
|---|---|---|---|---|---|---|---|---|---|
| Hydrogen (Gaseous) | Feedstock | 269 | 15 | 19 | 40 | 0.02 | 0.10 | 0.00 | 0.04 |
| | Fuel | 2305 | 43 | 364 | 375 | 0.02 | 0.21 | 0.07 | 0.24 |
| | Operation | 3474 | 0 | 0 | 0 | 0.00 | 0.00 | 0.01 | 0.00 |
| | Vehicle | 1030 | 203 | 80 | 84 | 0.31 | 0.12 | 0.05 | 0.37 |
| Grid-Independent SI HEV | Feedstock | 234 | 64 | 17 | 29 | 0.02 | 0.15 | 0.01 | 0.05 |
| | Fuel | 1002 | 445 | 68 | 71 | 0.13 | 0.14 | 0.02 | 0.10 |
| | Operation | 5310 | 5233 | 408 | 412 | 0.75 | 0.80 | 0.02 | 0.01 |
| | Vehicle | 840 | 146 | 65 | 68 | 0.31 | 0.10 | 0.04 | 0.30 |
| Grid-Dependent SI HEV | Feedstock | 230 | 69 | 17 | 31 | 0.03 | 0.14 | 0.06 | 0.05 |
| | Fuel | 1929 | 335 | 220 | 223 | 0.09 | 0.34 | 0.02 | 0.67 |
| | Operation | 4375 | 3545 | 273 | 276 | 0.50 | 0.54 | 0.02 | 0.00 |
| | Vehicle | 840 | 146 | 65 | 68 | 0.31 | 0.10 | 0.04 | 0.30 |

**TABLE D-6** GREET Energy and Emission Factors for Light-Duty Autos in 2020

| | | Total Energy | Petroleum | $CO_2$ | $CO_{2e}$ | VOC | $NO_x$ | $PM_{2.5}$ | $SO_x$ |
|---|---|---|---|---|---|---|---|---|---|
| | | BTU/VMT | BTU/VMT | g/VMT | g/VMT | g/VMT | g/VMT | g/VMT | g/VMT |
| RFG SI Autos (Conventional Oil) | Feedstock | 314 | 57 | 13 | 25 | 0.02 | 0.11 | 0.00 | 0.04 |
| | Fuel | 1125 | 410 | 64 | 69 | 0.12 | 0.11 | 0.01 | 0.06 |
| | Operation | 4753 | 4553 | 365 | 369 | 0.15 | 0.07 | 0.01 | 0.01 |
| | Vehicle | 633 | 126 | 47 | 50 | 0.21 | 0.08 | 0.03 | 0.14 |
| RFG SI Autos (Tar Sands) | Feedstock | 913 | 55 | 60 | 73 | 0.02 | 0.12 | 0.01 | 0.04 |
| | Fuel | 1177 | 410 | 65 | 70 | 0.12 | 0.11 | 0.01 | 0.06 |
| | Operation | 4753 | 4553 | 365 | 369 | 0.15 | 0.07 | 0.01 | 0.01 |
| | Vehicle | 633 | 126 | 47 | 50 | 0.21 | 0.08 | 0.03 | 0.14 |
| CG SI Autos (Conventional Oil) | Feedstock | 314 | 57 | 13 | 25 | 0.02 | 0.11 | 0.00 | 0.04 |
| | Fuel | 1125 | 410 | 64 | 69 | 0.12 | 0.11 | 0.01 | 0.06 |
| | Operation | 4753 | 4553 | 365 | 369 | 0.15 | 0.07 | 0.01 | 0.01 |
| | Vehicle | 633 | 126 | 47 | 50 | 0.21 | 0.08 | 0.03 | 0.14 |
| CG SI Autos (Tar Sands) | Feedstock | 913 | 55 | 60 | 73 | 0.02 | 0.12 | 0.01 | 0.04 |
| | Fuel | 1177 | 410 | 65 | 70 | 0.12 | 0.11 | 0.01 | 0.06 |
| | Operation | 4753 | 4553 | 365 | 369 | 0.15 | 0.07 | 0.01 | 0.01 |
| | Vehicle | 633 | 126 | 47 | 50 | 0.21 | 0.08 | 0.03 | 0.14 |
| RFG SIDI Autos (Conventional Oil) | Feedstock | 273 | 50 | 12 | 21 | 0.01 | 0.09 | 0.00 | 0.03 |
| | Fuel | 979 | 357 | 56 | 60 | 0.10 | 0.10 | 0.01 | 0.06 |
| | Operation | 4133 | 3959 | 317 | 321 | 0.15 | 0.07 | 0.01 | 0.01 |
| | Vehicle | 633 | 126 | 47 | 50 | 0.21 | 0.08 | 0.03 | 0.14 |
| RFG SIDI Autos (Tar Sands) | Feedstock | 794 | 48 | 52 | 63 | 0.02 | 0.10 | 0.01 | 0.03 |
| | Fuel | 1023 | 357 | 57 | 60 | 0.10 | 0.10 | 0.01 | 0.06 |
| | Operation | 4133 | 3959 | 317 | 321 | 0.15 | 0.07 | 0.01 | 0.01 |
| | Vehicle | 633 | 126 | 47 | 50 | 0.21 | 0.08 | 0.03 | 0.14 |

| | | | | | | | | | |
|---|---|---|---|---|---|---|---|---|---|
| Diesel (Low Sulfur) | Feedstock | 262 | 48 | 24 | 33 | 0.01 | 0.09 | 0.00 | 0.03 |
| | Fuel | 560 | 279 | 42 | 43 | 0.02 | 0.06 | 0.01 | 0.04 |
| | Operation | 3961 | 3961 | 313 | 317 | 0.06 | 0.08 | 0.02 | 0.00 |
| | Vehicle | 633 | 126 | 47 | 50 | 0.21 | 0.08 | 0.03 | 0.14 |
| Diesel (Fischer-Tropsch) | Feedstock | 254 | 17 | 20 | 32 | 0.02 | 0.07 | 0.00 | 0.04 |
| | Fuel | 2345 | 61 | 85 | 91 | 0.03 | 0.19 | 0.05 | 0.08 |
| | Operation | 3961 | 0 | 302 | 306 | 0.06 | 0.08 | 0.02 | 0.00 |
| | Vehicle | 633 | 126 | 47 | 50 | 0.21 | 0.08 | 0.03 | 0.14 |
| Diesel (Soy BD20) | Feedstock | 454 | 189 | -22 | -6 | 0.02 | 0.14 | 0.01 | 0.09 |
| | Fuel | 8561 | 66 | 18 | 11 | 0.23 | -0.05 | 0.00 | -0.07 |
| | Operation | 3961 | 3218 | 314 | 318 | 0.06 | 0.08 | 0.02 | 0.00 |
| | Vehicle | 633 | 126 | 47 | 50 | 0.21 | 0.08 | 0.03 | 0.14 |
| CNG | Feedstock | 352 | 19 | 25 | 52 | 0.03 | 0.09 | 0.00 | 0.05 |
| | Fuel | 332 | 14 | 27 | 29 | 0.00 | 0.03 | 0.01 | 0.05 |
| | Operation | 4527 | 0 | 269 | 275 | 0.12 | 0.07 | 0.01 | 0.00 |
| | Vehicle | 633 | 126 | 47 | 50 | 0.21 | 0.08 | 0.03 | 0.14 |
| E85 (Dry Corn) | Feedstock | 730 | 208 | -204 | -163 | 0.04 | 0.22 | 0.01 | 0.11 |
| | Fuel | 5621 | 265 | 61 | 70 | 0.22 | 0.35 | 0.03 | 0.09 |
| | Operation | 4753 | 1245 | 358 | 362 | 0.14 | 0.07 | 0.01 | 0.00 |
| | Vehicle | 633 | 126 | 47 | 50 | 0.21 | 0.08 | 0.03 | 0.14 |
| E85 (Wet Corn) | Feedstock | 737 | 399 | -198 | -146 | -0.12 | 0.34 | 0.02 | 0.19 |
| | Fuel | 5770 | 182 | 200 | 211 | 0.22 | 0.25 | 0.09 | 0.13 |
| | Operation | 4753 | 1245 | 358 | 362 | 0.14 | 0.07 | 0.01 | 0.00 |
| | Vehicle | 633 | 126 | 47 | 50 | 0.21 | 0.08 | 0.03 | 0.14 |
| E85 (Herbaceous) | Feedstock | 477 | 169 | -252 | -217 | 0.04 | 0.16 | 0.01 | 0.03 |
| | Fuel | 3122 | 163 | 5 | 15 | 0.17 | 0.35 | 0.02 | 0.00 |
| | Operation | 4753 | 1245 | 358 | 362 | 0.14 | 0.07 | 0.01 | 0.00 |
| | Vehicle | 633 | 126 | 47 | 50 | 0.21 | 0.08 | 0.03 | 0.14 |

*continued*

**TABLE D-6** Continued

| | | Total Energy BTU/VMT | Petroleum BTU/VMT | CO₂ g/VMT | CO₂ₑ g/VMT | VOC g/VMT | NOₓ g/VMT | PM₂.₅ g/VMT | SOₓ g/VMT |
|---|---|---|---|---|---|---|---|---|---|
| E85 (Corn Stover) | Feedstock | 342 | 170 | −239 | −237 | 0.02 | 0.11 | 0.01 | 0.06 |
| | Fuel | 2805 | 163 | 5 | 14 | 0.17 | 0.34 | 0.02 | 0.00 |
| | Operation | 4753 | 1245 | 358 | 362 | 0.14 | 0.07 | 0.01 | 0.00 |
| | Vehicle | 633 | 126 | 47 | 50 | 0.21 | 0.08 | 0.03 | 0.14 |
| E10 (Dry Corn) | Feedstock | 350 | 70 | 8 | 22 | 0.02 | 0.12 | 0.01 | 0.04 |
| | Fuel | 1200 | 379 | 58 | 61 | 0.12 | 0.11 | 0.01 | 0.06 |
| | Operation | 4753 | 4446 | 365 | 368 | 0.15 | 0.07 | 0.01 | 0.01 |
| | Vehicle | 633 | 126 | 47 | 50 | 0.21 | 0.08 | 0.03 | 0.14 |
| E10 (Wet Corn) | Feedstock | 351 | 87 | 9 | 24 | 0.00 | 0.13 | 0.01 | 0.05 |
| | Fuel | 1212 | 372 | 70 | 73 | 0.12 | 0.10 | 0.02 | 0.06 |
| | Operation | 4753 | 4446 | 365 | 368 | 0.15 | 0.07 | 0.01 | 0.01 |
| | Vehicle | 633 | 126 | 47 | 50 | 0.21 | 0.08 | 0.03 | 0.14 |
| E10 (Herbaceous) | Feedstock | 328 | 67 | 4 | 17 | 0.02 | 0.11 | 0.00 | 0.04 |
| | Fuel | 981 | 370 | 53 | 56 | 0.11 | 0.11 | 0.01 | 0.05 |
| | Operation | 4753 | 4446 | 365 | 368 | 0.15 | 0.07 | 0.01 | 0.01 |
| | Vehicle | 633 | 126 | 47 | 50 | 0.21 | 0.08 | 0.03 | 0.14 |
| E10 (Corn Stover) | Feedstock | 317 | 67 | 5 | 16 | 0.02 | 0.11 | 0.00 | 0.04 |
| | Fuel | 953 | 370 | 53 | 56 | 0.11 | 0.10 | 0.01 | 0.05 |
| | Operation | 4753 | 4446 | 365 | 368 | 0.15 | 0.07 | 0.01 | 0.01 |
| | Vehicle | 633 | 126 | 47 | 50 | 0.21 | 0.08 | 0.03 | 0.14 |
| Electric | Feedstock | 126 | 41 | 9 | 20 | 0.02 | 0.04 | 0.09 | 0.03 |
| | Fuel | 2061 | 53 | 283 | 285 | 0.00 | 0.23 | 0.01 | 0.53 |
| | Operation | 1358 | 58 | 0 | 0 | 0.00 | 0.00 | 0.01 | 0.00 |
| | Vehicle | 645 | 122 | 50 | 52 | 0.21 | 0.08 | 0.03 | 0.23 |

| | | | | | | | | | | |
|---|---|---|---|---|---|---|---|---|---|---|
| Hydrogen (Gaseous) | Feedstock | 153 | 8 | 11 | 23 | 0.01 | 0.04 | 0.00 | 0.02 |
| | Fuel | 1233 | 21 | 201 | 207 | 0.01 | 0.08 | 0.04 | 0.08 |
| | Operation | 1964 | 0 | 0 | 0 | 0.00 | 0.00 | 0.01 | 0.00 |
| | Vehicle | 792 | 160 | 62 | 65 | 0.21 | 0.09 | 0.04 | 0.29 |
| Hydrogen (Liquid) | Feedstock | 153 | 8 | 11 | 23 | 0.01 | 0.04 | 0.00 | 0.02 |
| | Fuel | 3122 | 101 | 355 | 367 | 0.03 | 0.23 | 0.09 | 0.37 |
| | Operation | 1964 | 0 | 0 | 0 | 0.00 | 0.00 | 0.01 | 0.00 |
| | Vehicle | 792 | 160 | 62 | 65 | 0.21 | 0.09 | 0.04 | 0.29 |
| Grid-Independent SI HEV | Feedstock | 207 | 38 | 9 | 16 | 0.01 | 0.07 | 0.00 | 0.02 |
| | Fuel | 740 | 270 | 42 | 45 | 0.08 | 0.07 | 0.01 | 0.04 |
| | Operation | 3127 | 2996 | 240 | 244 | 0.11 | 0.06 | 0.01 | 0.00 |
| | Vehicle | 645 | 122 | 50 | 52 | 0.21 | 0.08 | 0.03 | 0.23 |
| Grid-Dependent SI HEV | Feedstock | 187 | 41 | 10 | 18 | 0.02 | 0.06 | 0.04 | 0.03 |
| | Fuel | 1289 | 201 | 137 | 140 | 0.05 | 0.14 | 0.01 | 0.23 |
| | Operation | 2618 | 2029 | 161 | 163 | 0.07 | 0.04 | 0.01 | 0.00 |
| | Vehicle | 645 | 122 | 50 | 52 | 0.21 | 0.08 | 0.03 | 0.23 |

TABLE D-7 GREET Energy and Emission Factors for Light-Duty Trucks 1 in 2020

| | | Total Energy BTU/VMT | Petroleum BTU/VMT | $CO_2$ g/VMT | $CO_{2e}$ g/VMT | VOC g/VMT | $NO_x$ g/VMT | $PM_{2.5}$ g/VMT | $SO_x$ g/VMT |
|---|---|---|---|---|---|---|---|---|---|
| RFG SI Autos (Conventional Oil) | Feedstock | 373 | 68 | 16 | 29 | 0.02 | 0.13 | 0.01 | 0.04 |
| | Fuel | 1338 | 488 | 76 | 82 | 0.14 | 0.13 | 0.02 | 0.08 |
| | Operation | 5652 | 5414 | 434 | 438 | 0.18 | 0.10 | 0.02 | 0.01 |
| | Vehicle | 730 | 140 | 54 | 57 | 0.31 | 0.08 | 0.04 | 0.15 |
| RFG SI Autos (Tar Sands) | Feedstock | 1086 | 66 | 71 | 87 | 0.02 | 0.14 | 0.01 | 0.04 |
| | Fuel | 1399 | 488 | 77 | 83 | 0.14 | 0.13 | 0.02 | 0.08 |
| | Operation | 5652 | 5414 | 434 | 438 | 0.18 | 0.10 | 0.02 | 0.01 |
| | Vehicle | 730 | 140 | 54 | 57 | 0.31 | 0.08 | 0.04 | 0.15 |
| CG SI Autos (Conventional Oil) | Feedstock | 373 | 68 | 16 | 29 | 0.02 | 0.13 | 0.01 | 0.04 |
| | Fuel | 1338 | 488 | 76 | 82 | 0.14 | 0.13 | 0.02 | 0.08 |
| | Operation | 5652 | 5414 | 434 | 438 | 0.18 | 0.10 | 0.02 | 0.01 |
| | Vehicle | 730 | 140 | 54 | 57 | 0.31 | 0.08 | 0.04 | 0.15 |
| CG SI Autos (Tar Sands) | Feedstock | 1086 | 66 | 71 | 87 | 0.02 | 0.14 | 0.01 | 0.04 |
| | Fuel | 1399 | 488 | 77 | 83 | 0.14 | 0.13 | 0.02 | 0.08 |
| | Operation | 5652 | 5414 | 434 | 438 | 0.18 | 0.10 | 0.02 | 0.01 |
| | Vehicle | 730 | 140 | 54 | 57 | 0.31 | 0.08 | 0.04 | 0.15 |
| RFG SIDI Autos (Conventional Oil) | Feedstock | 325 | 59 | 14 | 26 | 0.02 | 0.11 | 0.00 | 0.04 |
| | Fuel | 1164 | 424 | 66 | 71 | 0.12 | 0.12 | 0.02 | 0.07 |
| | Operation | 4915 | 4708 | 377 | 381 | 0.18 | 0.10 | 0.02 | 0.01 |
| | Vehicle | 730 | 140 | 54 | 57 | 0.31 | 0.08 | 0.04 | 0.15 |
| RFG SIDI Autos (Tar Sands) | Feedstock | 944 | 57 | 62 | 75 | 0.02 | 0.12 | 0.01 | 0.04 |
| | Fuel | 1217 | 424 | 67 | 72 | 0.12 | 0.12 | 0.02 | 0.07 |
| | Operation | 4915 | 4708 | 377 | 381 | 0.18 | 0.10 | 0.02 | 0.01 |
| | Vehicle | 730 | 140 | 54 | 57 | 0.31 | 0.08 | 0.04 | 0.15 |

| | | | | | | | | | |
|---|---|---|---|---|---|---|---|---|---|
| Diesel (Low Sulfur) | Feedstock | 311 | 57 | 28 | 39 | 0.02 | 0.10 | 0.00 | 0.04 |
| | Fuel | 666 | 331 | 50 | 51 | 0.02 | 0.07 | 0.01 | 0.05 |
| | Operation | 4710 | 4710 | 372 | 376 | 0.06 | 0.12 | 0.02 | 0.00 |
| | Vehicle | 730 | 140 | 54 | 57 | 0.31 | 0.08 | 0.04 | 0.15 |
| Diesel (Fischer-Tropsch) | Feedstock | 302 | 20 | 23 | 38 | 0.03 | 0.08 | 0.00 | 0.05 |
| | Fuel | 2789 | 72 | 101 | 109 | 0.03 | 0.23 | 0.06 | 0.09 |
| | Operation | 4710 | 0 | 359 | 363 | 0.06 | 0.12 | 0.02 | 0.00 |
| | Vehicle | 730 | 140 | 54 | 57 | 0.31 | 0.08 | 0.04 | 0.15 |
| Diesel (Soy BD20) | Feedstock | 540 | 224 | -26 | -7 | 0.02 | 0.16 | 0.01 | 0.11 |
| | Fuel | 10179 | 78 | 21 | 13 | 0.28 | -0.06 | 0.00 | -0.09 |
| | Operation | 4710 | 3827 | 373 | 377 | 0.06 | 0.12 | 0.02 | 0.00 |
| | Vehicle | 730 | 140 | 54 | 57 | 0.31 | 0.08 | 0.04 | 0.15 |
| CNG | Feedstock | 418 | 23 | 30 | 62 | 0.03 | 0.11 | 0.00 | 0.06 |
| | Fuel | 395 | 17 | 33 | 34 | 0.00 | 0.03 | 0.01 | 0.06 |
| | Operation | 5383 | 0 | 319 | 326 | 0.15 | 0.10 | 0.02 | 0.00 |
| | Vehicle | 730 | 140 | 54 | 57 | 0.31 | 0.08 | 0.04 | 0.15 |
| E85 (Dry Corn) | Feedstock | 868 | 247 | -243 | -193 | 0.05 | 0.27 | 0.01 | 0.13 |
| | Fuel | 6683 | 316 | 72 | 83 | 0.26 | 0.41 | 0.04 | 0.11 |
| | Operation | 5652 | 1480 | 426 | 430 | 0.17 | 0.10 | 0.02 | 0.00 |
| | Vehicle | 730 | 140 | 54 | 57 | 0.31 | 0.08 | 0.04 | 0.15 |
| E85 (Wet Corn) | Feedstock | 876 | 475 | -236 | -174 | -0.14 | 0.40 | 0.02 | 0.22 |
| | Fuel | 6861 | 217 | 237 | 251 | 0.26 | 0.30 | 0.10 | 0.16 |
| | Operation | 5652 | 1480 | 426 | 430 | 0.17 | 0.10 | 0.02 | 0.00 |
| | Vehicle | 730 | 140 | 54 | 57 | 0.31 | 0.08 | 0.04 | 0.15 |
| E85 (Herbaceous) | Feedstock | 567 | 201 | -300 | -258 | 0.05 | 0.19 | 0.01 | 0.03 |
| | Fuel | 3712 | 194 | 6 | 17 | 0.20 | 0.42 | 0.03 | 0.00 |
| | Operation | 5652 | 1480 | 426 | 430 | 0.17 | 0.10 | 0.02 | 0.00 |
| | Vehicle | 730 | 140 | 54 | 57 | 0.31 | 0.08 | 0.04 | 0.15 |

continued

**TABLE D-7** Continued

| | | Total Energy BTU/VMT | Petroleum BTU/VMT | CO$_2$ g/VMT | CO$_{2e}$ g/VMT | VOC g/VMT | NO$_x$ g/VMT | PM$_{2.5}$ g/VMT | SO$_x$ g/VMT |
|---|---|---|---|---|---|---|---|---|---|
| E85 (Corn Stover) | Feedstock | 407 | 202 | −284 | −281 | 0.02 | 0.13 | 0.01 | 0.08 |
| | Fuel | 3336 | 194 | 6 | 17 | 0.20 | 0.40 | 0.03 | 0.00 |
| | Operation | 5652 | 1480 | 426 | 430 | 0.17 | 0.10 | 0.02 | 0.00 |
| | Vehicle | 730 | 140 | 54 | 57 | 0.31 | 0.08 | 0.04 | 0.15 |
| E10 (Dry Corn) | Feedstock | 417 | 84 | 10 | 26 | 0.02 | 0.14 | 0.01 | 0.05 |
| | Fuel | 1426 | 451 | 69 | 72 | 0.14 | 0.13 | 0.02 | 0.07 |
| | Operation | 5652 | 5287 | 434 | 437 | 0.18 | 0.10 | 0.02 | 0.01 |
| | Vehicle | 730 | 140 | 54 | 57 | 0.31 | 0.08 | 0.04 | 0.15 |
| E10 (Wet Corn) | Feedstock | 417 | 104 | 10 | 28 | 0.01 | 0.15 | 0.01 | 0.06 |
| | Fuel | 1442 | 442 | 84 | 87 | 0.14 | 0.12 | 0.02 | 0.07 |
| | Operation | 5652 | 5287 | 434 | 437 | 0.18 | 0.10 | 0.02 | 0.01 |
| | Vehicle | 633 | 126 | 47 | 50 | 0.21 | 0.08 | 0.03 | 0.14 |
| E10 (Herbaceous) | Feedstock | 390 | 80 | 5 | 21 | 0.02 | 0.13 | 0.01 | 0.04 |
| | Fuel | 1166 | 440 | 64 | 66 | 0.14 | 0.13 | 0.02 | 0.06 |
| | Operation | 5652 | 5287 | 434 | 437 | 0.18 | 0.10 | 0.02 | 0.01 |
| | Vehicle | 730 | 140 | 54 | 57 | 0.31 | 0.08 | 0.04 | 0.15 |
| E10 (Corn Stover) | Feedstock | 376 | 80 | 6 | 19 | 0.02 | 0.13 | 0.01 | 0.05 |
| | Fuel | 1134 | 440 | 64 | 66 | 0.14 | 0.12 | 0.02 | 0.06 |
| | Operation | 5652 | 5287 | 434 | 437 | 0.18 | 0.10 | 0.02 | 0.01 |
| | Vehicle | 730 | 140 | 54 | 57 | 0.31 | 0.08 | 0.04 | 0.15 |
| Electric | Feedstock | 150 | 48 | 11 | 23 | 0.03 | 0.05 | 0.10 | 0.03 |
| | Fuel | 2450 | 62 | 337 | 339 | 0.01 | 0.28 | 0.01 | 0.63 |
| | Operation | 1615 | 69 | 0 | 0 | 0.00 | 0.00 | 0.01 | 0.00 |
| | Vehicle | 840 | 146 | 65 | 68 | 0.31 | 0.10 | 0.04 | 0.30 |

| | | | | | | | | | |
|---|---|---|---|---|---|---|---|---|---|
| Hydrogen (Gaseous) | Feedstock | 181 | 10 | 13 | 27 | 0.01 | 0.05 | 0.00 | 0.03 |
| | Fuel | 1466 | 26 | 238 | 246 | 0.01 | 0.10 | 0.04 | 0.09 |
| | Operation | 2335 | 0 | 0 | 0 | 0.00 | 0.00 | 0.01 | 0.00 |
| | Vehicle | 1030 | 203 | 80 | 84 | 0.31 | 0.12 | 0.05 | 0.37 |
| Hydrogen (Liquid) | Feedstock | 182 | 10 | 13 | 27 | 0.01 | 0.05 | 0.00 | 0.03 |
| | Fuel | 3712 | 121 | 422 | 437 | 0.03 | 0.27 | 0.10 | 0.44 |
| | Operation | 2335 | 0 | 0 | 0 | 0.00 | 0.00 | 0.01 | 0.00 |
| | Vehicle | 1030 | 203 | 80 | 84 | 0.31 | 0.12 | 0.05 | 0.37 |
| Grid-Independent SI HEV | Feedstock | 246 | 45 | 10 | 19 | 0.01 | 0.08 | 0.00 | 0.03 |
| | Fuel | 880 | 321 | 50 | 54 | 0.09 | 0.09 | 0.01 | 0.05 |
| | Operation | 3718 | 3562 | 285 | 289 | 0.13 | 0.08 | 0.02 | 0.00 |
| | Vehicle | 840 | 146 | 65 | 68 | 0.31 | 0.10 | 0.04 | 0.30 |
| Grid-Dependent SI HEV | Feedstock | 222 | 49 | 11 | 22 | 0.02 | 0.07 | 0.04 | 0.03 |
| | Fuel | 1533 | 239 | 163 | 166 | 0.06 | 0.17 | 0.01 | 0.28 |
| | Operation | 3113 | 2413 | 191 | 194 | 0.09 | 0.06 | 0.01 | 0.00 |
| | Vehicle | 840 | 146 | 65 | 68 | 0.31 | 0.10 | 0.04 | 0.30 |

TABLE D-8 GREET Energy and Emission Factors for Light-Duty Trucks 2 in 2020

| | | Total Energy | Petroleum | $CO_2$ | $CO_{2e}$ | VOC | $NO_x$ | $PM_{2.5}$ | $SO_x$ |
|---|---|---|---|---|---|---|---|---|---|
| | | BTU/VMT | BTU/VMT | g/VMT | g/VMT | g/VMT | g/VMT | g/VMT | g/VMT |
| RFG SI Autos (Conventional Oil) | Feedstock | 436 | 80 | 19 | 34 | 0.02 | 0.15 | 0.01 | 0.05 |
| | Fuel | 1564 | 570 | 89 | 96 | 0.16 | 0.16 | 0.02 | 0.09 |
| | Operation | 6605 | 6327 | 507 | 511 | 0.22 | 0.14 | 0.02 | 0.01 |
| | Vehicle | 730 | 140 | 54 | 57 | 0.31 | 0.08 | 0.04 | 0.15 |
| RFG SI Autos (Tar Sands) | Feedstock | 1269 | 77 | 83 | 101 | 0.03 | 0.16 | 0.01 | 0.05 |
| | Fuel | 1635 | 570 | 90 | 97 | 0.16 | 0.16 | 0.02 | 0.09 |
| | Operation | 6605 | 6327 | 507 | 511 | 0.22 | 0.14 | 0.02 | 0.01 |
| | Vehicle | 730 | 140 | 54 | 57 | 0.31 | 0.08 | 0.04 | 0.15 |
| CG SI Autos (Conventional Oil) | Feedstock | 436 | 80 | 19 | 34 | 0.02 | 0.15 | 0.01 | 0.05 |
| | Fuel | 1564 | 570 | 89 | 96 | 0.16 | 0.16 | 0.02 | 0.09 |
| | Operation | 6605 | 6327 | 507 | 511 | 0.22 | 0.14 | 0.02 | 0.01 |
| | Vehicle | 730 | 140 | 54 | 57 | 0.31 | 0.08 | 0.04 | 0.15 |
| CG SI Autos (Tar Sands) | Feedstock | 1269 | 77 | 83 | 101 | 0.03 | 0.16 | 0.01 | 0.05 |
| | Fuel | 1635 | 570 | 90 | 97 | 0.16 | 0.16 | 0.02 | 0.09 |
| | Operation | 6605 | 6327 | 507 | 511 | 0.22 | 0.14 | 0.02 | 0.01 |
| | Vehicle | 730 | 140 | 54 | 57 | 0.31 | 0.08 | 0.04 | 0.15 |
| RFG SIDI Autos (Conventional Oil) | Feedstock | 379 | 69 | 16 | 30 | 0.02 | 0.13 | 0.01 | 0.04 |
| | Fuel | 1360 | 496 | 78 | 83 | 0.14 | 0.14 | 0.02 | 0.08 |
| | Operation | 5743 | 5502 | 441 | 445 | 0.22 | 0.14 | 0.02 | 0.01 |
| | Vehicle | 730 | 140 | 54 | 57 | 0.31 | 0.08 | 0.04 | 0.15 |
| RFG SIDI Autos (Tar Sands) | Feedstock | 1103 | 67 | 72 | 88 | 0.03 | 0.14 | 0.01 | 0.04 |
| | Fuel | 1422 | 496 | 79 | 84 | 0.14 | 0.14 | 0.02 | 0.08 |
| | Operation | 5743 | 5502 | 441 | 445 | 0.22 | 0.14 | 0.02 | 0.01 |
| | Vehicle | 730 | 140 | 54 | 57 | 0.31 | 0.08 | 0.04 | 0.15 |

| | | | | | | | | | |
|---|---|---|---|---|---|---|---|---|---|
| Diesel (Low Sulfur) | Feedstock | 363 | 66 | 33 | 46 | 0.02 | 0.12 | 0.01 | 0.04 |
| | Fuel | 778 | 387 | 58 | 60 | 0.02 | 0.09 | 0.01 | 0.06 |
| | Operation | 5504 | 5504 | 435 | 439 | 0.07 | 0.17 | 0.02 | 0.00 |
| | Vehicle | 730 | 140 | 54 | 57 | 0.31 | 0.08 | 0.04 | 0.15 |
| Diesel (Fischer-Tropsch) | Feedstock | 353 | 23 | 27 | 44 | 0.03 | 0.09 | 0.00 | 0.06 |
| | Fuel | 3259 | 84 | 118 | 127 | 0.04 | 0.27 | 0.08 | 0.11 |
| | Operation | 5504 | 0 | 420 | 424 | 0.07 | 0.17 | 0.02 | 0.00 |
| | Vehicle | 730 | 140 | 54 | 57 | 0.31 | 0.08 | 0.04 | 0.15 |
| Diesel (Soy BD20) | Feedstock | 631 | 262 | −30 | −8 | 0.03 | 0.19 | 0.01 | 0.13 |
| | Fuel | 11895 | 92 | 25 | 16 | 0.32 | −0.07 | 0.00 | −0.10 |
| | Operation | 5504 | 4472 | 436 | 440 | 0.07 | 0.17 | 0.02 | 0.00 |
| | Vehicle | 730 | 140 | 54 | 57 | 0.31 | 0.08 | 0.04 | 0.15 |
| CNG | Feedstock | 489 | 27 | 35 | 72 | 0.04 | 0.13 | 0.00 | 0.07 |
| | Fuel | 461 | 20 | 38 | 40 | 0.00 | 0.04 | 0.01 | 0.07 |
| | Operation | 6290 | 0 | 373 | 381 | 0.18 | 0.14 | 0.02 | 0.00 |
| | Vehicle | 730 | 140 | 54 | 57 | 0.31 | 0.08 | 0.04 | 0.15 |
| E85 (Dry Corn) | Feedstock | 1014 | 288 | −284 | −226 | 0.06 | 0.31 | 0.01 | 0.15 |
| | Fuel | 7810 | 369 | 85 | 97 | 0.31 | 0.48 | 0.05 | 0.13 |
| | Operation | 6605 | 1730 | 498 | 502 | 0.20 | 0.14 | 0.02 | 0.00 |
| | Vehicle | 730 | 140 | 54 | 57 | 0.31 | 0.08 | 0.04 | 0.15 |
| E85 (Wet Corn) | Feedstock | 1024 | 555 | −275 | −203 | −0.17 | 0.47 | 0.02 | 0.26 |
| | Fuel | 8018 | 253 | 278 | 293 | 0.31 | 0.35 | 0.12 | 0.18 |
| | Operation | 6605 | 1730 | 498 | 502 | 0.20 | 0.14 | 0.02 | 0.00 |
| | Vehicle | 730 | 140 | 54 | 57 | 0.31 | 0.08 | 0.04 | 0.15 |
| E85 (Herbaceous) | Feedstock | 663 | 234 | −351 | −302 | 0.05 | 0.22 | 0.01 | 0.04 |
| | Fuel | 4337 | 227 | 7 | 20 | 0.23 | 0.49 | 0.03 | 0.00 |
| | Operation | 6605 | 1730 | 498 | 502 | 0.20 | 0.14 | 0.02 | 0.00 |
| | Vehicle | 730 | 140 | 54 | 57 | 0.31 | 0.08 | 0.04 | 0.15 |

continued

**TABLE D-8** Continued

| | | Total Energy BTU/VMT | Petroleum BTU/VMT | $CO_2$ g/VMT | $CO_{2e}$ g/VMT | VOC g/VMT | $NO_x$ g/VMT | $PM_{2.5}$ g/VMT | $SO_x$ g/VMT |
|---|---|---|---|---|---|---|---|---|---|
| E85 (Corn Stover) | Feedstock | 475 | 236 | -332 | -329 | 0.03 | 0.15 | 0.01 | 0.09 |
| | Fuel | 3898 | 227 | 7 | 20 | 0.23 | 0.47 | 0.03 | 0.00 |
| | Operation | 6605 | 1730 | 498 | 502 | 0.20 | 0.14 | 0.02 | 0.00 |
| | Vehicle | 730 | 140 | 54 | 57 | 0.31 | 0.08 | 0.04 | 0.15 |
| E10 (Dry Corn) | Feedstock | 487 | 98 | 11 | 31 | 0.03 | 0.16 | 0.01 | 0.06 |
| | Fuel | 1667 | 527 | 81 | 84 | 0.17 | 0.15 | 0.02 | 0.08 |
| | Operation | 6605 | 6178 | 507 | 511 | 0.22 | 0.14 | 0.02 | 0.01 |
| | Vehicle | 730 | 140 | 54 | 57 | 0.31 | 0.08 | 0.04 | 0.15 |
| E10 (Wet Corn) | Feedstock | 488 | 121 | 12 | 33 | 0.01 | 0.17 | 0.01 | 0.07 |
| | Fuel | 1685 | 517 | 98 | 101 | 0.17 | 0.14 | 0.03 | 0.08 |
| | Operation | 6605 | 6178 | 507 | 511 | 0.22 | 0.14 | 0.02 | 0.01 |
| | Vehicle | 633 | 126 | 47 | 50 | 0.21 | 0.08 | 0.03 | 0.14 |
| E10 (Herbaceous) | Feedstock | 456 | 93 | 6 | 24 | 0.03 | 0.15 | 0.01 | 0.05 |
| | Fuel | 1363 | 514 | 74 | 78 | 0.16 | 0.15 | 0.02 | 0.07 |
| | Operation | 6605 | 6178 | 507 | 511 | 0.22 | 0.14 | 0.02 | 0.01 |
| | Vehicle | 730 | 140 | 54 | 57 | 0.31 | 0.08 | 0.04 | 0.15 |
| E10 (Corn Stover) | Feedstock | 440 | 93 | 7 | 22 | 0.02 | 0.15 | 0.01 | 0.05 |
| | Fuel | 1325 | 514 | 74 | 78 | 0.16 | 0.15 | 0.02 | 0.07 |
| | Operation | 6605 | 6178 | 507 | 511 | 0.22 | 0.14 | 0.02 | 0.01 |
| | Vehicle | 730 | 140 | 54 | 57 | 0.31 | 0.08 | 0.04 | 0.15 |
| Electric | Feedstock | 176 | 56 | 13 | 27 | 0.03 | 0.06 | 0.12 | 0.04 |
| | Fuel | 2863 | 73 | 394 | 396 | 0.01 | 0.32 | 0.01 | 0.74 |
| | Operation | 1887 | 80 | 0 | 0 | 0.00 | 0.00 | 0.01 | 0.00 |
| | Vehicle | 840 | 146 | 65 | 68 | 0.31 | 0.10 | 0.04 | 0.30 |

| | | | | | | | | | | |
|---|---|---|---|---|---|---|---|---|---|---|
| Hydrogen (Gaseous) | Feedstock | 232 | 13 | 16 | 34 | 0.02 | 0.06 | 0.00 | 0.03 | |
| | Fuel | 1875 | 33 | 305 | 314 | 0.02 | 0.13 | 0.05 | 0.12 | |
| | Operation | 2989 | 0 | 0 | 0 | 0.00 | 0.00 | 0.01 | 0.00 | |
| | Vehicle | 1030 | 203 | 80 | 84 | 0.31 | 0.12 | 0.05 | 0.37 | |
| Hydrogen (Liquid) | Feedstock | 233 | 13 | 16 | 34 | 0.02 | 0.06 | 0.00 | 0.03 | |
| | Fuel | 4750 | 154 | 540 | 559 | 0.04 | 0.35 | 0.13 | 0.56 | |
| | Operation | 2989 | 0 | 0 | 0 | 0.00 | 0.00 | 0.01 | 0.00 | |
| | Vehicle | 1030 | 203 | 80 | 84 | 0.31 | 0.12 | 0.05 | 0.37 | |
| Grid-Independent SI HEV | Feedstock | 305 | 56 | 13 | 24 | 0.02 | 0.10 | 0.00 | 0.04 | |
| | Fuel | 1094 | 399 | 62 | 67 | 0.11 | 0.11 | 0.01 | 0.06 | |
| | Operation | 4619 | 4424 | 354 | 358 | 0.20 | 0.11 | 0.02 | 0.01 | |
| | Vehicle | 840 | 146 | 65 | 68 | 0.31 | 0.10 | 0.04 | 0.30 | |
| Grid-Dependent SI HEV | Feedstock | 272 | 59 | 14 | 27 | 0.02 | 0.09 | 0.05 | 0.04 | |
| | Fuel | 1835 | 295 | 193 | 197 | 0.08 | 0.20 | 0.01 | 0.32 | |
| | Operation | 3821 | 2995 | 238 | 240 | 0.13 | 0.07 | 0.02 | 0.00 | |
| | Vehicle | 840 | 146 | 65 | 68 | 0.31 | 0.10 | 0.04 | 0.30 | |

## Mobile6.2 Heavy-Duty Truck Energy and Emissions Factors

The operational factors for heavy-duty vehicles were determined with Mobile6.2 and are shown in Table D-9 and Table D-10. Default Mobile6.2 values were used for these vehicles.

## GREET and Mobile6.2 Comparison

The GREET vehicle operation factors can be compared against Mobile6.2's to evaluate the accuracy of particular vehicles. GREET assumes default emission factors for conventional gasoline and diesel vehicles and the operating conditions of the vehicles is not transparent. Mobile6.2 is designed to model emissions from conventional fuel vehicles and low level ethanol blends and provides the ability to adjust many vehicle operation and fuel characteristics in determining emission factors. Table D-11 and Table D-12 compare the GREET default conventional gasoline and diesel vehicle emissions against Mobile6.2. The lack of transparency in the vehicle and operating characteristics used to generate GREET factors results in some difficulty in verification using Mobile6.2. In 2005, GREET assumes low-sulfur concentrations of 26 ppm in gasoline and 200 ppm in conven-

**TABLE D-9** Mobile6.2 Energy and Emission Factors for Heavy-Duty Vehicles in 2005

|          | Total Energy | $CO_2$ | VOC   | $NO_x$ | $PM_{2.5}$ | $SO_x$ |
|----------|--------------|--------|-------|--------|------------|--------|
|          | Btu/VMT      | g/VMT  | g/VMT | g/VMT  | g/VMT      | g/VMT  |
| HDGV2A   | 12500        | 888    | 1.79  | 4.13   | 0.07       | 0.05   |
| HDGV2B   | 12500        | 888    | 1.79  | 4.13   | 0.07       | 0.05   |
| HDGV3    | 13587        | 963    | 2.47  | 4.71   | 0.08       | 0.06   |
| HDGV4    | 14205        | 1005   | 5.30  | 5.90   | 0.07       | 0.06   |
| HDGV5    | 15823        | 1124   | 3.10  | 5.40   | 0.06       | 0.06   |
| HDGV6    | 15823        | 1119   | 2.91  | 5.30   | 0.07       | 0.07   |
| HDGV7    | 17123        | 1217   | 3.43  | 6.09   | 0.07       | 0.07   |
| HDGV8A   | 18382        | 1296   | 4.05  | 6.79   | 0.00       | 0.00   |
| HDDV2A   | 10195        | 795    | 0.23  | 3.99   | 0.12       | 0.01   |
| HDDV2B   | 10195        | 795    | 0.23  | 3.99   | 0.12       | 0.01   |
| HDDV3    | 11250        | 879    | 0.25  | 4.44   | 0.13       | 0.01   |
| HDDV4    | 12921        | 1004   | 0.31  | 5.41   | 0.11       | 0.01   |
| HDDV5    | 13316        | 1036   | 0.32  | 5.68   | 0.25       | 0.01   |
| HDDV6    | 15000        | 1176   | 0.47  | 7.99   | 0.26       | 0.01   |
| HDDV7    | 17400        | 1354   | 0.58  | 9.94   | 0.33       | 0.01   |
| HDDV8A   | 20077        | 1561   | 0.56  | 12.89  | 0.36       | 0.02   |
| HDDV8B   | 21048        | 1647   | 0.66  | 15.10  | 0.36       | 0.02   |

**TABLE D-10** Mobile6.2 Energy and Emission Factors for Heavy-Duty Vehicles in 2030

|  | Total Energy | $CO_2$ | VOC | $NO_x$ | $PM_{2.5}$ | $SO_x$ |
|---|---|---|---|---|---|---|
|  | Btu/VMT | g/VMT | g/VMT | g/VMT | g/VMT | g/VMT |
| HDGV2A | 12376 | 876 | 0.35 | 0.18 | 0.02 | 0.02 |
| HDGV2B | 12376 | 876 | 0.35 | 0.18 | 0.02 | 0.02 |
| HDGV3 | 13298 | 945 | 0.76 | 0.23 | 0.02 | 0.02 |
| HDGV4 | 13298 | 949 | 0.82 | 0.21 | 0.02 | 0.02 |
| HDGV5 | 15625 | 1107 | 0.91 | 0.24 | 0.02 | 0.02 |
| HDGV6 | 15432 | 1090 | 0.90 | 0.24 | 0.02 | 0.02 |
| HDGV7 | 16779 | 1191 | 0.95 | 0.27 | 0.03 | 0.02 |
| HDGV8A | 17606 | 1255 | 1.00 | 0.28 | 0.00 | 0.00 |
| HDDV2A | 10038 | 785 | 0.10 | 0.25 | 0.01 | 0.01 |
| HDDV2B | 10038 | 785 | 0.10 | 0.25 | 0.01 | 0.01 |
| HDDV3 | 11154 | 873 | 0.12 | 0.26 | 0.02 | 0.01 |
| HDDV4 | 12794 | 998 | 0.14 | 0.41 | 0.02 | 0.01 |
| HDDV5 | 13182 | 1030 | 0.15 | 0.44 | 0.02 | 0.01 |
| HDDV6 | 15000 | 1169 | 0.19 | 0.47 | 0.02 | 0.01 |
| HDDV7 | 17400 | 1352 | 0.23 | 0.58 | 0.03 | 0.01 |
| HDDV8A | 19773 | 1544 | 0.26 | 0.64 | 0.03 | 0.02 |
| HDDV8B | 20714 | 1616 | 0.29 | 0.75 | 0.03 | 0.02 |

**TABLE D-11** Comparison of Emission Factors (g/VMT) for a Light-Duty Gasoline Automobile in 2005

|  | VOC Exhaust | VOC Evap | $NO_x$ | $PM_{2.5}$ Exhaust | $PM_{2.5}$ TBW | $SO_x$ |
|---|---|---|---|---|---|---|
| GREET | 0.15 | 0.07 | 0.3 | 0.008 | 0.007 | 0.01 |
| Mobile6.2 | 0.27 | 0.87 | 0.8 | 0.005 | 0.007 | 0.02 |

**TABLE D-12** Comparison of Emission Factors (g/VMT) for a Light-Duty Diesel Automobile in 2005

|  | VOC Exhaust | $NO_x$ | $PM_{2.5}$ Exhaust | $PM_{2.5}$ TBW |
|---|---|---|---|---|
| GREET | 0.09 | 0.3 | 0.07 | 0.007 |
| Mobile6.2 | 0.33 | 1.3 | 0.15 | 0.007 |

tional diesel. GREET further specifies sulfur contents for low-sulfur diesel. Outside of fuel sulfur levels, vehicle emission factors are fixed based on inputs and assumptions from 1990 through 2020.

The differences between GREET and Mobile6.2 emission factors are most likely due to the variations in vehicle operation and fuel input parameters. These differences could be from cold start and warm running, fuel vapor pressure, summer or winter fuel mix, and vehicle model assumptions. While the Mobile6.2 factors tend to be larger than the GREET factors. The GREET factors are assumed to be reasonable, given the uncertainty in vehicle and fuel parameters and that they are within the bounds of Mobile6.2 estimates for the year.

### EPA Mobile6 Ammonia Emissions Factors

Ammonia emissions, which ultimately contribute to particulate formation, are evaluated by APEEP but not included in the default transportation damage assessment. GREET does not evaluate ammonia emissions but Mobile6.2 does for a subset of vehicle and fuel combinations included in GREET. Table D-13 summarizes the Mobile6.2 ammonia emission factors for 2005 and 2030.

For light-duty gasoline vehicles, the ammonia factors are about 0.1 g/VMT; for light-duty diesel vehicles, they range from 0.01 to 0.03 g/VMT for both years. The heavy-duty gasoline and diesel vehicle factors are 0.05 and 0.03 g/VMT.

### Applying GREET Feedstock and Fuel Production Factors to Heavy-Duty Vehicles

Feedstock and fuel production factors from GREET are used to supplement the Mobile6.2 heavy-duty-vehicle operational emissions. Because Mobile6.2 evaluates only the operational phase of heavy-duty vehicles, there is a need to supplement this component with feedstock and fuel production requirements so that results are commensurate with light-duty vehicles evaluated in GREET. To do this, the GREET feedstock and fuel production factors from reformulated gasoline and low-sulfur diesel light-duty vehicles are used. Using the energy content of gasoline or diesel consumed during vehicle operation, the corresponding GREET feedstock and fuel production factors are prorated and assessed to the heavy-duty vehicles. This procedure is done across all of the energy and emissions factors for each of the heavy-duty vehicles assessed with Mobile6.2.

Heavy-duty vehicle-manufacturing factors are not included in the assessment. Unlike feedstock and fuel production processes that are specific to a fuel (which is the same for both light- and heavy-duty vehicles), vehicle-

**TABLE D-13** Mobile6.2 Ammonia Emissions (g/VMT)

|         | 2005  | 2030  |
|---------|-------|-------|
| HDGV2B  | 0.045 | 0.045 |
| HDGV3   | 0.045 | 0.045 |
| HDGV4   | 0.045 | 0.045 |
| HDGV5   | 0.045 | 0.045 |
| HDGV6   | 0.045 | 0.045 |
| HDGV7   | 0.045 | 0.045 |
| HDDV2B  | 0.027 | 0.027 |
| HDDV3   | 0.027 | 0.027 |
| HDDV4   | 0.027 | 0.027 |
| HDDV5   | 0.027 | 0.027 |
| HDDV6   | 0.027 | 0.027 |
| HDDV7   | 0.027 | 0.027 |
| HDDV8A  | 0.027 | 0.027 |
| HDDV8B  | 0.027 | 0.027 |
| LDGV    | 0.100 | 0.102 |
| LDGT1   | 0.100 | 0.102 |
| LDGT2   | 0.097 | 0.102 |
| LDGT3   | 0.097 | 0.102 |
| LDDV    | 0.007 | 0.007 |
| LDDT    | 0.027 | 0.027 |
| LDDT12  | 0.007 | 0.007 |

manufacturing processes are unique. There is no known information that estimates the energy requirements and resulting emissions of manufacturing heavy-duty gasoline and diesel fuels of different classes. As a result, this component was excluded from the assessment.

## COUNTY-LEVEL DAMAGE CALCULATIONS

The vehicle feedstock, fuel, operation, and manufacturing per VMT emission factors are used in conjunction with APEEP county unit-damage factors to determine county resolution total damages. For each of the life-cycle components, particular assumptions were made in performing the calculations. APEEP has county-level pollutant-unit damages for all states except Alaska and Hawaii. For every county, APEEP reports ground levels and various heights of emission-unit damages (dollar per metric tonne emitted) for VOCs, $NO_x$, $PM_{2.5}$, $SO_2$, and $NH_3$ (ammonia).

### Feedstock Production Damages

The location of feedstock production and associated emissions is not clear for the various fuel energy inputs. From crude oil to corn to coal, the identification of feedstock production locations is not transparent. Feedstock can be produced internationally (for example, conventional crude oil from overseas or tar sands crude oil from Canada) or domestically (for example, coal or corn), and transport of raw energy inputs can occur along the fuel production pathway. The difficulty of estimating feedstock production and transport locations resulted in the assignment of these emissions to the county where travel occurs. The feedstock emissions are assessed the lowest level above ground-level height in APEEP.

## Fuel Production Damages

Fuel production damages are assessed to particular geographic regions based on petroleum refinery and ethanol plant locations. PADD (Petroleum Administration for Defense Districts) regions are used to identify five geographic areas of the United States for petroleum production and consumption statistics. The regions are East Coast, Midwest, Gulf Coast, Rocky Mountain, and West Coast and serve as a common resolution for petroleum data. The U.S. Energy Information Administration reports petroleum refinery locations and production capacity (EIA 2009b). Using these locations, associated counties could be determined for assessment of APEEP damage factors for conventional fueled vehicles. Without knowing which refinery produces the fuel for a VMT in another county, PADD resolution was used to assess fuel production unit damages. For each PADD, a weighted-average APEEP fuel factor (further referred to as $APEEP_{FUEL}$) was determined from the percentage of PADD fuel production capacity for each refinery and the corresponding county. The result produced five $APEEP_{FUEL}$ pollutant damage factors, one for each PADD. The $APEEP_{FUEL}$ damage factors were assessed to each county in the United States based on its PADD location. The fuel production life-cycle emissions were used in conjunction with the $APEEP_{FUEL}$ factors to determine fuel production damages for each county given a specific vehicle's per VMT emissions.

A PADD-based resolution approach was also used for ethanol fuel production. Using ethanol refinery locations (RFA 2009), $APEEP_{FUEL}$ factors were determined for ethanol production for each of the five PADD regions. Given the mix of ethanol in the fuel (10% or 85%), this fraction was multiplied by the $APEEP_{FUEL}$ ethanol factor and the remainder by the $APEEP_{FUEL}$ gasoline factor. For example, for an E10 vehicle operating in a county in PADD 1, 10% × $APEEP_{FUEL,ETHANOL}$ and 90% × $APEEP_{FUEL,GASOLINE}$ are added and assessed to that county. This mixed $APEEP_{FUEL}$ factor for each pollutant is then multiplied by the corresponding fuel production emissions for an E10 vehicle. Similar to feedstock production, the fuel production APEEP factors are based on the lowest level height above ground level.

For electric vehicles, power-plant emissions were assumed to occur according to petroleum PADD locations. Given the complexity of modeling electricity-generation emissions associated with specific driving locations, fuel production emissions were assigned to the petroleum production locations within relevant PADD regions.

## Vehicle Operation Damages

Vehicle operation VMT are based on county populations, which are assumed to be a reasonable metric for disaggregation of total state VMT. Given the GREET and Mobile6.2 per VMT emissions factors, VMT estimates are needed for each U.S. county to determine total emissions in that county. State-level VMT is available but not any higher resolution (FDA 2008). Using U.S. census population estimates, state-level VMT is disaggregated to each county by the fraction of population. These county VMT are then multiplied by the GREET and Mobile6.2 vehicle operational emission factors to determine total emissions for each county. The emissions of each pollutant are then joined with the APEEP ground-level-pollutant county factors to determine total damages.

## Vehicle-Manufacturing Damages

PADD regions are used to aggregate vehicle-manufacturing APEEP costs, similar to fuel production. Census data were examined for information on vehicles, parts, and tire manufacturing facilities (including number of facilities, employee counts, and county). The census data details the location and employee count for over 8,000 facilities. For each county in the United States, the total number of employees from these industries was determined. A weighted-average vehicle-manufacturing APEEP factor (further referred to as $APEEP_{MANUFACTURING}$) was determined for each PADD based on the percentage of employees and the APEEP factor foreach county in a PADD. Again, this process was done because of the lack of information that identifies whether a vehicle is driven in a particular county where it was manufactured. The PADD-based approach assumes that for a vehicle driven in a particular county, the manufacturing took place in that county's PADD and the weighted-average $APEEP_{MANUFACTURING}$ factor is applied.

## Total Life-Cycle Damages

Total damages are determined from feedstock production, fuel production, vehicle operation, and vehicle-manufacturing factors. This assessment was performed for each vehicle and fuel combination. Given a specific vehicle and fuel combination, the feedstock, fuel, operation, and manufacturing emission factors (in grams of VOC, $NO_X$, $PM_{2.5}$, and $SO_2$ per VMT) are multiplied by the APEEP county and pollutant factors (dollar damages per gram of pollutant, which may be weighted averages for the PADD region). Furthermore, the APEEP factors are reported in dollars of damage of emission for mortality, morbidity, and other damages (for example, agricultural or visibility impairment). For each vehicle and fuel combination, the life-cycle emission factors are joined with the APEEP pollutant damage factors for mortality, morbidity, and other to determine total damages for each county. The result is a mortality-morbidity-and-other dollar damages for each county and each vehicle type (light-duty autos, truck 1, and truck 2) in both 2005 and 2030.

### Damages Related to Electric Vehicles and Grid-Dependent Hybrids

For the vehicle-manufacturing component and the fuel feedstock (for example, coal or natural gas) component of the life cycles of electric vehicles (EVs) and grid-dependent hybrid vehicles (GD-HVs), the GREET model's estimates of emissions per VMT were paired with results from the APEEP model, a process that provided estimates of the physical health and other nonclimate-change-related effects and monetary damages per ton of emissions that form criteria air pollutants. However, the allocation of electric-utility-related damages to the vehicle operations and electricity production components of the life cycles were approximated by applying a GREET-generated kWh/VMT and applying that to the estimated average national damages per kWh from the electricity analysis presented in Chapter 2.

The committee used 1.59 cents/kWh for 2005 and 0.79 cents/kWh for 2030 for the damages due to producing (not consuming) electricity for both EVs and GD-HEVs. Those values were obtained by determining the aggregate marginal damages for coal-fired and natural gas plants based on their shares of net generation and the average marginal damages for each type of plant. For example, for 2005: [0.485 (coal share of net generation) × 3.2 cents/kWh] + [0.213 (natural gas share of net generation) × 0.16 cents/kWh] = 1.59 cents/kWh.

We estimated the fuel (electricity generation) component damages based on the damages associated with producing electricity at the rate of 0.52 kWh/VMT, and the fuel damages for 2005 were calculated as follows: 0.52 kWh/VMT × 1.59 cents/kWh = 0.83 cents/VMT. For 2030, the estimate for fuel damage is 0.31 cents/VMT. For the vehicle operation component, we estimated damage associated with a 10% loss of electricity over transmission and distribution lines (for example, 0.05 kWh/VMT for 2005) (DOE 2009).

A similar approach was used for estimating the electricity-related damages for GD-HEVs. However, no more than 35% of energy supplied to GD-HEVs was estimated to come from the grid.

## REFERENCES

Argonne National Laboratory. 2009. The Greenhouse Gases, Regulated Emissions, and Energy Use on Transportation (GREET) Model. U.S. Department of Energy, Argonne National Laboratory [online]. Available: http://www.transportation.anl.gov/modeling_simulation/GREET/ [accessed Oct. 12, 2009].

DOE (U.S. Department of Energy). 2009. Overview of the Electric Grid. Office of Electricity Delivery and Energy Reliability, U.S. Department of Energy [online]. Available: http://sites.energetics.com/gridworks/grid.html [accessed Sept. 4, 2009].

EIA (Energy Information Administration). 2006. Annual Energy Outlook 2007, With Projections to 2030. DOE/EIA-0383(2007). Energy Information Administration, Office of Integrated Analysis and Forecasting, U.S. Department of Energy, Washington, DC. February 2006 [online]. Available: http://tonto.eia.doe.gov/ftproot/forecasting/0383(2007).pdf [accessed Nov. 19, 2009].

EIA (Energy Information Administration). 2009a. Annual Energy Outlook 2009, With Projections to 2030. DOE/EIA-0383(2009). Energy Information Administration, Office of Integrated Analysis and Forecasting, U.S. Department of Energy, Washington, DC. March 2009 [online]. Available: http://www.eia.doe.gov/oiaf/aeo/pdf/0383(2009).pdf [accessed Apr. 22, 2009].

EIA (Energy Information Administration). 2009b. Ranking of U.S. Refineries. Energy Information Administration [online]. Available: http://www.eia.doe.gov/neic/rankings/refineries.htm [accessed Nov. 19, 2009].

EPA (U.S. Environmental Protection Agency). 2009. Mobile6 Vehicle Emission Modeling Software. U.S. Environmental Protection Agency [online]. Available: http://www.epa.gov/OMS/m6.htm [accessed Nov. 13, 2009].

FDA (Federal Highway Administration). 2008. Highway Statistics Series. Policy Information. U.S. Department of Transportation, Federal Highway Administration [online]. Available: http://www.fhwa.dot.gov/policy/ohpi/hss/index.cfm [accessed Nov. 13, 2009].

RFA (Renewable Fuel Association). 2009. Biorefinery Locations. Renewable Fuel Association, Washington, DC [online]. Available: http://www.ethanolrfa.org/industry/locations/ [accessed Nov. 13, 2009].

# E

# Supplemental Information on Land-Use Externalities from Biofuels: A Case Study of the Boone River Watershed

The committee uses the erosion productivity impact calculator (EPIC) model in conjunction with detailed field-level data for the Boone River Watershed to evaluate a number of "scenarios" where each scenario is associated with a different possible land use constituting different crops grown and different management practices on the land. To begin, a baseline land use corresponding to the 2005 cropping pattern and land management is developed, and an estimate of the externalities associated with the baseline is made. Then, various alternative land uses are proposed and evaluated using EPIC to predict the amount of nitrogen, phosphorous, sediments exported from each field as well as the amount of carbon sequestered. These levels can be aggregated to the watershed level and compared with the baseline. We also compute the amount of biofuels that the new land use can produce so that the externalities can be considered relative to the amount of acreage used to grow the feedstock or to the amount of fuel produced or both. Finally, using values from the literature, we monetize the externality end points.

The land-use conditions we evaluate include a baseline that represents 2005 cropping patterns and land use and the following counterfactual scenarios:

1. *Continuous corn*: The existing corn acreage that rotates with soybeans is converted to continuous corn—a change to about 90% of the acreage. As there is very little Conservation Reserve Program land or idle land in this watershed, this change is the main way in which planting decisions in this watershed can respond to increased demand for corn usage via ethanol.

2. *Corn stover*: The stover is removed from the baseline acreage and used to produce ethanol. We consider three possible rates of removal: 50%, 80%, and 100%.

3. *Continuous corn and corn stover*: This scenario is a combination of the first two; all corn and soybean rotations are changed to continuous corn, and then stover removals of 50%, 80%, and 100% are simulated.

4. *Switchgrass*: Switchgrass acreage is randomly placed on the baseline acreage from the baseline in percentages of 25%, 50%, 75% and, on the complete watershed, 100%. A nitrogen fertilizer rate of 123 kg/ha was simulated for the switchgrass, a rate consistent with optimal rates reported by Vogel et al. (2002) and Heggenstaller et al. (2009) for Iowa switchgrass biofuel production.

Our analysis draws heavily from the model and data sources developed by Gassman (2008), and we refer the interested reader to that document for substantially greater details on the data sources, collection methods, and assumptions. Here, we outline the basics of the model and summarize the externality estimates from the model. The Boone River Watershed covers over 500,000 acres in north central Iowa. Figure E-1 shows its location along with the Upper Mississippi River Basin and the state. The watershed is dominated by corn and soybean production, which together account for nearly 90% of its land use. The watershed is also characterized by intensive livestock production; land-applied manure from these livestock operations and commercial fertilizer applications are the primary sources of nutrients to the watershed stream system. However, manure applications were not accounted for in these simulations.

A key source of land-use data for the Boone simulations is a field-level survey of cropping patterns and conservation practices undertaken by C. Kiepe, formerly with the USDA-Natural Resources Conservation Service, who visually inspected all the fields (common land units) in the Boone watershed during the spring of 2005. These highly detailed spatially explicit data provide the basic information to populate the EPIC model. Table E-1 summarizes the cropping pattern observed: The region is almost entirely in a 1-year rotation of corn and soybeans. A few acres are in continuous corn or pasture, and a few are enrolled in the Conservation Reserve Program and are planted in a perennial cover. Additional data sources include soils information from the Soil Survey Geographic Database and the Iowa Soil Properties and Interpretations Database, climate data from NOAA and the Iowa Environmental Mesonet, topographic information from the Iowa Digital Elevation Model, and livestock operations from the Iowa Department of Natural Resources. Extensive additional details on these and other data sources used to populate the model can be found in Gassman (2008).

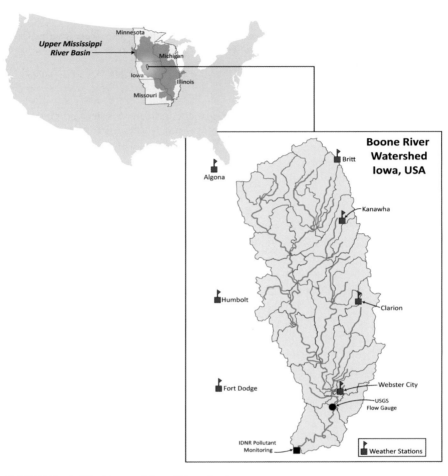

**FIGURE E-1** The Boone River Watershed.

**TABLE E-1** Boone River Watershed Baseline Cropping Pattern

|                                               | Acres   | Percent of Watershed |
|-----------------------------------------------|---------|----------------------|
| Corn-soybean rotation                         | 474,000 | 89                   |
| Continuous corn rotation                      | 21,000  | 4                    |
| Pasture                                       | 16,000  | 3                    |
| Conservation Reserve Program                  | 13,000  | 2                    |
| Other (mixture of other rotations and alfalfa)| 9       | <1                   |
| Total                                         | 533,000 | 100                  |

# REFERENCES

Gassman, P.W. 2008. A Simulation Assessment of the Boone River Watershed: Baseline Calibration/Validation Results and Issues, and Future Research Needs. Ph.D. Thesis, Iowa State University, Ames, IA.

Heggenstaller, A.H., K.J. Moore, M. Liebman, and R.P. Anex. 2009. Nitrogen influences productivity and resource partitioning by perennial, warm-season grasses. Agron. J. 101(6):1363-1371.

Vogel, K.P., J.J. Brejda, D.T. Walters, and D.R. Buxton. 2002. Switchgrass biomass production in the Midwest USA: Harvest and nitrogen management. Agron. J. 94(3):413-420.